CONN'S
Biological Stains

A Handbook of Dyes, Stains and Fluorochromes for Use in Biology and Medicine

10th Edition

Purchasers of the printed version will be entitled to free online access for a trial period, after which access will be limited to subscribers. The online version will have regular updates. Purchasers of the print version should register by sending an email to Conns@bios.co.uk. Notification will be sent out when the online version is released.

CONN'S
Biological Stains

A Handbook of Dyes, Stains and Fluorochromes for Use in Biology and Medicine

10th Edition

Edited by

Richard W. Horobin
Institute of Biomedical and Life Sciences, University of Glasgow, UK

and

John A. Kiernan
Department of Anatomy and Cell Biology, University of Western Ontario, London, Ontario, Canada

Published for the Biological Stain Commission
by BIOS Scientific Publishers, Oxford, UK

© Biological Stain Commission, 2002

First published 1925
Second edition 1929
Third edition 1936
Fourth edition 1940
Fifth edition 1946
Sixth edition 1953
Seventh edition 1961
Eighth edition 1969
Ninth edition 1977
Tenth edition 2002

A CIP catalogue record for this book is available from the British Library.

ISBN 1 85996 099 5

BIOS Scientific Publishers Ltd
9 Newtec Place, Magdalen Road, Oxford OX4 1RE, UK
Tel. +44 (0)1865 726286. Fax +44 (0)1865 246823
World Wide Web home page: http://www.bios.co.uk

Production Editor: Andrea Bosher
Typeset by Saxon Graphics Ltd, Derby, UK
Printed by Cromwell Press, Trowbridge, UK

Contents

How to use this book

How to find entries for individual dyes and fluorochromes

Within any major chemical grouping, for which see Chapter 3, non-ionic compounds are treated first, followed by anionic, then zwitterionic, cationic, and finally reactive dyes. Within each subgroup the entries are set out in increasing order of the dye's formula weight. This arrangement usually places compounds with similar properties near to one another in the text. The Index provides an even simpler route to substances whose names, or Colour Index designations, are known.

How to find information about a named dye

Find the major entry for a dye or fluorochrome, as above. In each entry, when available and appropriate, information is provided on the following topics, in the sequence given:

Usual dye name
Colour Index designations
Synonyms
Empirical formula and formula weight

Absorption and emission maxima
Qualitative indicator properties and pK values
Metal complexation behavior
Solubilities, in various solvents

Sketch of chemistry of dye with a structural formula

Staining applications in biomedicine – routine, and less common. Microscopic stains are emphasized, but stains for flow cytometry are also noted.
Other biomedical applications – e.g. macro stain, or use as a marker, tracer or assay reagent

Industrial uses

Purity of commercial batches – homogeneity, dye content (including the minimum required for Commission certification), availability of HPLC or thin layer chromatographic methods, note of Commission assay methods.

Stability – of dry powder & staining solutions; light fastness

Special hazards

How to find the stain you need

Now suppose you are asking yourself a question such as, "Which dye could I use to counterstain nuclei yellow, after doing this blue immunostain?" Does this book help here? Not directly; this is not a staining manual. The book will, however, inform you where to go for such advice: to the listing of practical sources found under the heading **Literature of biological stains**, at the end of Chapter 1.

Contributors list

Dr Brian Bracegirdle, Cold Aston Lodge, Cold Aston, Cheltenham GL54 3BN, UK

Dr Richard W Horobin, Division of Neuroscience & Biomedical Systems, Institute of Biomedical & Life Sciences, The University of Glasgow, Glasgow G12 8QQ, UK

Dr Frederick H. Kasten, Department of Anatomy & Cell Biology, James H. Quillen College of Medicine, East Tennessee State University, P.O. Box 70582–0582, Johnson City, TN 70582, USA

Dr John A. Kiernan, Department of Anatomy & Cell Biology, The University of Western Ontario, London, Canada N6A 5C1

Dr Hans O. Lyon, Department of Pathology, Hvidovre Hospital, University of Copenhagen, Kettegard Alle 30, DK-2650 Hvidovre, Denmark

Dr David P. Penney, Biological Stain Commission, Department of Pathology and Laboratory Medicine, University of Rochester Medical Center, Box 626, Rochester, NY 14642–0001,USA

Dr James M. Powers, Department of Pathology and Laboratory Medicine, University of Rochester Medical Center, 601 Elmwood Avenue, Rochester, NY 14642–0001, USA

Dr Juan C. Stockert, Facultad de Ciencia, Departamento de Biologia, Universidad Autonoma de Madrid, Cantoblanco, E-28049 Madrid, Spain

Dr Mark Wainwright, Senior Research Fellow, Department of Colour Chemistry, The University, Leeds LS2 9JT, UK

Professor Dr Dietrich Wittekind, Anatomische Institut der Universitat, Albertstrasse 17, D-78000 Freiburg, Germany

Preface

There have been nine previous editions of *Biological Stains*, published from 1925 to 1977. In each of these books, chapters concerning general aspects of dyes and other stains were followed by a compendium of more detailed accounts of individual compounds, grouped in chapters on the basis of chemical structure. The tenth edition has a similar plan, although it is a multiauthor work, for which all the chapters have been newly written. The average reader is assumed to be a biologist or from a biomedical field; but not a specialist in analytical, organic, physical or dyestuff chemistry, and not an expert in animal or plant histology, pathology or toxicology. The book is a source of information about dyes, fluorochromes and other colorants; and an account of their uses. It does not contain technical instructions, for which the reader must consult the manuals and articles cited in the text.

Overview. The first two chapters provide respectively a sketch of staining in biological laboratories and the history of staining for microscopy. Chapter 3 explains the categorization of dyes and fluorochromes, which differs from that in previous editions, along with some principles that govern the spelling of informal names of dyes. Applications, including those outside the laboratory, are reviewed in Chapter 4; and this is followed by a chapter reviewing the mechanisms of attraction and attachment of dyes to biological substrates. Examples and mechanisms of staining and labeling by covalent attachment are discussed in Chapter 7, and dyes used as indicators are reviewed in Chapter 8.

Conn's Biological Stains is published for the Biological Stain Commission, an independent non-profit organization in the USA that tests and certifies many stains, especially those used in histopathology. Standardization is also attempted elsewhere and by other organizations; this is described in Chapter 6, together with an account of the impurities that occur in dyes – the raison d'être of the Biological Stain Commission. Most of the remaining 20 chapters contain accounts of particular dyes and fluorochromes, and the final chapter relates to the methods used by the assay laboratory of the Biological Stain Commission.

Included and excluded items. Dyes and fluorochromes have been chosen for inclusion on the basis of being mentioned in recent published literature, both staining manuals and research papers. Only substances of known chemical composition are discussed. Consequently some proprietary colored and fluorescent reagents used by research workers are not included. One of us discussed this issue with a major supplier, with his persuasive efforts proving inadequate when faced with commercial pressures. The ninth edition of *Conn's Biological Stains* (Lillie, 1977) included several dyes, some of questionable identity, that had not been available for many years, and also some newer ones that never came to be used as stains. Most of these items have been omitted from the tenth edition to make room for newer compounds, especially fluorescent labels and probes, currently widely used in laboratories.

The electronic version. To keep this book a reasonable size, about half the recent material that was unearthed from the literature has been omitted. We have also omitted entries for compounds that have been little used in recent years, and we do not include many older but nevertheless useful references to traditional uses of dyes. A complete and fully searchable version of *Conn's Biological Stains* will be available on a web site to purchasers of the printed book. This electronic version will cover more compounds, give more examples of older and current uses and will be updated at regular intervals. Purchasers of the printed version will be entitled to free online access for a trial period, after which access will be limited to subscribers. The online version will have regular updates. Purchasers of the print version should register by sending an email to Conns@bios.co.uk. Notification will be sent out when the online version is released.

Disclaimers. The inclusion of a substance in this book does not constitute a commendation of its efficacy for any purpose. Note that dyes and fluorochromes often vary in composition (see Chapter 6), and consequently there are inconsistencies in physicochemical data such as solubility and absorption maxima reported in the literature. The values given here should be taken as indicative, not authoritative.

For the case of solubility an additional, and specific, caveat should be given. Dye molecules stick to themselves, as well as to cells and tissues; resulting in ready formation of colloidal suspensions and related forms. This makes measurement of solubilities a technically difficult, indeed a conceptually ambiguous, task. Solubility data deserve especial skepticism even in the, unusual, circumstance of their relating to pure dye samples.

Toxic, allergic and other hazards are mentioned for some, but not all items. Although the great majority of dyes do not pose serious hazards, laboratory workers must always follow good laboratory practices and obey the rules set out by local safety officers and committees.

Most of the dyes discussed in this book are or formerly were items of commerce, and many of their names are or were legally protected terms. Biological and biomedical end-users and writers routinely ignore this. They will refer, for instance, to Texas red without noting that this is a legally protected term owned by Molecular Probes Inc., or to dye names containing the words coomassie or sirius without regard to the Ciba-Geigy or Bayer corporations respectively. Failure to mention the owner of a trademark does not imply any disrespect for the originators of such names, without whom there would be no biological stains.

Acknowledgements

As editors, we recognize the contribution of Dr Fred Kasten, past President of the Biological Stain Commission, who started the process of revising the ninth edition of *Conn's Biological Stains*. We also appreciate the support of the Trustees of the Commission during this book's six-year gestation.

Assistance of various kinds, including spectral measurements, have been provided by many people, including: Russ Allison (University of Cardiff), Mike Barer (University of Leicester), Dan Belliveau (University of Western Ontario), Dick Dapson (Anatech Inc.), Dave Goldman (Research Software Design, Portland, Oregon; author of the PAPYRUS bibliographic database program), Andi Grantham (California), Floyd Green (formerly of Aldrich Inc.), John Griffith (University of Leeds), Bryan Hewlett (McMaster University), Lamar Jones (Histological Consultants Inc.), Andrew Moran (University of Sheffield), Debbie Rai (University of Stirling), and Juan Stockert (Free University of Madrid), members of the staff of Aldridge Chemicals Inc. (Milwaukee, WI), Amersham Pharmacia Biotech (Amersham, UK), and Molecular Probes Inc. (Eugene, OR), and Mrs Denise Currie, Mrs EA Gordon and other librarians at the Universities of Glasgow and Strathclyde. Mark Wainwright (University of Leeds) and Tom Wickersham (formerly of Aldrich Inc.) provided valuable on-going guidance to the complexities and peculiarities of dyestuff chemistry. We thank all of them, and apologize to those others excluded only by the failures of our notebooks.

May 2002   R.W.H.
                                                                                                      J.A.K.

Dedications

From RWH:

To Dorothy, long after time;
To Rose, not before time;
and to The Trustees, hopefully just in time.

From JK:

To my family, especially Tessa, Julia, Edward,
Susan, Jeffrey and Philip, with love and appreciation.

The Biological Stain Commission

Biological Stain Commission is a non-profit corporation, governed by a board of Trustees, that has been testing and certifying dyes for use in biology since 1922. Among the objectives of the Commission are the education of users about sources of reliable dyes and fluorochromes, and the publication of information about these substances and about traditional, improved and new staining methods and histochemical techniques. These objectives are met by publishing books and a journal. The first edition of *Biological Stains* by Harold J. Conn, a founder of the Commission, was published in 1925. The Commission's bimonthly journal was entitled *Stain Technology* from 1925–1990 and has been *Biotechnic & Histochemistry* since 1991. The Commission's laboratory is located in Rochester, NY. The web site at **www.biostains.org** may be consulted for more information.

Introduction to dyes and stains

F.H. Kasten

The first 8 chapters of *Conn's Biological Stains* review several aspects of the dyes and other coloring agents used in botanical, biomedical, microbiological and zoological laboratories. This first chapter introduces the reader to the terminology, chemical natures, and some other properties of colored and otherwise visible or potentially visible substances. Difficulties in obtaining correctly labeled and sufficiently pure compounds are pointed out, and attempts to address these problems are summarized. Chapter 1 concludes with lists of some major books and journals that cover the large field of dyes, dyeing, biological staining and histochemistry.

THE NATURE OF BIOLOGICAL STAINS AND THEIR USES

The purpose of staining is to add color to plant or animal tissues. This is done sometimes to increase the visibility of objects (or regions within an object) that can be seen with the unaided eye. The application of an iodine solution to a piece of food is a traditional test for starch, which forms a blue complex with iodine. In gynecological practice, iodine applied to a mucous membrane identifies and delimits areas of epithelium that contain glycogen. The red–brown iodine–glycogen complex contrasts with the deep yellow color imparted by iodine to almost everything else. There are many other macroscopic tests of this type, but most of the rich variety of staining methods is due to the need to reveal finer structural details of tissues and the more precise positions of chemical substituents when viewed with a microscope.

Colored compounds and dyes

The great majority of substances used in biological staining are dyes. Almost all dyes are organic compounds of the aromatic series and are therefore derivatives of benzene, C_6H_6. The six carbon atoms of benzene are joined to form a ring. The C–C bonds in the ring can be illustrated in different ways, as shown in Figure 1.1.

Kekulé's representation, with alternating single and double bonds, is used in the structural formulae of aromatic ring systems in this book because it shows

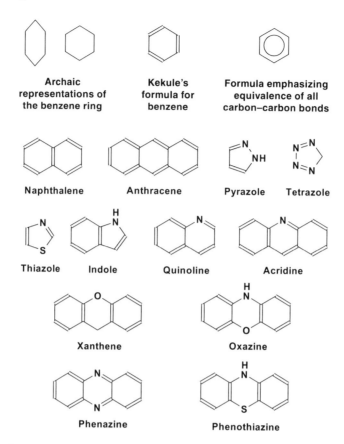

Figure 1.1 Representations of benzene (above) and some other aromatic ring systems found in dye molecules.

the arrangements of atoms and bonds responsible for color more clearly than the other conventions. Alternating double and single bonds are said to be **conjugated**. The electrons that form the bonds are shared by the whole conjugated system (delocalized), and a structural formula showing single and double bonds in certain positions represents only one extreme form that the bonds may transiently assume. Aromatic compounds commonly contain two or more fused rings (Figure 1.1), and one or more of the carbon atoms may be replaced by another element (a heteroatom) such as nitrogen or sulfur. Some examples that form parts of dye molecules are shown in the lower part of Figure 1.1.

Chromophores and auxochromes

Benzene, naphthalene and similar aromatic hydrocarbons absorb radiation from the ultraviolet but not from the visible part of the spectrum and are therefore not colored. In order to absorb visible light the aromatic rings must form part of a larger molecule, known as a **chromogen**, in which one possible arrangement of the conjugated bonds is not entirely aromatic. This results in an uneven distribution of electric charge in the molecule. Examples of chromogens are given in Figure 1.2. The part of a chromogen responsible for the property of color is called a **chromophore**.

The notion of chromophores and chromogens was developed in 1876 by O.N. Witt, 21 years before the discovery of the electron. It is now known that ultraviolet and visible light are absorbed when photons interact with electrons, and modern theories of color formation are consequently far more complicated than indicated above (Harms, 1965; Horobin, 1982). A simple descriptive account of the interaction of light with conjugated systems of electrons in organic molecules is provided later in this chapter.

The colored compounds shown in Figure 1.2 are not dyes; they would not stain a fabric or tissue. In order to be a **dye** a colored compound must be able to impart its color to other substances. This property is achieved by adding to the molecule one or more substituent groups known as **auxochromes**, which are capable of joining the chromogen to the substance being dyed. Auxochromes also modify (and usually increase the intensity of) the color of the chromogen.

The simplest auxochromes are side-chains that can form positive or negative ions. Examples are amine groups (which are protonated in acidic media), quaternary nitrogen atoms (which are positively charged at any pH), sulfonic acid groups (negatively charged when dissolved in water at any pH), and carboxyl groups (which ionize to yield negative ions in neutral or alkaline conditions). Most solid dyestuffs are salts in which the auxochrome ions are balanced by sodium or chloride. The charged dye ion is attracted to oppositely charged ions that form part of the material being dyed. For example, negatively charged dye ions, applied from an acidic solution, bind to protonated amino groups of proteins, including those of cytoplasm and collagen in sections of animal tissues, or the keratin and sericin of which wool and silk are respectively composed. Attraction of opposite charges is one of several mechanisms that can serve to bind a dye to a substrate. The mechanisms of staining are discussed in detail in Chapter 5.

Light absorption by organic molecules

Absorption of ultraviolet or visible light is a function of the energy required to promote electronic transitions within the molecule and there is now a sufficient

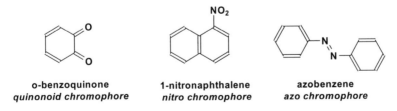

	NO$_2$	
o-benzoquinone	1-nitronaphthalene	azobenzene
quinonoid chromophore	*nitro chromophore*	*azo chromophore*

Figure 1.2 Some colored compounds (chromogens) that are not dyes.

knowledge base to enable the prediction of light absorption for a given molecule, normally using molecular orbital calculations. The Parr–Parriser–Pople (PPP) method is perhaps the best known of these, but an excellent qualitative treatment is given in the standard work on this subject by Griffiths (1984).

The presence of functionalized aromatic features (heteroatoms, auxochromes), allows the absorption of visible light. The absence of part of the perceived reflectance or transmittance spectrum gives the impression of a certain color or shade. For example, removal of the red component of white light gives the sensation of blue color, with the shade of blue depending on the exact type of red light (i.e. the wavelength range) absorbed. It should be remembered here that there is a reciprocal relationship between wavelength and energy.

Electronic absorption in organic molecules is allowed due to the presence of vacant electronic orbitals (levels). Normally, the absorption of light energy allows the promotion of an electron from the highest occupied molecular orbital (HOMO) to the lowest unoccupied molecular orbital (LUMO, vacant). The energy required for this transition, that is, the gap between the HOMO and LUMO, depends on the structure of the molecule. There are other transitions that occur, and the more complex the molecule, the greater is the number of absorption bands. The present discussion must be limited to the longest wavelength of absorption (λ_{max}) and for deeper treatments the reader is referred to the texts of Griffiths (1984) and Zollinger (1991, 1999).

As a simple example, anthracene (Figure 1.2) is an aromatic molecule of the benzenoid class, containing three linear-fused benzene rings. The lowest energy absorption (i.e. the longest wavelength) of anthracene occurs in the ultraviolet range (300 nm). This absorption can be lowered in energy by the inclusion in the ring system of a nitrogen atom in the central benzene ring, giving the acridine molecule (Figure 1.1), which has a lowest energy absorption of 340 nm. The introduction of the ring nitrogen changes the electronic structure of the aromatic system, narrowing the gap between the energy levels.

The attachment of other chemical groups or atoms to the exterior of the ring system can also affect the electronic transition, the magnitude depending on the nature of the group and also its position in the ring. Where there is electronic communication (conjugation) between the group and a ring heteroatom or another, remote, group the effect will be significant. Important groups in this respect are usually strongly electron releasing (such as amino) or withdrawing (such as nitro). Such groups are often termed auxochromes.

Applying this second criterion to the acridine system, electron-releasing dimethylamino groups ($-N(CH_3)_2$) fixed in positions 3- and 6- of the ring are both strongly conjugated with the ring nitrogen and lower the lowest energy electronic transition still further, to 488 nm. The resulting dye is acridine orange (see Chapter 17).

The use of heteroatom substitution and/or auxochromes is essential in tailoring dyes to fit staining requirements. Both these factors also affect fluorescence properties. In addition, an auxochromic moiety can be produced with associated desirable physicochemical properties such as increased lipophilicity, or tethered reactive groups for chemical attachment to macromolecules such as proteins or nucleic acids.

Dyes, stains and colorants

The question is often asked, 'What are dyes and how do they differ from fluorochromes, dyestuffs, colorants, biological stains, and pigments?' We have already given a brief chemical explanation of what a dye is and stated that it is an organic aromatic molecule containing the requisite groups that provide visible color and permit molecular binding to a material. The analogous word fluorochrome refers to a dye that is capable of fluorescing, though some fluorochromes absorb only in the ultraviolet and are therefore not colored. 'Dye' is synonymous with 'dyestuff'. The latter word is probably derived from the German word *Farbstoff* ('color substance'), which is translated as dye. *Farbstoff* dates back to the products of the great German dye industries of the late 19th century and has been used widely ever since. It is a matter of custom as to whether the words dye or dyestuff are used. Dyestuff is popular in industrial circles and dye is favored by histologists and cytologists. Examples of dyes are the natural dye hematoxylin and the synthetic dye crystal violet.

The word colorant is applicable to any substance that is capable of coloring or staining a material. Baker (1958) argued that this word should be the term of choice for biologists when referring to any substance that stains tissues. In spite of the cogent argument made by this influential writer, the word colorant is still not in vogue among biologists and histotechnologists.

The word stain, employed as a noun, includes not only dyes but all the other colorants used in biological work, and some of these compounds are not colored. For example, Schiff's reagent is used in the Feulgen method for DNA (Kasten, 1964) and in the periodic acid–Schiff (PAS) test for polysaccharides and glycoproteins (McManus, 1946). Fluorescent brightening agents are chemically similar to dyes but their absorption is largely in the near ultraviolet. Their various uses in biology include vital fluorescent staining of calcification in tissues (Rahn and Perren, 1970) and detection of fungi in material that is also conventionally stained to show bacteria (Ruchel and Schaffrinski, 1999). Dyes or fluorochromes are produced as final products in the techniques of enzyme histochemistry, immunohistochemistry, and *in situ* hybridization

Inorganic reagents

Other colorless colorants include osmium tetroxide and silver nitrate, which are readily reduced to insoluble osmium dioxide (black) and metallic silver (brown or black), in a finely divided or colloidal state. Gold may also be deposited in tissues, nearly always by reduction of the tetrachloroaurate ion of $HAuCl_4$ or $NaAuCl_4$ (salts colloquially called 'gold chloride'). The color of colloidal gold may be violet, red or black, depending on the average particle size (Frens, 1973). These and other inorganic reagents are used in many histochemical and traditional staining methods. The older methods are often called metal impregnations, from the belief that the soluble compound selectively permeated certain parts of the block or section of tissue and was then reduced *in situ* to the visible metal or lower oxide (Gatenby and Beams, 1950; Gray, 1954).

The adjective argentaffin means possessing the capacity to reduce simple silver ions (Ag$^+$) or complex ions such as Ag(NH$_2$)$^+$, in the dark and without the aid of a chemical reducing agent. This is a property of some substances, including serotonin (5-hydroxytryptamine) and ascorbic acid that occur at high concentration in certain cells (Lillie and Fullmer, 1976; Pearse, 1985). An argyrophilic (or argyrophil) object in a tissue also reacts with deposition of silver atoms, but these are too small and too widely separated to interrupt the passage of light. They resemble, in this respect, the specks of silver that form the latent image in an exposed but undeveloped photographic film. A chemical reducing agent (or exposure to bright light in a few methods) is required to produce a black deposit of colloidal metallic silver. The chemical mixtures used to make argyrophilic materials visible are similar to photographic developers. They extract loosely bound (unreduced) silver ions from nearby parts of the tissue and reduce them to the metal at the sites of previously formed invisible clumps of a few silver atoms. Each tiny cluster of silver atoms serves as a catalyst, and the silver particles grow until the original catalytic sites are big enough to be seen. This is an example of amplification: the original signal is too small to be detected, but it triggers other reactions that lead to a visible product.

The silver methods are the basis for the demonstration in light microscopy of reticular fibers, intercellular boundaries, enterochromaffin (argentaffin) cells, and the Golgi apparatus. Traditional neurological staining methods depend heavily on the use of silver salts selectively to impregnate different types of neuroglial cells and neurons and their cytoplasmic extensions. Examples of such neurological methods are those named after Golgi, Ramon y Cajal and Bielschowsky.

Electron stains

Salts of heavy metals, notably lead and uranium, are also used as electron stains, to impart additional electron density to thin sections prior to examination by electron microscopy. Electron stains are outside the scope of this book, but considerable information about them is available elsewhere (Glauert and Lewis, 1998; Hayat, 1975; Horobin, 1982; Lewis and Knight, 1977; Ogawa and Barka, 1992; Pearse, 1985; Polak and Van Noorden, 1997; Stoward and Pearse, 1991).

Pigments

Another group of colorants is that of the pigments, defined for industrial purposes as colored or white inorganic or organic substances consisting of small particles, which are insoluble in the medium they are dispersed in, and which remain insoluble throughout the coloration process. Pigments are used in paints, inks and cosmetics, for bulk coloring of plastics, cellulose fibers for the paper industry, and many other materials. Inorganic pigments include titanium dioxide (white), carbon (black), cadmium sulfide (yellow) and iron oxides (yellow, red) (Schiek, 1982). Organic pigments are also widely used in industry and include compounds representative of all the major chemical groups of dyes (Singer, 1982).

Organic and inorganic pigments are formed during various biological staining procedures. Examples are Prussian blue (in the Perls method for tissue iron) and various azo, formazan and indigo compounds in histochemical techniques for enzymes.

Dyeing and staining

Textile dye specialists, formerly called **dyers**, refer to the process of producing permanent colors on natural and synthetic fibers as **dyeing**. While some biologists also use this word for their applications, most prefer to describe the process of using a dye in solution as **staining**. Examples are hematoxylin and eosin staining (which colors the nuclei of cells blue–purple and other tissue components pink) and methyl green–pyronine staining, which imparts different colors to DNA and RNA. These are techniques for use with sections of fixed tissue.

Biological staining usually requires a prior fixation, which preserves the architecture of the tissue and also enhances its affinity for some colored compounds. This is not always the case, however, and nontoxic **vital staining** of living cells has a long history, beginning with the use of carmine by Trembley in the 18th century.

Vital staining is defined as the coloration of living cells or other components of tissues (Emmel and Cowdry, 1964). When this is done in a whole living animal or plant the term **intravital staining** has been used (Wilson, 1910). **Supravital staining** is the application of dyes to living cells or small pieces of tissue freshly removed from the body, which are believed to be alive at least during the early stages of the procedure (Cappell, 1929). This type of staining has been much used to study blood and other cell suspensions (Cunningham and Tompkins, 1938). Intravital fluorochroming of whole animals was introduced by Ellinger and Hirt (1929), who examined the microcirculation. In recent years fluorescent compounds have come greatly to outnumber conventional dyes in vital staining applications, and they are typically described merely as fluorescent probes.

Examples of vital stains dating from the 19th century and still in use are neutral red, which is taken up into endocytotic vesicles or vacuoles, Janus green B, which enters mitochondria, and methylene blue, which enters the thin cytoplasmic processes of neurons (Baker, 1958). These dyes can be used in either intravital or supravital techniques. A special type of intravital staining is the tracing of neural connections. Any of a wide variety of fluorescent dyes (Aschoff and Hollander, 1982; Kuypers *et al.*, 1977) or fluorescently or enzymatically labeled macromolecules (Kristensson *et al.*, 1971; Mesulam and Rosene, 1979; Trojanowski, 1983) is injected at a localized site. The tracer is taken up by synaptic terminals and transported retrogradely along axons to the cell bodies of neurons, in which it accumulates and is detected. Some tracers are transported anterogradely, from cell bodies at the site of the injection to the synaptic terminals of the axons.

Certain anionic dyes such as trypan blue pass through damaged but not through intact cell membranes and are then trapped in the cytoplasm. These dyes therefore stain dead but not living cells. The trypan blue dye-exclusion technique is a popular method used by cell culturists to quantify the proportion of live and dead cells in freshly isolated cells. Several fluorochromes, including propidium iodide, similarly enter dead cells but then bind strongly to the DNA in their nuclei. Living cells can be selectively stained by immersion in a solution containing an invisible compound that enters the cell and is changed to a fluorescent substance as a result of an enzyme-catalyzed reaction. Fluorochrome precursors of this kind include various esters and amides (acted upon by esterases or peptidases) (Haugland, 1995; Johnson, 1998) and certain tetrazolium salts that change into colored or fluorescent formazans when they accept electrons from coenzymes involved in cellular respiration (Sasaki and Passaniti, 1998; Ullrich *et al.*, 1996). Thus vital staining provides a variety of ways to determine the proportions of live and dead cells in suspensions and cultures.

NOMENCLATURE AND IDENTIFICATION OF DYES

The common names of dyes are meaningless and are not sufficient to identify particular compounds. An individual dye may have several synonyms, including the trade names, often long obsolete, of different manufacturers or suppliers. Conversely, the same common name may be applied to completely different dyes. Furthermore, a bottle may contain a substance completely different from the dye named on its label.

Here is a simple example of difficulty in accurate dye identification. The acridine dye known today as acriflavine (CI 46000) was called trypaflavine many years ago. This name does not appear in modern catalogs, so without more information at hand a dye user would find it difficult to replace an old, depleted bottle of trypaflavine under the old name. A listing of dye synonyms, such as is provided in the index of this book, might solve this particular problem but is not foolproof. Therefore, special systems of identification and numbering are employed to verify the identity and chemical structure of each dye.

The *Colour Index* (CI)

The Society of Dyers and Colourists (Bradford, UK) and the American Association of Textile Chemists and Colorists (Lowell, Massachusetts) have cooperated in identifying the chemical constitutions of most dyes. These are listed under various classifications in six volumes of the third edition of the *Colour Index* (1971–1975). Since then, three supplemental volumes have been published, the last in 1992. The first digit of a page number in the CI is the number of the volume. Copies of the *Colour Index* may be found as reference books in university, technical, industrial, and government libraries. The work is also available on CD-ROM as *The Colour Index International* (Society of Dyers and Colourists, 1996), which is updated at frequent intervals. Since 2001 the 4th edition of the *Colour Index* has been an online publication (information at http://www.colour-index.org).

Approximately 8000 individual dyes and pigments were initially identified chemically and each was assigned a CI generic name and a 5-digit CI constitution number to accompany its chemical structure. Both the CI number and the generic name provide an exact identification of a dye. The generic names are usually less important to biological users because they are derived from modes of application in textile dyeing (Acid dyes, Basic dyes, Direct dyes, Food dyes, Reactive dyes, Natural dyes, Solvent dyes, Pigments, etc.). The CI generic names are found in sequence under their classification groups in Volumes 1, 2, 3, 5, and 6 of the *Colour Index*. For example, methyl blue (CI 42780) has the assigned generic name Acid blue 93 (Vol. 1, p. 1331). In the absence of a common name for a dye, the CI generic name is used. CI Direct red 75 (Vol. 2, p. 2128) specifies an azo dye that is used as a biological stain.

Volumes 4 and 5 of the *Colour Index* are of special value to people who want to gain detailed information about dyes for biological staining. Volume 4 lists all dyes with assigned CI numbers in sequence, according to their chemical constitution. For example, all xanthenes are listed by number, starting at CI 45000 with acridine red 3B and CI 45005 for Pyronine G, and ending with CI 45555 (Fluorescent brightener 155, a compound formerly added to lubricating oils). A copy of a page from Volume 4 is shown in Figure 1.3. For each dye there is recorded, in addition to the CI number, the structural formula (with archaic representation of aromatic rings), methods of preparation, chemical and physical properties, references, patent numbers, and, where appropriate, common names. Volume 5 is useful if one has a commercial name for a dye and is looking for the CI number. All commercial names are listed alphabetically. For the example of trypaflavine, this name is found on p. 5787 of Vol. 5. Its CI Number is given as 46000. If one now looks in Volume 4 at CI 46000, details are given about acriflavine, the name now commonly used for this dye.

The Chemical Abstract Service (CAS) Registry

Precise identification of a dye may also be accomplished by reference to its Chemical Abstract Service Registry or CAS number. These are unique numbers assigned to compounds and mixtures recorded in the CAS System. The CAS number is not related to chemical structure or composition, but is assigned to each substance as it enters the registry. Every substance that has been cited in chemical, scientific, and technical literature and has been indexed in *Chemical Abstracts* since 1975 has been assigned a CAS number. This number is specific to all details of a compound. It is therefore possible to specify a substance by way of its CAS number, without regard to synonyms, trade names, or common names. For the dye acriflavine, the CAS number is 8048–52–0.

XANTHENE COLOURING MATTERS

The chromophore of the aminoxanthene dyes is the resonance-hybrid [structure] ⟷ [structure], where

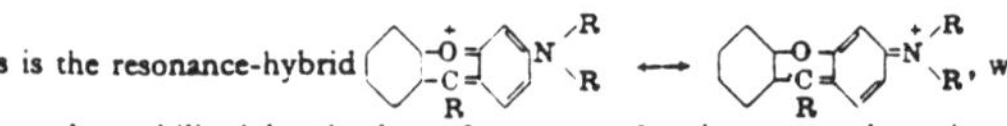

R = H, or alkyl, or aryl; the hydroxyxanthenes can be stabilised by the loss of a proton, forming an uncharged system in which the chromophore is the quinoid structure [structure]

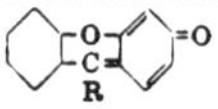

The dyes are prepared from xanthene derivatives with the usual auxochromes in *para*-position to the methane carbon atom. These derivatives are not obtained from xanthene itself, but by reacting together suitably chosen simple intermediates.

When R is an aryl radical, the dyes, although possessing the pyrone ring, have analogies with the triarylmethane class. The xanthene class is subdivided into amino, aminohydroxy, and hydroxy derivatives.

In general, the xanthenes are basic dyes which possess remarkably pure bright hues, and their solutions are strongly fluorescent. They dye wool and silk directly from weak acid baths, and cotton on a tannin mordant. Some of the hydroxy compounds are valuable mordant dyes.

Special Literature

Hewitt, *Dyestuffs derived from Pyridine, Quinoline, Acridine, and Xanthene*, Longmans, Green & Co, London, 1922
Fierz-David, *Künstliche Organische Farbstoffe*, Julius Springer, Berlin, 1926
Elderfield, *Heterocyclic Compounds*, Vol. 2, p. 419, John Wiley & Sons, New York, 1951
Venkataraman, *The Chemistry of the Synthetic Dyes*, p. 740, Academic Press, New York, 1952
Lubs, *The Chemistry of Synthetic Dyes and Pigments*, p. 291, Reinhold Publishing Corporation, New York, 1955

XANTHENE COLOURING MATTERS

(I) — AMINO-DERIVATIVES (FLUORENE COLOURING MATTERS)
- (a) Pyronines (C.I.45000–45020)
- (b) Succineins (C.I.45050)
- (c) Sacchareins (C.I.45070)
- (d) Rosamines (C.I.45090–45105)
- (e) Rhodamines (C.I.45150–45225)

(II) — AMINO-HYDROXY-DERIVATIVES (RHODOLS)

(III) — HYDROXY-DERIVATIVES (FLUORONE COLOURING MATTERS)
- (a) Hydroxy-phthaleins (C.I.45350–45460)
- (b) Anthrahydroxy-phthaleins (C.I.45500–45510)

(IV) — MISCELLANEOUS-DERIVATIVES

(I) — AMINO-DERIVATIVES (FLUORENE COLOURING MATTERS)

(a) PYRONINES

45000 — **Basic Dye**

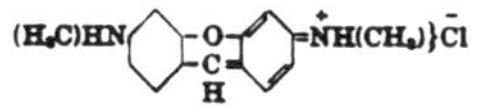

(H₃C)HN[structure]ṄH(CH₃)}Cl̄

Oxidise C.I.45005 with potassium permanganate

Discoverers — Bender and Kämmerer 1891
Acridine Red 3B
Leonhardt Co., BP 1231/92; USP 489625; FP 219023; GP 65282 (*Fr.* 3, 176)
Biehringer, *J. prakt. Chem.* 54 (1896), 235

Soluble in water (red with greenish yellow fluorescence)
Soluble in ethanol (red with greenish yellow fluorescence)
H₂SO₄ conc. — yellow with green fluorescence; on dilution — orange then red
Aqueous solution + NaOH — red ppt.

45005 — **Basic Dye**

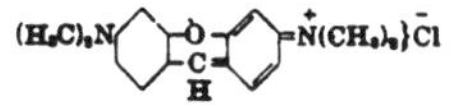

(H₃C)₂N[structure]Ṅ(CH₃)₂]Cl̄

Condense m-dimethylaminophenol with formaldehyde, dehydrate the product with sulfuric acid and oxidise with ferric chloride

Discoverer — Bender 1889
Pyronine G (By)
Bayer Co., BP 8673/89; FP 198785; GP 54190 (*Fr.* 2, 61)
Leonhardt Co., BP 13217/89, 18606/91; USP 445684; FP 200401; GP 58955, 59003, 63081, (*Fr.* 3, 92, 94, 93)
Gerber Co., GP 60505 (*Fr.* 3, 96)
BIOS 959, 10
Monit. sci. 4 [4] (1890), 751
Möhlan & Koch, *Ber.* 27 (1894), 2896
Biehringer, *Ber.* 27 (1894), 3299; *J. prakt. Chem.* 54 (1896), 217
Scott & French, *The Military Surgeon*, Nov. 1924

H₂SO₄ conc. — reddish yellow; on dilution — red
Aqueous solution + HCl — bright orange

Soluble in water (red with yellow fluorescence)
Soluble in ethanol (red with yellow fluorescence)

Figure 1.3 A typical page from the *Colour Index* showing CI numbers, textile uses, common names, chemical structures and other information. (Reproduced with permission, from Vol. 4 of 3rd edn, p. 4417.)

Substitution and misidentification of dyes

The label on a bottle does not always correctly name the contents. For example, an investigation of pyronine dyes from various commercial sources (Kasten, 1962) revealed widespread substitution of rhodamines for pyronine Y(G) (CI 45005), a dye widely used in a staining technique for nucleic acids. This situation is much improved today as a result of a purer pyronine Y(G) supplied by industry and the modern dye standardization procedures used by the Biological Stain Commission (BSC) for dye certification. Confusion and deception about the names of dyes used in biology were more widespread years ago but have occurred as recently as the 1990s (Penney and Powers, 1995), when some companies supplied unsatisfactory and unrelated products in place of light green SF yellowish (CI 42095). This dye is an ingredient of Papanicolau's stain, used worldwide for detection of malignant cells in smears.

An interesting example of beneficial dye substitution is seen today in the case of aniline blue WS (CI 42755). This anionic triphenylmethane dye is a component of many trichrome recipes to stain collagen. Aniline blue WS is often contaminated with other dyes, including methyl blue (CI 42780), which gives equally good staining results, but older samples varied in their dye content and staining properties. It is prohibitively expensive to purify aniline blue WS, but relatively easy to synthesize methyl blue of satisfactory purity, and the latter is now supplied by at least one major dye firm, as a certified dye but under the name of aniline blue (Green, 1990). The company is to be commended for being frank about this desirable substitution. Histologists and technologists need to realize that the two dyes are interchangeable in trichrome procedures. More than 20 names have been applied in the past to aniline blue WS, methyl blue and mixtures of these and similar dyes (Krause, 1926–1927; Lillie, 1977).

LITERATURE OF BIOLOGICAL STAINS

There are numerous sources of information about the chemistry of dyes and pigments. Useful references are the third edition of the *Colour Index* (Society of Dyers and Colourists, 1971–1992; 1996) and works by:

Fay (1911)
Bucherer (1914)
Cain and Thorpe (1923)
Schultz (1931–1939)
Fierz-David (1949)
Venkataraman (1952–1978)
Lubs (1955)
Allen (1971)
Rys and Zollinger (1972)
Bird and Boston (1975)
Venkataraman (1977)

Fabian and Hartman (1980)
Gordon and Gregory (1983)
McLaren (1986)
Waring and Hallas (1990)
Peters and Freeman (1991).
Zollinger (1991)
Zollinger (1999)

The previous editions of *Biological Stains* provide a historical view, reflecting as they do the available information and interests for early periods in this field. The first edition by H.J. Conn was published in 1925 and the ninth, by R.D. Lillie, appeared in 1977. Other books with information about stains and discussion of how they work include the following:

Mann (1902)
Becher (1921)
Krause (1926–1927)
Baker (1958, 1966)
Emmel and Cowdry (1964)
Harms (1965)
Gurr (1971)
Gabe (1976)
Lillie and Fullmer (1976)
Horobin (1982, 1988)
Clark and Kasten (1983)
Lyon (1991)
Hayat (1993)
Kiernan (1999).
Bancroft and Gamble (2002)

The *Merck Index* (Budavari, 1996) is a valuable compilation of information about thousands of compounds, including many that are dyes or other compounds used in staining. The catalogs and handbooks of some major chemical companies contain valuable information about dyes such as the *Sigma–Aldrich Handbook of Stains, Dyes and Indicators* (Green, 1990). A comprehensive source of data about fluorescent molecular probes is to be found in the *Handbook of Fluorescent Probes and Research Chemicals* from Molecular Probes, Inc. (Haugland, 1996). This company also has an updated CD-ROM that appeared in 2000. A volume that centers particular attention on the theories of fluorescence, fluorescent dyes and probes, and on molecular interactions is *Fluorescent Probes in Cellular and Molecular Biology* (Slavik, 1994). There are some comprehensive reviews of the history of fluorescent dyes/probes and fluorescence microscopy, as well as histological and cytological applications (Kasten, 1983, 1989, 1999). Other important books are available that focus on fluorescent dyes and probes and their applications in biology. These volumes include *Fluorescence Microscopy of Living Cells in Culture* (Part A edited by Wang and

Taylor, 1989; Part B edited by Wang, 1989), *Cell Structure and Function by Microspectrofluorometry* (edited by Kohen and Hirschberg, 1989), *Flow Cytometry* (edited by Darzynkiewicz and Crissman, 1990) and *Fluorescent and Luminescent Probes for Biological Activity: A Practical Guide to Technology for Quantitative Real-Time Analysis* (edited by Mason, 1999).

For those engaged practically in the application of biological stains in medicine, cell and molecular biology, histotechnology, histochemistry, and immunostaining, many books from the older and recent literature are worth knowing about. Some of the older volumes have historically valuable information and useful stain recipes that do not appear in more recently published books. The following list contains only a selection of the many books that contain instructions for carrying out staining procedures:

Chamberlain (1901)
Mann (1902)
Ehrlich *et al.* (1903)
Guyer (1906)
Langeron (1925)
Krause (1926–1927)
Mallory (1938)
Johansen (1940)
Gatenby and Beams (1950)
Jones (1950)
Gray (1954)
Ganter and Jollès (1969, 1970)
Lison (1960)
McManus and Mowry (1960)
Thompson (1966)
Luna (1968)
Romeis (1968)
Gabe (1976)
Berlyn and Miksche (1976)
Lillie and Fullmer (1976)
Lewis and Knight (1977)
Humason (1979)
Drury and Wallington (1980)
Pearse (1980, 1985)
Clark (1981)
Culling *et al.* (1985)
Boon and Drijver (1986)
Sheehan and Hrapchak (1987)
Polak and Van Noorden (1997)
Larsson (1988)
Stamp and Wright (1990)
Chayen and Bitensky (1991)
Armed Forces Institute of Pathology (1992)

Stoward and Pearse (1991)
Van Noorden and Frederiks (1992)
Sanderson (1994)
Taylor (1994)
Presnell and Schreibman (1997)
Horobin and Bancroft (1998)
Kiernan (1999)
Ruzin (1999)
Bancroft and Gamble (2002)

The works of Gatenby and Beams (1950), Jones (1950) and especially Gray (1954) provide comprehensive accounts of staining methods available at a time when histochemistry was in its infancy. The three volumes of the 4th edition of *Histochemistry: Theoretical and Applied* (Pearse, 1980, 1985; Stoward and Pearse, 1991) contain the most extensive available accounts of histochemical techniques of all kinds.

It should also be mentioned that there are more than 60 volumes published by Gustav Fischer on topics in histochemistry and cytochemistry under the general title of *Progress in Histochemistry and Cytochemistry*.

Several journals regularly contain articles dealing with the chemistry and industrial applications of dyes, and with the use of these compounds in histotechnology and histochemistry. A listing of the more important journals includes the following:

Acta Histochemica
Advances in Colour Science and Technology
American Journal of Clinical Pathology
Archives of Patholoqy and Laboratory Medicine
Biotechnic and Histochemistry (formerly *Stain Technology*)
Coloration Technology (formerly *Journal of the Society of Dyers and Colourists*)
Cytometry
Dyer
Dyes and Pigments
European Journal of Histochemistry (formerly *Basic and Applied Histochemistry*)
Histochemical Journal
Histochemistry and Cell Biology (formerly *Histochemistry*, and before that *Histochemie*)
Industrial and Engineering Chemistry
Journal of Clinical Microbiology
Journal of Clinical Pathology
Journal of Histochemistry and Cytochemistry
Journal of Histotechnology
Journal of Microscopy (formerly *Journal of the Royal Microscopical Society*)
Journal of Neuroscience Methods

The history of staining

B. Bracegirdle

BEFORE STAINS

When a child, or anyone else new to a microscope, first looks at a specimen, he puts it on the stage as best he can, shines a light on it, and views it by reflected light. This is exactly how Robert Hooke looked at most of the specimens he illustrated so magnificently in his book *Micrographia* of 1665. He went on to describe how he made what we would now call sections in various planes, how he achieved a kind of darkground illumination, and even how he cleared some of his specimens. It is a great pity that much of what he said was not understood and thus not acted upon. His readers may well have looked at the pictures and not his text!

We now know that the microscopical image must resolve required detail at the required magnification, but it must also provide sufficient contrast to allow the structure of the specimen to be observed. This contrast can nowadays be obtained physically (by optical means), and/or chemically (usually by coloring). It remains a salutary lesson, however, to take a piece of fresh tissue and observe it squashed very flat without staining, as would have been the norm until at least the 1850s. A surprising amount of detail can be made out, given a lot of patience and the odd tilt of the mirror.

Nonetheless, observers from the dawn of microscopy on must have wished for more contrast in transparent specimens, increasingly so as magnifications slowly climbed in the 18th- and rapidly in the 19th century. Leeuwenhoek followed Hooke as a microscopist, and used the smear technique with fresh specimens, among other techniques (most of which he kept to himself). However, scientific work after his in the 18th century (and on into the 19th) was based almost entirely on injection. Such injected preparations, often using several colors in different vessels, are quite amazing when viewed, but do not tell very much about the nature of the specimen apart from the course of the fine vessels. They took the average microscopical observer by storm, so amazing do they look, and for sheer effect are still worth a look if some can be found in old collections. A few 18th-century workers, such as Hill, avoided such techniques, and used cochineal as a stain. Hill also invented the microtome and used microincineration of plant sections to increase contrast. His work represented a real scientific advance, but it also failed to take root.

Much more than this sample of results was obtained before 1830, of course, but is outside the scope of this short chapter. More detailed accounts, including one of the development of the microtome, have been provided elsewhere (Bracegirdle, 1986; Clark and Kasten, 1983).

AFTER 1830

This date has been chosen because Canada balsam was introduced as a mountant about then. Soon afterwards specimens began to be cleared on the slide before mounting, and thus the basis of modern technique had been established, although it would be 50 years before it became fully developed. The use of stains at this time was just beginning to be accepted, and received its real impetus not from scientific work as such, but from the sale of preparations for the growing number of amateur microscopists; visibility of the specimen was paramount for such a purpose. From about 1840 a vast number of slides was produced (the 3 x 1 inch size was formally adopted late in 1839), and many still survive to illustrate the development of staining. Older collections are well worth the attention of the professional microscopist, for much has been reinvented several times.

Something must be said of techniques other than actual staining. The preparation of microscopic mounts is an involved procedure, and it is a relatively modern concept that the prepared specimen might bear little resemblance to the original condition, but that when a tissue is prepared in a wide variety of ways, a full idea of its make-up can be obtained. It can be difficult to be sure just what was done in earlier days, as methods were often only vaguely specified, reagents might have contained significant impurities, and their names might be quite different nowadays.

The method used to kill the animal providing the tissue can critically affect the result on the slide, and heavy narcotization before sacrifice remains unhelpful. The preservation of the tissue sample also affects the result. It was not until 1833 that chromic acid was first used as a hardening agent, although Malpighi had hardened brains for study by boiling them as early as 1666. Mercuric chloride was first used as a fixative in 1846, but not established until 1878. Acetic acid had been used, as vinegar, to preserve food, and alcohol was known to have preservative properties, but neither was used effectively in microscopy before the mid-19th century. Osmium tetroxide ('osmic acid') was introduced as a fixative in 1864, but did not find widespread use until 1882, and formaldehyde was not so used until 1893. References for all these are given by Bracegirdle (1986), but two important points have emerged. One is that the date of first use might well be very different from the date when the technique became widely used. The other is that most of the dates of first use are surprisingly recent.

NATURAL STAINS

Saffron was used by Leeuwenhoek in 1714 to color muscle fibers (van Leeuwenhoek, 1719). Such a use repeated nowadays shows that it would not have been very effective. It seems that Reichel (1758) used logwood, in simple unmordanted solution, to stain plant tissues. Hill (1770) introduced cochineal, and both hematoxylin and carmine remain in use even today.

Carmine was the only dye used for many years, but was not applied as a nuclear stain until Corti (1851) so used it. Unfortunately he published in French in an obscure German journal and his work was ignored. Von Gerlach (1858) used greatly diluted ammonia–carmine overnight, and rapidly established the technique among his students. Early workers thought that carmine stained everything, and even Beale (1860) was led to quite wrong conclusions as to the course of nerve fibers as a result: a salutary lesson even today.

Hematoxylin is an unsatisfactory stain in the absence of a mordant. It was mentioned for the first time after Reichel by Quekett (1848), whose book is an important landmark in the history of staining. It was Böhmer (1865) who first used logwood with an alum mordant, and this revolutionized the use of hema-toxylin. Some famous names in histology published their own mixtures containing this stain, but its ripening was not understood until the mechanism was described by Mayer (1891).

Many other natural stains – virtually every colored plant substance – were tried out over the years; a characteristic of such papers is their distinctly alchemical approach, and none was scientifically very useful.

ANILINE COLORS

The introduction of mauve, the first aniline dye, by Perkin in 1856 was a major innovation not only for microscopy but for the entire chemical industry. It was first used in staining by Beneke (1862), whose sample might well have been from a different source and had slightly different properties. Waldeyer (1863) used mauve, and also Paris blue and basic fuchsine, and in the same year Frey (1863) used aniline blue, and Roberts (1863) used picric acid. For a decade (apart from the introduction of indigocarmine by Chronszczewsky (1864)) the initial spate of aniline dyes as stains was over. However, in 1874 Ranvier (1874) used cyanin, Zuppinger (1874) introduced iodine violet, and Lieberkühn (1874) adopted synthetic alizarin.

The way was then clear. All histologists could scan the dyemakers' lists, and publish a paper, perhaps of indifferent quality, on the use of a new color. From this date any survey of the literature must be highly selective.

Methyl violet was introduced by Cornil (1875), in an important paper which also described for the first time the metachromatic effects of aniline dyes. Another important paper of that year was by Hermann (1875), which brought to the attention of biologists the principle of differentiating a stain. This was not the first description, for it had been first mentioned by Schweigger-Deidel and Gogiel in 1866, and again by Böttcher in 1868, but neither of these earlier investi-gators established the technique.

Ehrlich introduced at least 12 stains. These include safranine (Ehrlich, 1877), which had been available since 1859 and remains a valuable stain. He also surveyed the then available dyes, making some mistakes as to their composition. He introduced methyl blue, nigrosin, and acid fuchsin in 1879, neutral red in 1893, and janus green in 1898. His work with methylene blue (Ehrlich, 1881) was fundamental, especially for bacteriology.

Other significant stains introduced in this golden period included bismarck brown (Weigert, 1878), malachite green (van Beneden and Julin, 1884), light green and congo red (Griesbach, 1886), and the fat stains Sudan III (Daddi, 1896) and Sudan IV (Michaelis, 1901).

MULTIPLE STAINING

The first double stain was that of Schwartz (1867), who used successive solutions of picric acid and carmine. His paper has excellent color plates illustrating the effects obtained. The first to counterstain hematoxylin with an aniline dye was Poole (1875), who used aniline blue for the purpose. The following year saw the introduction of the still widely-used hematoxylin and eosin (Wissowzky, 1876). The first account of a triple stain was by Gibbes (1880), and Griesbach (1889) demonstrated quadruple staining in 1888. Such techniques are still in use for teaching, but it remains a truism that little is to be discovered merely by the use of extra colors on one tissue.

Be that as it may, some combinations have stood the test of time. Van Gieson's (1889) stain was originally worked out for nervous tissue but is now applied to connective tissue. Flemming's (1891) triple stain is still used for nuclear cytology, and Mallory's (1900) method is still valuable for showing collagenous and cytoplasmic fibers in contrasting colors.

The blood stains are important, not only for their practical utility but also for their scientific background. In their developed form they are combinations of basic with acidic dyestuffs, and were originated by Ehrlich (1879), with some understanding of their chemistry. Many others were to contribute to the perfection of such stains, culminating in Giemsa (1902a,b), who made the stains reliable. Bernthsen (1906) clarified the chemistry of the oxidation products of methylene blue present in these stains.

THE BEGINNINGS OF CYTOLOGICAL STAINING

The mitochondrion was one of the first organelles to attract attention, and was described under many different names before Benda (1901) described a procedure that rendered results easy to obtain. After refinement (Meves and Duesberg, 1908), the technique has remained useful into modern times. Golgi bodies attracted more attention still, because they were lost in routine fixation. Their very existence was fiercely disputed (Kirkman and Severinghaus, 1938) until use of the electron microscope proved that they were present.

The first silver impregnation technique (Krause, 1844) was used on pieces of skin. Improvements by von Recklinghausen (1860) brought this type of method into wider use.

Gold salts were first used by Cohnheim (1866). Osmium tetroxide was used by Golgi from 1873, and much improved in his rapid method (Golgi, 1886), which took only a week to carry through. His preparations (which survive in Pavia) are important in the history of histology. Ramon y Cajal developed the work further, especially for spinal ganglia, and his double method (Ramon y Cajal, 1891) was a considerable advance. Much mumbo-jumbo was written about these techniques, but Hill (1896) clarified the various formulae.

COLORING LIVING TISSUES

Although Trembley colored *Hydra* in 1744, the real interest in such work developed only from about 1880 (Certes, 1881; Ranvier, 1975). The modern development of true intravital staining took place when Bouffard (1906) injected acid dyes and Goldmann (1909) used a colloidal dye, pyrrhol blue. Subsequent workers placed much reliance on intravital techniques, especially in the development of histochemistry.

Although Girod-Chantrans used a variety of reagents and tests to classify plants as early as 1802, it was Raspail (1825) who founded histochemistry. He deliberately set out to apply histochemical tests in his researches, and consolidated the work in a famous essay (Raspail, 1830). The further history of histochemistry has been outlined by Petersen (1941), Sandritter (1964) and Pearse (1980)

Nothing has been said in this short summary about the development of infiltration and embedding methods, and the slow perfection of microtomy. However, by the end of the 19th century, much that would still be recognizable in a laboratory for histology had been developed. Possibly the most substantial work ever in the field of practical histology was the third edition of a book edited, in its first edition, by Ehrlich, and first published in 1903 (Krause, 1923). This first edition is a good summary of the state of histological work at the turn of the century, while the three volumes of the 1926–27 edition contain 2444 pages, and have copious references to each entry. This is some measure not only of the thoroughness of the authors, but of the development of the subject in a relatively few years.

The story may be brought further up to date by a series of editions of a substantial book edited originally by Arthur Bolles Lee, first published in 1885 (Lee, 1885), and reaching its 11th edition in 1950. These summarize most work of any note in the field, although coverage is less full in the last edition.

A few more general works on the nature of staining had appeared, prominent among which is that by Mann (1902). This seminal work gave a detailed scientific background of the mechanisms underlying the various techniques, as well as detailed notes on the different processes themselves.

INTO THE TWENTIETH CENTURY

By the end of the 19th century, the optics of the microscope had been developed to near perfection by the use of apochromatic objectives applied on a sound theoretical basis. Photographic recording of results had become less of an art and more of a science; most of the required mounting procedures were in place; some development of micromanipulation had taken place, and a few had looked at living material as well as fixed.

Evans and Schulemann (1915) introduced specialized acid dyes (trypan red, trypan blue, vital red, Evans blue) which remain of much importance, and vital staining was given a further boost in the 1920s. Simpson (1921) investigated the reactions of living leukocytes to more than 200 dyes, and Sabin (1923) analyzed human blood cells using neutral red and janus green B. Stockinger (1964) brought this work more up to date in a useful monograph on vital staining and fluorochroming of animal cells.

FLUORESCENCE MICROSCOPY

The fluorescence microscope, first used for histology and histopathology in the 1930s, is of the greatest importance in the 1990s. The techniques have developed beyond all recognition over the previous 25 years. Immunofluorescence, for localizing specific antigens relies on it, and both histo- and cytochemistry have been revolutionized by it. The distribution of macromolecules in the cell can be demonstrated using secondary fluorochromes, and a wide range of these is now available.

The technique was introduced by Heimstädt and Lehmann between 1911 and 1913 (Heimstädt, 1911). It was not to become widely used for well over another 40 years, although both Zeiss and Reichert offered suitable instruments by 1915. Almost all the important work before about 1940 was carried out in Germany; suitable reviews of what had been achieved are provided by Metzner (1931) and Hamperl (1943). All this work established the instrumentation, and especially the light sources and filter systems. The secondary fluorescence techniques were then applied in medicine and life sciences. The work caused much excitement in physiology, and a great deal was achieved before World War II intervened.

POST-1950 DEVELOPMENTS

Since about 1950 the world of light microscopy has been transformed at its cutting edges. Naturally, the larger part of histological microscopy remains that of routine coloring of sections. Less of this is now carried out in university departments for training medical and life sciences students, as histology now gets much less attention in premedical and other courses than it did until about 1985. (Interestingly, in the United Kingdom, it became an actual statutory requirement to include a considerable amount of practical and theoretical histology in the medical curriculum in 1886. It was to last for almost exactly a century.) Much routine histology is still carried out for pathological diagnosis, and this may well continue for many years yet. For those carrying out research, however, matters are now rather different.

Any historian working on the very recent past faces two difficulties. One is that there is a great abundance of material, making it impossible to appraise all that is to hand. The other is that, until some years have elapsed, it is impossible to decide what is going to be significant and what is not. These same difficulties apply to the history of staining; the literature is vast and the important techniques of tomorrow are impossible to forecast.

INCREASING PRECISION IN THE 1980s AND 1990s

Nucleic acids were discovered as long ago as 1869, in Tübingen, by Miescher (1871), who extracted them from pus and realized that they came from the nuclei of cells. From that observational feat came much of today's amazing precision in identifying the nature and geography of a wide range of cellular components. Miescher relied on gravity to recover his separated nuclei from other cell components; it would not be until 1934 that the centrifuge would be applied to cell fractionation. By the time of his death in 1895, Miescher recognized that he had founded biochemistry, but he never accepted that nuclein (chromatin) was the chemical foundation of heredity.

Zacharias (1881) had actually already demonstrated that chromatin, chromosomes, and nuclein were essentially one and the same, and others completed similar work. Kossel discovered adenine, thymine, and thymic acid, and he was awarded the 1910 Nobel prize in medicine for his work on the chemistry of cell nuclei. In spite of all this work, 47 years were to pass before it was categorically stated by Gulland (1938) that plant and animal nucleic acids were accepted as being one and the same, and that DNA and RNA were separate kinds found in both plant and animal cells.

Feulgen and Rossenbeck (1924) had pioneered the first cytochemical reaction for DNA and it was applied to the giant salivary gland chromosomes of some Diptera by King and Beams (1934). The Feulgen reaction obviously filled a great need for a reliable DNA/chromosome stain; by 1938 more than 400 references to the DNA nuclear reaction had been published, and it has remained a valuable tool through and beyond the 1950s.

Large numbers of workers have applied more modern techniques, and more especially a wide range of specific fluorochromes, to chromosome banding methods. Others have developed autoradiography, flow cytometry, and a range of other techniques to investigating with ever greater precision just what is where in the cell, and just what it is doing there. The electron microscope has been developed as an analytical tool, and various new kinds of light microscope now allow investigation of living cells more easily and precisely than ever before. The application of powerful electronics to imaging, image enhancement, and computer analysis has opened up still more possibilities. It would be premature here to analyze what might be achieved with all these even over the next 10 years.

3

Nomenclature and classification of dyes and other coloring agents

J.A. Kiernan

Most of this chapter is concerned with the chemical groups of dyes used in biology. At the end is a short section on more complex substances: probes and labeled compounds.

NAMING OF DYES

The formal chemical names of most dyes are so long that they are never used in conversation or writing. Instead, each dye has one or more informal or trivial names. In addition, many dyes have standardized *Colour Index* names based on their original industrial uses and colors.

Informal names

Most informal names include the color, and some also indicate a major property or the general class of compounds to which the dye belongs. For example, acridine orange is a dye that contains the acridine ring structure. Letters at the ends of names may indicate similar dyes with slightly different colors such as eosin Y and eosin B. Words in the informal names of dyes are often trade-names (e.g. coomassie, eriochrome, sirius) that bear no relation to chemistry or color but have entered the language of scientists through common usage. Often an informal name gives no information about the chemical constitution or colorant properties of a dye. The names methyl blue or Congo red might apply to almost any blue or red compound, and names like vesuvin and coriphosphine do not reveal even the colors.

Colour Index names and numbers

The *Colour Index* (CI) and its supplements, published by the Society of Dyers and Colourists (1971–1996), is an encyclopedic reference for dyes and other coloring agents. It provides for each dye a unique number, related to chemical structure, and a name based on the industrial method of application and color. Thus methylene blue, a cationic (basic) dye, is CI 52015 or Basic blue 9. Sirius red 4B (CI 28160) is named *Direct red 81* because it is one of the group of direct cotton dyes. The groups of colorants that form the basis of this naming system are summarized in Table 3.1. The chemical groups of dyes recognized in the CI are listed in Table 3.2.

Table 3.1 Application classes of dyes, as used in *Colour Index* names

Acid dyes	Colored anions, applied at low pH to wool, silk and nylon.
Azoic dyes	Insoluble azo dyes, formed within the substrate by reaction of a diazonium salt (**azoic diazo component**) with a naphthol (**azoic coupling component**).
Basic dyes	Colored cations. Used on polyacrylonitrile (acrylic) textiles, which have acid side chains. Formerly also on proteinaceous fibers and tannic acid-treated cotton.
Direct dyes	Acid dyes with large molecules that will bind to cellulose fibers directly.
Disperse dyes	Insoluble colored compounds in fine suspension, which enter hydrophobic materials (cellulose acetate, polyester) when applied in the presence of a 'plasticizer'.
Fluorescent brighteners	Absorb ultraviolet and emit blue; used to offset yellowing of white textiles.
Food dyes	Substances used to color foodstuffs.
Ingrain dyes	Temporarily solubilized colorants that change chemically and become insoluble after application to the substrate being dyed.
Mordant dyes	Dyes applied in conjunction with metals. The metal may be applied before, with or after the dye.
Natural dyes	Made or extracted from plants or animals.
Oxidation bases	Colorless amines and aminophenols that polymerize to brown or black insoluble substances when oxidized (usually by hydrogen peroxide). Used on fur and hair.
Pigments	Finely divided white, black or colored materials, insoluble in water and organic solvents. Used in paints and for mass coloration of plastics.
Reactive dyes	Colored compounds with side chains that react to form covalent bonds with the substrate. Used to provide very fast (resistant) dyeing of cellulose etc.
Solvent dyes	Colored compounds soluble only in hydrophobic solvents. Used for mass coloration of liquids and plastics.
Sulfur dyes	Cheap polymeric dyes made by heating aromatic amines and phenols with sulfur or sodium polysulfide. A soluble (reduced) form is applied to the textile and the insoluble polymer is then regenerated by oxidation.
Vat dyes	Applied to cloth as a soluble compound that is then oxidized to the insoluble, fully colored form of the dye. Used mainly on cotton.

Table 3.2 Constitution (Chemical) classes of dyes recognized in the *Colour Index*

Nitroso dyes	Indophenol dyes
Nitro dyes	Azine dyes
Azo dyes	Oxazine dyes
Azoic dyes	Thiazine dyes
Stilbene dyes	Sulfur dyes
Carotenoid dyes	Lactone dyes
Diphenylmethane dyes	Aminoketone dyes
Triarylmethane dyes	Hydroxyketone dyes
Xanthene dyes	Anthraquinone dyes
Acridine dyes	Indigoid dyes
Quinoline dyes	Phthalocyanine dyes
Methine dyes	Natural organic coloring matters
Thiazole dyes	Oxidation bases
Indamine dyes	Inorganic coloring matters

Choices of names and spellings

In this book, the informal names of dyes are those believed to be most widely used by biologists. Many synonyms are listed, as are the CI numbers and application names. For uniformity, all words in the informal names of dyes begin with lower-case letters unless they are obviously names of people or places. The names from the CI are treated in the same way as the generic and specific names of animals or plants, with an initial capital letter for the word denoting mode of application class, and lower case for the color. Thus CI 51180 is **Nile blue** or **Basic blue 12**.

Many words in the informal names of dyes end in –ine or –in, and these endings have often been used wrongly, especially for microscopical stains. The ending –ine is appropriate for an organic base or a compound derived from an organic base. The ending –in is used for compounds that are not bases or their derivatives. Thus, **acid fuchsine** receives a terminal –ine because it is basic fuchsine modified by addition of sulfonic acid groups. **Eosin** and **phloxin** on the other hand are heterocyclic phenols, not bases, so their names end in –in. Correct spellings are invariably used in modern works concerned with dye chemistry, such as those Gordon and Gregory (1983), Waring and Hallas (1990) and Zollinger (1991). In his classical text on staining, Baker (1958) reminded his readers of the rules that governed spelling of chemical and pseudo-chemical names. Other modern writers in the field, including Lillie (1977), Horobin (1982, 1988), Chayen and Bitensky (1991) and Green (1990), have been less consistent, and 'chemically incorrect' spellings abound in vendors' catalogs. The spellings of informal names in this 10th edition of *Conn's Biological Stains* are 'chemically correct' and are also the ones that have always been preferred in major American and British dictionaries such as *Webster's Encyclopedic* and the *Oxford English Dictionary*.

CLASSIFICATION OF DYES

Chemical structures determine the colors, properties and uses of dyes, and provide the only rational basis of a classification of these compounds. The numbering system and arrangement of dyes used in the CI and in *Conn's Biological Stains* are based primarily on the chemical structures of chromophores (Table 3.2). In this edition of *Conn* there is a stricter adherence to chemical structure than previously, so that chemically similar compounds are near to one another. For example, the 'natural dyes' in Chapter 17 of the 9th edition are now placed in groups that contain chemically similar synthetic dyes. Within each major category the dyes are arranged according to variations in the chromophore and the attached substituents. The works of Allen (1971), Venkataraman (1952–1978), Gordon and Gregory (1983) and especially Zollinger (1991) have guided the placement of compounds in the groups summarized in Table 3.3.

Table 3.3 Synopsis of the classification of dyes and other colored and fluorescent substances adopted in the 10th edition of *Conn's Biological Stains*

Group for 10th edition	Chapter in 10th edition	Chapter(s) in 9th edition	Group in 9th edition
Inorganic coloring agents	27	17	Natural dyes etc.
Nitroso and nitro dyes	9	5	Nitroso and nitro dyes, some indicators, and miscellaneous chromogenic reagents
Azo dyes:			
Monoazo dyes	10	6	Monoazo dyes and fluorochromes
Dis-, tris- and polyazo dyes	11	7	Dis-, tris- and polyazo dyes
Diazonium salts and their reaction products	12	8	Azoic diazo compounds, phenolic and other azo coupling reagents and some related chromogenic reagents
Tetrazolium salts and formazans	13	9	Tetrazolium salts and related chromogenic reagents
Arylmethane dyes:			
Diarylmethanes	14	10	Ketoneimines, diarylmethanes and hydroxyketones
Aminotriarylmethanes	14	11	Aminoarylmethanes
Hydroxytriarylmethanes	15	12	Hydroxytriarylmethanes etc.
Simple hydroxytriarylmethanes (rosolic acids)	15	12	Hydroxytriarylmethanes etc.
Phthaleins	15	12	Hydroxytriarylmethanes etc.
Sulfonphthaleins	15	12	Hydroxytriarylmethanes etc.

Table 3.3 Continued

Group for 10th edition	Chapter in 10th edition	Chapter(s) in 9th edition	Group in 9th edition
Xanthene dyes	16	13	Xanthenes and acridines
Acridine dyes	17	13	Xanthenes and acridines
Phenanthridines	17	16	Anthraquinone and other polycyclic dyes: phenanthrolines
Azine dyes	18	15	Quinone imine dyes etc.
Oxazine dyes	19	15	Quinone imine dyes etc.
Thiazine dyes	20	15	Quinone imine dyes etc.
Polyene dyes:	22		
Carotenoids	22	17	Natural dyes
Stilbene dyes and fluorescent brighteners	22	10	Ketoneimines, diarylmethanes and hydroxyketones
Other polyenes (Diphenylhexatrienes and benzofuranyl fluorochromes)	22	0	(None in 9th edn)
Polymethine dyes and fluorochromes:			
Cyanines and azamethines (Subgroups include cyanines, hemicyanines, streptocyanines, neutrocyanines, merocyanines, azacyanines and oxonols. An individual dye can belong to more than one of these subgroups.)	23	14, 15, 9	Quinoline, polymethine, and thiazole dyes; Quinone imine dyes; Tetrazolium salts and related chromogenic reagents
Cyanine-like fluorochromes (Benzimidazoles and indolenines) 25	23	16	Anthraquinone and other polycyclic dyes
Styryl dyes	24	0	(None in 9th edn)
Indamine and indophenol dyes	24	15	Quinone imine dyes
Quinophthalone dyes	24	14	Quinoline, polymethine, andthiazole dyes
Thiazole dyes	24	14	Quinoline, polymethine, and thiazole dyes
Anthocyanines	24	17	Natural dyes etc.
Coumarin dyes (and fluorescent brighteners)	24	0	(None in 9th edn)
Flavonoid and similar neutrocyanine dyes	24	10, 17	Hydroxyketones; Natural dyes

Table 3.3 Continued

Group for 10th edition	Chapter in 10th edition	Chapter(s) in 9th edition	Group in 9th edition
Carbonyl dyes:			
Indigoid and thioindigoid dyes and pigments	25	16	Anthraquinone and other polycyclic dyes
Naphthoquinone dyes	25	10, 17	Hydroxyketones; Natural dyes
Anthraquinone dyes	25	16	Anthraquinone and other polycyclic dyes .
		17	Natural dyes (carmine, tetracyclines)
Naphthalimide dyes and fluorochromes	25	10	Aminoketones etc.
Aza[18]annulene colorants: (includes porphyrins and phthalocyanines)	26	16	Anthraquinone and other polycyclic dyes (Natural ones not in 9th edn)
Other organic colorants	27	Various	(In various chapters of 9th edn)

    No classification of dyes can keep everyone happy because many compounds meet chemical criteria for inclusion in more than one group. The following account summarizes the general properties of each family of dyes.

1. Inorganic colorants

Iodine and silver nitrate are among several inorganic substances used to induce color in microscopical preparations. Colored products may also be formed by the reactions of colorless inorganic reagents, as when ferrocyanide ions react with ferric ions to form prussian blue.

2. Nitroso dyes

These result from the reaction of phenols with nitrous acid, which places a nitroso (–NO) group on a carbon atom *ortho* or *para* to the phenolic –OH. The C-nitroso compound is in equilibrium with a quinone oxime, as illustrated in **Background information** in Chapter 9. If the oxygen atom is *ortho* to the nitroso, the oxygens can form bonds to a metal atom. The resulting strongly colored compound may be a pigment or, if there is a solubilizing substituent (typically sulfonate $-SO_3^-$), a dye.

3. Nitro dyes

These yellow dyes contain one or more nitro ($-NO_2$) groups. The N–O and N=O bonds of the nitro group are equivalent because of resonance, and they are conjugated (alternating) with the C–C and C=C bonds of the aromatic ring. If the compound is a phenol, the aromatic compound exists in equilibrium with a quinonoid one.

4. Azo dyes

The azo group, –N=N–, is present in more commercially available dyes than any other chromophore. An azo group is formed when a diazonium ion, known as the **diazo component**, reacts with either a phenol or an amine, known as the **coupling component**. Diazonium ions are generated by the action of nitrous acid (from $NaNO_2$ and HCl) on aromatic primary amines at about 0°C. In the azo coupling reaction the end nitrogen atom of a diazonium ion displaces a hydrogen from the aromatic ring of a phenol or amine. Coupling components derived from benzene often couple in the *para* position, but naphthalene derivatives usually couple *ortho* to their hydroxyl or amine groups. The formation of diazonium salts and their coupling reactions are illustrated in the **Background information** of Chapter 10. Colored azo compounds can be synthesized by several methods other than the coupling of diazonium salts (Zollinger, 1961), but such reactions are not often commercially exploited.

Structure and properties

In the azo chromophore the bonds between the azo nitrogen atoms and the aromatic rings are angulated, so that *cis* and *trans* isomers are theoretically possible, as illustrated in **Background information** in Chapter 10. Azo dyes exist as *trans* isomers because of hydrogen bonding between an azo nitrogen atom and a hydrogen atom of a nearby polar substituent such as an *o*-hydroxyl or *o*-amino group:

chrysoidin
CI 11270, Basic orange 2

Sudan orange G
CI 11920, Solvent orange 1

A phenolic oxygen and an azo nitrogen atom may be suitably placed to form bonds to a metal ion, the presence of which brings about a major change in the color and other properties of the dye.

In most azo dyes the aromatic rings are benzene or naphthalene. Some include a heterocyclic ring, most commonly pyrazolone. The aromatic rings carry a wide range of substituent groups that determine the color and the dying properties. Many azo dye molecules contain two or more azo linkages. The number of azo groups forms the basis of segregation of this large group of dyes: **monoazo dyes** have one azo linkage, whereas **dis-**, **tris-**, **tetrakisazo dyes** have two, three, or four azo groups respectively. From their industrial applications and their properties as biological stains, azo dyes fall into seven groups.

(i) **Cationic or basic azo dyes.** These may have amine side chains or quaternary nitrogen atoms. The charged group may be part of the chromogen (delocalized cationic dyes) or attached by a nonconjugated chain of carbon atoms (pendant cationic dyes).

(ii) **Anionic or acid azo dyes.** These have sulfonic or carboxylic acid groups attached to their aromatic rings.

(iii) **Direct azo dyes.** These anionic dyes have large molecules (two or more azo linkages) that can assume a coplanar conformation. Direct dyes bind to cellulose (cotton, linen) by nonionic, noncovalent forces.

(iv) **Reactive azo dyes.** These have pendant side chains that can combine covalently with the substrate. Attachment to the hydroxyl groups of cellulose is a major commercial application.

(v) **Mordant azo dyes.** In these a hydroxyl group is adjacent to a ring carbon that is joined to either an azo nitrogen atom or a carboxyl group. Either arrangement can combine with a metal atom, such as chromium(III) to form a stable five- or six-membered chelate ring.

(vi) **Solvent azo dyes.** These lack sulfonic acid or ionized amino groups that would confer solubility in water. They are used for the mass coloring of hydrophobic materials.

(vii) **Azoic dyes.** These insoluble colored compounds are formed in or on the substrate by reaction of separately applied diazonium salts and coupling components. Most diazonium salts are chemically unstable, but some can be manufactured as dry powders. These stable compounds are called **azoic diazo components**. Some have two diazonium ions per molecule and are informally known as **tetrazonium salts**. Azo coupling reactions are exploited in many histochemical techniques, especially for detecting enzyme activities.

5. Tetrazolium salts and formazans

A tetrazolium cation contains a five-membered ring with four nitrogen atoms and one carbon. Reduction of this ion yields an insoluble colored formazan, with an open chain in which two pairs of nitrogen atoms are attached to the same carbon. Conjugated bonds extend through the formazan chromophore into the aromatic substituents (R, R′, R″ in the later formulae) derived from the original tetrazolium salt, as illustrated in **Background information** in Chapter 13.

A metal-complexing formazan, zincon, is used histochemically to detect zinc. The chromogenic reduction of tetrazolium salts is exploited in the study of enzymes that catalyze biological oxidations.

6. Arylmethane dyes

The general formula for this large group of dyes is:

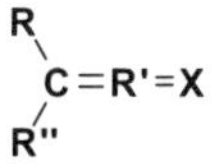

$$\begin{matrix} R \\ \diagdown \\ C=R'=X \\ \diagup \\ R'' \end{matrix} \qquad (\text{ X = N or O })$$

where R and R′ are benzene or naphthalene rings. R″ is an amino group in the diarylmethanes or another ring in the much more numerous triarylmethanes. The three bonds to the central carbon are all the same, and are quite unlike the C–H bonds of methane. For this reason, Zollinger (1991) prefers the name **aryl-methine dyes**. Other modern authors retain the traditional –methane suffix despite its chemical inaccuracy.

The most common substituents on the rings, which determine the properties of the dyes, are amine and hydroxyl groups. Thus, four families of arylmethane dyes are recognized.

(i) **Diarylmethanes.** This is a small group of cationic dyes with the general structure shown in **Background information** in Chapter 14.

(ii) **Aminotriarylmethanes.** The simplest member of this series is pararosaniline. Its colored cation can be correctly formulated in various ways. These are illustrated in Chapter 14, Figure 14.1. In other dyes of this group the rings may be naphthalene rather than benzene, the basic groups may be secondary or tertiary amines, or the basic groups may be out-numbered by sulfonate ions, giving anionic dyes. These dyes are notable for their bright, strong colors.

(iii) **Hydroxytriarylmethanes.** In these anionic dyes, at least two of the rings attached to the central carbon bear phenolic or quinonoid oxygen atoms. Three subgroups are recognized (CI; Lillie, 1977).

   (a) **Simple hydroxyarylmethanes** or **rosolic acids** are analogs of rosaniline in which oxygen replaces nitrogen. Thus, pararosolic acid has –OH or =O *para* to the central carbon atom. The generic structure of hydroxytri-arylmethanes is shown in the **Background information** of Chapter 15, Figure 15.1.

   (b) **Phthaleins** are triarylmethanes with *para* oxygen atoms in two rings. The third ring lacks a *para* oxygen but has a carboxyl group *ortho* to the central carbon. In a basic environment, this exists as a colored carboxylate anion. Acidic conditions suppress ionization, and the carboxyl group forms a lactone (internal cyclic ester) that incorporates the central carbon atom. The lactone, which does not contain the triarylmethane chromophore, is colorless (see Chapter 15, Figure 15.2).

   (c) **Sulfonphthaleins** have a sulfonic acid group *ortho* to the central carbon of the triarylmethane structure, and it is similarly able to form a lactone-like (sultone) linkage with the central carbon. An example is **phenol red**. The sultone form of this compound is illustrated in the appropriate dye entry in Chapter 15. The sultone forms of sulfon-phthaleins are white and exist only in the solid state.

31

7. Xanthene dyes

The chromophore in these dyes contains the planar skeleton of the oxygen-containing heterocyclic compound xanthene. In the general formula, shown in the **Background information** of Chapter 16 (Figure 16.1), R may be a hydrogen atom or an aliphatic or aromatic group, and A and D are nitrogen in the **aminoxanthenes** or oxygen in the **hydroxyxanthenes**.

In many of these dyes, R is a phenyl group with a one or more carboxyl or sulfonic acid side-chains. The structure is then that of a triphenylmethane with an oxa bridge joining two of its phenyl groups. Halogens commonly replace hydrogens in the xanthene and other parts of the molecule.

Xanthene dyes used for biological staining are yellow or red, and many are also fluorescent. The pyronines and rhodamines are examples of aminoxanthene dyes; fluorescein and the eosins are well known hydroxyxanthenes.

8. Acridine dyes

The structure of acridine resembles that of xanthene, except that the heteroatom is nitrogen, not oxygen. The acridines are strongly fluorescent yellow cationic dyes. Acridine orange and acriflavine are examples.

Phenanthridines are related compounds, with a chromophore that is an isomer of acridine. The phenanthridines include ethidium bromide, propidium iodide and related fluorochromes.

For general structures of acridine and phenanthridine dyes, see the **Background information** in Chapter 17 (Figure 17.1).

9. Azine dyes

The azine chromophore contains the skeleton of phenazine, which consists of two benzene rings linked by two nitrogen atoms. A delocalized positive charge and an alternation of aromatic and quinonoid structures constitute the chromophoric system, which is illustrated in the **Background information** of Chapter 18 (Figure 18.1).

A wide variety of substituents may be attached to one of the phenazine nitrogens, and one or both of the six-carbon rings in the general formula may be part of a larger system of fused rings. Neutral red and safranine O are azine dyes used as biological stains.

10. Oxazine dyes

In these dyes, a nitrogen and an oxygen atom make bridges between two six-carbon rings, as shown in the **Background information** in Chapter 19 (Figure 19.1).

Most oxazine dyes are either colored cations or metal-binding phenolic compounds. They include important pH indicators (e.g. litmus) and microscopical stains (e.g. cresyl violet and orcein).

11. Thiazine dyes

In the three-ring thiazine chromophore nitrogen and sulfur make bridges between two six-carbon rings. These are all cationic dyes, and electron delocalization results in partial positive charges being present on both nitrogen atoms and on the sulfur. Structures of various thiazine dyes are shown in the **Background information** of Chapter 20 (Figure 20.1).

These dyes are used for staining bacteria and blood cells, in biochemical studies of oxidation–reduction reactions, and as indicators (see **Redox indicators** in Chapter 8).

12. Polyene dyes and fluorochromes

The polyene chromophore is a simple chain of alternating double and single bonds. If the compound is a hydrocarbon the conjugated chain must contain at least 11 double bonds to bring its absorption spectrum into the visible range. This number is reduced to five if the chain has hydrophilic oxygen-containing groups (free or esterified carboxyl) at its ends. Aromatic and quinonoid rings also enhance visibility. The polyene group could include all dyes and fluorochromes, but its definition excludes compounds in which a positive or negative charge might exist at one end of the conjugated chain. This constraint consigns all but a few dyes and fluorochromes to other categories, leaving as polyenes the carotenoids, stilbenes and a few related fluorochromes.

(i) **Carotenoids.** These colorants have no aromatic rings in their molecules. More than 300 carotenoids occur in plants, and a few are synthesized industrially for coloring foods. The simplest member of the series is lycopene, named for its presence in *Lycopersicon*, the tomato:

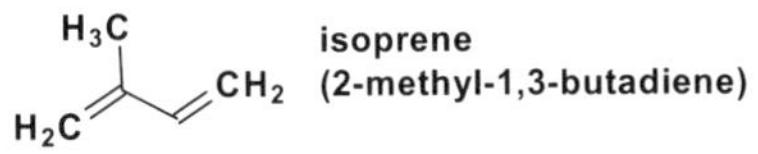

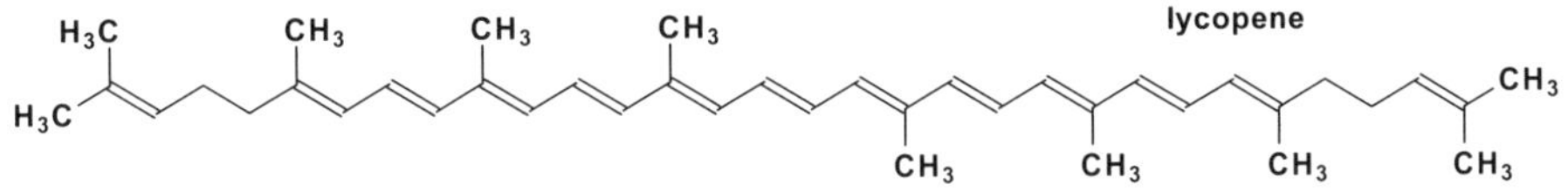

This formula shows the carotenoid structure, comprising eight isoprene units arranged so that there is a long conjugated chain in the middle part of the molecule (Kienzle and Isler, 1978). In other carotenoids the ends of the chains are folded into rings, which may be alicyclic or quinonoid, and may bear such substituents as =O, –OH and –OCOCH$_3$.

(ii) **Stilbene dyes and fluorescent brighteners.** The core stilbene structure is illustrated in the **Background information** of Chapter 22 (Figure 22.1). There are dyes manufactured in which the chromophoric system is extensively substituted with aromatic moieties.

33

This group includes some direct cotton dyes of uncertain composition and many compounds used as fluorescent brighteners. Some of the latter have been used as fluorochromes, especially in botanical microscopy.

(iii) **Other polyenes.** Certain fluorescent compounds used to trace transport within living cells are placed for convenience in this category. They include certain **diphenylhexatrienes** and the cationic **benzofuranyl** fluorochromes known as fast blue and true blue.

13. Polymethine dyes and fluorochromes

In these compounds the chromophore consists of one or more –CH= (methine) groups with an electron donor at one end of the chain and an electron acceptor at the other end. In most of these dyes the donors and acceptors are nitrogen atoms, at least one being within a heterocyclic ring. The hydrogen atom of a methine group may be replaced by another atom, such as carbon in a methyl group or as part of a ring. Related polymethine dyes have one or more –N= (aza, azamethine or indamine) rather than methine groups groups within the conjugated chain.

Dye chemists differ in their use of terms and placement of these dyes into groups (Brooker, 1966; Gregory, 1990). Here, nine major groups are recognized within the diverse family of polymethine dyes, on the basis of conspicuous chemical features such as ring systems. There are, however, many dyes that qualify for inclusion in more than one of the groups.

(i)   **Cyanines and azamethines.** The general formula for this group of dyes is:

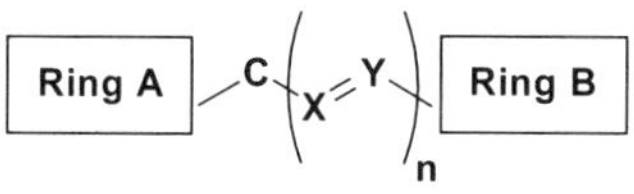

A and B are aromatic rings, of which at least one is heterocyclic with a positively charged (usually quaternary) nitrogen. Several subgroups are recognized, but they are not mutually exclusive. For example, a particular hemicyanine might also be an azamethine. The wavelength of maximum absorption increases with the number of bridging methine groups (n in the general formula).

Cyanine dyes proper have $n > 0$ and X and Y are carbon atoms in the general formula. Rings A and B, which contain the donor and acceptor nitrogen atoms, may be the same (symmetrical cyanines) or different (unsymmetrical cyanines). In a **hemicyanine** dye there is only one charged nitrogen-containing ring; the nitrogen at the other end of the polymethine bridge is not a part of Ring B. In a **streptocyanine** dye, the chain of methine groups is terminated at both ends by nitrogen atoms that are not in rings; A and B may be non-cyclic groups such as $-N(CH_3)_2$. The word **carbocyanine** is sometimes applied to dyes in which $n = 1$, but it is also used for cyanines that are not azamethine dyes (see below).

Neutrocyanines have a potentially anionic acceptor atom, such as oxygen in a carboxyl or phenolic hydroxyl group, and the molecule has no net charge.

The **flavonoids**, which are naturally occurring neutrocyanines, are here treated as a separate group of polymethine dyes; it includes some important biological stains. The **merocyanines** are synthetic neutrocyanine dyes.

In the **azamethine** dyes, either X or Y of the general cyanine formula is a nitrogen atom. Indamines and indophenols (see below) are azamethine dyes in which a single –N= forms the bridge between Rings A and B of the general cyanine formula.

An **oxonol dye** has methine (or azamethine) bridges between ring systems that contain oxygen atoms, such as ketone or hydroxyl groups, as the electron donor and acceptor.

If in the general cyanine formula $n = 0$, there is no methine bridge connecting Ring A with Ring B. Such a dye is sometimes called an **apocyanine**, because of its resemblance to a cyanine. The thiazole and quinophthalone dyes (see below) are examples of apocyanines.

Cyanines and azamethines have many industrial uses, especially in the manufacture of photographic materials. A few are used as biological stains and many as fluorochromes.

(ii) **Cyanine-like fluorochromes.** This arbitrary group contains two groups of cationic fluorochromes that bind strongly to nucleic acids, typically with a resultant increase in fluorescence. **Benzimidazoles** and **indolenines** include respectively the 'Hoechst' dyes and DAPI. These are fluorochromes that bind to DNA.

(iii) **Styryl dyes.** These have molecules containing a styryl group, usually conjugated with a tertiary amine group. The properties of the dyes are determined by the substituents (R, R', X and Y):

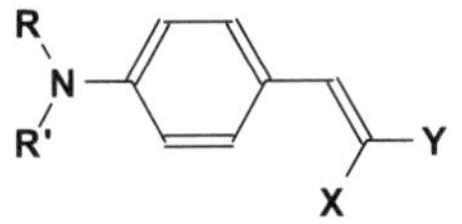

The structure resembles that of hemicyanines and other polymethines but the chromophoric system does not end in a heterocyclic ring. For this reason, Gregory (1990) places styryl dyes in a major class of their own. The group includes fluorescent probes of cellular activities. Note that certain dyes described in the biological literature as styryls do not fit the definition given here (see Chapter 24).

(iv) **Indamine and indophenol dyes.** In these azamethine dyes an aza or indamine (–N=) group forms a bridge between an aromatic and a quinonoid ring. In the indamines, nitrogen atoms terminate the conjugated chain in both ring systems, whereas an indophenol dye has the chain terminated by phenolic hydroxyl or a quinonoid carbonyl group at one end. Colored compounds with indamine and indophenol structures are the products of some histochemical reactions.

(v) **Quinophthalone dyes.** These, also known as quinoline dyes, have quinoline joined to phthalic anhydride. A methine and an azamethine

35

group are present in the center of the molecule. A few quinophthalones are used in biological research because they have greater affinity for cell membranes than for extra- or intracellular fluids.

(vi) **Thiazole dyes.** These, like the quinophthalones, are apocyanines. The methine that joins the two rings is partly included in benzothiazole.

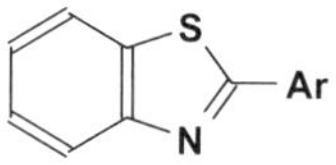

Ar is an aromatic ring, which may be benzene or another benzothiazole. The chromophoric system can be positively charged from protonation or methylation of the thiazole nitrogen. Some dyes in this group are made anionic by sulfonation. A few thiazole dyes, including primuline and the thioflavines, are used as fluorescent tracers and stains.

(vii) **Anthocyanines.** The aglycones (sugar-free phenolic components) of many glycoside pigments of plants are anthocyanines. These have a three-ring cationic structure that can be formulated with the positive charge on oxygen or carbon:

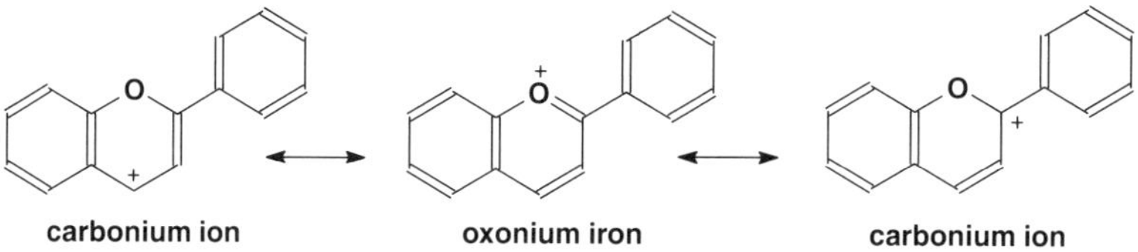

carbonium ion   oxonium iron   carbonium ion

The oxonium ion is the one usually depicted in structural formulae. This ring system has features in common with the coumarins and flavones.

(viii) **Coumarin dyes and fluorescent brighteners.** These fluorescent compounds contain coumarin, with a wide variety of substituents.

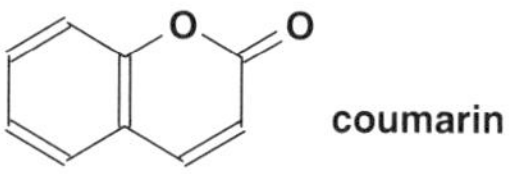

coumarin

Depending on the nature of the substituents, these compounds can also be classified as neutrocyanines or hemicyanines. They also have structural similarity to the flavonoids.

(ix) **Flavonoid dyes.** Flavone is a ketone that forms part of many colored compounds that occur in plants.

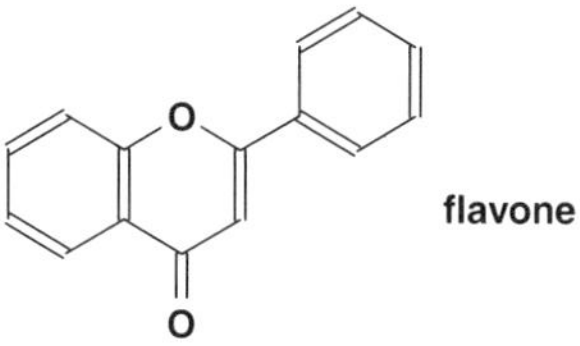

flavone

The arrangements of aromatic and quinonoid rings vary greatly. These dyes also fall within the definition of the **neutrocyanines**. Brazilein and hematein, formed by oxidation of brazilin and hematoxylin, are examples of flavone dyes. They have also been included in a group known as 'hydroxyketone' dyes (Allen, 1971).

14. Carbonyl dyes

The family of carbonyl dyes, recognized by Zollinger (1991), includes compounds in which two carbonyl groups form parts of a chain of conjugated bonds interposed between an electron donor and an electron acceptor. Four subgroups are recognized here.

(i) **Indigoid and thioindigoid dyes and pigments.** The chromophore is:

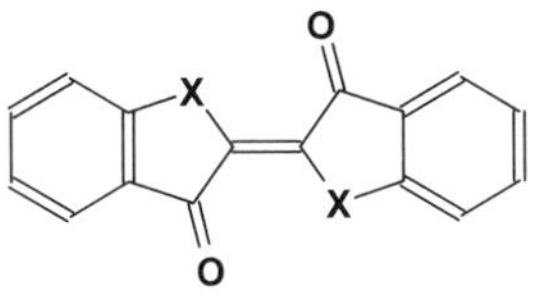

in which X is >NH in the indigo series (blue) and –S– in the thioindigos (red). Indigo and thioindigo are vat dyes: insoluble colored compounds formed by oxidation of soluble, colorless precursors. There are also dyes that are made soluble by adding sulfonate groups to the indigoid structure. Indigocarmine is used as biological stain, and insoluble indigoid compounds form in some reactions exploited in enzyme histochemistry.

(ii) **Naphthoquinone dyes.** Just as simple phenols can be oxidized to quinones, naphthols can be oxidized to naphthoquinones. Several naphthoquinones have been used as textile dyes (Tilak, 1971), and a few serve as stains or histochemical reagents.

(iii) **Anthraquinone dyes.** Anthraquinone dyes have molecules built around anthraquinone:

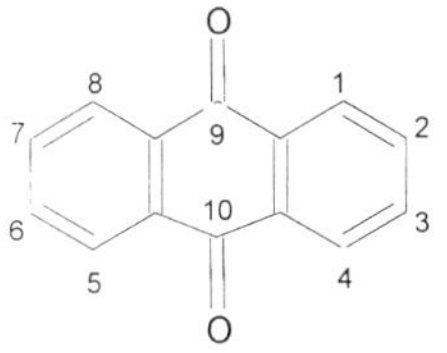

There can be many substitutions, and these may include junctions with other ring systems. This is the largest group of carbonyl dyes. The most notable ones in industry are vat dyes for cotton, disperse dyes (Table 3.1) and pigments (Greenhalgh, 1976).

Mordant dyes in this group include the natural colorants carmine and alizarin. The former has many uses as a histological stain. Synthetic anthraquinones include anionic, solvent and reactive dyes.

(iv) **Naphthalimide dyes and fluorochromes.** These were classified as 'aminoketones' by Lillie (1977). Here (following Zollinger, 1991) they constitute a small group within the carbonyl dyes. The central feature of the molecule is:

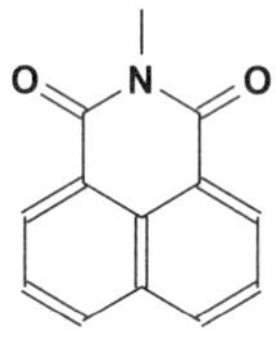

with various side-chains. Some fluorescent compounds of this type can be injected into living cells to study diffusion within the cytoplasm and through gap junctions into adjacent cells.

15. Aza[18]annulene colorants.

A ring of methine and azamethine linkages, included in and connecting four other ring systems, is the chromophore in many colored compounds of animals and plants, including the **porphyrins, cobalamins** and **chlorophylls. Phthalocyanine** and related compounds are synthetic dyes and pigments with large coplanar molecules containing a metal, most frequently copper. In dyes, solubilizing side chains are attached to the four benzene rings at the corners of the molecule. For core structures of porphyrins and phthalocyanines, see the **Background information** in Chapter 26.

16. Other organic colorants.

This miscellany contains substances that differ conspicuously in chemical structure from those in any of the other 15 groups.

PROBES AND LABELED COMPOUNDS

Some dyes are used to impart color or fluorescence to specific objects in living cells and tissues, such as cell membranes, DNA molecules, mitochondria and other organelles. Substances used for this purpose are known as **probes**. If a probe is strongly colored or (more usually) fluorescent, it may be directly visible. Examples are cationic phenanthridine and benzimidazole fluorochromes that attach to nucleic acids, diphenylhexatrienes and lipophilic cyanines used to label membranes of living cells, and dyes such as janus green B (monoazo), neutral red (azine) and methylene blue (thiazine) that are used in traditional vital staining techniques. Green fluorescent protein (GFP) provides an intrinsic fluorescent label for the progeny of cells infected or transfected with the gene that encodes the protein.

Other probes are invisible compounds, such as proteins or drugs, and these must be made visible by **labeling**. A simple example of a labeled probe is the protein phalloidin, which attaches specifically to actin filaments in cells. If phalloidin molecules are made to form covalent bonds with a fluorescent dye such as fluorescein, the resultant labeled probe will impart fluorescence to actin filaments.

Not all labeling agents are colored or fluorescent. Proteins, notably antibodies for use in immunohistochemistry and nucleic acid probes for *in situ* hybridization, are frequently labeled by covalent conjugation to a **histochemically demonstrable enzyme**. Horseradish peroxidase is popular for this purpose. Alkaline phosphatase is also widely used; glucose oxidase is somewhat less popular. The gene encoding β-galactosidase is often included in DNA sequences introduced experimentally into cells. The cells and their descendants can then be recognized by a staining method that localizes the activity of that enzyme. Two other invisible labeling agents are worth mentioning. It is easy to conjugate **biotin** to proteins, nucleic acids and other large molecules. This compound can be detected by virtue of its strong and specific affinity for **avidin**, a protein from egg-white that can be labeled with any fluorescent or enzymatic marker. **Digoxigenin**, a steroid of plant origin, can also be conjugated to large molecules. Commercially available antibodies are used for immunohistochemical detection of bound digoxigenin-labeled probes.

4

Applications of dyes, fluorochromes and pigments

M. Wainwright

Introduction

Use of chemicals for coloring of materials has deep roots: cave painting, staining of the skin, and dyeing of textile materials. In ancient times these were based on naturally occurring pigments and dyes. Woad is an oxidation product of an extract of the leaves of *Isatis tinctoria*, and the purple robes of ancient oligarchs were dyed with Tyrian purple extracted from marine mollusks. Available colorants were limited until the 19th century, when the discovery of synthetic aniline dyes caused an explosion of activity in the area of organic chemistry. This radically changed the clothing and textile industries and also, through scientists interested in microscopy, revolutionized the study of biology. Textile dyeing and biological staining are closely related techniques.

This chapter describes the application of dyes in the diverse areas where coloration is employed, and also addresses the question of how variations in the physicochemical make-up of dyes can lead to changes in their patterns of cellular localization or affinity for different substrates. Just as the science of medicinal chemistry grew out of vital staining, so a proper scientific approach to staining should be based on a fundamental appreciation of the physicochemical characteristics of dye molecules.

Biomedical applications: microimaging

The science of biological staining has been of immense importance in clinical diagnosis and pathology in human, veterinary and plant biology since the latter half of the 19th century. The main applications of this approach are the differential staining of fixed tissues, cells or organelles; the study of living cells and organisms; and the labeling of biomolecules to generate fluorescent probes. These topics constitute the focus of much of the remainder of the present book. Consequently this section merely summarizes the major biological and medical fields using the methods, mentioning the classic stains used. Note that the mode of action of many such stains is discussed in Chapter 5, whilst details on the dyes used may be found in the entries in the following chapters by use of the Index.

Clinical cytology

Identification of cell types in samples of dispersed cells constitutes a major area of diagnostic use of dyes. Although procedures in cytology have evolved over a considerable period, and immunostains are now widely used, the Papanicolaou stain remains a key method. For a scientific view of this, and other dyes used in cytology, see Boon and Drijver (1986). Note: all dyes involved – fast and light greens, eosin Y, and so on – are described in the entries in later chapters.

Cytogenetics and hematology

In addition to the widespread use of immunostaining in cytogenetics, chromosome banding is still carried out using Giemsa's stain; and entries for the constituent azure B and eosin Y dyes are provided in later chapters. Hematology, in which flow cytometry is now probably a major diagnostic technique in the counting and study of blood cells, also still uses Romanowsky–Giemsa staining of blood smears as a routine procedure; for a general account see Simmons (1997).

Histology and histopathology

Healthy and diseased tissue are regularly recognized by means of staining techniques using dyes. The most commonly used single method is still the hematoxylin and eosin oversight stain. For a general account see such texts as Bancroft and Gamble (2002). That text, and others such as those of Churukian (2000) and Kiernan (1999), also discuss the 'special stains' used in histology and histopathology. Entries for the dyes involved in such methods are given in later chapters of this book and may be found via the Index. The mode of action of many of these methods are discussed in Chapter 5.

Staining for botanical histology has to some degree developed independently; for an account of current procedures see Ruzin (1999). Again the dyes are described in the entries later in the book.

Microbiology

Diagnostic use is still made of the Gram stain; whose constituents (crystal violet, iodine, and safranine) are described in the later entries. This procedure is described in such texts as those of Barrow and Feltham (1993) and Seeley *et al.* (1991); its staining mechanism is discussed in Chapter 5.

Fluorescent stains, probes and labels

Fluorescence microscopy allows minute quantities of labeling reagents to be used because objects of interest are seen as bright sources of light on dark background. Of course such methods are used as highly sensitive alternatives to the types of staining already discussed. For instance, fixed chromosomes may be Q banded with quinacrine dyes rather than G banded with Giemsa stain; and acridine orange fluoresces green with DNA but red with RNA in fixed cells.

However, currently most **fluorescent probes** are applied to living cells or organisms. When, as is often the case, the cell remains viable after the staining process such **vital staining** permits monitoring of the structure and properties of live cells. Amongst many general accounts of such methods for detecting organelles such as mitochondria or the Golgi complex, or properties such as pH or concentration of intracellular calcium, see those of Celis (1998) and Mason (1999). To understand why dyes localize specifically in particular organelles or cellular regions it is necessary to compare chemical structure with activity; for a brief account of this QSAR approach see Chapter 5.

Another staining strategy uses fluorochromes as **fluorescent labels** of biological molecules which can bind selectively to structures of biological interest. Widely used fluorochrome systems include fluoroscein, rhodamine and BODIPY. Detector molecules include antibodies and nucleotides, for use in immunostaining and *in situ* hybridization respectively. For an extended account of such labels, see Chapter 7.

NONIMAGING BIOMEDICAL APPLICATIONS

Photodynamic therapy (PDT)

Photodynamic action is the combination of light and a photosensitizing compound to produce localized concentrations of free radicals or reactive oxygen species. Known for nearly a century, since Raab's experiments on the photodestruction of paramecium using acridine as the photosensitizing agent, this process is now the basis of the routine clinical treatment of skin cancer.

The physicochemical basis is that in some compounds, many of them dyes, photoexcitation leads via the singlet excited state to a population of reasonably long-lived triplet excited state molecules. When such a population, known as a photosensitizer, can transfer its electronic excitational energy to its surroundings, the compound is known as a photosensitizer. Transfer may occur either via redox reactions (e.g. hydrogen abstraction or electron transfer) or by direct energy transfer to ground state (triplet) oxygen. The latter results in the concomitant formation of highly reactive – and thus cytotoxic – short-lived singlet oxygen. This is illustrated in Figure 4.1. Photoactivation of a sensitizing dye *in situ* thus guarantees localized tissue damage.

In the past decade there has been a huge research effort in the synthesis and testing of new photosensitizers for clinical PDT (Bonnett, 1995; Wainwright, 1996). Much of this work has involved compounds closely related to porphyrins, such as chlorins, purpurins, verdins, and benzo- and naphtho-fused analogs such as phthalocyanines and naphthalocyanines. Such compounds offer advantages over earlier porphyrin derivatives in terms of improved photophysical and photochemical profiles. Phthalocyanine-type structures in particular can be readily altered to generate a wide range of chemically different compound, so

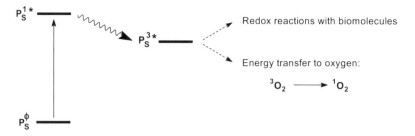

Figure 4.1 Electronic events underlying dye mediated photosensitization

modulating cell uptake and distribution. Moreover the central metal (or semi-metal) atom can also be varied, to alter the photosensitizing efficiency.

Very many other synthetic dyes have served as lead compounds in the search for improved photosensitizers. However naturally occurring photosensitizers are also of interest. For instance, perylenequinonoid pigments such as the hypo-crellins (one of which is illustrated below) are highly active against protein kinase C, a major target in anticancer and antiviral chemotherapy (Diwu *et al.*, 1994).

Photodynamic antimicrobial chemotherapy (PACT)

PACT utilizes organism-specific photosensitizers and visible or ultraviolet light in order to give a phototoxic response, normally via oxidative damage. This procedure's major current application is for the disinfection of blood products. Thus methylene blue has been routinely used by several European blood donation services since 1992 for the disinfection of plasma, particularly for viral inactivation. However, photobactericidal effects have been demonstrated, against a range of bacterial pathogens including refractory species such as *Helicobacter pylori* and methicillin-resistant *Staphylococcus aureus* (Millson *et al.*, 1996; Wainwright *et al.*, 1998). Various other clinical applications, including treatment of oral infections, are being developed.

Antimicrobial dyes

Dyes do of course possess antimicrobial properties even in the absence of light. It is often forgotten that the pharmaceutical industry has its roots in the testing of aniline dyes as stains for different cell types, including pathogens. At the forefront of this work was Paul Ehrlich and the German industrial textile dye producers such as Bayer. Extensions of such work led to the use of acriflavine in wound antisepsis in World War I, and subsequently to development of nonstaining antibacterial aminoacridines (e.g. aminacrine) and to antimalarials (e.g. mepacrine and chloroquine). These latter are still used in the tropics, and the acridine analog, amsacrine (AMSA), is effective against certain cancers (Denny *et al.*, 1982).

Clinical tracers and markers

Dyes such as the triarylmethane patent blue V are used to trace lymph pathways and lymph nodes for lymphadenectomy. They have been either injected directly into the lymphatics (Pump and Hirnle 1996) or intradermally over the primary tumor site (Borgstein *et al.*, 1997). Methylene blue and toluidine blue O have are used for preoperative surgical marking (Creagh *et al.*, 1995; Mashberg, 1983), and the latter dye is marketed as a marker for the delineation of oral carcinoma (Warnakulasuriya and Johnson, 1996).

BIOLOGY

Biochemistry: Staining electropherograms

Electrophoresis on gels or other layers is used extensively for separation of biomolecules, such as proteins and nucleic acids and their derivatives. Several staining methods are available for visualization of separated proteins. The most frequently used reagent is an acid dye, such as bromocresol purple, coomassie brilliant blue R250 or fast green FCF. Alternatively a reactive dye such as remazol brilliant blue R may be used to pre-stain the mixture (Saoji *et al.*, 1983). Glycoproteins have been stained with the periodic acid–Schiff procedure (Doerner and White, 1990.). Stains for nucleic acids include acridine orange and coriphosphine O, which allow the distinction between DNA and RNA to be made (Bruno *et al.*, 1996), and DAPI. The technique has been extended to the use of *bis*-intercalating dyes such as TOTO-1 and the *bis*-ethidiums (Zhu *et al.*,1994; Carlsson *et al.*, 1995). Further information on all dyes mentioned may be obtained in their entries later in this book.

Physiology: Cell viability

The trypan blue exclusion assay was for many years the benchmark for counting living cultured cells. However the method has been subject to various critiques, and new protocols have been devised. For instance a hemocytometer-based method using simultaneous staining with fluorescein diacetate and propidium iodide (Altman *et al.*, 1993). Nevertheless new trypan blue-based methods are still being reported, for example, its use in conjunction with terminal nick end labeling of DNA (TUNEL) to quantify toxicity in cell cultures (Perry *et al.*, 1997b).

Quite another strategy for assessing cell viability is to use the tetrazolium salt MTT. This colorless compound is reduced by the mitochondria of living cells, yielding a purple formazan. This can be assayed spectrophotometrically to give a quantitative measure of cell viability (Mosmann, 1983). See the later entry for chemical information on MTT.

APPLICATIONS IN NONBIOLOGICAL SCIENCES

Indicators

Prior to the advent of conductimetric and potentiometric methods, titrations were generally carried out using indicator dyes. Such titrations could involve the use of dyes in the measurement of pH changes, electron transfer reactions or metal ion concentrations. Each indicator method requires a change to occur in the indicator chromophore via the addition of protons, electrons or metal ions. A more detailed treatment of dyes used as indicators is given in Chapter 10.

Tracer dyes

Use of dyes as indicators includes indication of the presence of a liquid, for example in tracing water courses and drainage. Similarly fluorescent dyes may be flushed through pipelines to reveal leaks. Such usage occurs in geology, soil science and civil engineering. In these areas the dyes used are not required to change color due to changes in pH or ion gradient. Dyes employed include methylene blue, whose possible toxicity in this context has been assessed (Field *et al.*, 1995), and also fluorescein, nigrosine WS and rhodamine B; see their entries in later chapters for more detail.

DYEING OF TEXTILES AND RELATED MATERIALS

Textile dyeing or printing imparts color, or colored patterns, to a substrate. This is carried out by a wide variety of methods with a bewildering array of dyes. Different types of textile, fiber or other substrate, such as paper, can be classified according to their chemical make-up and this simplifies the choice of dyestuff. The following discussion therefore divides the various substrates and dyes into chemical subgroupings.

Substrates

Proteins

These include wool, silk and skin (leather); plus human hair (i.e. keratin) which may be dyed with similar reagents, albeit with toxicity safeguards. All protein molecules have side chains that are electrically charged at neutral pH.

Decreasing the pH of a solution in which the protein is immersed will cause protonation of side chains such as those of lysine and arginine residues. This is the principle used in dyeing wool with anionic (acid) dyes; the dye molecules are held on the fiber by ionic linkages. Proteins also contain negatively-charged carboxylate residues and nucleophilic thiol groups.

Cellulosics

Substrates such as cotton, viscose and paper all contain cellulose, which is a polymer of glucose. The linear character of the polymer makes it ideal for dyeing with long planar molecules. In addition the nucleophilic nature of the many hydroxyl groups facilitates the use of reactive dyes.

Synthetics

These are usually linear-chain polymers such as polyamides or polyesters, with polar groups linked by polymethylene or arylene groups. The linear nature of these synthetic polymers indicates the use of long, planar dye molecules. The absence of ionizable or nucleophilic side chains presents a different problem in dyeing such fibers, but the polar groups, amide (–CONH–) or ester (–CO–O–), allow the use of dyes that can form hydrogen bonds.

Dyes

Compounds used as dyes can be classified both by the type of chromophore (color-forming part of the molecule) and by the mode of dye application. The former aspect is addressed later in this book, in particular Chapters 10 and 11 for azo dyes, and Chapter 25 for carbonyl dyes. The classification of textile dyes into their application groups is done below.

Acid and direct dyes

Acid dyes are anionic, with relatively small aromatic systems, and form salt linkages to the fiber, typically through a sulfonate or carboxylate group. **Direct dyes** are also anionic, but have large, planar aromatic systems. Their uptake is in part mediated by dye–dye interactions within the fiber. Note that milling dyes though usually possessing large aromatic systems, and so both slow diffusing and of high affinity, are not planar.

Basic (cationic) dyes

These are usually small molecules, which carry either protonated (–N$^+$R$_2$H), quaternized (–N$^+$R$_3$) or pseudoquaternized (=N$^+$R$_2$) amino groups. Salt links are formed with anions in the fiber.

Both acid and basic dyes cover a wide range of chromophore types. For example, azo dyes giving a wide range of colors can be synthesized containing either acidic or basic functionality.

Disperse dyes

These small and nonpolar molecules are almost insoluble in water. They are applied to the substrate as suspensions or dispersions. Dye uptake is achieved through the dye 'dissolving' in hydrophobic regions of the polymers making up the fiber.

Vat dyes

Mostly insoluble and nonionic, vat dyes are converted to water-soluble anions on reduction, usually with sodium dithionite. This conversion is known as vatting. The soluble form of the dye is applied to the substrate in the usual way, following which the insoluble parent dye is regenerated in situ by atmospheric oxidation. The overall process is illustrated below for a typical vat dye of the indanthrene blue type.

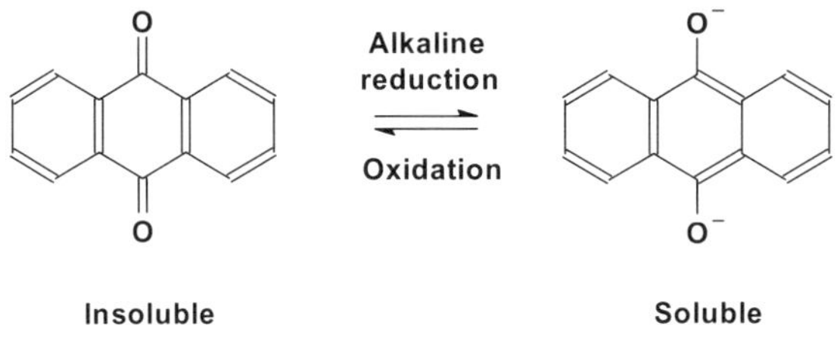

Insoluble   Soluble

Pigments

Particulate and insoluble, pigments are generally used as coloring agents for paints and surface coating media such as printing inks. They are also used in textile printing and for bulk coloration plastics and rubber.

 Some vat dyes have been used as pigments, but cheaper, purer and more efficacious materials have been developed. Pigments based on the phthalocyanine, quinacridone and diketopyrrolopyrrole ring systems represent major advances (Zollinger, 1991). Heterocyclic-fused and benzanelated analogs such as the near infrared-absorbing naphthalocyanines are also available, and are used in optical recording media (Gregory, 1991).

Mordant dyes

In the early history of textile dyeing affinity and color stability were sometimes improved by use of salts, typically of chromium, in the dyeing process. It was originally thought the salts opened up the textile fibers to the dyestuff; hence the name mordant (from the French *mordre* to bite). However, it is now considered that improved dyeing properties are due to chromium ions, introduced into the fibers, acting as complexing sites for dyes with the correct functionality. The chromium–dye complexes being much larger molecules than those of dyes themselves, and are consequently more difficult to remove from the fiber matrix.

 Many different classes of dye can form metal complexes. The presence of two adjacent polar groups (such as hydroxyl, amino, keto or carboxyl) in the dye molecule is a good indication of such behavior. Two examples of such dyes are illustrated below: the ancient anthraquinone dye alizarin (CI Mordant Red 11), present in madder (*Rubia tinctoria*); and the synthetic azo dye eriochrome blue black 2B (CI Mordant Black 3). The latter forms a complex using the two hydroxyl groups and the diazo (–N=N–) bridge.

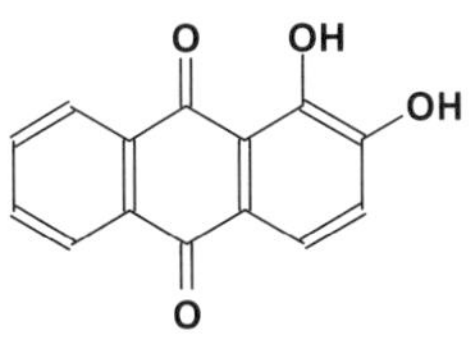

alizarin 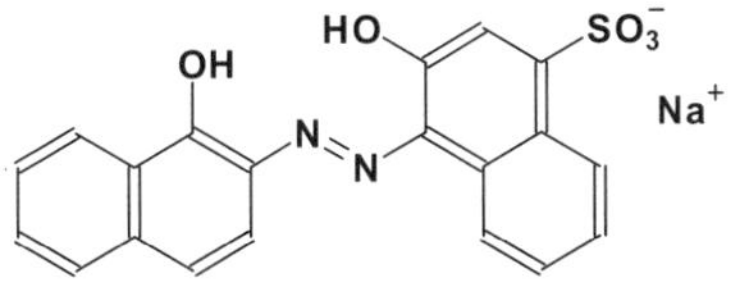   eriochrome blue black 2B

In metal ion complexation, coordination of an electron pair with a positively charged metal ion causes a decrease in the amount of electron density in the dye chromophore. This makes some dyes and fluorochromes suitable as indicators of the presence of specific metal ions. The 'jaws' of a complexing dye can be designed to fit metal ions of a certain radius, providing reagents of great utility in chemical analysis.

Reactive dyes

These involve joining the dye to the textile with a new, single covalent bond. Such strong attachments, known as reactive dyeing, became commercially available in the early 1950s. Attachment of the dye occurs by way of a highly reactive group tethered to the dye's chromophore, usually through an inert spacer. Electron-rich groups (nucleophiles) in the fiber attack the reactive group, forming a new covalent bond. Examples of reactive dyes are those with triazine and vinyl sulfone groups; see entries for procion and remazol dyes respectively, in later chapters.

NONCOLORANT INDUSTRIAL USES OF DYES AND FLUOROCHROMES

Photographic film

The use of dyes in photography arises from the fact that the silver halides used in photographic film absorb at wavelengths below 500 nm and have little sensitivity to yellow or red light. Sensitizing dyes are added to photographic emulsions to absorb long wavelength light and to transmit the associated energy to the silver halide crystals on which they are adsorbed. Cyanine dyes have been used for this purpose for over 50 years.

Precursors of dyes are present in color films, and the dyes are formed in the chemical reactions of development. Separation of green, blue and red in different halide layers allows the complete image to be formed by optical addition of partial images in the three primary colors.

Infrared absorbing dyes

Until the 1960s, dye research was concerned mainly with coloration of textiles, and infrared absorbing chromophores were a very small field of application. The advent of lasers brought great commercial possibilities for infrared absorbing dyes.

Generally such dyes are prepared by the extension of the aromatic portion of a chromophoric system, or the inclusion of electron releasing or withdrawing groups (auxochromes) at the correct sites. This leads to bathochromic shifts (absorption at longer wavelength and lower energy) in the electronic absorption spectrum of the resulting compound. For instance extension of the polymethine chain in cyanine dyes has such effects, as in the example given in Table 4.1.

Infrared absorbing dyes are used in optical recording and information storing devices such as compact and video disks, and in thermal imaging systems. The theory, synthesis and application of infrared absorbing dyes have been reviewed by Matsuoka (1990).

Photochromic dyes

Such compounds change color on irradiation, due to intramolecular rearrangement, the activation energy required being supplied by the incident light. The rearranged material may be stable for seconds, hours or days and is normally returned to its original state by a second exposure to light or by heating. There are many different types of photochromic compounds, the photoinduced structural isomerism varying in complexity from ring-opening associated with spiro-compounds to simple *cis/trans*-isomerization of azobenzenes (see introduction to Chapter 10). Uses of these interesting materials include sunlight-responsive lenses for sunglasses and components of optical data storage systems.

COLORATION OF COSMETICS, FOOD AND DRUGS

Natural and synthetic dyes are used to color food and medicines, both for consumer appeal and for color coding. Cosmetics and hair dyeing are ancient traditions constituting major commercial fields in the modern world. As all such colorants may end up in the human body following ingestion or absorption through the skin, they require more rigorous toxicological examination than textile dyes or biological stains. Many azo dyes, for example, are

Table 4.1 Effects of polymethine chain length on light absorption by a cyanine dye, N–ethylquinolinium–2–$(CH=CH)_n$–CH=(2–N–ethylquinoline)

$n =$	Absorption maximum (nm)
1	708
2	818
3	943

mutagenic (Chung and Cerniglia, 1992). Dyes used in livestock foodstuffs may also be passed on up the food chain, either intact or as potentially harmful products of metabolic conversion by the animals (Roybal *et al.*, 1996). For such reasons it has been emphasized that labeling of colorants used in pharmaceutical preparations is advisable because of potential side effects in compromised patients (Kumar *et al.*, 1993).

Finally consider dyes for hair, which are often generated *in situ* from aromatic amines. Such compounds are often toxic if ingested, but skin reactions are infrequent. The toxicology of these dye precursors has been thoroughly investigated (Johnson 1994a,b).

Mechanisms of biological staining

R.W. Horobin

THE BACKGROUND TO STAINING THEORY

Staining methods in current use have survived repeated methodological discontinuities. By the late 19th century stains using dyestuffs and silver salts were widely used. Later came cytochemistry, with chromogenic methods such as the periodic acid–Schiff procedure. Overlapping this was the development and application of enzyme histochemistry, followed by immunostaining and its related affinity–histochemical approaches. More recently *in situ* hybridization achieved prominence, together with a renaissance of live-cell staining. Technical innovation has been repeatedly followed by the largely empirical application of new procedures to biological and clinical problems. Methods from older methodological waves survive into each new age. Technical development of earlier methodologies continues, albeit at a slower pace, as fashions in staining move on.

Theory building – essential for standardization, selection and optimization of methods, and trouble-shooting technical failures – has lagged behind technical innovation and application. Concern with the scientific bases of staining procedures typically is marked in the early years of a new methodology, then diminishes. For instance, the physico-chemical principles underlying dyeing methods were discussed by Mann (1902) but no equivalent theoretical account was available until the 1950s. Earlier staining theories developed as if different staining methodologies were physicochemically unrelated. This fragmentation was well illustrated by the Mechanisms of Staining chapter in the previous edition of this book.

A few accounts have emphasized the single set of physicochemical factors operating in all methodologies. In English there are the scholarly books of John Baker (Baker, 1958), various monographs by Horobin (1982, 1988) and Horobin and Bancroft (1998) and Lyon (1991), and more recently a fascinating review by Prento (2001). Such accounts have a characteristic fate: they are widely cited, but their larger message is ignored.

Basic concepts and terminology

The purpose of biological staining is to generate information, most often to address the following questions.

(i)   What is this, that we see in the microscope?
(ii)   Where is this found, within the cell, tissue or organism?
(iii)   How much of this is there, or how many of them are present?

To provide answers a staining method must be sufficiently sensitive to detect the biological target. This often depends upon the amount of material present, and on the precise mode of action of the stain. Other things being equal fluorescent reagents and catalytic staining processes favor high sensitivity. Fluorescent reagents provide a superior signal to noise ratio in the microscope, whereas catalytic staining processes allow sensitivity to be increased by prolonging staining times. With most stains, selectivity is as important as sensitivity. Biological targets must often be stained differently from adjacent structures before identification is possible. Consequently an understanding of the mode of action of a staining method requires answers to two key questions:

(i)   Why does anything stain?
(ii)   Why doesn't everything stain in the same way?

This chapter provides answers to these questions. Although factors controlling sensitivity and selectivity are numerous, the same physicochemical phenomena are important whatever the staining methodology. Thus electric charges carried by dyes and biopolymers determine the selectivity of acid and basic dyes, and also influence the action of fluorescently labeled antibodies. The sizes of dye ions can control staining patterns, and the sizes of labeled antibody molecules critically influence sensitivity in the immunostaining of resin sections.

CONTROL OF STAINING BY TARGET CONCENTRATION AND SPECIMEN DENSITY

In the simplest situation staining intensity reflects the amount of target material present. Quantitative histochemistry and cytochemistry depend on such relationships, as discussed by Chayen and Bitensky (1991), Fricker *et al.* (1999) and Van Noorden and Gossrau (1991). Even when staining is not stoichiometric a correlation between staining intensity and quantity of target is commonplace. Variations in specimen concentrations are strongly influenced by local variations in specimen density. Thus eosin acts as a nonspecific protein stain in routine hematoxylin and eosin staining, and structures such as the cornified layer of the skin, mitochondria-rich cytoplasms, red blood cells, and protein secretion granules stain especially strongly. These structures contain different proteins, but all are physically dense.

A related situation arises when distribution of material varies although the total amount of target material present is the same. Consider a biopolymer such as glycogen which, after certain fixative regimens, is evenly distributed across the cell. However, other fixatives result in intracellular aggregation of glycogen. The former can result in 'weak background coloration' while the latter is recognized as 'intense granular staining'. For a dramatic illustration of this see Simson (1977).

CONTROL OF STAINING BY SPECIMEN GEOMETRY

Specimen geometry influences the staining process at a variety of scales. After block staining only superficial elements of the specimen are strongly stained. This is most marked with larger staining reagents such as labeled antibodies. However, variation in section thickness also affects penetration. Even with dyestuffs penetration of stains into the center of thick sections is relatively slow (Cooper *et al.*, 1988). This is most obvious with techniques that are strongly influenced by diffusion rate, such as picro-trichromes: the thicker the section the yellower the color. This is because the yellow dye is the smaller and faster diffusing one. In keeping with this, cell smears, typically being extremely thin, often require shorter staining times than sections stained in the same way. However, cells in the center of any cell masses present tend to stain more slowly than superficial cells, as illustrated by Boon and Drijver (1986, Plate 6.4).

Specimen preparation also affects section geometry, and hence stain penetration. Surface topography of cryosections is usually more uneven than paraffin sections, which are themselves rougher than resin sections. Freezing typically shatters cells and tissues, and paraffin embedding also produces tissue damage. Consequently, other things being equal, cryosections require shorter staining times than do paraffin sections, and resin-embedded material takes even longer to stain. Rate controlled staining patterns vary because of these effects.

Fixation also alters the microtopography of cells and tissues. Agents such as alcohol or Carnoy's fluid are coagulant, and shatter cells and tissues. Aldehyde fixatives produce more coherent specimens, and agents such as Zenker's fluid give rise to coherent and somewhat shrunken specimens. These effects have been summarized by Horobin (1988, Figure 14a–d). Shattering maximizes staining rate, whereas shrinking minimizes it; the largest effects occur with the largest (i.e. slowest diffusing) dyes present.

Innate biological geometry also influences staining patterns. Consider Schmorl's thionine stain for demonstrating canaliculi within bone matrix. These microchannels are more readily penetrated by the dye than is the surrounding dense collagen. Consequently only canaliculi are stained.

Even a failure to stain can form the basis of a biological staining technique, namely negative staining. Thus micro-organisms are often dense and impermeable, excluding large dyes such as nigrosine, but if a dye solution is allowed to dry in contact with a bacterium or fungus, the surface topography can be observed.

AFFINITY CONTROLLED STAINING

Accounts of affinity can adopt different emphases. One is to consider large cellular units, such as nuclei or fat droplets, each comprising vast numbers of molecules. In such a colligative approach underlying concepts are often thermodynamic, such as entropy. Much of DJ Goldstein's work (Goldstein, 1963a) was of this type. Alternatively accounts focus on the biomolecules constituting these structures, emphasizing intermolecular forces; Prento's review (2001) is an example. Most accounts, such as this present one, are perhaps more muddled.

Electrical interactions

Electrical or coulombic interactions are widely discussed, although sometimes described as electrostatic binding or salt links. These are involved in acid and basic dyeing of fixed tissues. Many routine staining systems utilize acid and basic dyes; for example, the Gram stain in microbiology; the hematoxylin and eosin stain in histopathology; the cytological Papanicolaou stain; and the Romanowsky–Giemsa stain in hematology.

Acid dyes contain colored anions, basic dyes colored cations. Interaction of these with tissue polyions is substantially influenced by the signs of the electrical charges on dye and biopolymer. In fixed tissues most proteins will carry an overall positive charge under acid conditions. When applied under such conditions, acid dyes give nonspecific protein staining. Tissue polyanions present in fixed tissue under weakly acid or neutral conditions include DNA, ribosomal RNA and anionic glycosaminoglycans. These species stain with basic dyes. The electric charge on dyes both contributes to affinity and controls staining selectivity. An elegant account of acid and basic dyeing in biological staining was provided by Baker (1958); for a restatement from a molecular viewpoint see Prento (2001).

Coulombic influences arise with all classes of staining reagent, not only dyes. Consider the various nondye reagents used to stain anionic carbohydrates present in mucins and on cell surfaces. The common feature of cationized ferritin, colloidal iron and the high iron diamine reagent is that they all contain or give rise to cationic staining species of large size.

Electrical interactions of ionic stains with specimens are influenced by reagents that modulate electrical effects. Differentiation of the Gram stain to remove excess crystal violet triiodide from stained micro-organisms provides an example. For this to take place at a technically controllable rate, differentiating solvents with low dielectric constants are used, such as acetone or ethanol. Stain–tissue electrical effects are also influenced by addition of colorless salts to the staining baths. Effects of such additions are complex. Staining may be reduced, as salt ions compete with dye ions for ionized biopolymers, but staining is sometimes increased, as predicted by the Gibbs–Donnan model (Bennion and Horobin, 1974). Specific ion effects also occur, and are said to underlie selective staining of different tissue polyions in critical electrolyte concentration methods (Scott, 1973).

Noncoulombic dye–tissue attractions

Binding of acid and basic dyes to tissue sections is aided by various noncoulombic forces, often loosely termed **van der Waals attractions**. These include short-range forces such as dipole and induced-dipole attractions, and dispersion forces. These are significant with reagents possessing large aromatic systems (Horobin and Bennion, 1973; Prento, 2001). Methods for demonstrating elastin, such as Miller's stain and Congo red, illustrate the role and importance of van der Waals attractions. Such dyes are applied from solutions containing an electrolyte and an organic solvent, both of which inhibit electrical attractions. Moreover basic dyes such as Victoria blue are applied at low pH, and acid dyes such as Congo red at high pH, respectively inhibiting acid or basic dyeing of most proteins. Elastin, however, is an unusual protein, containing few ionizable amino acid residues, and it remains stainable due to van der Waals attractions (Horobin and Flemming, 1980). Another illustrative case is provided by the formazans produced in dehydrogenase demonstrations. Bis-formazans have large aromatic systems. Consequently they bind well to tissue proteins and are well localized intracellularly. Stained sections may be processed through routine clearing solvents without dye loss. In contrast, monoformazans have smaller aromatic systems, give poor localization, and need aqueous mountants (Seidler, 1991).

Other noncoulombic specimen–stain interactions can contribute to affinity. **Hydrogen bonding** involves formation of weak bonds between hydrogen and adjacent electronegative atoms, usually oxygen or nitrogen. Water forms such bonds, which compete with stain–specimen bonds. Hydrogen bonding is therefore unlikely to contribute to staining in aqueous solutions. When staining glycogen with Best's carmine the dye is dissolved in a salty, largely organic solution, and hydrogen bonding occurs between carminic acid and glycogen (Horobin and Murgatroyd, 1971). If sections stained with Best's carmine are inadvertently rinsed in water, all staining of glycogen is immediately lost.

Charge transfer phenomena involve partial transfer of charge between electron donor and electron acceptor molecules, with the resulting complex often being a strikingly different color from that of the original species (Atkins, 1997). Such effects arise in Romanowsky–Giemsa staining, when azure B and eosin Y molecules act as electron acceptors and electron donors respectively. They interact by charge transfer in chromosomes, nuclear chromatin and certain other sites, resulting in formation of a complex with an intense magenta color (Zanker, 1981). Another method involving charge transfer is staining of plant tissues by toluidine blue. A green complex forms between the electron accepting dye and electron donating phenolic biopolymers present in lignins.

The staining phenomenon termed **metachromasia** involves a single chemical species; in histochemistry a basic dye gives rise to more than one staining color. Metachromasia arises when tissue substrates – usually in the form of highly charged, flexible biopolymers – provide templates that encourage dye aggregation. In such aggregates dye ions bind together by short range forces, giving a new electronic entity with new spectral properties. Metachromasia thus contributes to increased staining affinity (Goldstein, 1963a).

Hydrophobic bonding and other hydrophobic effects

Metachromasia is also favored by another source of affinity, **hydrophobic bonding.** This involves the coming together of hydrophobic groups previously dispersed within an aqueous environment. In metachromasia the aromatic rings of basic dyes constitute the hydrophobic domains. This phenomenon occurs widely, not just during metachromasia, and can involve any dye or staining reagent with hydrophobic groupings associating with hydrophobic groups in the specimen. This takes place only in aqueous systems (Horobin and Bennion, 1973). The widespread contribution of hydrophobic bonding to the affinities of routine stains is shown by the routine use of 70% alcohol as a differentiating solvent. This solution, due to its content of organic solvent, disrupts hydrophobic bonding, and releases dyes from the tissues.

Staining of lipids by dyes such as Sudan black or oil red does not involve hydrophobic bonding because the solutions are largely nonaqueous. Instead these lipid stains involve partitioning processes driven by the increase in reagent entropy that arises when dyes are distributed between two phases (solution and lipid) rather than being restricted to one. Selectivity is due to dye being miscible with lipids but not with hydrated proteins, which make up most of the nonlipid portions of the specimens (Horobin, 1981; 1982, p. 73).

REACTIVITY CONTROLLED STAINING

Some reactions are best considered from the perspective of organic or metal-coordination chemistry. Catalytic reactions, both enzymatic and nonenzymatic, are also important in certain staining procedures.

Stains and staining involving covalent bond formation

The periodic acid–Schiff (PAS) stain for polysaccharides, and the Feulgen stain for DNA, involve formation of covalent bonds between stain and specimen. These bonds contribute to staining affinities, the chemistry of formation defining the selectivity. Both procedures use Schiff's reagent, which combines with aldehyde groups to form a colored product. The different selectivities of the two methods arise from the different reactions used to generate aldehydes. In the PAS method specimens are treated with an oxidizing agent that generates aldehydes only from the sugar residues in polysaccharides. In the Feulgen technique specimens are subjected to acid hydrolysis, which generates aldehydes only from DNA. Other procedures involving formation of stain–specimen covalencies include the thiol-reactive dye mercury orange, and the protein and polysaccharide-reactive procion dyes. For a more extended account see Chapter 7.

Other widely applied methods involve formation of polar covalent bonds (also termed coordinate bonds) between an organic grouping and a metal ion. Examples include staining of tissue calcium ions by forming chelates with alizarin red S; as illustrated in Chapter 7. The same principle has been applied in enzyme histochemistry, where calcium phosphate deposits are produced by alkaline phosphatase in a Gomori-type procedure. These can be visualized, and the enzyme located, by formation of a fluorescent complex with calcein blue (Murray and Ewen, 1992).

Attachment of metal ions to dyes by polar covalencies occurs also in mordant or metal-complex dyes. The best known is the 'hematoxylin' used in hematoxylin and eosin staining. This contains a mixture of cationic chelates formed between aluminum ions and hematein (Bettinger and Zimmermann, 1991). Other metal-complex dyes include alcian blue 8G, iron–eriochrome cyanine R and gallocyanine–chrome alum. Some complexes are pre-formed and applied to the specimen as such, others are formed *in situ* from the precursor dye plus a source of metal ions. Chemical structures of certain pre-formed complexes are given in the dye entries later in this book; for example, Chapter 24 for aluminum hematein complexes, and Chapter 19 for gallocyanine chrome alum.

Such stains have been considered to act by mordanting, which is formation of covalent bonds between the metal ions and tissue groupings such as the phosphate ions of DNA (Baker, 1958; Sandritter *et al.*, 1966). This would account for stain retention during dehydration and counterstaining. Formation of such bonds with tissue amino groups might also explain nuclear staining seen with some mordant dyes even after extraction of DNA (Clark, 1969; Clark and Meischen, 1978). An additional, not necessarily exclusive, viewpoint (Marshall and Horobin, 1973) follows from the hydrophilic character and large aromatic systems of some cationic metal-complex dyes. Hydrophilicity limits stain loss into dehydrating alcohols; aromatic systems contribute to affinity by way of van der Waals attractions.

Influence of catalytic processes on staining

Catalytic staining increases sensitivity because prolonged reaction times will amplify color intensity. One methodology exploiting catalysis is silver staining. Specimens are treated with solutions of various silver salts, which are reduced to form visible deposits of metallic silver. Catalytic silver staining variants are extremely sensitive, and the accumulated silver deposits sometimes visualize entities themselves too small to resolve by light microscopy. Silver staining has been investigated by many workers over many years; for a critical account see Gallyas (1979a).

Romanowsky–Giemsa stains also involve catalysis. These procedures, used in hematology and cytogenetics, involve an initial basic dyeing of chromatin of cell smears or chromosome spreads by azure B. A charge transfer complex with eosin is then formed, following which additional basic dye is bound by the DNA (Sumner, 1980). This provides intense staining of such tiny entities such as chromosome bands.

Catalytic processes are central to enzyme histochemistry, as reviewed by Stoward and Pearse (1991) and Van Noorden and Fredericks (1992). Enzymic catalyzed reactions provide both sensitivity and selectivity. Sections or smears are incubated in substrates that are stable except when the relevant enzyme is present. This limits staining to enzymatic sites. Enzymes, however, vary in their substrate selectivity. Some phosphatases, for instance, are poorly selective and catalyze hydrolysis of most organic phosphates; others catalyze hydrolysis only of a specific substrate.

THE SPECIMEN AS MATRIX – *IN SITU* PIGMENT FORMATION

Retention of stain in sections or smears can occur even in the absence of dye–tissue bonding. Insoluble pigments can be generated by strong stain–stain forces, and mechanical factors then retain the pigments within the specimen, which serves as a chemically passive matrix.

Such stains include organic pigments formed as final reaction products in enzyme histochemistry, for example, azo dyes, formazans, and indigo derivatives. Stain–stain binding here involves van der Waals attractions, hydrophobic bonding, and hydrogen bonding (Burstone, 1962; Holt and Sadler, 1958; Seidler, 1991). Nonenzymic staining may also involve organic pigment formation, as with formation of insoluble thionine picrate in Schmorl's picro-thionine technique for bone canaliculi. Stain–stain factors include electrical attractions, van der Waals forces and hydrophobic bonding.

Other enzyme histochemical final reaction products are inorganic pigments, such as lead sulfide formed in Gomori-style methods. Nonenzymic stains generating inorganic pigments include the Prussian blue reaction for ferric iron. Pigments such as this potassium ferric ferrocyanide are stabilized largely by intense coulombic attractions between the ionic constituents.

Silver stains, producing microcrystals of metallic silver, are a further example. These deposits are insoluble because of strong metal–metal bonds, not metal–tissue attractions. These internal attractions between the metal atoms differ from those previously described; they result from silver cations being surrounded by a sea of free electrons.

RATE CONTROLLED STAINING

Control of staining patterns by staining rates remains intuitively unconvincing to many, despite having been extensively demonstrated and discussed a century ago (Mann, 1902). In fact rates of access of stains to cellular and tissue targets can differ, as can rates of reaction with targets, and rates of loss from the targets. All these differences can influence staining patterns.

Influence of rate of stain access

The most dramatic rate effect is when stains clearly fail to enter cellular or tissue targets in the time available. An example is the failure of labeled antibodies to penetrate thick specimens such as vibratome sections (Piekut and Casey, 1983). Indeed labeled antibodies do not readily penetrate the surface membranes of cells, a problem addressed by slightly-damaging treatments such as freezing and thawing. Stain exclusion also arises with smaller molecules. In stripping film autoradiography, for example, the gelatin layer inhibits staining of cell nuclei with propidium iodide (Giordano *et al.*, 1985). Unappreciated stain exclusion is more confusing. Cell counts, for instance, are commonly underestimated when short staining times or large, slowly diffusing, dyes are used with thick neuroanatomical sections (Cooper *et al.*, 1988). Stain exclusion by resin embedding media is discussed later.

These examples are obviously problems or artifacts. Some methods, however, can intentionally control stain diffusion. Several routine techniques are influenced by rate control of stain entry. An example is selective staining of mucus, goblet cells and glycosaminoglycan (GAG)-rich structures by basic dyes such as alcian blue 8G. If staining times are lengthened, additional polyanions such as nuclear DNA also stain (Goldstein and Horobin, 1974). Basic dyes that selectively color mucus and GAGs are all large (i.e. slowly diffusing) but nonselective dyes are small. The structures rich in acid carbohydrates have been shown to be faster staining than nuclei or basophilic cytoplasms (Friedberg and Goldstein, 1969; Goldstein, 1963a).

Selectivity of trichrome stains is also influenced by rate of dye uptake. These methods use sequences or mixtures of acid dyes to give selective staining of collagen fibers or other protein-rich entities. Collagen fibers have been shown to stain faster than other structures (Baker, 1958; Horobin, 1974), and small acid dyes typically diffuse faster than large ones (Baker, 1958; Seki, 1932). When dye sequences are used, the last acid dye to be applied usually stains the collagen. That this is a rate effect controlled by the ready penetrability of the collagen is indicated by the fact that the sequence-control holds whether the last dye has smaller or larger anions than the earlier dyes (cf. methods summarized by Lillie and Fullmer, 1976, pp. 697–698). When dye mixtures are used, usually the larger (more slowly diffusing) dye stains the collagen fibers, again in keeping with the rate control model (Horobin and Flemming, 1988). This is, however, not the whole story, and for evidence of the role of nonrate effects with picro-Sirius red see Prento (1993).

Influence of rate of staining reactions

Regardless of how quickly a reagent reaches the target, the staining pattern may be controlled by the rate of reaction with the target. A well explored case is the acid hydrolysis step in the Feulgen technique for DNA. If hydrolysis time is too short, few aldehyde groups form and staining is weak. If hydrolysis time is too long, DNA breaks into small fragments that diffuse from the specimen, again giving weak staining. Such reaction–rate effects are especially important when the Feulgen reaction is used quantitatively (Kjellstrand, 1980).

Influence of rate of stain loss

Selectivity of many staining methods is influenced by rate of loss of dye. Used intentionally this is termed **destaining, differentiation** or **displacement.** In the

Gram stain differentiation is required to distinguish the bacterial classes. Initial treatment with crystal violet colors both Gram-positive and Gram-negative species. Following conversion of dye to a poorly soluble salt, to retard subsequent loss, the bacterial smear is destained (differentiated) in an organic solvent. Dye is extracted from thin-walled Gram-negative bacteria, leaving only the thick-walled Gram-positive organisms stained (Beveridge, 2001).

CONTROL BY PRE-STAINING MODIFICATION OF THE SPECIMEN

All specimen preparation steps can influence staining. Bench workers distinguish intentional pre-staining modification of specimens from artifacts on operational, not physicochemical, grounds. Illustrative examples of problems and artifacts due to inadvertent pre-staining specimen modifications follow.

Fixation influences tissue permeability and, as noted above, staining patterns. Reactivity too is fixative dependent in many cases. Other preparative steps influencing staining patterns include acid decalcification (Yoshika *et al.*, 1972), section thickness (Lison, 1955), and chemical dehydration of tissues (Beckmann and Dierichs, 1982).

Histochemical **blockades and extractions** entail deliberate alteration of the specimen to prevent certain materials or structures from staining, and are carried out as control measures. If a particular compound is to be revealed, extraction of this compound from the specimen, or its chemical transformation should eliminate staining. If it does not, some error has occurred or unacknowledged complexity exists. For instance, glycogen is extracted using an amylase solution, and enzymes are inactivated by inhibitors. For general accounts of such tactics see Horobin (1982, pp. 210–214) or Lillie and Fullmer (1976, their index entry 'blockade').

STAINING RESIN SECTIONS – MECHANISMS AND EXPLANATIONS

Resin embedding involves infiltrating specimens with liquid monomers which are then converted to solid polymers ('resins' or 'plastics') *in situ*.

Infiltration of tissues by resins

Routine histotechnological resins vary in hydrophilicity/hydrophobicity and in the degree of crosslinking, but all monomers are viscous and hydrophobic fluids. Monomers of water-miscible resins are less hydrophobic than those of epoxy or methylmethacrylate resins but are not hydrophilic.

Glycolmethacrylate has a water-miscible monomer, and has been widely applied in light microscopy. The polymer is slightly hydrophobic and in most formulations it is weakly crosslinked and carries no ionic substituents. Another resin used in light microscopy, especially for mineralized bone, is methylmethacrylate. This is not significantly crosslinked, and is appreciably hydrophobic.

The epoxy resins used for electron microscopy, and which are stained as semithin sections for preliminary inspection by light microscopy, are highly crosslinked and appreciably hydrophobic. Polymerization of epoxy resins is routinely achieved by heating the monomers; whereas glycolmethacrylate and its analogs may be polymerized at or below room temperature. Heating reduces antigenicity and enzymic activity in the embedded tissues.

Resin monomers infiltrate biological specimens unevenly. Some regions become resin-rich while others contain little resin (Eneström and Kniola, 1994; Horobin, 1983). The patterns of differential infiltration by the viscous, hydrophobic monomers are predictable and much the same for all commonly used resins (Horobin, 1983). To summarize, the following materials are routinely poorly infiltrated:

(i)   Dense structures, e.g. elastic fibers, red blood cells and protein secretion granules
(ii)   Hydrated structures, e.g. cellulose cell walls, collagen fibers and other polysaccharide- and glycosaminoglycan-rich materials

Overall, infiltration is slower at lower temperatures, with more viscous infiltration media, and if tissue is incompletely dehydrated.

Exclusion of stains by resin embedding media

Resins that are not crosslinked, such as methylmethacrylate, are routinely removed by solvent extraction prior to staining, and therefore have no more influence on staining than does the paraffin wax of de-waxed sections. Glycolmethacrylate and epoxy resin formulations are sufficiently crosslinked to prevent removal. Celloidin (cellulose nitrate) also falls into this category because celloidin sections are typically stained with the hydrophobic embedding medium still in place. Staining with resin present will now be considered.

Many stains act more slowly on resin sections than on cryostat or paraffin preparations, because resins limit access of reagents to the specimen. Increases in resin crosslinking usually decrease staining rates (Gerrits *et al.*, 1990). Increases in plasticizer content, however, make the resin more permeable and can increase staining (Frater, 1981).

Some structures stain rapidly even in the presence of crosslinked resins, on account of the differential infiltration effect previously noted. Poorly infiltrated material stains well, for example, goblet cell mucin with alcian blue. Staining sometimes looks crisper in resin sections than in paraffin, probably in part because the more completely infiltrated cellular regions exclude staining reagents.

Partial infiltration can have more dramatic effects on staining, especially with specimens embedded in crosslinked and hydrophobic resins. Some routine oversight stains for epoxy-embedded material use sequences of water-soluble basic dyes such as basic fuchsine and methylene blue. Only resin-free structures stain strongly, and the resin-free and low-density structures (e.g. GAG-rich materials) are selectively stained by the second dye applied. Consequently color patterns are reversed if one worker stains with a red dye followed by a blue, but another uses the blue dye followed by the red (compare Aparicio and Marsden, 1969 with Huber *et al.*, 1968).

Sizes of staining reagents are also important in such rate controlled staining. With one crosslinked epoxy resin no staining reagent whose molecular weight exceeded 500 Da gave significant coloration (Horobin and Proctor, 1990). With a glycolmethacrylate resin small dyes (< 550 Da) penetrated resin rapidly so that staining methods for paraffin sections needed no significant modification. Large dyes (> 1000 Da) stained only the poorly infiltrated structures such as goblet cell mucin, and dyes of intermediate size penetrated the resin but could not easily be washed out (Gerrits *et al.*, 1990).

Are water miscible resins different?

Unlike hydrophobic resin monomers the water miscible materials may be directly applied to hydrated specimens. Moreover routine water miscible resins do not need heat curing, thereby avoiding the denaturation of enzymes and protein antigens that occurs with standard epoxy resins. In other respects differences are not marked. Lipid extraction artifacts are similar for water miscible and epoxy resins (Cope and Williams, 1969). Patterns of monomer infiltration are also similar (Horobin, 1983). Moreover water miscible resins do show an affinity for many lipophilic stains (Horobin *et al.*, 1992). This can give background staining, as with aldehyde fuchsine, and can also inhibit the entry of lipophilic reagents into sections.

STAINING LIVING CELLS AND ORGANISMS

Staining of living organisms, a technique of some antiquity (Kasten, 1999), has been transformed by commercial availability of fluorescent reagents and fluorescence microscopes. A wide variety of cellular components and processes may be probed in this way (Haugland, 2001; Mason, 1999).

Mechanisms of such staining are different from those for fixed tissues and cryosections. Consider the widely used cationic fluorochromes bisbenzimide, DiI, $DiOC_6(3)$, neutral red, and rhodamine 123. Applied to cryo- or fixed sections all stain cell nuclei and other polyanion-rich sites by acting as fluorescent basic dyes. With live cells, however, DiI is a fluorescent stain of the plasma membrane, $DiOC_6(3)$ is a probe for the endoplasmic reticulum, neutral red for lysosomes, and rhodamine 123 for mitochondria; only bisbenzimide stains nuclei in living cells.

Peculiarities of live cells and organisms

To understand staining of live cells several new factors must be considered. Some concern cell structure and function, one concerns the stains themselves.

Live cells are highly compartmented. A fluorochrome may be applied to a cell from an external solution or be inserted in some way. In both instances some cellular regions may be inaccessible. For example, if a fluorochrome has high affinity for the first structure it reaches, for instance the plasma membrane, then the dye will move no further. DiI has such an affinity for membrane lipids. Compare this with section staining where materials from the interior of cells are exposed on the surface of the slice. Other probes mentioned above have lower membrane affinities and pass through by passive diffusion. Inside the cell, reagents are again faced with a set of discrete structures, some being membrane bound in eukaryotes.

Dyes and fluorochromes are also moved into and around live cells by physiological processes such as endocytosis. Compounds too hydrophilic to penetrate the plasma membrane passively (e.g. Lucifer yellow CH) may be taken up by fluid phase pinocytosis. Compounds binding to plasma membrane constituents (e.g. DiI, and Evans blue which binds nonspecifically to proteins) may be internalized by adsorptive pinocytosis. Weak bases (e.g. neutral red) may accumulate in regions of low pH such as lysosomes due to ion-trapping, where pH gradients are generated by ATP-dependent proton pumps (Holtzman, 1989). Uptake of methylene blue into nerve fibers requires the activity of a neuronal Na/K pump (Kiernan, 1974). Rhodamine 123 accumulates in respiring mitochondria, aided by the surface negative potential on these organelles. Some probes stain mitochondria, however, even in cryosections, due to precipitation as salts of cardiolipin, a particularly hydrophobic phospholipid of the mitochondrial inner membrane (Röding *et al.*, 1986).

Fluorochromes are applied at low external concentrations. Within cells organelles compete for the compounds. With higher dye concentrations or longer times, additional structures may stain. Neutral red accumulates in the Golgi complex and nuclei after first staining lysosomes, and $DiOC_6(3)$ first accumulates in mitochondria before moving into its principal target, the endoplasmic reticulum.

QSAR decision rules for staining of living cells

The role of simple physicochemical factors can be identified even in living systems. Compounds that accumulate in a given structure, by a given mechanism, show similar patterns of electric charge, of acid or base strength, of hydrophilicity/lipophilicity, of overall and conjugated system size, and so on. For instance, small lipophilic cations of strong bases (e.g. Rhodamine 123) accumulate in mitochondria; whereas small hydrophilic cations of weak bases (e.g. neutral red) accumulate in lysosomes. Such relationships can be expressed numerically as quantitative structure activity relations (QSAR). Decision rules based on QSAR, including some relating to the fluorescent probes already mentioned, are summarized in Table 5.1. For additional decision rules, definitions of parameters used, and a general discussion of this subject, see Horobin (2001).

Table 5.1 QSAR decision rules for fluorescent probe–live cell accumulation

Accumulation site; uptake mechanism	QSAR decision rule*
Endoplasmic reticulum	$AI > 4$; $5 > \log P > 0$; $CBN < 40$
Lysosomes, by ion-trapping	$Z > 0$; $\log P > -5$; $pK = ca\ 7$; $CBN < 40$
by fluid phase pinocytosis	$\log P << 0$; $CBN < 40$
Mitochondria, by potential/precipitation	$Z > 0$; $5 > \log P > 0$; $CBN < 40$
Nuclear chromatin, by DNA binding	$pK > 10$; $-5 > \log P > 0$; $LCF > 17$; $CBN < 40$
Plasma membrane, by lipid partitioning	$\log P > 8$
by protein binding	$CBC > 40$

* AI is the amphipathic index, CBN the conjugated bond number, LCF the largest conjugated fragment value, log P the logarithm of the octanol–water partition coefficient, pK a measure of the base strength, and Z the ionic charge

6

Dye purity and dye standardization for biological staining

H. Lyon

Through the years many authors have called for pure or at least less impure dyes (Conn, 1922a,b; Horobin, 1969, 1980; Lillie, 1944a; Lyon *et al.*, 1982, 1994; Schulte, 1994). Horobin (1980) claims that without standardized dyes, standardized procedures are impossible. As automation of biological staining methods and data analysis become more widespread, the need for procedures which are both standardized and rational increases, as does the need for standard stains of known composition.

Commercial dyes are rarely pure and are seldom produced primarily for biological staining; most are made for dyeing and printing of textiles. Manufacturers and industrial users of such dyes need cheap products that give reproducible results, and are not interested in chemical purity as such. As initially manufactured, these dyes may contain, in addition to colored products, unconsumed reactants, intermediate products and also inorganic salts added to precipitate the dye from solution. After production a concentrated, dry dyestuff is 'standardized' or 'cut' by mixing with a suitable diluent to obtain a strength acceptable to the dyer, but not necessarily to the biologist. Dispersing agents may also be added, and finally even other dyes might be included to standardize the shade. Only a few dyes are produced exclusively as biological stains. Examples are methylene violet Bernthsen and brilliant cresyl blue. Marshall (1979a) analyzed a number of commercially available and allegedly pure azure dyes; the weight loss on drying varied between 4.00% and 10.65%, and the amount of sulfated ash varied between 0.29% and 20.38%. High performance liquid chromatography (HPLC) showed that most batches contained some methylene blue and its demethylated derivatives, and only one batch (of azure B) was of acceptable quality.

TYPES OF IMPURITY

Three types of impurities are encountered in commercial dyes (Egerton *et al.*, 1967). These are: (i) diluents, such as sodium chloride, starch, and in the case of water-insoluble dyes, dispersing agents, (ii) substances, other than the required dye, that have formed during manufacture. These may arise from side-reactions, the use of impure intermediates, or may be isomers of the main component, and (iii) dyes of different constitution and color added for shading. Some examples follow.

Wrong dye present

Samples sold as pyronine G (Horobin, 1973) and pyronine Y (Lyon *et al.*, 1982) sometimes contain rhodamine B and rhodamine 6G, but none of the dye named on the label. Another kind of mislabeling is encountered with methyl green (CI 42585). Most dyes sold under this name are ethyl green (CI 42590), not methyl green, but this particular mislabeling is without any practical consequences (Jakobsen *et al.*, 1984a). Further examples of wrong labels are given by Lyon *et al.* (1994). Acetyl Sudan black B, advocated as a lipid and lipoprotein stain, cannot be synthesized by the methods given in the literature, and dye sold as acetyl Sudan black B is merely Sudan black B with a different label (Lauder and Beynon, 1989). Mowry *et al.* (1980) found that certain batches of basic fuchsine labeled CI 42500 for pararosaniline actually contained rosaniline (CI 42510) and were unusable for making aldehyde fuchsine (Proctor and Horobin, 1987).

Colored impurities can have deleterious effects on staining. For example, Kasten (1967) found certain acridine orange batches that gave false-positive staining for RNA.

Variation in amount of major dye present

Many staining methods are rate controlled, and overstaining occurs if sections are left in the dyebaths too long. The consequence of variable dye concentration is that for each batch the staining time, or the degree of differentiation, or the pH of the staining solution must be adjusted, to obtain the required staining results. With manual staining, this is irritating but rather trivial. With automatic or semiautomatic staining machines, it becomes an unacceptable complication (Horobin, 1980).

Additional dyes present

A classical example of this is methyl green (CI 42585), which always contains crystal violet (CI 42555) (Kurnick, 1955a,b). The last-named dye is extracted, either into chloroform or by filtration through polyamide, to obtain stable solutions of green dye (Andersen *et al.*, 1986; Jakobsen *et al.*, 1984a; Trevan and Sharrock, 1951). There are differing reports as to the harmful (Kasten, 1967) or harmless (Horobin, 1973) consequences of the presence of crystal violet in methyl green.

The azure dyes, used in hematology, provide another example, which is discussed at length in Chapter 21.

Kasten *et al.* (1962) have described the presence of primary amine dye impurities in batches of pyronine B and pyronine Y. Of 19 batches labeled pyronine B. Seventeen of 36 batches that contained some pyronine Y also contained primary amine dye impurities.

The typical problems arising from the presence of colored impurities are the occurrence of additional nonspecific staining and of unwanted shades of staining (Horobin, 1969, 1980). This subject has been discussed in relation to Sudan black B (Pfuller *et al.*, 1977), gallocyanine-chrome alum (Horobin and Murgatroyd, 1968, 1978b), and Romanowsky stains (Curtis and Horobin, 1975; Marshall, 1978b).

Colorless materials present

Water in dyes has been largely neglected despite its obvious analytical importance and the fact that many dyes are hygroscopic (Horobin, 1969). Cain and Thorpe (1933) indicate that the most common colorless impurities in commercial dyes are dextrin and sodium chloride, introduced during manufacture.

In commercial pyronine Y samples, Lyon *et al.* (1982) showed that the pH of aqueous 0.05% solutions ranged from 3.53 to 6.40. Further, that they contained 0.2–1.7 mM Fe^{3+}, 0–75 mM Na^+, and 2–175 mM K^+. The content of dextrin, a common diluent of dyes, varies widely. It is more than 50% in some samples of alcian blue and may give rise to falsely intense stainings (Horobin and Goldstein, 1972). The dextrin contents of pyronine Y samples, however, had little influence on methyl green–pyronine staining, as long as the actual dye content was adjusted to a standard value (Horobin, 1980). Even the presence of over 80% dextrin had no effect.

Dispersing agents are often added to dyes with low water solubility. They also may enter staining solutions from poorly rinsed detergent-washed glassware (Horobin, 1980), producing a wide variety of effects including precipitation of dye, reduced staining intensity without precipitation, and gross alterations of the staining pattern. Bennion *et al.* (1975) added a cationic surfactant to solutions of azure A or toluidine blue and achieved selective staining of RNA. Similarly, Meloan and Puchtler (1978) showed that adding mesitol WLS to Congo red gave selective and staining of amyloid in paraffin sections. Mesitol WLS is an anionic reserving agent used in dyeing of wool–cotton and polyamide–cotton fabrics. It prevents binding of direct dyes by wool and polyamides, but does not interfere with the dyeing of cellulose.

Inorganic salts are commonly present in dyes used as biological stains, and can decrease or increase equilibrium staining intensities (Horobin and Goldstein, 1974; Scott, 1967; Scott and Willet, 1966), and these changes can selectively affect different structures (Clemens and Toepfer, 1968; Horobin and Goldstein, 1972). Usually the effect of adding salt to a dyebath is to decrease the staining intensity, either generally (Singer, 1952) or selectively as in the 'critical electrolyte concentration' methods (Scott, 1967). Salts can also slow down dyeing by increasing dye aggregation.

Zinc chloride is often present in cationic dyes such as thiazines and acridine orange, and can quench the fluorescence of adsorbed dye (Stoward, 1967) and modify the metachromasia of various materials (Thompson, 1966).

TECHNICAL INVESTIGATIONS AND ASSESSMENTS OF DYES AND
THEIR IMPURITIES

With any unknown dye sample, the general principle is first to separate, then to identify and finally to assay the colored and colorless components. Information is required on the identities and amounts of the significant colored and colorless components of the sample. The term significant relates to biological staining and is not an absolute term (Horobin, 1980).

Separation of dye components

The standard methods of separation include extraction, precipitation, filtration, recrystallization, distillation, electrophoresis, and adsorption (chromatography). For details of these procedures, the reader is referred to standard texts, for example, Freifelder (1982); Mathews and van Holde (1996); van Holde (1985) and Williams and Wilson (1976).

Identification of dye components

Once separation has been achieved, identification can be attempted. If pure samples are available as standards, identification may be attempted by adding such samples to the analytical sample, and chromatographing the mixture (Horobin, 1980). A general outline of dye-identification procedures has been presented by Brown (1969). Usually, identification does not have to be carried out blindly because the common colorless impurities are known. The dyestuffs literature will suggest probable colored impurities. Methods for identifying the separated colored components of commercial dye samples include spectroscopy, chromatography and classical microanalysis (Horobin, 1980).

For details regarding molar extinction coefficient, absorption, transmittance, absorbance, and various forms of spectroscopy, see the standard texts, such as Burrin (1976). The techniques can be used for any dye and in addition for many colorless components. Rules for the absorption spectra of dyes were published by Lewis (1945). Marshall (1975) derived rules for the visible absorption spectra of halogenated fluorescein dyes. Porro *et al.* (1963) and Porro and Morse (1965) have presented charts and tables of absorbance and fluorescence data for a series of fluorescent dyes and reagents.

Dyes, like any other organic molecules, possess atomic linkages with characteristic absorption peaks in the infrared. Infrared spectrometry can often be extremely valuable (Brown, 1969). If samples of pure standards are available, absolute identification can be achieved by comparison (Horobin, 1980). Nash *et al.* (1963) have described a technique for the recovery of compounds from TLC strips for infrared analysis.

Mass spectroscopy involves bombardment of the sample with a stream of electrons in vacuum in order to ionize it. This produces an ionized form of the parent compound and ionized fragments. These are all positively charged ions and are separated on the basis of their different mass/charge ratios. This technique can only be used for the more volatile components. The sample may be tiny (10^{-6} to 10^{-9} g) but it must be pure. Interpretation of the analytical information obtained is quite complex, but with sufficient background information absolute identity of component may be achieved (Horobin, 1980). Heiss and Zeller (1969a,b) have studied the mass spectrometric fragmentation of sulfur containing dibenzo-heterocycles by exact mass measurements and details of the mass spectrum of pyronine Y were given by Jakobsen *et al.* (1983).

Nuclear magnetic resonance (NMR) depends on atomic nuclei behaving as small magnets. In a strong magnetic field they absorb radiation in the radiowave region of the electromagnetic spectrum (Burrin, 1976). NMR spectroscopy is a practical identification method for organic compounds. The technique is not critically dependent on component purity unless paramagnetic materials are present. For small dye molecules interpretation is straightforward, and with sufficient background information, absolute identity can be achieved (Horobin, 1980).

The required sample size for elemental microanalysis is a few mg for the determination of carbon, hydrogen, and nitrogen in simple organic compounds, though a pure sample is required. The analytical information obtained is easy to interpret and can sometimes with sufficient background data give absolute identity (Horobin, 1980). These techniques are suitable for the identification of simple inorganic compounds such as sodium sulfate or boric acid, typical colorless constituents of dyes. Marshall (1976a) found methanol-insoluble residues in commercial samples of erythrosins, fluorescein, phloxin and rose Bengal, and revealed by inorganic semimicroanalysis these were carbonates and halides.

Assaying the amount of dye components

With a pure dye sample, spectrophotometric assay is easy, requiring only a determination of the absorbance at the absorption maximum if the molar extinction coefficient of the dye is known. However, this latter requirement presents a formidable problem. The molar absorbances of many dyes are not available, or are unreliable for dyes that have never been adequately purified. In these cases Lyon *et al.* (1994) advise purification using HPLC. In a series of papers, the European Committee for Clinical Laboratory Standards (ECCLS), Subcommittee on Reference Materials for Tissue Stains (SRMTS) (1992a–i) has published re-evaluated molar extinction coefficients for a number of biological dyes. Further, in the predecessor of the present book (Lillie, 1977) spectrophotometric assay methods are described for auramine O, methyl violet 2B, crystal violet, pyronine Y, pyronine B, ethyl eosin, neutral red, safranine O, thionine, and methylene violet Bernthsen.

Spectrophotometric assay procedures have been described for several biological stains, including xanthene dyes (Emery *et al.*, 1950), aminoaryl-methane dyes (Emery and Stotz, 1953), toluidine blue O (Apgar and Patel, 1969), carminic acid (Marshall and Horobin, 1974a), hematoxylin and hematein (Marshall and Horobin, 1974b), and dyes used in methyl/ethyl green–pyronine Y staining (Jakobsen *et al.*, 1984a,b).

Lyle and Tehrani (1979) separated a number of water-soluble dyes by TLC. They made a comparison of elution followed by absorption spectrophotometry, densitometry directly on the chromatogram, and reflectance spectrophotometry on the removed dye spot including adsorbent. Densitometry gave the most consistent results. Schumacher and Adam (1994) determined the dye content of a number of commercial samples of alcian blue by spectrophotometry following the recommendations of Schenk (1981); dye content varied between 39% and 73%.

Lubrano *et al.* (1977) used HPLC to quantitate batch variations in commercial samples of thiazine dyes, thiazine eosinates, and Romanowsky-type blood stains. Large variations were observed between most of the samples of each type. Nakada *et al.* (1980) have shown that methylthymol blue and methylxylenol blue can be purified by HPLC.

Emery *et al.* (1950) described color acid precipitation assays of anionic xanthene dyes; the assay works well with pure dyes but does not distinguish between the members of the group. Maurina and Deahl (1943) have described a gravimetric assay for methylene blue using either potassium or sodium perchlorate to obtain a quantitative yield of methylene blue perchlorate.

The majority of dyes have hitherto been assessed titrimetrically by reduction with titanous chloride or oxidation with ceric sulfate, chloramine T, or iodine (Horobin, 1980; Lillie, 1977). These methods require homogeneous samples to obtain useful results. The equipment is inexpensive and easy to operate for the oxidation methods. Reduction with $TiCl_3$ is more complicated, requiring anaerobic conditions, but the results are fairly accurate (Horobin, 1980; Lillie, 1977; see also Chapter 28). Giles and Greczek (1962) have reviewed redox titration methods for assaying dyes.

Drying of dis-azo dyes for long periods at 110°C showed that these retained 1–2% of water even under these conditions. When the dyes were then placed in air of 100% humidity, they took up water avidly. For example, chlorazol sky blue FF contained 50% water (Horobin, 1969). Carminic acid lost between 6.0–8.2% and carmine 7.6–15.9% of weight, on drying (Marshall and Horobin, 1974a).

The common 'standardizing' additive in commercial samples, dextrin, may be easily estimated by extracting the dye with ethanol leaving dextrin and any inorganic salt as a residue (Cain and Thorpe, 1933). Inorganic salts such as NaCl, Na_2SO_4 or carbonates may then be estimated by simple chemical methods. Metals can be determined by atomic emission spectroscopy (Chavan, 1976; Jakobsen *et al.*, 1984b; Lyon *et al.*, 1982; Marshall and Lewis, 1975). If the absence of dextrin has been checked chromatographically, solvent extraction with chloroform, 2-propanol, ethanol, or methanol is an excellent method for estimating the salt content of dye samples (Horobin, 1968, 1970; Horobin and Murgatroyd, 1968, 1969).

Traditional semimicroanalysis can be used for determining the content of inorganic components in dyes. Arshid *et al.* (1953) have suggested a method for obtaining quantitative yields of nitrogen azo dyes; their method is unsuitable, however, for cyanine and water-insoluble anthraquinone dyes.

Purification of dyes

The task of purifying large quantities of dyestuffs is technically far more difficult than purifying a few milligrams to a few grams of dye, with the greatest diffi- culty being financial, not technical. If users were able and willing to pay enough, then pure dyes would become available (Horobin, 1980).

The useful preparative chromatographic methods involve thin layers or columns of various adsorbents. Preparative thin-layer chromatography (TLC) techniques are described by Horobin and Murgatroyd (1968), Stutz *et al.* (1968) and, for really large- scale work, by Visser (1967). Large-scale chromatographic purification of dyes has usually been carried out in columns packed with media that make use of adsorption (Kennedy *et al.*, 1970), gel filtration (Levinson *et al.*, 1977; Lohr *et al.*, 1974), or ion-exchange (Gupta *et al.*, 1967; Lohr *et al.*, 1975). HPLC can also be used to purify dyes (Lyon *et al.*, 1994). A technical problem of preparative column chromatography is recovery of the dyes from their dilute solutions in the eluted fractions. Techniques for concentration are described by Horobin (1980) and Shapeiro (1961).

Recrystallization to purify dyes on a large scale is widespread and has been discussed in general terms by Giles and Greczek (1962). The tendency of dyes to precipitate in colloidal form can greatly hinder filtration, but high yields are possible, such as 99.98% of acridine orange from hexane or 96.65% from chloro- form (Zanker, 1973). Chavan (1976) purified anionic dyes by recrystallization, and Jakobsen *et al.* (1984a) and Andersen *et al.* (1986) obtained fluoroborate salts of methyl and ethyl green in high purity.

Electrophoresis is primarily an analytical tool rather than a means of preparing samples (Mathews and van Holde, 1996), but it has been utilized for purification and separation of water-soluble dyes on the basis of charge and ionic weight. If polymers are not formed, this is a quite effective method for small quantities of dyes.

THE STANDARDIZED STAINING METHOD APPROACH

Ever since dyes were first used for staining cells and tissues, it has been common practice to test new batches by using them under the same conditions as the previous batch. That this did not always lead to the expected results is evident from the numerous variations, published and unpublished, that are to be found for most staining procedures.

Using standard staining methods for assessing stains

The above situation has naturally led to the assumption that it is possible to assess a dye by its performance in a staining procedure that has been more or less standardized. This view has led the Biological Stain Commission to base its assessment of dyes not only on physico-chemical characteristics but also on performance in standardized staining techniques, many of which have been published (Churukian, 2000; Clark, 1981; Lillie, 1977; see also Chapter 28). Wittekind (1985) claims that it is possible to evaluate dyes including their staining performance solely on the basis of their physico-chemical characteristics.

PRACTICAL RESPONSES TO STAIN IMPURITY

Practical responses to the impurity of dyes are to call for quality control of the manufacturing process of dyes and to introduce a rigorous standardization of reagents and staining procedures. Preferably, both these approaches should be taken.

Commercial quality control

Quality control of dyes makes it mandatory that there should be a continuous surveillance of all steps in the production processes. This should lead to the production of the pure dye uncontaminated with unconsumed reactants, intermediate products, byproducts, or inorganic salts used to precipitate the dye from solution. Further, no dispersing agents or diluents such as dextrin or sodium sulfate should be present.

According to Conn (1946), the only source of dyes available for biologists in the early days of staining was the regular dye industry. Sometimes these products were standardized crudely as to color, but never as to chemistry. In 1880, Dr Georg Grubler set up a company in Germany for marketing aniline dyes for biologists. The company bought up large batches that Grubler had previously tested for microscopic use. In that way an empirical standardization of dyes was achieved with surprisingly little batch-to-batch variation (Lillie, 1977). In World War I the temporary unavailability of German stains forced Britain and the United States to develop their own sources (French, 1926a; Lillie, 1977; Scott, 1924a,b). In recent years, some of the larger dye manufacturers have set up programs for quality control. An example is the Certistain (E. Merck, Darmstadt, Germany) project. The firm has introduced an internal quality assessment scheme for dyes it produces. This means that each dye undergoes a specified testing procedure defined by the firm itself. The published procedures are quite similar to those used by the Biological Stain Commission (see below under Third-party quality control). There can be little doubt that these dyes represent an advance towards better products. They do not claim to be pure dyes; the user may expect a good and uniform quality, but must trust the firm for having carried out the testing procedures with care (Schulte, 1994).

Third-party quality control

Inconsistent dye quality in the USA after World War I led to the foundation of the Commission on Standardization of Biological Stains in 1922 (Editorial, 1926). This was later renamed the Biological Stain Commission (Lillie, 1977). Working together with manufacturers and dealers the Commission endeavors to see that the supply of stains available in the USA is of the highest quality (Lillie, 1977). This is done on a voluntary basis, with companies submitting samples of their dye batches to the Commission's laboratory for testing. If the dye passes the tests, the Commission issues a special certification label with the certification number of the dye batch. Unfortunately, this assessment does not ensure a supply of pure dyes. Certified dyes have to meet certain standards of dye content and purity and they must work satisfactorily in certain biological tests. The history of the Biological Stain Commission was written by Conn (1980a,b; 1981a,b,c), and a paper on the goals and recent status of The Commission was published by Mowry (1980).

Standardization of reagents, protocols, and documentation

The standardization of reagents, particularly of dyes, has already been discussed above under Identification of dye components, and reference has been made to this edition of *Conn's Biological Stains*, its predecessor (Lillie, 1977), and to the papers from the European Committee for Clinical Laboratory Standards (1992a–i). The availability of dye lots with high dye content and little colored impurity has improved in recent years but there is still a need for standards. Wilson (1992) notes that, without standards, the individual researcher may still need to test each batch to determine if it is satisfactory.

According to Boon and Wittekind (1986), the principal aim of standardizing staining methods is to render their application reproducible and therefore reliable. This is of the utmost importance when dyes are used for automated cell pattern recognition (Wittekind, 1985; Wittekind and Schulte, 1987a). Any preparatory step – from cell sampling to mounting of the stained slide – will somehow affect the structure of the cell and ultimately lead to the production of a particular staining pattern which, in strict terms, is an artifact (Lyon *et al.*, 1991). The aim of standardization in cell and tissue preparation is to make this artifact reproducible. It seems logical both from practical and theoretical points of view to choose as standard the staining pattern that best fulfils the requirements of the observer, be this human or a computer. The preparatory technique yielding this staining pattern is therefore defined as the standard technique (Lyon *et al.*, 1991).

Standardization of a staining protocol requires consideration of all steps in the procedure: slide preparation, dyes and other reagents, staining techniques, and documentation (Lyon *et al.*, 1991). Before carrying out a standardized staining protocol, two factors should be carefully evaluated. The first is the accessibility of the staining sites in the cells or tissue, and the second factor is the staining intensity of the observed cell or tissue area. The latter applies regardless of whether the stained product is viewed by direct microscopy or by image analysis techniques. Altman (1980) puts this very aptly by asking, 'How thick is your section?' In cytology this is the thickness of the intact cells; in histology it is the thickness of the sections of tissue. The latter is not identical to the setting on the microtome, and must therefore be determined in each case (Hoyer *et al.*, 1991).

The current state of the world in this area

In 1974, the Biological Stain Commission arranged a conference on the importance of dye purification and standardization in biomedicine (Mowry and Kasten, 1975). It was concluded that the dyes used in cytology and biomedicine were not fine chemicals but usually mixtures of variable and often undefined composition. Sometimes they were even incorrectly labeled. The situation has undoubtedly improved somewhat since 1974, but the lack of standardization is still a problem when cytochemists and cell biologists seek the physico-chemical bases for the special affinities shown by cell and tissue components for particular dyes under controlled conditions. This is not exclusively a problem with dyes; it also includes such reagents as antibodies and nucleic acid probes. Another conclusion of the 1974 conference was that the facts concerning inadequately standardized staining reagents were not well known and deserved increased attention.

Important work is taking place in the field of immunohistochemistry. In 1991 an Immunohistochemical Steering Committee was formed and a planning document published (Taylor, 1992a). This was shortly followed by a report on the proposed format for package inserts for immunohistochemistry products (Taylor, 1992b). In 1997 the NCCLS published a proposed guideline on quality assurance for immunocytochemistry (NCCLS, 1997).

Pure dyes are still difficult or impossible to purchase. Approximations are BSC Certified Stains, Merck Certistains, or similar products. A working group under the European Committee for Standardization (CEN), Technical Committee 140 for In Vitro Diagnostic Devices, has prepared a standard for the 'labeling' of dyes. The major implications are that dyes and other *in vitro* diagnostic reagents used in biology (antibodies, nucleic acid probes etc.) shall contain precise and detailed information on contents and applications. Only if the manufacturer indicates that the dye or other reagent is for *in vitro* diagnostic work in laboratory medicine, does it fall under the rules of the standard. Grizzle and Mowry (1994) give an excellent overview of the problems associated in defining whether a reagent is *in vitro* diagnostic or not. Furthermore, The US Department of Health and Human Services, Food and Drug Administration (FDA) has published proposed rules to classify or reclassify analyte-specific reagents (FDA, 1996a) and to classify or reclassify immunohistochemical reagents and kits (FDA, 1996b).

7

Reactive staining reagents and fluorescent labels

J.C. Stockert

Dye and nondye reagents can directly attach to biological components by strong covalent bonds. Such reactive dyes and fluorescent reagents represent an interesting category of microscopic stains for fixed or living cells and tissues. Fluorescent labels are also attached to biomolecules, to aid analysis of their interaction with biological substrates.

Direct cell and tissue staining

Classic histochemical examples of reactive staining include demonstration of aldehyde groups with Schiff's reagent, and the detection of thiols by mercurial dyes. Likewise, reactive fluorochromes (FITC, TRITC) and nonfluorescent reagents (dansyl chloride, fluorescamine) can bond covalently to tissue proteins, which are visualized by the fluorochrome emission or the development of fluorogenic reactions. In addition to staining of fixed tissues, such reactions can be applied to living material. The formation of fluorescent complexes between inorganic cations and chelating reagents (e.g. FURA-2 for Ca^{2+}) is an example.

Drawbacks of reactive staining include nonspecific binding of reagents, variable accessibility of target groups, possible noncovalent binding, and the necessity for suitable conditions for covalent bonding (solvent, pH, temperature, ionic strength, reagent concentration). Previous treatment of the substrate (fixation, processing) can also affect subsequent staining. Without critical consideration of these factors, results and interpretations from reactive staining can be uncertain.

Contrary to what is often said, reactive staining does not improve light fastness; photofading depends on the chemistry of the dye. The important advantage of reactive staining is washing fastness; covalent bonding makes dyes highly resistant to extraction by water or organic solvents.

Biomolecular labeling *in vitro*

Isolated substances of biological origin (proteins, polysaccharides, nucleic acids, lipids) may be covalently labeled with reactive fluorophores, and used to analyze uptake and accumulation in living cells, or to reveal specific antigens or nucleic acids. An advantage of fluorescent labeling and fluorogenic reactions is that fluorescence is detected more sensitively than absorption by a factor of 10^4 or higher.

For biomolecular labeling procedures, important factors are the availability of target tissue components, and the binding specificity of fluorescent probes. Binding and detection of labeled biomolecules often require additional techniques (improvement of cell permeability, signal amplification), and then controls and stardardization are necessary. Phototoxic effects can occur when fluorescent probes are applied to living cells.

BONDING MECHANISMS TO BIOLOGICAL MATERIALS

Reactive stains and labels display several bonding mechanisms. Typical reactions are discussed below under the four major categories of organic, photo-, organometallic and coordination chemistry.

Organic chemistry

A well-known reactive moiety is the isothiocyanate group (–N=C=S). In particular fluorescein isothiocyanate (FITC) and tetramethylrhodamine isothiocyanate (TRITC) are common fluorescent labels for antibodies and other proteins. The isothiocyanate reacts with amines to form thioureides (–NH–CS–NH–), and can also react with hydroxyls and thiols. Consequently lysine, tyrosine, and cysteine may be labeled in proteins. Other isothiocyanate-substituted dyes used to label proteins include xanthenes such as Eosin Y (EITC) and rhodamine B (RITC) and various acridines, anthracenes and pyrenes. Fluorochromes containing reactive groups other than isothiocyanate are also employed for molecular labeling. Examples are iodoacetamido, dichlorotriazinyl, maleimido and succinimidyl esters, of which the last is the most widely applied. FITC and TRITC remain the most popular compounds for preparing fluorescent conjugates. A visual summary of organic reactive groups involved is given in Figure 7.1.

Photochemistry

Azido (–N=N=N) substituents in aromatic rings activate under ultraviolet light and bind to nucleophilic groups such as amines, phenolic hydroxyls and thiols. Simple aryl azides are initially photolyzed to electron-deficient nitrenes:

$$\text{aryl–N=N=N} \rightarrow \text{aryl–N}^+ + \text{N}_2$$

These undergo ring expansion to yield reactive dehydroazepines. On account of their light sensitivity, azido derivatives must be used under red light illumination.

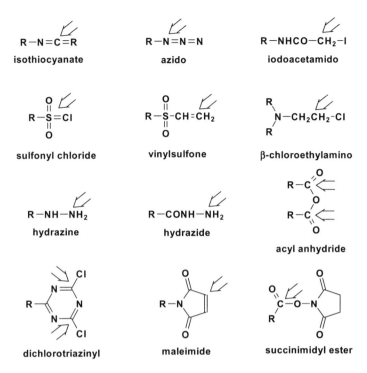

Figure 7.1 Some organic reactive groups found in dyes and reagents applied in reactive staining and labeling methods. Bonding atoms are arrowed.

Thus a photoactivatable analog of biotin contains the 4-azido-2-nitrophenyl group, which photolyzes and forms a stable linkage with DNA, RNA, and proteins (Foster *et al.*, 1985). Photoactivatable fluorophores are nonfluorescent compounds that become fluorescent after UV irradiation. Photoactivation of proteins labeled with caged 2-nitrobenzylfluorescein is an example of this, illustrated in Figure 7.2. UV light triggers an efficient photocleavage of the caged label, generating the actual fluorophore (Mitchinson *et al.*, 1994).

Organometallic chemistry

In organometallic compounds a metal atom is directly bound to carbon. Organomercurials contain mercury with a bound anion (C–Hg–X, with X being a halogen, OH, CN, or S). The Hg^{2+} ion mercurials ($R–Hg^+$) have strong affinity for thiols, with which they form mercaptides:

$$R–Hg–X + HS–Protein \rightarrow R–Hg–S–Protein + XH$$

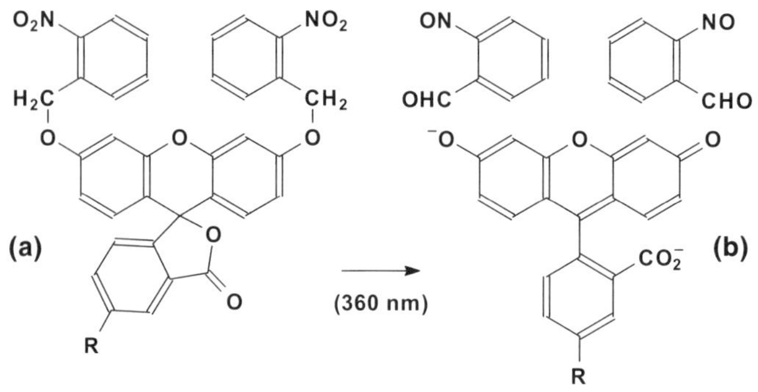

Figure 7.2 Photoactivation of caged 2-nitrofluorescein lactone (**a**) to give fluorescein and nitrosobenzaldehyde (**b**). R represents the labeled biomolecule.

Since the Hg–S bond is essentially covalent, with a very low dissociation constant, stable compounds form when $HgCl_2$, mercuric acetate, or organomercurials react with fixed tissues containing thiol groups. For instance the dye mercury orange was synthesized for specific staining of tissue thiols. The C–Hg bond, though stable, can be cleaved by reduction or electrophilic attack, as by treatment with alcoholic solutions of thiol compounds. The possibility of nonspecific bonding must be also taken into account, such as reaction with heterocyclic rings in DNA and RNA, as discussed by Horobin and Flemming (1982).

Coordination chemistry

Coordination occurs when a polar covalent bond forms between a metal cation and an electron-donating –OH, C=O, $-NH_2$, –N=N– or $-COO^-$ group. Bi- or multidentate ligands form stable five- or six-membered chelate rings. Coordination chemistry is involved in several staining and labeling methods.

Fluorescent Ca^{2+} indicators including Fura-2, Quin-2, and Indo-1 share nearly identical octa-coordinate binding sites. These salts and their acetoxymethyl (AM) esters are popular for studying intracellular Ca^{2+} concentration and mobilization from calcium-binding proteins. Carboxyl ionization is necessary for chelation and enhancement of the fluorescence. The affinity for Ca^{2+} is higher than for Mg^{2+}, because the latter ion is too small to contact all the ligand groups simultaneously. The charge on Na^+ or K^+ ions is not enough to fold the carboxylic chains into the cation-binding pocket.

To allow entry into the cell, these calcium indicators are often applied as AM esters. Cytoplasmic enzymes de-esterify these, liberating the carboxyl groups, and allowing chelation to occur. Fluorescein- or rhodamine-tetracarboxylates, either free or caged with photoreactive groups, are also used for this purpose.

An extensive, albeit uncritical, bibliography of fluorescent Ca^{2+}-indicators is provided by Haugland (1996).

REACTIVE STAINS FOR FIXED TISSUES

Schiff's reagent

The structure of Schiff's reagent and its reactions with aldehydes are discussed in Chapter 14. Here it will merely be noted that this colorless derivative of basic fuchsine reacts with tissue aldehydes to yield colored alkylsulfonate derivatives (Gill and Jotz, 1976; Nettleton and Carpenter, 1977).

As long ago as 1924 Feulgen and Rossenbeck described the use of Schiff's reagent for microscopic identification of DNA, but unfortunately the second author is seldom mentioned. McManus (1946) introduced the PAS reaction for polysaccharides. Measurements of Feulgen-stained DNA are frequent in biomedical studies. The DNA content of a cell identifies its position in the cell cycle, and Feulgen staining contributes to the analysis of proliferation kinetics. The assessment of altered DNA patterns is also useful in pathology and cyto-genetics. Specificity and applications of the Feulgen and PAS methods are discussed in histochemical textbooks (Kiernan, 1999; Pearse, 1985; Thompson, 1966).

Vinyl ether (or acetal) phospholipids can release Schiff-reactive aldehydes (**plasmal reaction**), as can the atmospheric oxidation of C=C bonds of unsatu-rated lipids (**pseudoplasmal reaction**). Aldehydes naturally present or chemi-cally produced in elastin, collagen, mast cells and lignin are also capable of reacting with Schiff's reagent (Lillie and Fulmer, 1976; Thompson, 1966). Alcoholic solutions of a phosphotungstic acid–Schiff complex proved useful for detection of water-soluble polysaccharides (Bedi and Horobin, 1976).

Schiff-type reagents and related compounds

Many dyes and fluorochromes can be treated with bisulfite to provide reagents that will combine with aldehydes (Kasten, 1959, 1963). These **Schiff-type** or **pseudo-Schiff reagents** can be used for Feulgen and PAS staining. The best substitutes for Schiff's reagent are made from thionine (Bancroft and Gamble, 2002; Van Duijn, 1956) and, for fluorescence staining, acriflavine (Clark, 1981).

Primary amines react rapidly with aldehydes, and stable imines are formed in a nonaqueous acidic environment (Lillie, 1962). The basic fuchsine dyes are primary amines, and an alcoholic solution of basic fuchsine with added HCl is a useful substitute for Schiff's reagent in the detection of hydrolyzed DNA or periodate-oxidized mucosubstances (Horobin and Kevill-Davies, 1971).

Mercurials

Azo and hydroxyxanthene mercurials are used in histochemistry. **Mercury orange** is a standard method for demonstration of tissue –SH groups. Low dye concentration, apolar solvents, and prolonged treatment are the best conditions for staining. Greater sensitivity can be obtained by observation using fluorescence microscopy (Cowden and Curtis, 1970). Fluorescent mercurials such as **fluorescein mercuric derivatives** and **mercurochrome** reveal thiols and reduced disulfides in chromosomes and other intracellular structures (Cowden and Curtis, 1974; de la Torre *et al.*, 1990).

Caution is needed in the interpretation of mercurial stains. Nucleic acid binding of mercurials has been reported (Horobin and Flemming, 1982; Takeuchi and Maeda, 1979). Glutaraldehyde fixation increases the affinity of mercurochrome for chromatin, and DNA–dye binding was also suggested by Novello and Brauer (1986). A variety of other nonthiol mercurial staining is possible.

Another problem of mercurial staining is to determine what proportion of thiol groups is detected. This depends on their availability after protein fixation, and their intentional or unintentional generation by reduction of disulfides. Histochemical staining with mercurials requires careful standardization and controls to avoid misleading results (Horobin and Flemming, 1982; Pearse, 1985).

Reactive dyes as tissue stains

The **textile reactive dyes**, used industrially to improve wet fastness especially of cellulosic materials, can form covalent bonds with chemical groups of objects examined microscopically (Lillie, 1977). The main advantage of such a reactive dye is that the stained preparation resists extraction of colorant by water and other solvents.

Some of the reactive dyes used for microscopic staining are mono- or dichlorotriazines (H and M procions, and cibacron dye ranges); vinylsulfones (remazol range); or dichloroquinoxalines (levafix E range); with azo or aminoanthraquinone chromophores. In fixed tissues, protein structures stain with procion dyes (Schwenke and Geyer, 1966). Remazol dyes, including remazol brilliant blue R, stain collagen, elastin, basement membranes, and erythrocytes (Salthouse *et al.*, 1971). Glial fibers (intermediate filaments of reactive astrocytes) can be visualized with levafix dyes (Waldrop and Puchtler, 1975).

In addition to these textile colorants, a number of other reactive dyes form covalent bonds with tissue sections. The alkylating alpha-chloroethylamine group is present in quinacrine mustard, a derivative of the antimalarial drug quinacrine. Its acridine ring can intercalate into DNA, where its basic nitrogens form ionic bonds with DNA phosphates, and the alkylating groups can bond covalently to guanine. Quinacrine mustard binds to DNA about 20 times more strongly than quinacrine. Ethanol or dilute acids extract quinacrine but not quinacrine mustard from DNA complexes (Modest and Sengupta, 1973). Quinacrine mustard was introduced by Caspersson *et al.* (1968) for demonstrating fluorescent banding patterns in plant and animal chromosomes. The dye is still used for visualizing specific chromosome bands (Q-banding), where staining is based on the brighter fluorescence of DNA regions rich in adenine and thymine (Bickmore and Craig, 1997; Sumner, 1990).

Reactive microscopic staining of proteins in fixed cells can be accomplished using dye reagents such as FITC and SITS (Benjaminson and Katz, 1970; Cornelisse and Ploem, 1976), or dinitrofluorobenzene (Tas *et al.*, 1980). Likewise, fluorescent protein labeling of fixed cell populations by FITC, RITC, TRITC, and SITS has applications in flow cytofluorometry (Stohr *et al.*, 1978).

Chelating dyes and reagents for metal detection

Coordinative and ionic forces result in the formation of metal complexes with bi- or multidentate ligands. These contain resonance-stabilized chelate rings.

Such structures occur for instance with anthraquinones, flavonols and tetracyclines where oxygens are in *peri* positions. Calcium-chelating anthraquinone dyes include alizarin red S and nuclear fast red; which have been applied to demonstrating calcification in fixed tissues (Bancroft and Gamble, 2002; Chaplin and Grace, 1976). Alizarin red S finds particular use for staining fetal skeletons in whole mounts, as reviewed by Klymkowsky and Hanken (1991). The structure of the calcium complex is shown in Figure 7.3a. Chelating flavones and flavonols with *peri* hydroxyl and ketone groups, especially morin, also form fluorescent metal complexes, as shown in Figure 7.3b. Indeed, the first fluorescence-based analysis was performed in 1867 using the Al^{3+} chelate of morin, and this dye is still used for detection of aluminum and calcium in tissue sections (Bancroft and Cook, 1994; Platt *et al.*, 2001).

Other chelating stains include the dihydroxyazo dyes, notably eriochrome blue black B and eriochrome black T which have also been used as calcium stains (Meloan *et al.*, 1973).

Nondye reagents for organic functional groups

There are chromogenic and fluorogenic reactions that form covalent bonds with many organic functional groups of tissue biomolecules and reagents that are not dyes. Some examples follow, listed alphabetically.

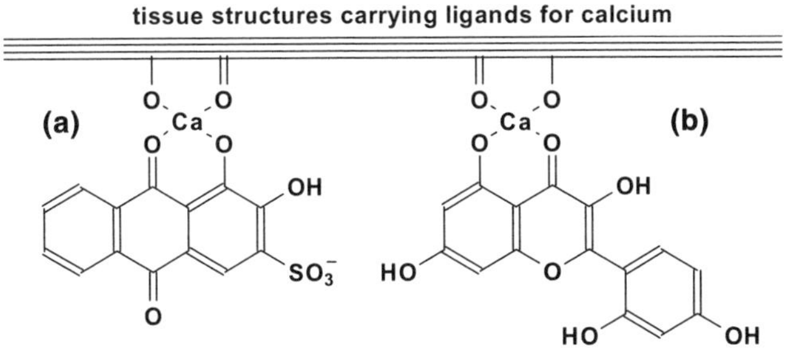

Figure 7.3 Structure of the alizarin red S (a) and morin (b) complexes with Ca^{2+} in tissue components.

Dansyl chloride

This compound (1-dimethylaminonaphthalene-5-sulfonyl chloride) reacts with primary and secondary amino groups to give fluorescent sulfonamides:

$$-SO_2Cl + -H_2N \rightarrow -SO_2-NH- + HCl$$

Under histochemical conditions, dansyl chloride selectively reveals terminal and lysine amino groups of proteins (Ringertz, 1968; Rosselet and Ruch, 1968); and the detection of chromosomal and extracellular proteins has been described (Burkholder, 1982; Utakoji and Masukuma, 1974; Vidal, 1980). Dansyl fluorescence intensity is influenced by the tissue environment and the mounting medium; apolar media and hydrophobic sites induce stronger emission and a shift to shorter emission wavelengths (Chen, 1967). This feature makes the dansyl fluorophore an important structural probe for macromolecules.

Diazonium salts

Aromatic diazonium salts (aryl–NN^+) or their bases (aryl–N=N–OH) couple through electrophilic attack to tissue components yielding colored products (Abrahart, 1977; Nachlas *et al.*, 1959). Direct chromogenic reactions with fast red B, fast red TR, fast red GG, fast garnet GBC and diazosafranine O were used by Lillie (1977) for the histochemical detection of bilirubin and indolamines. In the coupled tetrazonium reaction, aromatic side-chains of proteins are demonstrated by azo coupling (Pearse, 1985).

Fluorescamines

Fluorescamine (4-phenylspiro-(furan-2(3H)-1'-phthalan)-3,3'-dione), and **BMDPF** (2-methoxy-2,4-diphenyl-3(2H)-furanone) react selectively with primary amines, yielding highly fluorescent products (Weigele *et al.*, 1972). These compounds are customarily shown as 2-hydroxypyrrolinones, but the actual fluorophores are pyrrolones derived from their pseudobases. The reaction has histochemical application for amino groups (Cuellar *et al.*, 1991; Stockert and Trigoso, 1993), and appears useful for detection of neoplastic cells (Parry *et al.*, 1982).

NBD halides

The 4-chloro and fluoro derivatives of NBD (7-nitrobenzo-2,1,3-oxadiazole) form fluorescent products with primary and secondary amines:

$$(NBD)-Cl + -NH_2 \rightarrow (NBD)-NH- + HCl$$

NBD is a commonly used fluorophore for labeling peptides (Sai *et al.*, 1998) and phospholipids (Gong *et al.*, 2000).

Osmium tetroxide (OsO₄)

OsO_4 is well known as a fixative, but is also used as a stain (Hayat, 1975, 1981). The reaction with carbon–carbon double bonds (–C=C–) involves initial formation of cyclic Os(VI) esters; these hydrolyze giving black OsO_2 and a *cis*-glycol derivative. In addition to unsaturated lipids, various structures including the pyrimidine 5,6-double bonds of single-stranded nucleic acids contribute to tissue osmiophilia (Chang *et al.*, 1977). Such reactions are illustrated in Figure 7.4. As the reaction of OsO_4 with native DNA is sterically hindered by basepair stacking, OsO_4 is a probe for pyrimidine unstacking (Paul and Ferl, 1993). RNA pyrimidines and terminal riboses also react with OsO_4 (Daniel and Behrman, 1976); and osmiophilic ribonucleoproteins (in nucleoli, Balbiani rings, chromatid cores and kinetochores) can be detected using OsO_4 and *p*-phenylenediamine (Antonio *et al.*, 1996; Stockert, 1977).

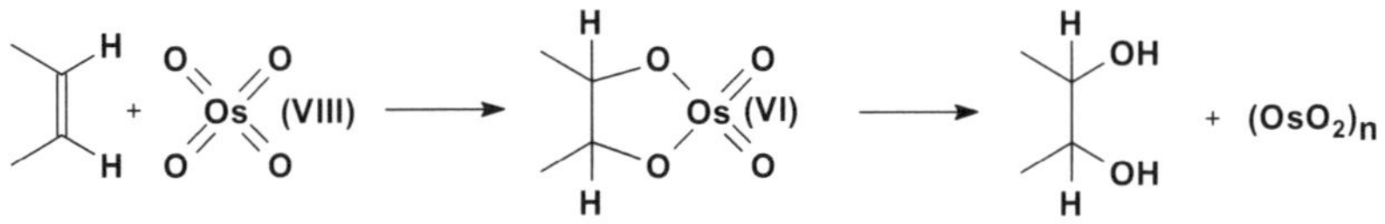

Figure 7.4 Reaction of a carbon–carbon double bond with osmium tetroxide, yielding black osmium dioxide as a final reaction product.

Phenanthrenequinone

A fluorogenic reaction occurs between 9,10-phenanthrenequinone and arginine, with formation of a 2-aminophenanthroimidazole derivative. The method is used for histochemical demonstration of arginine in proteins and peptides (Ekman *et al.*, 1980). Cleavage of the guanidino group of arginine may occur (Yamada and Itano, 1966) but the fluorescent product remains bound to tissues, allowing microscopic localization.

REACTIVE STAINS FOR LIVING CELLS AND ORGANISMS

Reactive dyes and fluorochromes for vital staining are of increasing interest, especially as cell markers, tracers of neuronal connections, and labels for growing bones. The possibility of cellular damage due to direct or photosensitized toxicity must be taken into account, and the latter effect certainly occurs during observation of living materials.

Isolated cells

SITS (4-acetamido-4'-isothiocyanatoostilbene-2,2'-disulfonic acid) was introduced as a fluorescent label for the erythrocyte membrane (Maddy, 1964), on the basis that hydrophilic substituents would reduce membrane permeability, and the isothiocyanate group would bind to membrane proteins. SITS has also been used for neuronal tracing (Schmued and Fallon, 1985). However, SITS and related compounds have been shown to be impure and unstable (Jakobsen and Horobin, 1989 and cited work). It is dimeric material present in some commercial products that is probably responsible for fluorescent neuronal tracing (Horobin *et al.*, 1987).

Cells with fluorescently labeled plasma membranes may be used for studying cell–cell and cell–matrix interactions. For long-term tracking, a label must be stable and not disturb the functions of the living cell. Protein-reactive reagents (e.g. FITC and TRITC) are used for labeling cell membranes (Edidin, 1989; Horan *et al.*, 1990). Often these substances become internalized, and sometimes labels leak from the cells.

Tissues and organisms

Reactive chlorotriazine and vinylsulfone derivatives of azo or azopyrazolone dyes (e.g. procion navy blue M3RS and remazol black B) have been applied *in vivo* for staining connective tissue fibres and calcification sites (Henry, 1968; Maltha *et al.*, 1977). Procion yellow M4R has been used as a fluorescent filler for neurons and their processes, especially in invertebrates (Gregory, 1973; Payton, 1970). The dye binds to protein amino groups in cell cytoplasms (Flanagan *et al.*, 1974). All these covalently bonded dyes resist extraction during such subsequent procedures as decalcification, tissue processing, and embedding.

The fluorescent naphthalimide dyes lucifer yellow CH and lucifer yellow VS were first synthesized for labeling neurons (Steward, 1978). They have reactive hydrazine (CH) or vinylsulfone (VS) groups, which react with carbonyls and amines, respectively. Lucifer yellow CH shows stronger emission than procion yellow M4R, and remains covalently bound to labeled cells. After iontophoretic injection it passes through gap junctions, revealing cells that are electrically-coupled (Steward, 1981). Applications include labeling of photoreceptors, motor neurons, cortical neural networks, and various embryonic cell-types (Bach *et al.*, 1993). Lucifer yellow CH can also be taken into cells by endocytosis (Racoosin and Swanson, 1994).

Fluorescent labels incorporated into calcifying tissues can provide time-markers for the study of bone and tooth development. Ca^{2+}-chelating fluorochromes (alizarin red S, calcein, calcein blue, tetracyclines, xylenol orange) are administered systemically to growing animals. The occurrence of different emission colors with this set of dyes allows polychromatic sequential labeling (Olerud and Lorenzi, 1970).

FLUORESCENT LABELS AND PROBES

Fluorescent labeling of biomolecules has many important applications, and these, together with the chemistry of fluorescent labels, have been reviewed by

Haugland (1996). When labeling, choosing suitable fluorophores and reactive groups is critical.

Ideal and available fluorophores

Ideally a fluorophore has satisfactory optical and physicochemical properties, and is also nontoxic, easy to manipulate, and available without impurities at low cost. Real compounds have only subsets of these features, and choosing a label always involves compromise. Moreover the chemical structures of newer fluorescent labels are not always revealed.

Perhaps for these reasons the traditional fluorescent labels remain widely used. These include xanthenes (FITC, TRITC, sulforhodamine B), dansyl and coumarin compounds, and NBD chloride. More recently Texas red, cascade blue (a derivative of 1-hydroxy-3,6,8-pyrene-trisulfonate), and BODIPY labels have become extensively applied. Information on compounds of all these classes is available in the entries in later chapters.

Fluorescent or phosphorescent complexes containing Zn, Pd, Ru, Tb, and Eu have also been developed for use as biomolecular labels (Hemmila, 1989; Papkovsky, 1991). Ru(II) bound to O-diimines shows a characteristic emission, and analytical and microscopic uses of fluorescent Ru(II) complexes are known (Bertolesi *et al.*, 1995; DiCesare *et al.*, 1993). Luminescence of Tb^{3+} and Eu^{3+} chelates has applications in microscopic, cytochemical, and analytical studies (Evangelista *et al.*, 1991). These lanthanide chelates constitute a new generation of labeling fluorophores, suitable for nucleic acid hybridization (Bush *et al.*, 1992; Prat *et al.*, 1991). An inorganic phosphor, Eu^{3+}-activated yttrium oxysulfide, can be bound to antibodies and visualized by time-resolved fluorescence microscopy (Beverloo and Tanke, 1991). This detects the persisting emission of a phosphorescent label but not the nonspecific autofluorescence of fixed tissues.

Fluorescent labels can be replaced with diachromic colored deposits by **photochemical conversion**. This involves irradiation in the fluorescence microscope in the presence of diaminobenzidine (DAB). Oxidation of the DAB yields brown, insoluble, osmiophilic products at the fluorescently labeled sites. Photochemical conversion was first described for lucifer yellow CH, and subsequently for other fluorescent dyes and labels such as acridine orange, DAPI, ethidium bromide, and FITC- and TRITC-conjugates (Sandell and Masland, 1988). Fluorescent neuronal tracers also photo-oxidize DAB, making ultrastructural correlations possible (Schmued and Snavely, 1993). Eosin Y-labeled antibodies have proved particularly effective for photo-oxidation of DAB (Deerinck *et al.*, 1994).

Labeled biomolecules used as probes

The introduction of **fluorescent labeled antibodies** stimulated the development of other fluorescent labeling strategies. A wide variety of labeled proteins (e.g. enzymes, hormones and lectins) and polysaccharides are now used in many biomedical fields.

Fluorescent *in situ* hybridization (FISH) is based on labeling a nucleic acid probe with a fluorophore that can be directly observed, or with a reporter molecule that is indirectly detected by immunocytochemistry. Current fluorescent labels for nucleic acid probes are NBD chloride, tetramethylrhodamine isothiocyanate, and Texas red (see Haugland, 1996). Less used labels are 7-methoxycoumarin, eosin Y, carbocyanines, and BODIPY. Labeling is achieved by direct chemical bonding or enzymatic incorporation of labeled precursors. Simultaneous detection of multiple probes is possible by using different fluorophores. Multicolor FISH facilitates the recognition of genetic rearrangements, such as the Philadelphia chromosome in interphase nuclei of patients with chronic myeloid leukemia (Arnoldus *et al.*, 1990). Chromosome painting techniques with specific DNA probes allow the multicolor analysis of chromosomes and chromosomal aberrations (Tucker *et al.*, 1994).

Fluorescent lipids have been prepared by labeling with anthracene, BODIPY, carbocyanine, coumarin, fluorescein, naphthalene, NBD chloride, and pyrene fluorophores. NBD-ceramide and BODIPY-cholesteryl ester are examples (Johnson *et al.*, 1991). Such fluorescent lipids are vital stains for the Golgi apparatus, being subsequently transported to the plasma membrane by vesicle-mediated processes (Pagano and Martin, 1994). Labeled membrane lipids are also used in studies of transport, signal transduction, and selective uptake of lipoproteins (Bach *et al.*, 1993; Furlong *et al.*, 1992; Reavent *et al.*, 1996).

8

The use of dyes and fluorochromes as indicators

M. Wainwright

INTRODUCTION

The electronic nature of light absorption allows the use of dyes to indicate physical or chemical changes. In solution chromophores, and fluorophores, are surrounded by solvent molecules. A change in the properties of the solvent system, such as a drop in temperature or increase in pH, may alter the configuration of the solvating molecules and cause a dynamic event in the arrangement of the electron cloud of the chromophore. If this is significant a color change can occur.

Many different events can lead to the changes in the electron cloud, and they can be measured at macro- and microscopic levels. For instance protonation of an auxochromic moiety in a dye molecule will remove a lone pair of electrons from the conjugated system. If donation of the lone pair to the chromophoric system is essential to one of its absorption bands, the protonated dye may absorb at a longer wavelength. If both the neutral and protonated molecules are chemically stable and there is a sufficient difference in the wavelengths of the respective absorption bands, the dye has the potential to be used as an indicator of acid–base behavior.

Other types of chemical groups may alter the electronic state or stable solvation of a dye. For example, a dye containing chelating groups will form complexes with metal ions of the correct size. Chelation can lead to decreased electron density in the chromophore if the chelating groups are directly attached to it, because chelation involves electron donation to the metal ion.

In biology acid–base and metal ion indicators are used to measure protonic equilibria and metal ion concentrations in microenvironments. Potentiometric probes report fluctuations in the transmembrane potentials of living cells. Ion-specific indicators are employed to measure local concentrations of cell-signallers such as Ca^{2+} and Mg^{2+}. Generally, indicators used as biological probes have fluorescent properties, to allow their detection at very low concentrations.

Whereas acidimetric and metallochromic titrations are reversible by the simple addition or removal of protons or metal ions, redox indicators undergo more drastic changes in structure on being reduced or oxidized. However, the redox indicator behavior remains reversible, and there must be a well-defined difference in color between the two forms.

Major applications of indicators

Indicators may be used in either single or multiphase systems, with the complexity of the system governing the technical difficulty of data collection. For instance, use of phenolphthalein in acid–base titrations is straight-forward, requiring only simple mixing of components and a visual check on the color change.

However, monitoring the change in potential across a cell membrane requires several refinements, both in sample preparation and in measuring changes in fluorescence. Results are also influenced by distribution of indi-cators within and outside cells. To predict the distribution of an indicator a thorough knowledge of its properties is required. Thus the lipophilicity of a protonated indicator is usually lower than that of the corresponding neutral species. With fluorescent indicators the wavelength and intensity of emission of each form will be different. Such pH-distribution behavior can be used to demonstrate particular intracellular compartments, such as mitochondria and lysosomes. The basis of the differential uptake is in part related to the degree of polarity of the molecule in its different forms. A discussion of the differences in uptake between polar and nonpolar molecules is given in Chapter 5.

Indicators for analysis of ions in solution may be immobilized on a polymeric membrane, and the response to the ion monitored via a change in fluorescence or of potential gradient. Such optical or solid state sensors are now in common use to detect pH changes, to identify different cations and anions, and even to distinguish between enantiomers (Vogtle and Knops, 1991).

pH indicators

The use of an organic molecule to indicate the acidity of a solution depends on the molecule's protonation or deprotonation. Typically this is facilitated by heteroatoms with free lone pairs of electrons. In bases these are donated to protons in the environment; or in acids a proton is lost to water molecules. Typical equi-libria for these types of indicators are shown below. Table 8.1 contains pK_a values for many of the dyes and fluorochromes mentioned in this book.

The ionizable groups in indicators often facilitate absorption of light. For example, a delocalized *p*-system may be conjugated to a carboxylic acid group,

For acids: $HA + H_2O \rightleftharpoons A^- + H_3O^+$

For bases: $B + H_2O \rightleftharpoons BH^+ + HO^-$

which also contains p orbitals. The two electronegative oxygen atoms of a carboxylic acid group tend to pull electron density away from the chromophore to which it is attached. Deprotonation of the acid generates the carboxylate anion which is considerably more electron-rich, and donates electron density back into the chromophore. This change in electron density at the chromophore due to the change in the auxochrome leads to a shift of the maximum wavelength of absorption, usually into the visible or, if already in the visible region, towards the red end of the spectrum (bathochromic shift). The use of acidic, or basic, substituents in the design of indicator dyes is thus a straightforward concept, with the rider that the system must be conjugated to show the desired effect on pH change.

The effect of deprotonation on a chromophoric system can be seen in indicators such as fluorescein. Here the neutral molecule is protonated in both its carboxylic acid and phenolic residues. On loss of the more acidic carboxylic proton, the neutral molecule can either form a lactone or exist as the free carboxylate ion. In a more alkaline medium the less acidic phenolic proton is also lost. The various forms of fluorescein are shown in Figure 8.1.

For efficient visual pH differentiation, as needed in simple titrations, the differences in the absorption characteristics of the various ionized and neutral species must be clearly visible. When using spectrophotometry it is possible to distinguish species with separate but closer absorption maxima.

However, when indicators are used to give information concerning the pH of living cells, the distribution of the different dye species also becomes important.

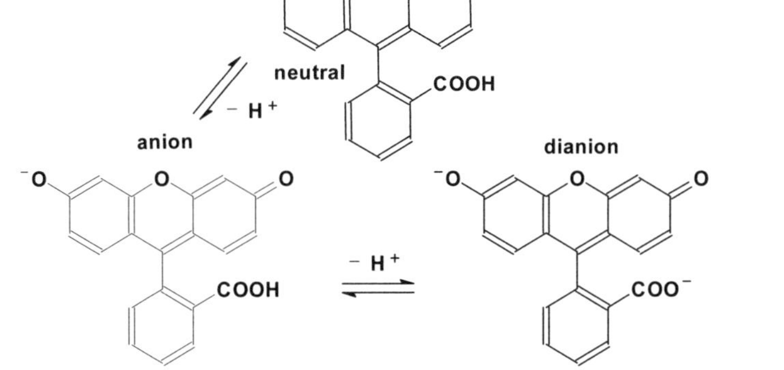

Figure 8.1 Ionic and uncharged fluorescein species.

Lipophilicity is a key parameter in controlling both intra/extracellular distribution, and distribution between organelles within the cell.

An example of the importance of the former is again provided by fluorescein. This dye, introduced as its membrane permeant diacetate which is converted to fluorescein by intracellular esterases, has been used to measure intracellular pH of living cells. However, it leaks rapidly from the cell. Carboxyfluorescein diacetate on the other hand yields carboxyfluorescein within the cell. This dye, due to its greater anionic character, exhibits lower efflux and is therefore widely used for intracellular pH measurements (Declerck *et al.*, 1994). For a brief account of the role of lipophilicity in controlling intracellular distribution see Chapter 5.

Metal ion indicators

A metal ion in solution exists in a stabilizing cage of solvent (typically water) molecules. This is due to attraction between the point concentration of cationic charge (M^{n+}) and the electron donating lone pairs of the oxygen atom in the water molecule. The nucleophilic nature of the interaction of water with metal ions is parallelled by other nucleophile-containing molecules. Some molecules have more than one nucleophilic center, suitably spaced, and can act as bidentate ligands or chelating agents. Ionic radii are important in determining the 'jaw' size and the sizes of metal ions that can be chelated.

In a metal ion-indicator, charge withdrawal, that is displacement of electron density towards the metal ion, following chelation leads to significant changes in the electron cloud of the indicator molecule. This produces a measurable change in the absorption or emission spectrum of the indicator. A simple example of this is provided by chrome Bordeaux B, illustrated below, which contains carboxylic acid and phenolic moieties. Coordination of the carboxylate anion of this dye to a metal ion such as zinc causes a color change from yellow to violet (Bishop, 1972, p. 238) because the electron density of the carboxylate group decreases when a metal ion is coordinated.

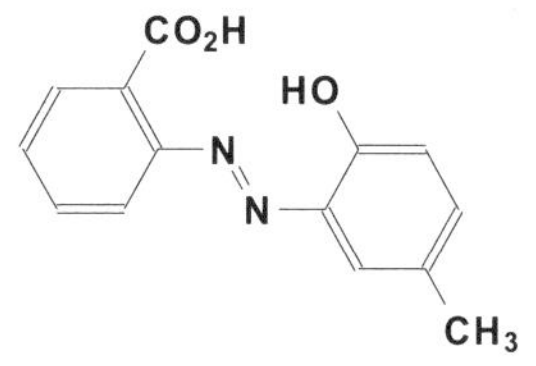

The size and charge of the metal ion also effect the magnitude of change in electron distribution of the coordinating molecule. A small, strongly positive cation (e.g. Al^{3+}) will be more electron withdrawing than a larger cation of lower charge such as Ca^{2+}. In addition, the size of the cation can inhibit chelation and this can form the basis of differential analytical methods. Generally the binding selectivity of Group I and II cations is sterically governed, whereas transition metal ions may bind more strongly due to the electronic involvement of the metal's *d* orbitals, if these are partly filled.

Calcium indicators are widely used in the study of Ca^{2+} in cellular responses. A probe for calcium must be highly selective because other mono- and divalent cations, notably sodium, potassium and magnesium, occur in the same regions. Even in the presence of large excesses of these other ions it is possible to detect low levels of calcium with fluorescent probes such as calcium green-1 and fura-2; these probes are described in entries in Chapters 16 and 27 respectively.

Such multidentate ligands have chelating groups organized in three dimensions. These compounds can enclose a metal ion completely. This allows the chelating agent to be highly dependent on the size and charge of the metal ion, and so extremely selective. Three-dimensional arrays are difficult to prepare, and fewer examples are available of this class than of chelating agents previously discussed. The principle of charge distortion due to ion chelation is, however, similar.

Redox indicators

Addition of hydrogen to the delocalized *p*-systems of dyes usually interrupts or destroys the delocalization, leading to loss of visible light absorption. In many dye classes the reduced (leuco) form can lose its added hydrogen by a reverse process, regaining its intact *p*-system and its color. Such behavior is typical of the dyes used as redox indicators.

A well-understood example is provided by methylene blue, whose cation can be chemically reduced to leucomethylene blue, as illustrated below. That the leuco compound is less conjugated is demonstrated by the positive charge on the dimethylamino group in methylene blue being able to be delocalized onto the other dimethylamino group, whereas the positive charge caused by the protonation of a dimethylamino group in leucomethylene blue is localized to the auxochrome. In keeping with this, the absorption maximum of methylene blue in aqueous media is 656 nm, but that of the colorless leuco form is 330 nm.

The change from blue to colorless may be brought about by any reducing agent of sufficient power. Comparing redox couples to a standard (such as a hydrogen electrode) provides a ranked order of electrode potentials, thereby simplifying the choice of a redox indicator for a particular reaction. The ease of reduction of dyes based on the phenazine and phenothiazine chromophores, has led to their use in redox electrodes. Here the redox dye is polymerized or

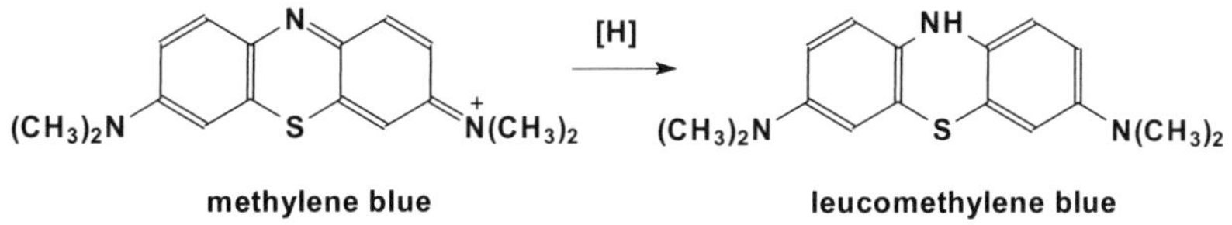

methylene blue　　　　　　　leucomethylene blue

otherwise immobilized on the surface of the electrode. Polymerized methylene blue has been employed in an electrode for assaying dehydrogenases (Cosnier and LeLous, 1996), and poly(neutral red)-coated carbon fiber microelectrodes are of use in the determination of neurotransmitters such as dopamine (Sun *et al.*, 1998).

A major application of redox indicators is in cell viability studies, for instance in the screening of cytotoxic compounds such as potential anticancer agents. In particular the use of MTT (methylthiazolyldiphenyl tetrazolium; see entry in Chapter 13) in cytotoxicity assays has revolutionized the field, permitting automated analyses using microculture plates with multiple cell lines. MTT is reduced by the mitochondria of live cells to an insoluble purple formazan compound, and can thus be used to give counts of viable cells (Mosmann, 1983).

The use of dyes to show cellular reduction is obviously not restricted to the tetrazolium salts. For example, methylene blue is reduced by the surfaces of endothelial cells (Merker *et al.*, 1997) whereas more lipophilic analog dyes are likely to be reduced inside the cell, for example, at mitochondria. Dyes may also be used to probe isolated redox systems *in vitro* by removing them from the cellular environment. Thus NADPH-dependent dehydrogenases such as horse liver alcohol dehydrogenase may be assayed using a blue triazine dye template as an artificial coenzyme (Burton *et al.*, 1996).

Potential-sensitive indicators

Polarity variation can perturb the electron cloud of a fluorophore. Changes of sufficient magnitude lead to changes in electronic energy levels, and manifest as displacement of emission wavelengths or altered fluorescence intensity.

Dyes used in the measurement of membrane potential can be classified as fast or slow, depending on the speed of their responses to changes in membrane potential. The fast class includes merocyanine 540, with a response time in the millisecond range. These dyes are used to measure cell surface potential, as in patch-clamp experimentation (Huser J. *et al.*, 1996). Slow potentiometric dyes are usually found in the oxonol, carbocyanine and rhodamine families. They have response times of seconds or even minutes and are used to monitor slower fluctuations in trans-membrane potential (Plasek and Sigler, 1996). For further information on these dyes, see entries in Chapters 16, 23 and 24.

Table 8.1 Properties of pH indicators

Class and Name	CI number	pH range	Color change
Nitro and nitroso compounds			
Picric acid	10305	0.0–1.0	colorless–yellow
Martius yellow	10315	2.0–3.2	colorless–yellow
2,6-Dinitrophenol		2.4–4.0	colorless–yellow

2,4-Dinitrophenol		2.8–4.4	colorless–yellow
Phenacetolin		3.0–6.0	yellow–red
2,5-Dinitrophenol		4.0–5.6	colorless–yellow
Isopicramic acid		4.1–5.6	pink–yellow
o-Nitrophenol		5.0–7.0	colorless–yellow
p-Nitrophenol		5.6–7.6	colorless–yellow
6,8-Dinitro-2,4-(1H,3H)-quinazolinedione		6.4–8.0	colorless–yellow
m-Nitrophenol		6.8–8.6	colorless–yellow
Ethyl-*bis*(2,4-dinitrophenyl)-acetate		7.5–9.1	colorless–blue
Nitroamine		11.0–13.0	colorless–orange/brown
2,4,6-Trinitrotoluene		11.5–13.0	colorless–orange
1,3,5-Trinitrobenzene		11.5–14.0	colorless–orange
2,4,6-Trinitrobenzoic acid		12.0–13.4	colorless–yellow

Azo dyes

2-(*p*-Dimethylaminophenyl)azopyridine		0.4–1.8	yellow–red
Metanil yellow	13065	1.2–2.3	red–yellow
4-Phenylazodiphenylamine		1.2–2.5	red–yellow
p-Methyl red		1.2–3.4	red–orange
Benzopurpurin 4B	23500	1.2–4.0	violet–red
Tropaeolin OO	14270	1.4–2.6	red–yellow
Fast garnet GBC base	11160	1.4–2.8	orange yellow
Alizarin yellow R	14030	1.9–3.3	red–yellow
Benzyl orange	22195	1.9–3.3	red–yellow
m-Methyl red	13020	2.0–4.0	red–yellow
4-(*m*-Tolyl)azo-*N*,*N*-dimethyl-aniline		2.6–4.8	red–yellow
Oil yellow II	11020	2.9–4.0	red–yellow
Methyl orange	13025	3.0–4.4	red–yellow
Ethyl orange		3.0–4.5	red–yellow
Hessian purple N	24865	3.0–4.6	blue/violet–red
N-Phenyl-1-naphthyl-aminoazobenzene-*p*-sulfonic acid		3.0–5.0	violet–red/orange
Congo red	22120	3.0–5.2	blue/violet–red
4-(4'-Dimethylamino-1'-naphthyl)-azo-3-methoxy-benzenesulfonic acid		3.5–4.8	violet–yellow

Table 8.1 continued

Class and Name	CI number	pH range	Color change
p-Ethoxychrysoidine		3.5–5.5	orange / red–yellow
α-Naphthyl red	11350	3.7–5.0	red–yellow / orange
Chrysoidine (chrysoidineY)	11270	4.0–7.0	orange–yellow
1-Naphthylaminazobenzene- *p*-sulfonic acid		4.2–5.8	pink–orange
Methyl red	13020	4.2–6.3	red–yellow
2-(*p*-Dimethylamino- phenyl)azopyridine		4.4–5.6	red–yellow
Ethyl red		4.5–6.5	red–yellow
Propyl red		4.6–6.6	red–yellow
N-Phenyl-1-naphthyl- aminoazo-*o*-carboxybenzene		5.0–7.0	violet–yellow
Nitrazol yellow	14890	6.0–7.2	yellow–blue
Brilliant yellow	24890	6.4–8.0	yellow–red / orange
Brilliant yellow S	13085	7.4–8.6	yellow–orange / red
Orange II	15510	7.4–8.6	yellow–orange
Propyl o-naphthyl orange		7.4–8.9	yellow–red
Orange I	14600	7.6–8.9	yellow–purple / red
Hessian Bordeaux	24860	8.0–9.0	blue–red
Diazo violet		10.1–12.0	yellow–violet
α-Naphthol violet		10.1–12.1	yellow–violet
Orange II	15510	0.2–11.8	orange–red
Alizarin yellow R	14030	0.2–12.0	yellow–red
Alizarin yellow GG (Metachrome yellow 2RD)	14025	0.2–12.0	colorless–yellow
Chrome orange GR	26520	0.5–12.0	yellow–red
Sulfone acid blue R	13390	0.5–12.0	blue–red
Lanacyl violet BF	13375	1.0–13.0	violet–orange
Tropaeolin O	14270	1.1–12.7	yellow–red / orange
Orange G	16230	11.5–14.0	yellow–pink
Aminoarylmethane dyes			
Crystal violet	42555	0.0–1.8	yellow–blue
Methyl violet B	42535	0.0–2.0	yellow–blue
Malachite green	42000	0.0–2.0	yellow–green
Brilliant green	42040	0.0–2.6	yellow–green
Ethyl violet	42600	0.0–3.5	yellow–blue

Methyl violet 6B	42535	0.1–1.5	yellow–blue / green
Methyl violet	42535	0.1–3.2	yellow–violet
Ethyl / methyl green	42590 / 42585	0.2–1.8	yellow–blue
Basic fuchsine	42500	1.2–3.0	purple–red
Methyl violet 6B	42535	1.5–3.2	blue–violet
Patent blue V	42045	1.8–3.0	yellow / green–blue
Alkali blue	42750 / 42765	9.4–14.0	blue–red
Aniline blue	42780 / 42755	10.0–13.0	blue–pink
Basic fuchsine	42500	11.6–14.0	red–colorless
Malachite green	42000	11.6–14.0	green–colorless
Acid fuchsine	42685	12.0–14.0	red–colorless

Phthalein and sulfonphthalein dyes

o-Naphthol benzein		0.0–0.8	green–yellow
Cresol red (o-cresol red)		0.2–1.8	red–yellow
m-Cresol purple		0.5–2.5	red–yellow
Xylenol blue (p-xylenol blue)		1.2–2.8	red–yellow
Thymol blue		1.2–2.8	red–yellow
Pentamethoxy red		1.2–3.2	red / purple–colorless
Hexamethoxy red		2.6–4.6	red / purple–colorless
Tetrabromophenolphthalein, ethyl ester, K⁺ salt		3.0–4.2	yellow–blue / violet
Bromophenol blue		3.0–4.6	yellow–blue
Bromophenol blue, K⁺ salt		3.0–4.6	yellow–blue
Tetraiodophenol-sulfonphthalein		3.0–4.8	yellow–blue
Bromochlorophenol blue		3.2–4.8	yellow–blue
Bromocresol green		3.8–5.4	yellow–blue
Chlorocresol green		4.0–5.6	yellow–blue
Chlorophenol red		5.0–6.6	yellow–red
Heptamethoxy red		5.0–7.0	red–colorless
Bromocresol purple		5.2–6.8	yellow–violet
Sulfonaphthyl red		5.2–6.8	red–yellow
Bromophenol red		5.2–7.0	yellow–red / purple
Dibromophenol-tetrabromo-phenol-sulfonphthalein		5.6–7.2	yellow–purple
Bromothymol blue		6.0–7.6	yellow–blue

97

Table 8.1 continued

Class and Name	CI number	pH range	Color change
Aurin	43800	6.8–8.2	yellow–red
Phenol red (phenol- sulfonphthalein)		6.8–8.4	yellow–red
o-Cresol benzein		7.2–8.6	yellow–red
Cresol red (o-cresol red)		7.2–8.8	yellow–red
α-Naphtholphthalein		7.3–8.7	yellow–green/blue
m-Cresol purple		7.4–9.0	yellow–purple
Xylenol blue (p-xylenol blue)		8.0–9.6	yellow–blue
Thymol blue		8.0–9.6	yellow–blue
Phenoltetrachlorophthalein		8.2–9.4	colorless–purple
Cresolphthalein (o-cresolphthalein)		8.2–9.8	colorless–red
α-Naphtholbenzein		8.2–10.0	yellow/orange–blue
Phenoltetraiodophthalein		8.3–9.7	colorless–pink
Phenolphthalein		8.3–10.0	colorless–red
Thymolphthalein		9.3–10.5	colorless–blue
Xanthene dyes			
Eosin Y	45380	0.0–3.0	yellow–fluorescent red
Erythrosin B	45430	0.0–3.5	yellow–fluorescent red
Erythrosin	45430	2.2–3.6	orange–red
Gallein	45445	3.8–6.6	yellow–red
Gallein	45445	10.6–13.0	pink–violet
Azine and oxazine dyes			
Brilliant cresyl blue	51010	0.2–1.0	green–blue
Resazurin		3.8–6.5	orange–yellow
Lacmoid (resorcein)		4.2–5.6	red–blue
Lacmoid (resorcein)		4.4–6.4	red–blue
Litmus		4.5–8.3	red–blue
Azolitmin		5.0–8.0	red–blue
Neutral red	50040	6.8–8.0	red–yellow
Nile blue 2B	51185	7.2–8.6	blue–pink
Nile blue (Nile blue A)	51180	10.0–11.0	blue–purple/red
Alizarin yellow GG	14025	10.0–12.0	violet–orange
Brilliant cresyl blue	51010	12.0–12.4	blue–red

Polymethine dyes

Hematoxylin*	75290	0.0–1.0	red–yellow
Quinaldine red		1.4–3.2	colorless–red
Hematoxylin*	75290	5.0–6.0	yellow–violet/blue
Pinachrome		5.8–7.8	colorless–red
Indo-oxine		6.0–8.0	red–blue
Quinoline blue		6.6–8.6	colorless–blue
bis-5-Bromovanillidenecyclo- hexanone		7.2–8.6	yellow/green– orange/red
bis(2'-Hydroxystyryl)ketone		7.3–8.7	yellow–green
Curcumin	75300	7.4–8.6	yellow–red
bis(4-hydroxy-3-ethoxy- benzylidene)-cyclohexanone		8.0–10.2	yellow–red
Curcumin	75300	10.2–11.8	violet–orange
Thiazol yellow G	19540	11.0–13.0	yellow–red

Carbonyl dyes (anthraquinones and indigoids)

Alizarin blue (alizarin blue B)	67410	0.0–1.6	pink–yellow
Alizarin orange	58015	2.0–4.0	yellow/orange–yellow
Alizarin red S	58005	3.7–5.2	yellow–purple
Carminic acid	75470	4.8–6.2	yellow–purple
Alizarin orange	58015	5.0–6.5	yellow–purple/red
Alizarin	58000	5.5–6.8	yellow–red
Alizarin blue (alizarin blue B)	67410	6.0–7.6	yellow–green
Rufianic acid	58055	8.5–10.0	orange–purple
Alizarin red S	58005	10.0–12.0	blue/red–yellow
Alizarin	58000	10.1–12.1	red–purple
Rufianic blue	58055	10.5–13.0	purple–blue
Alizarin blue SWR	58605	11.0–12.0	green–blue
Indigocarmine	73015	11.6–14.0	blue–yellow

* This refers to hematein, the oxidation product of hematoxylin.

9

Nitroso and nitro dyes

R.W. Horobin

NITROSO DYES

Nitroso dyes used as biological stains contain a nitroso group in an *ortho* position to a phenolic hydroxyl. As such compounds exist as equilibrium mixtures of tautomers, as shown below for naphthol green Y, they can also be regarded as quinonoximes.

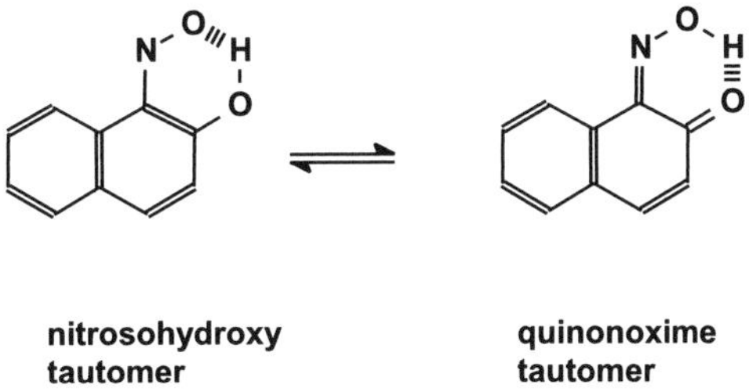

**nitrosohydroxy
tautomer**

**quinonoxime
tautomer**

The *ortho*-substitution pattern results in nitroso dyes being readily able to chelate a variety of transition metal ions, yielding colored metal coordination products.

Naphthol green B

CI 10020, CI Acid green 1
$C_{30}H_{15}N_3O_{15}S_3Na_3Fe$, FW 878

Absorption maximum: 714 nm in water.
Solubilities: 10% in water; insoluble in ethanol.

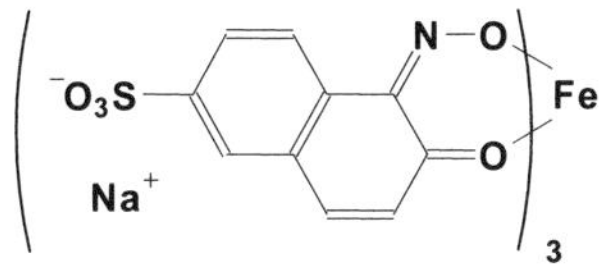

Naphthol green B is the sodium salt of a 3:1 dye–iron complex of the monosulfonated derivative of the nitroso dye **naphthol green Y** (CI 10005)). Naphthol green B has a large anion, and so acts as an acid dye.

Applications: This dye has been used in a variety of polychrome stains applied to animal tissues (Lillie, 1945; Mollier, 1938; Volkmann and Strauss, 1934), and also in a study of the modes of action of such polychrome methods (Horobin and Flemming, 1988).

Industrially the dye has been used for coloring wool and nylon, paper, anodized aluminum, and soap.

Purity of commercial batches: These can contain around 50% dye.

Stability: Aqueous solutions are stable for at least 2 years in the laboratory; the dye is extremely fast to light.

NITRO DYES

Nitro groups are present in many dyes, but usually not as essential parts of the chromophoric system. The structural characteristic of nitro dyes as narrowly defined is the presence of a nitro group in the *ortho* position to an electron donor. For most nitro dyes used as biological stains, this donor is a phenolic hydroxyl. The strongly electron accepting character of the nitro group results in ionization of these hydroxy groups at relatively low pH, hence their being used as acid (i.e. anionic) dyes.

Picric acid

CI 10305
Synonyms: Trinitrophenol
$C_6H_3N_3O_7$, FW 229

Absorption maximum: 354 nm in water.
Solubilities: 1.3% in water; 9% in ethanol; soluble in xylene.

pH of a 1% aqueous solution is 1.9: the dye decolorizes below pH 1.

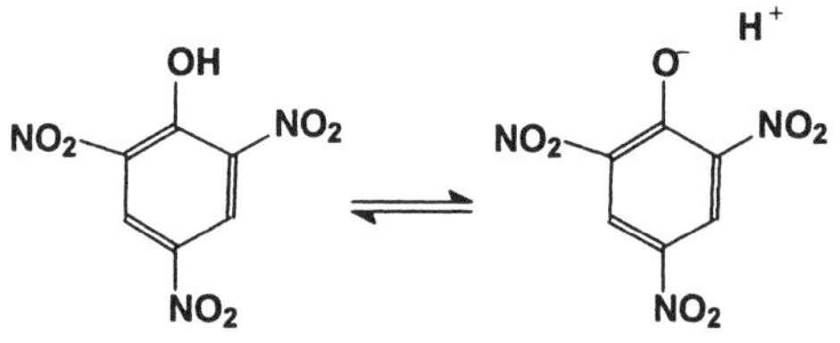

A moderately strong acid, in aqueous solutions present largely as a hydrophilic anion; as illustrated. As the free acid tautomer is weakly hydrophobic, alcohol may extract picric acid from stained tissues. Picric acid has a small aromatic system, resulting both in a low tissue affinity and a high rate of tissue penetration.

Applications: Picric acid is employed in many polychrome stains in conjunction with larger acid dyes: classically with acid fuchsine in the Van Gieson procedure, or its modern versions in which the red dye is sirius red F3B (Sweat *et al.*, 1964). To appreciate the variety of such methods, including picro-indigocarmine (Jullien, 1872), see the overview by Lillie and Fullmer (1976, p. 694).

The dye is used as a general cytoplasmic stain, contrasting with other acid or basic dyes. Such applications are found in botany (Jacobs *et al.*, 1996), as well as the combined counterstain and acid differentiator of hematoxylin staining found in animal histology. Other staining applications exploit the fast diffusion rate of this small acid dye, for example, Schmorl's (1934) picro-thionine stain for bone canaliculi.

There are also nonstaining applications in histotechnology. It is a component of fixative agents; for some currently used examples see Bancroft and Gamble (2002). Underlying this is the dye's ability to precipitate proteins; and indeed other substances such as cationic dyes (Stark *et al.*, 1969). Neutral fixative solutions containing formaldehyde and picric acid (Stefanini *et al.*, 1967) are much used for subsequent immunochemistry because they are reputed to conserve antigenicity (Accini *et al.*, 1974), and provide structural preservation superior to that attainable with simple neutral buffered formaldehyde (Kiernan, 1985). At neutral pH, picrate anions do not precipitate proteins, so their beneficial effects action must be due to some other property (Kiernan, 1999, p. 33). The acidic character of picric acid is utilized in decalcification, differentiation of dyes such as aluminum–hematoxylin, and in the removal of formalin pigment.

Industrially picric acid has been used as a textile dye; and has other uses such as etching copper and manufacture of colored glass; it was formerly used as a military explosive.

Purity of commercial batches: The dye can be manufactured in a pure state. There is a thin-layer chromatographic method (Graham, 1968). Due to the instability of the dry powder, solid picric acid is sold moistened with 35% of water.

Stability: The dry powder is flammable and may explode due to heat or mechanical shock. Aqueous solutions are stable indefinitely. Light fastness of stained sections is good.

Hazards: Picric acid is a toxic irritant, which can be absorbed through the skin.

Martius yellow

CI 10315, CI Acid yellow 24
Synonyms: Manchester yellow
$C_{10}H_5N_2NaO_5$, FW 256

Absorption maximum: 420 nm in water, 432 nm in methanol
Solubilities: The sodium salt is soluble at 4.4% in water; 0.2% in ethanol. The calcium salt, sometimes termed naphthol yellow, is much less soluble in water.

Martius yellow is usually the sodium salt of 2,4-dinitro-1-naphthol, although the ammonium and the poorly water-soluble calcium salts have also been sold. This small compound is a moderately strong acid, being used in biological staining as the anionic.
Who was Martius? The dye's second discoverer (in 1864), it was first discovered by Ganahl in 1856.
Applications: Martius yellow is a cytoplasmic stain in several trichromes used in human and animal histology: for example, Gabe's (Gabe, 1976) and the MSB method for fibrin (Lendrum *et al.*, 1962), where M indicates the use of Martius yellow. This latter method is widely used, for instance, in forensic pathology to assess mode of injury (Ali, 1992). Similar methods have been used in botany (Muller, 1912; Nebel, 1931; Vaughan, 1914). Martius yellow has been used as a counterstain, for example, to periodic acid–Schiff stain for demonstrating parasitic amoebae (Hulman and Taylor, 1987).
Purity of commercial batches: Dye contents are up to 95%.

Naphthol yellow S

CI 10316, CI Acid yellow 1
Synonyms: the free acid is sold as flavianic acid.
$C_{10}H_4N_2Na_2O_8S$, FW 358

Absorption maxima: for an aqueous solution of the salt 428 and 392 nm in water, for flavianic acid 336 nm.
Solubilities: 4.4% in water; 0.2% in ethanol.

This small sulfonated dye is anionic under the conditions of use. The commercially available free acid (flavianic acid) is less soluble in water though more soluble in ethanol. In practical staining methods the free acid will also be present as an anion.

Applications: Naphthol yellow S or flavianic acid is used in various polychrome procedures, for instance in combination with acid fuchsine (Lillie *et al.*, 1967) or with trypan blue (Clark, 1981, p. 115) for animal and human tissue. A variant has been applied to investigate smears of spermatozoa (Bryan, 1970), which application has become a standard method (Christensen *et al.*, 1996). The dye is used as a counterstain for blue and red dyes, yielding strongly polychromatic results (Schipper *et al.*, 1991; Siegel, 1967), and as a substitute for Martius yellow in Gabe's (1976) one-step trichrome method. Naphthol yellow S is also applied in a botanical polychrome (Hoffmeister, 1953) and as a counterstain, for instance in Feulgen or silver methods (Mellerowicz *et al.*, 1993).

Naphthol yellow S is applied histochemically for quantitation of proteins; for a critical review see Tas *et al.* (1980). This application remains in general use (Erokhina *et al.*, 1992). Note, however, that the dye is not particularly light fast.

Industrial applications include dyeing of furs, leather, paper and wool.

Purity of commercial batches: These can contain around 75% of dye. A thin-layer chromatographic method is available (Tas *et al.*, 1980).

10

Monoazo dyes

R.W. Horobin

BACKGROUND INFORMATION

Issues of nomenclature

Azo dyes used as biological stains contain an azo group (–N=N–) linking benzene, naphthalene or aromatic heterocylic rings. This chapter considers dyes containing single azo groups; Chapter 11 describes dyes with multiple groups. Certain azo dyes in this present chapter contain chromophores in addition to their azo group, for example, alcian yellow also contains thiazole chromophores, and Janus green is an azo–azine dye. The term 'azo' – used by biologists, dye chemists and indeed all practical dye users – is disapproved of by specialists in chemical nomenclature. The Chemical Abstracts index guide comments that whenever possible 'azo compounds are indexed as derivatives of diazene, HN:NH'.

Physicochemical properties of significance for staining

Azo group stereoisomerism

Azo dyes can exhibit *trans/cis* (or E/Z in current nomenclature) isomerism; as shown below for azobenzene. The *cis* stereoisomer may be obtained by illuminating the *trans* form; the isomers differ in color. Such phototropy can result in fading or changes in hue, reversible in the dark, and so is a potential problem in quantitative work. Azo dyes used as biological stains exist predominantly as their *trans* isomers, and are shown as such below.

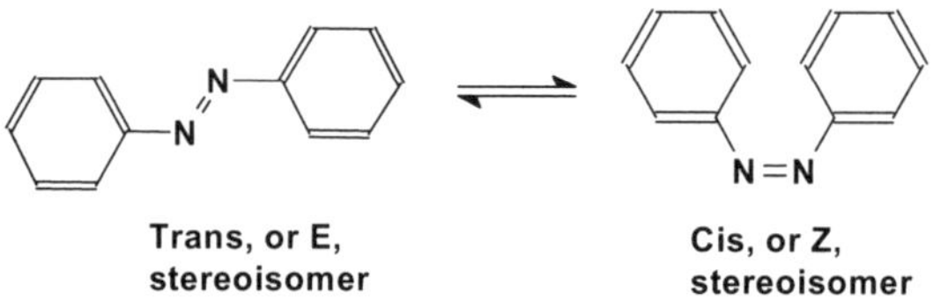

Trans, or E,
stereoisomer

Cis, or Z,
stereoisomer

Acid–base properties

The azo group is weakly basic, and may be protonated yielding an azonium cation, as illustrated below for azobenzene.

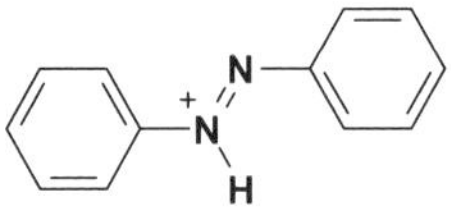

Protonation of an aminoazo dye can either give an azonium tautomer or can instead occur at the amino group giving the ammonium tautomer, as illustrated below for butter yellow.

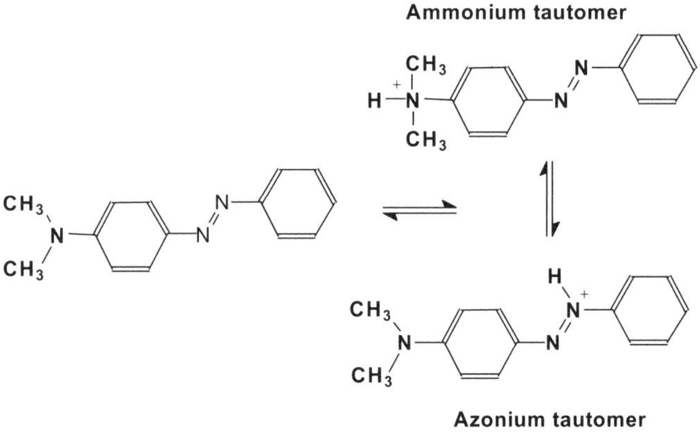

Relative amounts of the azonium/ammonium tautomers are influenced both by solvent acidity and also by electronic, steric and hydrogen bonding effects within the dye itself; as briefly reviewed by Gordon and Gregory (1983, p. 113). The pK_a of the azo group is typically around 2 (Zollinger, 1961, 1991, p. 132), as are many aromatic amines. However, some dye amines retain their positive charge in the pH range used for biological staining. This may result in staining solutions containing mixtures of species of differing electric charges and, in some cases, colors. For examples of the application of such effects in biology see Congo red in Chapter 11.

Most basic azo dyes carry quaternary nitrogen substituents, and so are cationic under most staining conditions. Such cationic charges are usually delocalized, but the Kekulé structures used below do not show this, only depicting extreme electron distributions.

A related issue is the acid strength of the hydroxy group in hydroxyazo dyes. As a necessary preliminary the **azo-hydrazone tautomerism** in such compounds is illustrated in Figure 10.1. Such azo/hydrazone tautomers can interconvert rapidly, both in solution and on dyed substrates. Note that *ortho-* but not *para*-hydroxyazo tautomers can involve strong intramolecular hydrogen bonding. Because the color, solubility and tendency to fade of azo/hydrazone tautomers differ, such equilibria are potentially significant for quantitative biological staining.

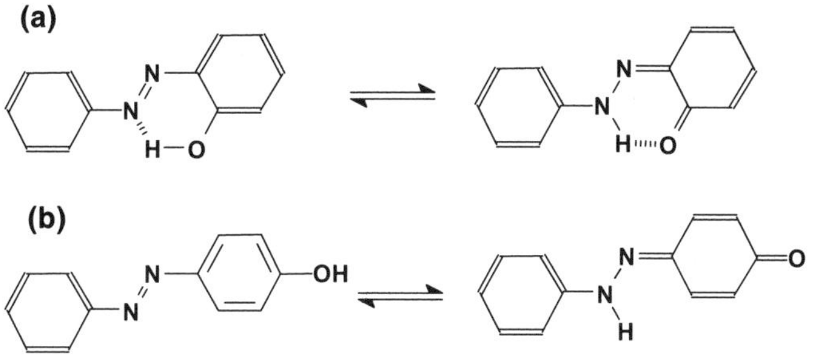

(a)

(b)

Figure 10.1 Azo–hydrazone tautomerism occurring with isomeric hydroxyazo dyes. Note that the *ortho*-isomer (structure **a**) permits strong hydrogen bonding, this is not the case for the *para*-isomer (structure **b**).

The relative amounts of azo and hydrazone tautomers are influenced by the hydrophilicity/lipophilicity of the solvent or biological substrate, as well as by the chemistry of the dye itself. See Gordon and Gregory (1983, p. 96) for a short review. For most dyes the favored tautomer is unknown, and may vary from one tissue component to another. This book routinely shows hydroxyazo and aminoazo dyes as the azo tautomers.

Nevertheless the azo/hydrazone equilibrium does influence the acidity of the hydroxy groups. This is because the strong hydrogen bond present in *ortho*-hydroxyazo compounds (Figure 10.1) causes them to be much weaker acids. This is of practical significance because the stronger acids are likely to be present as anions, not neutral species, under biological staining conditions, influencing acid dyeing. Moreover the neutral and ionized species will differ in color, once again a potential problem for quantitative investigations.

As indicated above, hydroxy groups in many azo dyes can deprotonate and gain negative charge even in neutral staining solutions; as occurs for instance with chromotrope 2R. Most acid azo dyes used as biological stains are substituted by sulfonate groups (e.g. orange G and tartrazine). These are fully ionized over the pH range used in biological staining. Consequently the sign of the electric charge carried by such dyes is not dependent on the pH of the staining solution. This has considerable practical significance because the sign of the charge largely controls the typical acid or basic dye staining patterns (see Chapter 5).

Hydrophilic/lipophilic properties

Ionization also strongly influences dye hydrophilicity/lipophilicity, and hence staining patterns. This is so even with nominally nonionic dyes such as those used for partition-based lipid staining, as was suggested by Michaelis as long ago as 1901. However as noted above, but contrary to Michaelis' view, there are no simple rules predicting ease of ionization of nonionic hydroxyazo dyes.

Other technically significant phenomena influenced by hydrophilicity/lipophilicity are dye losses during post-staining rinses, dehydration, clearing and mounting. Such losses occur because many hydrophilic dyes are soluble in aqueous solutions and mounting media, though not in many organic solvents or media. In contrast, lipophilic dyes are soluble in nonpolar organic solvents and mountants, but not in water. Although ionized dyes used in biological staining are more hydrophilic than nonionized species, hydrophilicity/lipophilicity is influenced by all the substituents, not merely the ionic ones. Thus the presence of methyl, naphthyl or phenyl substituents on an azo acid dye will decrease its hydrophilicity, and perhaps increase its alcohol solubility. The azo group itself is weakly lipophilic, even though it can form intra- and inter-molecular hydrogen bonds.

Synthesis of monoazo dyes

Azo dyes used as biological stains are synthesized by a two-step process. First, an aromatic or heteroaromatic amine is converted to a diazonium salt. In the second step, this is linked ('azo coupled') to a nucleophilic substrate. This latter is usually an aromatic structure carrying electron donating substituents such as hydroxy or amino groups, but can be an enol species. Examples of these possibilities are shown in Figure 10.2. Although diazonium salts are highly reactive, and hence chemically unstable, some can be stabilized for storage. This is described in Chapter 12.

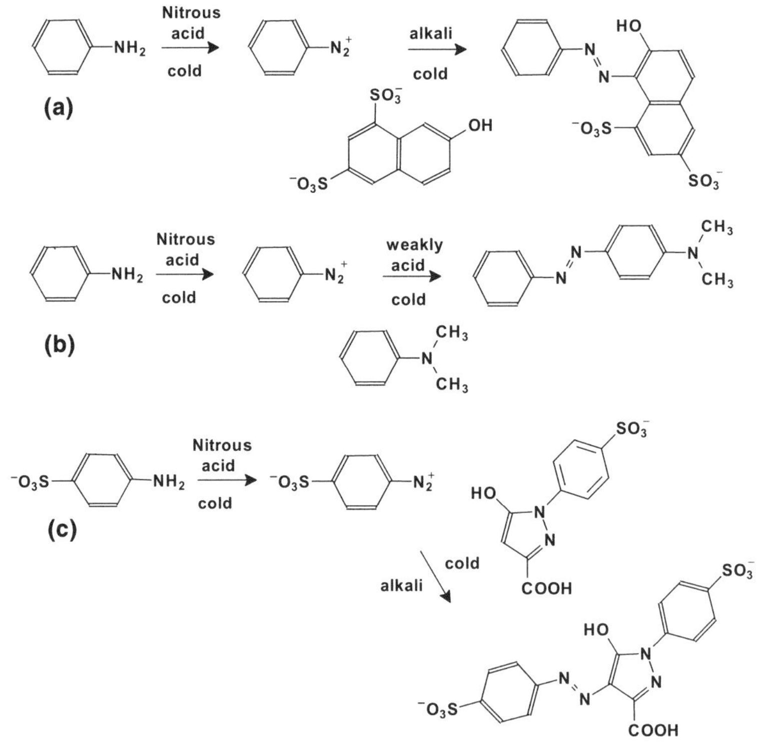

Figure 10.2 – Azo dye synthesese involving azocoupling between diazonium salts and: **(a)** a hydroxy-activated naphthalene, yielding orange G; **(b)** an amino-activated benzene, giving rise to oil yellow; and **(c)** an enol-activated heterocycle producing tartrazine.

ANIONIC DYES

Orange II

CI 15510, CI Acid orange 7
Synonyms: tropaeolin OOO
$C_{16}H_{11}N_2NaO_4S$, FW 350

Absorption maximum: 483 nm in water.
Color changes orange to red between pH 10.2 and 11.8; *pK value* 11.4
Solubilities: 11% in water; 0.2% in ethanol.

Applications: Orange II a useful substitute for orange G when a stronger yellow is required for contrast purposes. Bergonzini (1891) replaced orange G with the dye in Biondi's polychrome stain; Ebbinghaus (1902) used it for staining keratin in skin sections; French (1926a) used orange II in combination with eosin and azure C in a general tissue stain; and Kalter (1943) included orange II as a component of his polychrome stain, in combination with fast green FCF, crystal violet and safranine O. More recently, orange II has been used for the quantitative histochemical assay of protein (Oud *et al.*, 1984).

Industrially, orange II is used to dye and print nylon, silk and wool textiles; and to stain chromed leather.
Purity of commercial batches: Dye batches may contain up to 90%; the minimum dye content for Commission certification is 85% anhydrous dye. Thin-layer chromatographic assay methods for orange II are available (Penner, 1968). See Chapter 28 regarding Commission assay.
Stability: If in a sealed container in the dark, this dye is stable on storage; the dye is light fast.

Metanil yellow

CI 13065, CI Acid yellow 36
$C_{18}H_{14}N_3NaO_3S$, FW 375
Absorption maximum: 414 nm in methanol. The dye changes color from red to yellow between pH 1.5 and 2.7.

Solubilities: 5.4% in water; 1.5% in ethanol; slightly soluble in xylene.

Due to its sulfonate substituent, this small dye is present in aqueous solutions as a slightly lipophilic anion, except when strongly acidified.

Applications: Metanil yellow has been used as the cytoplasmic stain in various polychrome procedures, such as Masson's (1929) method; which usage was further developed by Lillie (1945). The dye has also been used as a cytoplasmic counterstain, for example, in Gridley's demonstration of fungi in histopathological sections (Clark, 1981); and in the application of Bielschowsky's silver stain to demonstrate nucleolar organizer region-associated protein (White *et al.*, 1994). In combination with periodic acid–Schiff (PAS) metanil yellow has been used to demonstrate ciliary muscle connections (Schachar, 1996). A metanil yellow–PAS–toluidine blue combination has been used to identify parietal cells in gastric glands (Tepperman *et al.*, 1989) and in disaggregated stomach epithelium (Barr *et al.*, 1989). Variant methods are applicable to resin embedded sections (Chappard *et al.*, 1996; Quintero-Hunter *et al.*, 1991).

Industrial uses include textile dyeing; coloring paper, lacquers and polishes; staining wood; and in the manufacture of heavy metal salts for pigments.

Purity of commercial batches: Commercial batches can contain up to 70% dye. A thin-layer chromatographic method is available (Chiang and Lin, 1969).

Stability: moderately light fast. Solutions of metanil yellow slowly turn brown.

Chromotrope 2R

CI 16570, CI Acid red 29, Mordant blue 80
$C_{16}H_{10}N_2Na_2O_8S_2$, FW 468

Absorption maximum: 510 nm in water, shoulder at 530 nm.
pK values of the two hydroxy groups are 6.3 and 8.9.
Metal complexation behavior: Can form a blue complex with chromium.
Solubilities: 19.3% in water; 0.2% in ethanol; insoluble in xylene.

Chromotrope 2R   Chromotrope 2B

A hydrophilic anionic dye of moderate size. The peri-substituted dihydroxy moiety results in formation of an additional anion in neutral solutions, and the ability to form metal coordination complexes.

Applications: Chromotrope 2R has been used as a cytoplasmic counterstain in histopathology, e.g. to celestin blue nuclear staining (Lendrum, 1935) and as a substitute for eosin in hematoxylin and eosin. The latter is recommended by Gabe's manual (1976, p. 228) in the context of a critique of this routine oversight stain, which may be recommended as a highpoint of the polemicists' art. Chromotrope 2R has also been used in various trichromes, for example, in a variant of Gomori's trichrome stain (Garvey *et al.*, 1996).

Further applications include use in selective stains for eosinophils (Ohno *et al.*, 1992), and protozoal spores in stool smears (Bretagne *et al.*, 1993); and as a counterstain for aldehyde–fuchsine in insect histology (Ewen, 1962). The dye is also used in several stains for tissue embedded in resins, both hydrophobic (Dougherty and King, 1984) and water miscible (Chu *et al.*, 1995; Lopez and Kornegay, 1991).

Industrially chromotrope 2R has been used for dyeing textiles and chromed leather, and for coloration of resins.

Purity of commercial batches: Samples with dye contents up to 75% are sold, and such batches may contain a single major component. A paper chromatographic analytical method is available (Rosenthal *et al.*, 1965). *Stability:* this dye is moderately light fast.

A related dye, **Chromotrope 2B (CI 16575, CI Acid red 176)**, is also shown above. This differs only in the presence of a nitro substituent; which results in a bluer shade and increased light fastness. The dye has been used in biology as an acid dye with similar properties to chromotrope 2R.

Orange G

CI 16230, CI Acid orange 10
Synonyms: wool orange 2G
$C_{16}H_{10}N_2Na_2O_7S_2$, FW 452

Absorption maximum: 475 nm in water.
The dye changes color from yellow to pink between pH 11.5 and 14.0, the hydroxy substituent having a *pK value* of 12.8.
Solubilities: 10.9% in water; 0.2% in ethanol; insoluble in xylene.

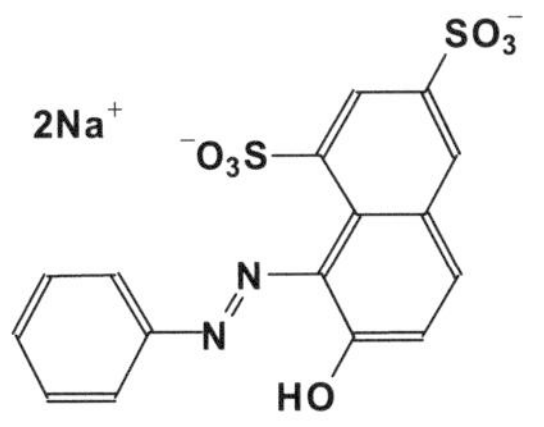

This moderately sized hydrophilic dye is anionic under practical staining conditions, due to its two sulfonate substituents.

Applications: Orange G is a component in many biological stains, typically as a background or cytoplasmic stain. One common application is in clinical cytology, in some Papanicolaou variants (1941; for a review see Boon and Drijver, 1986). In histopathology, orange G is a constituent of the Mallory connective tissue stain and of many of its modifications, such as Heidenhain's AZAN procedure. For a summary account of such stains see Kiernan (1999); for a recent application see Kikui and Miki (1995).

The dye has many other applications, for example, as a component of a stain to identify rice endosperm protein bodies (Barber *et al.*, 1991); and for staining resin sections, both polyester (Bryant and Watson, 1967) and epoxy (Reempts and Borgers, 1975).

The dye has other laboratory uses in biology. For instance in the assay of food and feedstuff proteins (Perlnemolnar *et al.*, 1985), and as a tracer in tracking pulmonary blood flow in rabbits (Luchtel *et al.*, 1991).

Industrially, orange G is used as a dye for textiles and chromed leather; and also to color paper, wood, inks, and pencils.

Purity of commercial batches: Material containing a single major component, and a dye content of up to 95% is available. Commission certified batches contain at least 80% anhydrous dye. A thin-layer chromatographic method is available (Frodyma and Frei, 1969). Intriguingly, a paper chromatographic method for orange G was devised in the 19th century (see Kornhauser, 1954, for an English summary). For the Commission's assay procedures see Chapter 28.

Stability: This dye is stable when stored in a sealed container in the dark; and is moderately light fast.

Ponceau 2R

CI 16150, CI Acid red 26
Synonyms: ponceau de xylidine, xylidine ponceau 3RS
$C_{18}H_{14}N_2Na_2O_7S_2$, FW 480

Absorption maximum: 503 nm, shoulder at 388 nm, in water.
Solubilities: 5.0% in water; 0.1% in ethanol; insoluble in xylene.

This is a moderately sized, moderately hydrophilic anionic dye. There is some confusion in the literature over the identity of 'ponceau de xylidine'. Gurr (1960)

114

suggested this name applies to the dis-azo dye CI 27000; described in previous editions of this book as **wool fast scarlet R**.

Applications: It was probably the present dye, CI 16150, which was used as a histological counterstain in the Masson (1911) method, and by Foot (1933) and others who modified Masson's procedures. In any event it is the dye used in modifications of the Papanicolaou clinical cytology stain (as reviewed by Boon and Drijver, 1986); to stain the protein in smears of wheat flour (Flint and Moss, 1970); and to distinguish the *ortho-* and *para-*cortex of wool fibers (Chen and Fujishige, 1996). Ponceau 2R has also been used with tissues embedded in epoxy resin to stain protein granules (Gori, 1977).

Industrial applications – has been used for dyeing textiles and chromed leather; and as a colorant of inks, paper and wood

Purity of commercial batches: Samples of dye content up to 70%, containing a single major constituent, are available. There is a thin-layer chromatographic assay method (Proctor and Horobin, 1985).

Stability: This dye is moderately light fast.

Hazards: A cancer suspect agent if chronically absorbed or ingested.

Crystal scarlet

CI 16250, CI Acid red 44
Synonyms: brilliant crystal scarlet 6R, crystal ponceau 6R
$C_{20}H_{12}N_2Na_2O_7S_2$, FW 502

Absorption maximum: 510 nm in water.
Solubilities: 5.0% in water; 2.3% in ethanol; very slightly soluble in xylene. Note: there is some disagreement over solubility data, perhaps reflecting confusion over dye identity.

This is a disulfonated, weakly hydrophilic dye of moderate size.

Applications: This dye is a component of the MSB fibrin-staining polychrome, introduced by Lendrum *et al.* (1962), and still favored by European histopathologists. The dye has also been used for investigating pancreatic secretion granules (Grauman and Arnold, 1969). Crystal ponceau 6R has been used as a wool dye, and as a food dye in some countries.

Purity of commercial batches: May contain up to 80% dye. There is a thin-layer chromatographic method available (Hayes *et al.*, 1973).

Stability: Crystal scarlet has moderate light fastness.

Tartrazine

CI 19140, CI Acid yellow 23, Food yellow 4
$C_{16}H_9N_4Na_3O_9S_2$, FW 534

Absorption maximum: 425 nm in water.
Solubilities: 6.0% in water; 0.1% in ethanol; soluble in cellosolve, insoluble in xylene. Note: a considerably higher aqueous solubility is indicated by Green (1990).

This is a hydrophilic polyanionic dye of moderate size. A dispute concerning its identity was resolved by spectroscopic investigations (Jones *et al.*, 1963), which showed it has the enol form illustrated.

Applications: Used in the phloxine–tartrazine procedure for cellular inclusions such as viral aggregates (Bancroft and Gamble, 2002) and eosinophil granules (Clark *et al.*, 1996). The dye has been used in other polychrome stains, for example to substitute for orange G in a variant of Mallory's trichrome (Lillie, 1965, p. 236), and in a methyl violet–tartrazine mixture to stain cryptosporidia in fecal smears and mucosal scrapings (Milacek and Vitovec, 1985).

Industrial uses include dyeing leather, nylon, paper and textiles; and the production of writing inks and wood stains. Tartrazine is also used for the coloration of anodized aluminum, soap and casein resins. A pure grade of tartrazine (CI Food yellow 4) is used as a food dye, and for coloration of cosmetics and pharmaceuticals.

Purity of commercial batches: Dye contents are up to 60%. Since this is a food dye, samples of 90% or higher dye content are also available. Some specimens have only a single major component. There is a thin-layer chromatographic method (Penner, 1968).

Stability: The dye is hygroscopic, readily picking up moisture from the atmosphere. Its light fastness is moderate.

Amaranth

CI 16185, CI Acid red 27
Synonyms: azorubin S
$C_{20}H_{11}N_2Na_3O_{10}S_3$, FW 605

Absorption maximum: 521 nm in water.
Solubilities: 7.2% in water; 0.1% in ethanol.

Amaranth is a strongly hydrophilic anionic dye of moderate size.

Applications: There have been occasional reports of this dye being used in biological staining, for example, Curtis (1905) who found the dye a selective collagen tissue stain in a picric acid mixture, and by Chambers (1935) for the vital staining of embryonic chick kidney. Gabe has recommended it as a nuclear stain in a one-step trichrome stain, notable for its ability to yield crisp staining even with 20 μm sections (Gabe, 1976, p. 217).

Industrially amaranth is used for textile dyeing and printing; for dyeing leather; and for coloring wood and paper coatings. Prior to 1977 amaranth was widely used for coloring food products, pharmaceuticals and cosmetics. At the time of writing (1997) it is a permitted food dye, with restrictions, in the European Union.

Purity of commercial batches: Dye contents are up to 80%. A paper chromatographic method is available (Nursten and Williams, 1972).

Stability: Light fastness of this dye is moderate.

ZWITTERIONIC DYES

Methyl red

CI 13020, CI Acid red 2

Nonionized form: $C_{15}H_{15}N_3O_2$, FW 269; hydrochloride FW 306, sodium salt FW 291

Absorption maxima: Nonionized form 410 nm in methanol, hydrochloride 493 nm in acidified methanol, sodium salt 437 nm in methanol.

Methyl red changes color from red to yellow between pH 4.4–6.0; *pK values* are 2.3 and 5.0.

Solubilities: The hydrochloride is slightly soluble in water and ethanol (0.2% and 0.5% respectively); the sodium salt is soluble in water and ethanol (7.0% and 4.0% respectively); the free form is barely soluble in water (0.01%), slightly soluble in ethanol (0.2%), but soluble in benzene.

Under usage conditions this small dye can exist in several states of protonation, some of which differ in color as well as charge sign. The dye is sold in the form of the hydrochloride, as a sodium salt, and as a 'free' form. It is this latter whose structure is routinely drawn as the nonionic species shown above. However, this may in fact be zwitterionic, as indicated by its low solubility in polar solvents.

Applications: Methyl red has rarely been used for staining, although Carter (1933) did report its application as a vital stain for protozoa. However, this dye is widely used in pure and applied biology. Schrader (1994) used it to study the ability of earthworms to modify their environmental pH. It is still used in preparation of agar plates in microbiology (Rugsaseel *et al.*, 1993). In food science, methyl red has been used as a component in an indicator to check pork quality during cold storage (Yoon *et al.*, 1994), and as a tracer for assessing penetration of foods into packaging (Ducruet *et al.*, 1996).

Purity of commercial batches: dye content of analytical reagent 95% or higher. A thin-layer chromatographic analysis method is available (Kirchner, 1971).

Stability: A wide variety of reducing agents cause rapid fading of aqueous dye solutions.

Hazards: Harmful if ingested. Mutagenic if chronically ingested.

Methyl orange

CI 13025, CI Acid orange 52
Synonyms: orange III
$C_{14}H_{14}N_3NaO_3S$, FW 327

Absorption maximum: 507 nm in aqueous acid.
In the pH range 3.0–4.4, this dye changes color from red to yellow; its *pK value* is 3.7.
Solubilities: 0.5% in water; 0.1% in ethanol; trivially soluble in xylene.

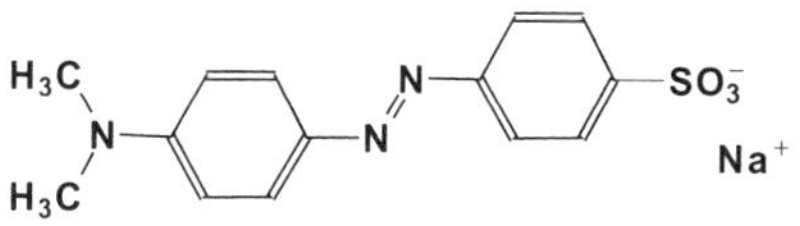

This small monosulfonated dye is sold as the sodium salt, as illustrated. When used as a pH indicator, it is transformed into a zwitterion of a different color.

Applications: Methyl orange has seen little application as a biological stain. However, Krause (1926/1927, p. 2305) cited Pfeffer as using it as a pH indicator in cell sap; Newcomer (1938) used it, under the synonym gold orange, as a counterstain to crystal violet in staining pollen tubes; and Lopez (1946) used it as a component of a polychrome histological stain.

Methyl orange has other biomedical applications, including use in film dosimeters (Chung and Miller, 1994) and as a reagent for the assay of bromide ions (Hasty *et al.*, 1981). Methyl orange is also used as a solution indicator, such as in assays of oxalic and malonic acids (Sagi *et al.*, 1992).

Purity of commercial batches: Dye contents are up to 95%, the minimum dye content for Commission certification being 85%. The dye may be analyzed by thin-layer chromatography, on silica layers using n-butanol–acetic acid–water (1:1:1 by volume) as developing solvent. For Commission assay procedures see Chapter 28.

CATIONIC DYES

Janus green B

CI 11050
Synonyms: diazine green S
$C_{30}H_{31}N_6Cl$, FW 511

Absorption maxima: 630 nm, with a subsidiary peak at 395 nm, in 50% aqueous ethanol.
Solubilities: 5.2% in water; 1.1% in ethanol.

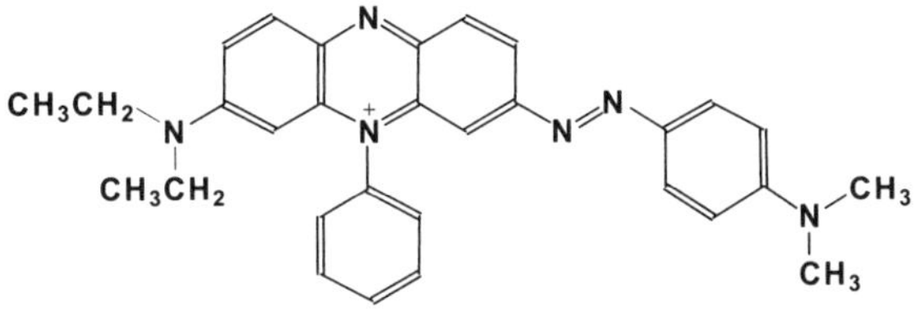

This lipophilic cationic dye is of moderate size, with a large conjugated system for a basic dye. The dye contains both azo and azine chromophores.
Applications: The most important biological staining application of Janus green B is the demonstration of mitochondria in living cells. This was first described by Michaelis (1900), and later developed by Bensley (1911). It has since been widely applied, for instance to blood cells (Sabin, 1929), and more recently to the cornea (Vandelft *et al.*, 1983), for viability testing of yeasts (Sano *et al.*, 1993), and as a vital wholemount stain to aid the dissection and neurophysiology of insect nervous systems (Yack, 1993). Other staining applications are the histological staining of embryonic tissues (Faris, 1924), staining coccidia oocysts in fecal specimens (Crough and Becker, 1931) and as a counterstain in a modified Ziehl–Neelsen procedure (Lillie, 1944a).

Janus green has a wide variety of other applications in biology, including use as a staining reagent in chromatographic analysis of phosphoinositides (Dale, 1988), and as an aid for the differentiation of brewery yeast species (Simpson *et al.*, 1992). The dye is also used to identify isolated and cultured guinea pig parietal cells (Giebel *et al.*, 1995), and to assess corneal epithelial cell viability following toxic insult (Weidner and Sillman, 1997). Other recent laboratory applications include contrast enhancement of latent fingerprints (Kempton and

Rowe, 1992) and the determination of anionic sites on crystalline silica, in cancer research (Saffiotti *et al.*, 1994).

Purity of commercial batches: Most commercial samples contain two major components, as well as minor or trace compounds. Commercial batches contain up to 65% dye, with the minimum dye content for Commission certification being 50%. A thin-layer chromatographic analytical method is available (Marshall, 1976c). See Chapter 28 for Commission assay procedures.

Stability: This dye is stable when stored in a sealed container in the dark.

Alcian yellow G

CI 12840, CI Ingrain yellow 1
$C_{40}H_{46}Cl_2N_8S_4$, FW 838

Absorption maxima: 395 nm in water, 388 nm in methanol. Emission maximum: if excited between 340–380 nm alcian yellow emits between 428–440 nm depending on concentration.

Solubility: 0.2% in water; 1% in ethanol.

Alcian yellow is a large, weakly hydrophilic cationic dye, with an extended conjugated system encompassing both the azo and thiazole chromophores. The solubilizing cationic isothiouronium cations are readily split off by dilute alkali, acidic oxidants or nucleophilic tissue groupings to yield an insoluble pigment.

Applications: Alcian yellow has been used in conjunction with alcian blue to distinguish between different types of glycosaminoglycans (Maxwell, 1963), and to demonstrate neurosecretory material (Ezzughayyar, 1993). Alcian yellow has also been used as a counterstain in a rapid method for the demonstration of *Helicobacter pylori* in gastric biopsies (Leung *et al.*, 1996). As a fluorochrome of fixed histological specimens the dye acts as a typical basic dye, coloring nuclei, basophilic cytoplasms and glycosaminoglycan-rich structures (Stockert *et al.*, 1989). Industrially alcian yellow was used in textile dyeing for the ingrain coloration of cotton.

Purity of commercial batches: Commercial batches can contain up to 70% of a single colored component. A thin-layer chromatographic assay method is available (Horobin and Goldstein, 1972).

Stability: The dry powder is stable over long periods, but solutions readily decompose yielding an insoluble derivative in the presence of alkali or oxidants; light fastness is moderate.

   Alcian green (CI Ingrain green 2) is a mixture of alcian yellow and alcian blue 8G (Horobin and Goldstein, 1972).

DYES FORMING COVALENT BONDS WITH TISSUE OR CO-SOLUTES

Bromo-PADAP

Synonyms: Br-PADAP, bromopyridylazo-diethylaminophenol,
2-(5-bromo-2-pyridylazo)-5-(diethylamino)phenol
$C_{15}H_{17}BrN_4O$, FW 349

Absorption maximum: 443 nm in methanol.
Metal complexation behavior: Forms intensely colored complexes with a wide variety of metal ions.
Solubilities: Only very slightly soluble (0.1%) in water, but dissolves readily if acidified or a little alcohol is added; soluble in ethanol (2%), methanol and many other organic solvents.

This is a small monoazo dye. The basic substituent protonates forming a cation under acid conditions. The phenolic hydroxyl ionizes under weakly alkaline conditions, and the resulting anion – together with the ring nitrogen and azo group – provides a chelation site suitable to bind many metal ions.
Applications: Demonstration of metal ions in tissue sections has been achieved using bromo-PADAP (Sumi *et al.*, 1983a,b, 1999). The dye has been widely used as an analytical reagent for metal ions, see the review by Marczenko (1986). Recent analytical applications of biomedical significance include assay of trace amounts of iodide in laver and table salt (Sun *et al.*, 1997) and of trace amounts of lead, cadmium and copper in drinking water (Lu *et al.*, 1998). The dye is also used in the assay of biochemicals, such as ascorbic acid in fruit juices and pharmaceutical preparations (Ferreira *et al.*, 1997).
Purity of commercial batches: Dye content can be up to 97%.

Mercury orange

Synonyms: Bennett's sulfhydryl reagent, p-chloromercuriphenyl-azo-α-naphthol. red sulfhydryl reagent.
$C_{16}H_{11}ClHgN_2O$, FW 483

Excitation at 450–490 nm, emission at 600 nm.
Solubilities: Soluble in water; 12% in methanol; also soluble in xylene.

This moderately sized nonionic dye carries a chloromercuric group, which can form covalent links with thiols or other nucleophils and suitably reactive aromatic rings.

Applications: Mercury orange was introduced into histochemistry by Bennett in 1951 and became a standard method for the light microscopic demonstration of cysteine, glutathione and free thiol groupings in tissues. It is still used in this way (Allen, 1993; Pearce *et al.*, 1997; Thomas *et al.*, 1995) despite doubts concerning specificity (Horobin and Flemming, 1982), sensitivity (Weise, 1980), and localization of low molecular weight adducts (Chieco and Boor, 1983). It has occasionally been used to stain thiols for transmission electron microscopy (Mundkur, 1964). Other laboratory uses of mercury orange include the flow cytometric assay of intracellular nonprotein thiols (O'Connor *et al.*, 1988).

Purity of commercial batches: Chromatographic analysis (Horobin, 1971) of a commercial sample showed that several colored components were present.

Procion yellow M4R

CI Reactive orange 14
Synonyms: procion yellow MX4R
$C_{19}H_8Cl_2N_8Na_3O_9S_2$, FW 696

Excitation at 488 nm, emission at 530 nm.
Solubility: 6% in water.

This is a fluorescent, hydrophilic acid dye of moderate size. Procion yellow carries a dichlorotriazinyl reactive group which forms covalent bonds with nucleophilic groups such as hydroxyl and amino present in tissues or solvents.

122

In some biological studies authors state merely that they used 'procion yellow'. This lack of precision is unfortunate, because there are several procion yellows of differing chemistry.

Applications: Procion yellow M4R was first used in biology as a label for newly formed bone (Goland and Engel, 1963). However, it has subsequently been widely used as an intracellular marker of neurones (Kater and Nicholson, 1973; Stretton and Kravitz, 1968). Once introduced into the cell, for instance by iontophoresis, the dye's hydrophilic character prevents loss through cell membranes, and its reactive character prevents loss during eventual tissue processing (Flanagan *et al.*, 1974); although there may be interneuronal transfer (Zieglansberger and Reiter, 1974) perhaps via gap junctions because the dye molecules are not excessively large. This remains a widely used procedure: see the combined labeling and intracellular electrophysiological recording of retinal cells described by Negishi *et al.* (1997).

Other cell types have also been labeled with procion yellow, for example, cockroach ganglia (Gregory, 1973), retinal cone cells (Petry and Bassi, 1991), and retinal glial cells (Reichenbach *et al.*, 1995). The dye continues to be used in other staining methods, such as investigation of changes in arterial and venous walls following damage (Berlin *et al.*, 1992), and the assessment of membrane damage in sarcolemmal cells following hypoxia (Bannister and Publicover, 1995). Procion yellow M4R has also been used for block staining prior to resin embedding (Payton, 1970).

Industrially procion yellow M4R has been used for textile dyeing.

Purity of commercial batches: Several components may be present, some could be nonreactive hydrolysis products. A thin-layer chromatographic analytical method is available (Sivitz *et al.*, 1973).

Stability: Dry powder is moderately stable, but aqueous solutions will decompose yielding a nonreactive acid dye on standing; light fastness is good.

Hazards: Reactive dyes may be irritants to skin and eyes, and may be skin sensitizers.

11

Dis-, tris- and polyazo dyes

R.W. Horobin

R.W. Horobin

BACKGROUND INFORMATION

Various generic features of azo dyes – acid–base strengths and isomerism – were discussed in Chapter 10. Only features peculiar to dyes with two or more azo groups are considered below.

Extent of conjugation

In commercially available dyes the extent of conjugation typically increases in the sequence dis-, tris- and polyazo; all have more extensive aromatic systems and usually more extended conjugation than analogous monoazo dyes. However even within a given class, such as the widely available dis azo anionic dyes, significant variations in extent of conjugation occur. Thus amido black 10B and biebrich scarlet have planar conjugated systems of moderate size; and such dyes are used industrially as general purpose colorants for wool. Dye such as trypan blue and sirius fast red F3B on the other hand have planar conjugated systems of considerable size; and such compounds can be used for direct dyeing of cellulosic textiles. Finally, dyes such as milling red B and sulfone cyanin 5R also have large aromatic systems, but since these are markedly nonplanar the extent of conjugation is limited; and such colorants find industrial application as milling dyes for wool, lacking affinity for cellulosic materials.

Impurities of dis and higher azo dyes

Because such dyes are usually assembled by repeated azocoupling, impurities are more common than with monoazo dyes. Thus with disazo dyes impurities commonly include monoazo dyes, as well as disazo isomers, and also disazo derivatives arising from impurities in starting reactants. These variations have been well documented for biological stains (Lloyd and Beck, 1963; Proctor and Horobin, 1985; Rosenthal *et al.*, 1965).

Safety issues

Benzidine and various of its carcinogenic derivatives are intermediates in the manufacture of many dis- and polyazo dyes. For the safety of workers in the dyestuffs industry, these compounds should be synthesized, diazotized and coupled entirely within enclosed apparatus (Allen, 1971). Moreover although the dyestuffs derived from such intermediates are not themselves carcinogenic, there is a possibility of bacterial reduction of these dyes, yielding the carcinogens. For these reasons few benzidine-derived dyes are manufactured on a large scale in industrial countries.

ANIONIC DISAZO DYES

Sudan III

CI 26100, CI Solvent red 23
Synonyms: Sudan G, fat ponceau G
$C_{22}H_{16}N_4O$, FW 352

Absorption maxima: 507 nm in toluene, with a subsidiary peak at 354 nm; 503 nm in ethanol.
Solubilities: 0.2% in ethanol; 0.3% in xylene; 1.3% in ethylene glycol; insoluble in water.

Sudan III is a strongly lipophilic disazo dye, with no substituents that ionize under staining conditions.
Applications: This dye has long been used as a stain for fats and other hydrophobic materials, such as suberin in vascular plants; see Daddi (1896) and Bugnon (1919) respectively. Sudan III is still used for such staining tasks, as shown by recent papers concerned with the staining of fat droplets in cryosections (Ballardini *et al.*, 1994) and of suberized abscission layers in plants (Adaskaveg, 1995). The dye will stain fat in other types of preparation, for example, blood smears (Gulati and Hyun, 1996) and sputum cytology (Vejar and LeCerf, 1997). Sudan III has also been used for the demonstration of other hydrophobic materials, for example, wear particles of polymethyl methacrylate bone cement in tissues (Lintner *et al.*, 1982), and pitch extractives in wood pulp (Schafer and Roffael, 1996). Sudan III is also used in biomedical laboratory procedures, for instance diagnostically to demonstrate fat in feces (Iacono *et al.*, 1991).

Industrially applications include coloration of oils, fats and waxes; of organic solvents; and polystyrene resins. The dye is used to color contact lens; and in some countries has been used in cosmetics.

Purity of commercial batches: A single major component plus many minor impurities are usually present. Dye content is up to 85%; the minimum dye content for Commission certification is 75%. A thin-layer chromatographic analytical method is available (Lansink, 1968); see Chapter 28 for Commission testing procedures.

Stability: Dry dye is stable when stored in a sealed container; and the dye's light fastness is good.

Sudan IV

CI 26105, CI Solvent red 24.

Synonyms: scarlet red, scarlet R, fat ponceau.

$C_{24}H_{20}N_4O$, FW 380

Absorption maxima: 520 nm in ethanol, with a subsidiary peak at 357 nm.

Solubilities: 0.3% in ethanol; 2.5% in ethylene glycol; also soluble in xylene to 3.5%; virtually insoluble in water.

Sudan IV is a strongly lipophilic disazo dye, carrying no substituents ionized under staining conditions. There are two methyl substituents on the aromatic rings.

Applications: The dye was introduced as a fat stain for animal tissues, applied from aqueous alcoholic solutions, at the beginning of the 20th century by both Michaelis and Herxheimer. Later its use from supersaturated isopropanol was recommended by Lillie, and from glycols by Chiffelle and Putt and others; for a brief review see Lillie and Fulmer (1976, pp. 565–568). Such methods remain in current use, for example, for staining fat droplets in cultured preadipocytes (Marko *et al.*, 1995) and lipid in atheroschlerotic lesions (Boerboom *et al.*, 1997). Sudan IV has also been used to stain fats and other hydrophobic materials such as suberin in plants (Rawlins, 1933). The dye is still used by botanists, for example, for the demonstration of oil droplets in coconut haustoria (Krishnankutty *et al.*, 1990) and for the identification of the outer lipid-rich layer of pollen grains of alfalfa (Henning and Teuber, 1992).

Less traditional staining applications include identification of liquid fats in food (Flint, 1994, p. 38) and assessment of fat-transplants in plastic surgery (Fagrell *et al.*, 1996).

Industrially applications include coloration of oils, waxes, shoe polish, and greases; oily-resin lacquers and varnishes; acrylic, cellulose acetate and poly-styrene resins; and as a wood stain.

Purity of commercial batches: One major and one minor red components are usually present, plus many minor components of various colors. Commercial samples contain up to 80% dye, which is the minimum dye content for Commission certification; see Chapter 28 for the Commission testing proce-dures. A thin-layer chromatographic method is available (Lansink, 1968).

Stability: The dry dye is stable on storage in sealed containers; and the dye's light fastness is good.

Oil red O

CI 26125, CI Solvent red 27
Synonyms: Sudan red 5B
$C_{26}H_{24}N_4O$, FW 409

Absorption maxima: 518 nm in toluene, subsidiary peak at 359 nm.
Solubilities: 0.7% in ethanol; 3.5% in xylene; also soluble in ethylene glycol to 2.5% and water to 0.4%.

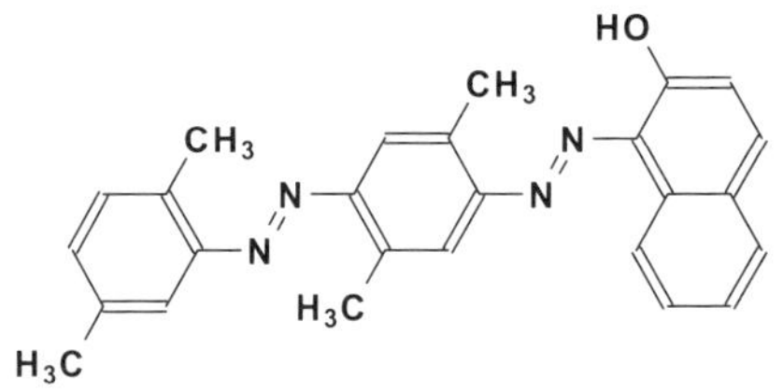

Oil red O is a superlipophilic disazo dye, carrying no substituents that ionize under staining conditions. There are four methyl substituents on the aromatic rings.

Applications: This dye is recommended as a routine fat stain both in the Herxheimer and the supersaturated isopropanol techniques (Bancroft and Gamble, 2002). Recent applications include identification of fat in food products (Flint, 1994), investigation of the differentiation of adipocytes in culture (Sorisky *et al.*, 1996), and study of aortic fatty streaks (Rong *et al.*, 1997). Oil red O has also been used to stain lipid in epoxy resin sections (Eyden *et al.*, 1991).

Demonstration of other hydrophobic materials can also be achieved using oil red O, for example, latex droplets in plants (Jayabalan and Shah, 1986) and polyethylene wear particles released from prostheses (Basle *et al.*, 1996). Less routine staining applications of this dye include use as a whole-mount stain for identifying laticifers in plants (Inamdar *et al.*, 1987) and for visualizing lipids in nematodes (Stamps and Linit, 1995).

Other biomedical laboratory applications of oil red O include staining lipid in flow cytometry (Whitman *et al.*, 1991) and the quantification of lipids in cultured cells (Ramirez-Zacarias *et al.*, 1992). Neutrophil functioning has been assessed by measuring the uptake of oil droplets stained with this dye in a microtiter plate system (Rosen *et al.*, 1991).

Industrial uses include coloration of oils, fats and waxes; also varnishes and cosmetics.

Purity of commercial batches: One major and one minor red dye, plus numerous minor components of different colors, are usually present. Dyes contents are up to 85%; the dye content of most Commission certified batches is in the range 75–80%. A thin-layer chromatographic analytical method is available (Lansink, 1968). For details of the Commission test procedures, see Chapter 28.

Stability: Light fastness is good.

Sudan black B

CI 26150, CI Solvent black 3
$C_{29}H_{24}N_6$, FW 457

Absorption maxima: 598 nm in ethanol, with a shoulder at 415 nm.
Solubilities: 3% in ethanol; also soluble in ethylene glycol to 1% and in xylene to 2.5%; trivially soluble in water.

Sudan black B is a superlipophilic disazo dye, carrying secondary amine substituents which only ionize under acid conditions, not in routine staining solutions. Commercial batches all contain two isomers, as illustrated (Pfuller *et al.*, 1977). Note: most catalogs and texts ignore the existence of isomer I.

Applications: This dye was first proposed as a myelin stain by Lison and Dagnelie (1935), as a fat stain in animals by Gerard (1935), and later as a stain for bacterial lipid by Burdon *et al.* (1947) and for botanical specimens (Gahan, 1984). It was also found that a preliminary bromination step permitted Sudan black to stain free cholesterol (Bayliss and Adams, 1972). The dye can be used as a substrate for demonstration of myeloperoxidase-containing leukocyte granules (Lillie and Burtner, 1953). All these methods are in current general use.

Additional staining applications include identifying lipids within insect cuticles (Majumdar *et al.*, 1996), in cryosection-autoradiographs (Moldovan *et al.*, 1994), glycol methacrylate sections of brain (Gerrits *et al.*, 1992) and epoxy resin sections (Sire and Vernier, 1980). The lipid staining properties of Sudan black B have also been applied to the diagnostic fiber-typing of skeletal muscle

(Dux *et al.*, 1981) and to the staining of myelinated nerves in amphibian and reptile whole-mounts (Filipski and Wilson, 1985). Other hydrophobic substances demonstrable with Sudan black B include suberin in plants (Mould and Robb, 1992), and polyurethane–polylactide prostheses in glycol methacrylate embedded specimens (Hoeksma *et al.*, 1988).

Other biomedical applications include estimation of venule diameter with Sudan black-stained oil droplets (Sadurski *et al.*, 1994), in gel electropherograms (Shibusawa *et al.*, 1995), and assessment of epididymal fluid movement by observing movement of Sudan black-stained droplets of mineral oil (Yamamoto *et al.*, 1995).

Industrial uses have included coloration of organic solvents, printing inks, cellulose lacquers, and oils, fats and waxes.

Purity of commercial batches: All samples contain two major blue–black components (isomers I and II, see above) plus a variety of minor constituents which vary from batch to batch. Brown polyazo products form in staining solutions exposed to air. A thin-layer chromatographic analytic method is available (Lansink, 1968); for Commission assay procedures see Chapter 28.

Stability: Staining solutions slowly form polyazo derivatives on standing, and should be discarded after 3–4 weeks; light fastness is fair.

Acetyl Sudan black has been stated to be a superior alternative. However commercial samples are Sudan black B itself (Marshall, 1977) and chemical investigations (Lauder and Beynon, 1989) showed that acetylation processes in the literature were ineffective. Acetyl Sudan black is indeed 'a nonexistent reagent' (Horobin and Proctor, 1989).

ANIONIC DISAZO DYES

Biebrich scarlet

CI 26905, CI Acid red 66
Synonyms: biebrich scarlet water soluble, ponceau B
$C_{22}H_{14}N_4O_7S_2Na_2$, FW 556

Absorption maximum: 505 nm in methanol.
Solubilities: 4% in water; 0.2% in ethanol; insoluble in xylene.

Biebrich scarlet is an anionic, hydrophilic, disulfonated disazo dye; of moderate overall size, as is its conjugated system. Note: a nonionic disazo fat stain, Sudan

IV, is occasionally termed 'biebrich scarlet red', leading to confusion when authors refer to this latter dye as 'biebrich scarlet'.

Applications: Biebrich scarlet has been widely used for cytoplasmic staining, as briefly reviewed by Gurr (1960). The Shorr stain (1941), which involves the use of biebrich scarlet, is still used in diagnostic cytology; and the dye is also a component of various polychromes (Garvey *et al.*, 1993). Spicer and Lillie (1961) used alkaline solutions of biebrich scarlet to selectively stain basic proteins. This dye was chosen because its color does not change over a wide range of pH. Biebrich scarlet has also been used to stain epoxy resin sections (Moxey and Yeomans, 1976), as a component of one-step fixative-stains for fish eggs (Klinger and Van den Avyle, 1993), and to facilitate counting of eosinophils in a hemocytometer (Brattig *et al.*, 1993).

Industrial uses have included textile dyeing and the coloration of paper.

Purity of commercial batches: One major and one minor red component are routinely present; commercial samples contain up to 50% dye. Biebrich scarlet can be analyzed by thin-layer chromatography on cellulose using cellosolve acetate–acetic acid–water (5:4:32 by volume) as developing solvent; a paper chromatographic method is also available (Rosenthal *et al.*, 1965).

Stability: Light fastness is moderate.

The structure of an isomeric dye, **woodstain scarlet** or **brilliant crocein MOO** (CI 27290, Acid red 73), is given above. Like biebrich scarlet, this has been used as a component of histological polychrome stains (Movat, 1955; Wismar, 1966).

Amido black 10B

CI 20470, CI Acid black 1
Synonyms: naphthol blue black, naphthalene black 12B, pontacyl blue black S
$C_{22}H_{14}N_6Na_2O_9S_2$, FW 617

Absorption maximum: 618 nm in water.
Solubilities: 3% in water; 3.2% in ethanol; insoluble in xylene. There is disagreement in the literature over solubility data, perhaps reflecting batch variation.

Amido black 10B is an anionic, hydrophilic disulfonated disazo dye. It is of moderate overall size, as is its conjugated system. Incomplete azo coupling in the second synthetic step gives rise to a red monoazo dye of smaller size.

Applications: This dye was introduced as the collagen staining component of Van Gieson-type trichromes (Lillie, 1954; Puchtler and Sweat, 1964), and as a hemoglobin stain by Puchtler's group (Puchtler and Sweat, 1962; Puchtler *et al.*, 1964).

Amido black has since been used for the quantitation of proteins in tissue sections (Nohammer, 1983), and this technique has been adopted for assessing cultured cell numbers (Ciapetti *et al.*, 1996). Amido black has subsequently been applied as a total-protein stain (Wojcik *et al.*, 1996). Another application of the dye has been as a nucleolar stain in cytology and histology (Bedrick, 1970; Mundkur and Greenwood, 1968; Wood and Green, 1958); and it is still used in this way cytologically (Boon and Drijver, 1986). The dye can also be used to stain total protein (Pihakaski-Maunsbach and Walles, 1990) and nucleoli (Burgauer and Stockert, 1975) in epoxy resin sections. Amido black has been recommended as a counterstain for periodic acid–Schiff staining in botanical material (Ruzin, 1999).

Other applications of amido black 10B in biomedicine include protein assays, for example, in milk and the related staining of proteins on gel electropherograms (Shilina *et al.*, 1997) and on polymer membranes, following transfer from electropherograms (Sanchez *et al.*, 1997) or extraction from cells (Dieckmann-Schuppert and Schnittler, 1997). The dye has also been used as a tracer of apoplastic (Jacobsen *et al.*, 1992) and vascular (Schurr *et al.*, 1996) fluid movement in plants.

Industrial uses include textile dyeing and printing; dyeing of leather; the coloration of paper, anodized aluminum, of resins; and as a component of wood stains and writing inks.

Purity of commercial batches: Samples routinely contain one major, one minor, and several trace components; some batches contain very little amido black. Dye contents are up to 80%. A thin-layer chromatographic analytic method is available (Nettleton *et al.*, 1986).

Stability: Light fastness is good.

Congo red

CI 22120, CI Direct red 28
Synonyms: Congorot, Kongoröt
$C_{32}H_{22}N_6Na_2O_6S_2$, FW 697

Absorption maximum: 497 nm in alkaline water (containing 0.01% w/v sodium carbonate). When excited at 500 nm, Congo red often fluoresces red, but the emission wavelength is dependent on the substrate to which dye is bound. Congo red changes from blue to red between pH 3.0–5.0; its *pK value* is 4.1.

Solubilities: 5% in water; 0.2–0.8% in ethanol; also soluble to 2.5% in ethylene glycol; insoluble in xylene. Addition of acid to aqueous solutions yields a blue precipitate. This is a hydrophilic disulfonated dye with a large, planar aromatic system; there is a lipophilic domain between two sulfonated naphthyl end units.

Applications: First synthesized by Böttiger in 1884, by 1886 Congo red was already being applied as a biological stain for axons (Griesbach, 1886) and cellulose (Klebs, 1886); and by 1888 as a stain for embryonic material (Schaffer, 1888). Congo red has also been used as a cytoplasmic contrast stain, as briefly surveyed by Emig (1941).

Congo red was used as an elastic fiber stain as early as 1925 (Matsuura, 1925), with subsequent applications of this kind including its use as a fluorescent elastin stain (Hospelhorn *et al.*, 1988), and as a single-dye polychrome method exploiting its pH indicator properties (Davies and Young, 1982). Congo red is still used as a routine elastic stain, for example, a recent demonstration of elastic fibers within tendons (De Carvalho and Vidal, 1995). Sloper (1954, 1955) used Congo red in conjunction with phosphotungstic acid, to stain neuro-secretory material in the pituitary gland. Congo red is also still used to demonstrate polysaccharides such as cellulose, see a recent fluorescence method (Verbelen and Stickens, 1995) and a procedure used with plant root hairs (Webster and Stone, 1994).

Long used to stain amyloid in histopathology (Bennhold, 1922), Congo red is recommended in current histopathology manuals for this purpose (Bancroft and Gamble, 2002). It is widely used for demonstrating amyloid plaques resulting from Alzheimer's disease (Hsiao *et al.*, 1996). The green birefringence of Congo red-stained amyloid greatly enhances the sensitivity of the method (Wolman and Bubis, 1965).

Congo red also has a history of application in the study of unicellular organisms, both pro- and eukaryotic, for instance as a negative stain for bacteria and spirochetes (Benians, 1916; Cumley, 1935; Maneval, 1934), for protozoa (Merton, 1932), and for yeasts (Gutstein, 1937). Recent examples of this type include demonstration of certain strains of *E. coli* living within meal-worms (McAllister *et al.*, 1996), and the fluorescent staining of chitin in fungal cell walls (Matsuoka *et al.*, 1995). Other microscopic staining applications of interest include selective staining of eosinophils (Grouls and Helpap, 1981), testing the viability and acrosomal status of mammalian sperm (Kovacs and Foote, 1992), and the assessment of damage to starch grains in food products (Adler *et al.*, 1995).

A small selection from the wide variety of nonmicroscopic applications include use in plate counts to assess aerobic cellulose decomposers in soil samples (Suyama *et al.*, 1993), demonstration of xylenases on gel electro-pherograms (Breccia *et al.*, 1995), and counting and identification of inverte-brates recovered from sampling devices in the field (Brinkman and Duffy, 1996).

Congo red has long been used as a pH indicator in test papers and in solution; as well as in bacteriological starch and agar gels. Currently the dye is also used as a pH indicator diagnostically, in the endoscopic evaluation of gastritis (Iseki *et al.*, 1997), and as a component in fiber optic and related sensors (Egami *et al.*, 1997).

Industrial applications – Congo red has been used as a textile dye; and for coloration of paper. Due to the use of benzidine in the traditional manufacturing process, the dye is no longer manufactured on a large scale in the major industrialized countries.

Purity of commercial batches: One major red component and a minor yellow product are routinely present. Dye contents are up to 85%, with the minimum dye content for Commission certification being 75%. A thin-layer chromatographic analytical method is available (Raban, 1963). See Chapter 28 for the Commission's assay methods.

Stability: The dry powder is stable when stored in sealed containers; light fastness is poor.

Ponceau S

CI 27195, CI Acid red 112
Synonyms: fast ponceau 2B
$C_{22}H_{12}N_4Na_4O_{13}S_4$, FW 761

Absorption maxima: 520 nm in water, with a subsidiary peak at 352 nm.
Solubilities: 1.4–4.0% in water; 0.1–1.2% in ethanol; also soluble in ethylene glycol to 4.7%; insoluble in xylene. Variations in solubilities reported may reflect batch variation in degree of sulfonation.

This is an anionic, very hydrophilic disazo dye of moderate size.

Applications: The dye was proposed as a substitute for acid fuchsine in the Van Gieson procedure by Curtis as long ago as 1905. However it appears to have been little used in this way until several authors (Leach, 1946; Lillie, 1945; Ruth, 1946) noted its merits, including its greater resistance to fading. The dye is also recommended in botanical staining manuals as a stain for sieve plate elements (Berlyn and Miksche, 1976).

Ponceau S is widely applied to stain proteins on gel electropherograms (Tormey and O'Brien, 1993), and has been used as a stain for proteins adsorbed onto nitrocellulose membranes (Goldring and Ravaioli, 1996). The dye has been used industrially for the acid dyeing of wool.

Purity of commercial batches: Dye content can be up to 75%.
Stability: Light fastness is moderate.

Arsenazo III

Synonyms: 2,2'-(1,8-dihydroxy-3,6-disulfonaphthyl-2,7-bisazo)bisbenzene-arsonic acid

$C_{22}H_{12}As_2N_4Na_6O_{14}S_2$, FW 776

Absorption maxima: the free acid has a maximum at 560 nm in methanol; the disodium salt a maximum of 533 nm in water. In water the calcium complex has an absorption maximum of 652 nm at pH 7.5.

pK values: arsenazo III shows eight protonation equilibria, whose pK values are spread over the range < 1.3–11.0.

Metal complexation behavior: The dye can be used as a selective calcium chelator at neutral pH; but under suitable conditions also forms complexes with cadmium and various actinide and lanthanide ions.

Solubilities: The sodium salt is 2.0% soluble in water, and 0.3% in ethanol; the free acid form is soluble in ethanol and slightly soluble in water.

A disazo dye of substantial size, which carries two sulfonate and two arsonic acid substituents. It is anionic except under extremely acid conditions. The fully ionized species is strongly hydrophilic, the free acid form very much less so.

Applications: Arsenazo III has occasionally been used to assay neuronal calcium in single cells using microscopy (Miyakawa *et al.*, 1992). However, in nonimaging mode the dye is a routine metallochromic indicator and assay reagent for cellular calcium and related ions. Recent examples include estimation of calcium release from the endoplasmic reticulum (Gilchrist *et al.*, 1997) and of cadmium transfer across the plasma membrane (Tu *et al.*, 1996).

Purity of commercial batches: Dye contents can be over 95%, although some batches contain several minor dye impurities. A thin-layer chromatographic analytical method is available (Nemcova *et al.*, 1986).

Trypan blue

CI 23850, CI Direct blue 14
Synonyms: diamine blue 3B
$C_{34}H_{24}N_6Na_4O_{14}S_4$, FW 961

Absorption maxima: 607 nm in methanol, 588 nm in water; emission when bound to protein is in the red.

Solubilities: 1% in water; 7.2% in ethylene glycol; trivially in ethanol; insoluble in xylene.

This is a large, very hydrophilic, tetrasulfonated anionic dye. It has a large, planar aromatic system; with a lipophilic domain sandwiched between sulfonated naphthyl end-units.

Applications: Traditionally trypan blue was used as a component of polychrome stains. Curtis (1905) recommended it as a collagen stain in a Van Gieson procedure, and that application has continued. It is used as a blue alternative to Congo red in Puchtler–Bennhold stains for amyloid (Puchtler *et al.*, 1962). Other tissue staining applications include demonstration of fungi (Tisserant *et al.*, 1993), of starch in food products (Flint, 1994), and as an oversight stain for monolayers of cultured cells (Perry *et al.*, 1997a,b). Trypan blue has also been applied as a quenching agent to suppress unwanted autofluorescence, both in fluorescent-antibody staining of cryosections (Benne *et al.*, 1997) and in flow cytometry (Mosiman *et al.*, 1997).

Trypan blue has been widely used as a vital stain since the early 20th century (see Chapter 17 in Clark and Kasten, 1983). It is excluded by most living cells, but can be taken into phagocytes and certain other cells. Current applications of this kind include wide use for viability testing, for cells as diverse as frozen sperm (Sanchez *et al.*, 1995) and aortic muscle cells exposed to antifungal agents (Osaka *et al.*, 1997). In embryology trypan blue has been applied as a fluorescent tracer of cell populations (Callebaut and Vakaet, 1981), and for imaging the action of the chick cardiovascular system (Kosaki *et al.*, 1997). When trypan blue binds to proteins, notably albumin, the resulting complex emits red fluorescence. This property has been exploited in studies of exudation from blood vessels in the injured central nervous system (Baskaya *et al.*, 1997; Hamberger and Hamberger, 1966).

Trypan blue also has numerous applications in biomedicine not involving microscopy, a few examples of which follow. In experimental pathology the dye has been used to assess arterial endothelial barrier dysfunction (Berman and Martin, 1993), and creatine kinase release as a marker of irreversible injury of myocytes (Huser *et al.*, 1996b). In microbiology, trypan blue is applied as a constituent of agar gels for the enumeration of yeasts and molds (Hart *et al*, 1991). In pharmacology the dye has also been used as a model drug to assess the *in vivo* permeability of the colonic mucus layer (Szentkuti, 1997).

Oncologists have used the dye as a tumor promoter modulating permeability of lysosomal membranes (Kozlowska *et al.*, 1995), and toxicologists have used the dye to assess the validity of a novel procedure for eye irritancy testing (Hagino *et al.*, 1991).

Industrial application of trypan blue has been discontinued, because of use of dimethylbenzidine in the manufacturing process, and the dye is no longer produced on a large scale in the major industrialized countries.

136

Purity of commercial batches: Typically there is one major blue constituent and one or two reddish purple components; dye contents usually fall in the range 25–80%. A thin-layer chromatographic analytic method is available (Raban, 1963).

Stability: Light fastness is poor.

Evans blue

CI 23860, CI Direct blue 53

$C_{34}H_{24}N_4Na_4O_{14}S_4$, FW 961

Absorption maximum: 611 nm in water: when excited in the green, the emission maximum is around 680 nm. The dye changes color near pH 10.

Solubilities: 6.5% in water; 0.5% in ethanol; soluble in ethylene glycol to 5%; insoluble in xylene.

This is a large, very hydrophilic, tetrasulfonated anionic dye. It has a large, planar aromatic system; with a lipophilic domain sandwiched between sulfonated naphthyl end-units.

Applications: Histological staining using Evans blue has included demonstration of specific pituitary cell types (Landing and Hall, 1956). More recently the dye has been applied as a fluorochrome, for example, as a general stain for proteins (Marttin *et al.*, 1997), and as an immunostaining and *in situ* hybridization counterstain (Holland *et al.*, 1996). It has also been used in fluorescence microscopy as a quenching agent to suppress unwanted autofluorescence (de la Lande and Waterson, 1968).

Evans blue has long been used as a vital stain. Recent examples include assessment of cell viability, for example, of yeast in frozen dough (Autio and Mattila-Sandholm, 1992), and of bud viability following application of herbicides (Becker *et al.*, 1997). Other vital staining applications include selective staining of neutrophil plasma membranes (Hed *et al.*, 1983). Evans blue, mixed with bovine albumin, was the first fluorescent tracer to be used in studies of retrograde axonal transport in the nervous system (Kristensson *et al.*, 1971). Later, the dye alone was injected, for tracing neuronal projections in the brain (Kuypers *et al.*, 1977).

Evans blue is also extensively applied in biomedicine in procedures not involving microscopy. A few examples follow. In botany Evans blue has been used in spectrophotometric viability assays of cells in suspension or in leaf disks (Baker and Mock, 1994). Physiologists have used the dye as a marker. Examples include assessment of protein leakage from microvasculature of airways (Sakamoto *et al.*, 1994), and of the occurrence of mucociliary transport (Tamaoki *et al.*, 1997). The dye is also often used to trace fluid flows, for example, of material within dentinal tubules (Vongsavan and Matthews, 1991), and of lymph within tonsils (Belz and Heath, 1995). Evans blue also finds extensive application in experimental pathology, for example, to mark sites of experimental injuries to the eye lens (Uga *et al.*, 1994), and to mark dystrophic muscle fibers (Matsuda *et al.*, 1995).

There are also clinical applications: for example, as a surgical marker dye (Shoemaker *et al.*, 1996), and to evaluate mucus clearance from endotracheal tubes (Trawoger *et al.*, 1997). Evans blue was once routinely used to assess patient's blood volume (Gregerson and Shiro, 1938). However, patients' lack of enthusiasm for blue-toned skin currently restricts this application to research settings, for example, to assess changes in serum volume after bed rest (Levine *et al.*, 1997).

Purity of commercial batches: Typically one major blue component plus a minor red/purple contaminant; with dye contents up to 85%. There is a paper chromatographic analytical procedure available (Lloyd and Beck, 1963).

Stability: Light fastness is poor.

Chicago blue 6B

CI 24410, CI Direct blue 1
Synonyms: chlorazol sky blue FF, Niagara blue 6B, pontamine sky blue 6B
$C_{34}H_{24}N_6Na_4O_{16}S_4$, FW 993

Absorption maximum: 618 nm in water, with possible variation due to concentration and salt effects.
Metal complexation behavior: The dye forms copper complexes of low water solubility and high light fastness.
Solubilities: 4.8% in water; soluble to 4.8% in ethylene glycol; very slightly (0.1%) soluble in ethanol; insoluble in xylene.

This is a large, very hydrophilic, tetrasulfonated anionic dye. It has a large, planar aromatic system; with a lipophilic domain sandwiched between sulfonated naphthyl end-units.

Applications: Chicago blue 6B has been applied as a selective collagen stain in Masson trichrome and Van Gieson methods (Lillie, 1945). The dye seems to have been used in place of benzo sky blue in various staining applications. As these two dye compounds are physicochemically similar such confusion is of little practical concern. Chicago blue 6B has been used to quench background autofluorescence prior to fluorescence-labeled immunostaining (Beesley, 1993). A more recent biomedical application of Chicago blue 6B is the induction of localized photodamage to endothelial cells of the microvasculature (Nishimura *et al.*, 1989).

The dye is no longer used industrially, due to use of dimethoxybenzidine in the manufacturing process.

Purity of commercial batches: Typically contain one major blue component; dye contents are in the range 20–50%. A paper chromatographic analytic method is available (Lloyd and Beck, 1963).

Stability: Light fastness is poor.

CATIONIC DISAZO DYES

Bismarck brown Y

CI 21000, CI Basic brown 1
Synonyms: Bismark brown Y, vesuvine
The dihydrochloride is: $C_{18}H_{20}N_8Cl_2$, FW 419

Absorption maxima: 643 nm in water; 457 nm in 50% aqueous ethanol containing 5% 1 M hydrochloric acid.

pK value: 5.

Solubilities: 1–5% in water; 1–3% in ethanol; also soluble to 7% in ethylene glycol; the cationic basic dye is insoluble in xylene, although the free base is highly soluble.

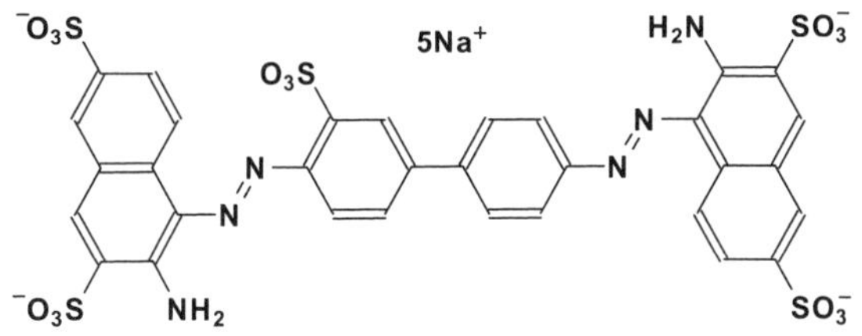

The overall size, and the size of the conjugated system, of this cationic disazo dye are moderate. Commercial samples are mixtures of various isomers and homologs, a major component is illustrated above. Probably because of these factors, reports of solubilities and colors vary somewhat.

Who was Bismarck? This dye was named to honor Bismarck, but many sources anglicize the spelling to 'Bismark'.

Applications: Bismarck brown Y has been widely used as a contrast stain, for example, as a component of the Birch–Hirscheld's procedure for demonstrating amyloid. The dye is still sometimes applied in this way, for instance as a metachromatic mucin and cartilage stain in contrast to methyl green (Clark, 1981; Lillie and Fullmer, 1976, p. 635). The dye was a component of the original Papanicolaou cytology stain (Papanicolaou, 1941). It is found in some current formulations, although it has been suggested that it plays no part in coloration (Marshall *et al.*, 1979).

In botanical work Bismarck brown is recommended by staining manuals for demonstration of cellulose cell walls (Berlyn and Miksche, 1976; Jensen, 1962). In histochemistry, the dye has been used to prepare a pseudo-Schiff reagent

(Kasten, 1958). In microbiology, it has been applied to stain bacteria in contrast to gentian violet in a modified Gram stain; to methyl or crystal violet in the Ljubinsky and Neisser stains for diphtheria organisms; and to carbol-fuchsine in stains for acid–fast bacteria. Bismarck brown Y has also been used as a bulk stain (Weissenberg, 1937); and for the vital staining of vertebrate embryos (Vogt, 1925), stenostonum (Carter, 1933), and protozoa (Gray, 1973).

Industrially Bismarck brown Y has been used as a textile dye; and also to color paper, leather and wood; and to manufacture pigments. The free base is a solvent dye (CI 21000:1, Solvent brown 41) used to color oils, fats and waxes; including shoe polishes.

Purity of commercial batches: Several components may be present. Overall dye content may be up to 50%; the minimum dye content for Commission certification is 45%. A thin-layer chromatographic analytic method is available (Rettie and Haynes, 1964). See Chapter 28 for Commission test procedures for this dye.

Stability: Chemical modification occurs when aqueous solutions are boiled. Light fastness is poor.

The dye **Bismarck brown R** (CI 21010, Basic brown 4) differs only in the presence of two aromatic methyl groups. When substituted for CI 21000 as a section stain, it gives very similar outcomes, albeit of a redder hue. Bismarck brown R has also been used to optimize staining of proteins in acrylamide gels (Choi *et al.*, 1996).

ANIONIC POLYAZO DYES

Chlorazol black E

CI 30235, CI Direct black 38
Synonyms: pontamine black E
$C_{34}H_{25}N_9Na_2O_7S_2$, FW 782

Absorption maxima: In water there are two broad maxima around 500–504 nm and 574–578 nm; in ethanol there is a broad maximum centered around 598–602 nm.
Solubilities: 6% in water; 0.1% in ethanol; 5% in ethylene glycol; insoluble in xylene.

This disulfonated anionic trisazo dye is of moderate size and hydrophilicity. It has a large, planar aromatic system.

Applications: Chlorazol black E was first popularized as a one-bath oversight stain in animal and plant histology and cytology by Cannon (1937), and the dye is still occasionally used in this way (Bameul, 1992; Fyson and Oaks, 1992). In a later paper, Cannon (1950) reported the selective staining of elastic fibers, which application had been previously noted by Levine and Morrill (1941) and subsequently by various others (Goldstein, 1963b). There are other diverse applications of chlorazol black E for biological staining. For instance, it is recommended in botanical staining manuals for demonstration of nucleoli (Berlyn and Miksche, 1976), and is also used in diagnostic microbiology to stain *Giardia* cysts in stool specimens (Alles *et al.*, 1995), and in hematology to stain granulocytes (Kass, 1981). Chlorazol black E has been used to stain starch in wheat-flour food products (Flint and Moss, 1970).

Specimens other than sections are also stained, for instance, in staining polychaete larvae as whole mounts (Hermans, cited in Gray, 1973). Indeed the dye can be used together with lactophenol for the staining, clearing and mounting of a variety of botanical, mycological and zoological specimens (Gurr, 1960). The dye has been applied as a fixative-stain of intestinal parasites (Bullock, 1980) and as a vital stain for macrophages (Baker, 1941).

Purity of commercial batches: Typically one major black component, accompanied by minor impurities of varied colors, for example, green, red and yellow. There is a paper chromatographic analytic method available (Rosenthal *et al.*, 1965). For the Commission's approach to assessing chlorazol black E samples, see Chapter 28.

Stability: Light fastness is good.

Sirius red F3B

CI 35780, CI Direct red 80
Synonyms: chlorantine fast red 7BLN or 5BRL, durazol brilliant red B, pontamine fast red 7BNL, sirius fast red. This must not be confused with Sirius red 4B (CI 28160, Direct red 81).
$C_{45}H_{26}N_{10}Na_6O_{21}S_6$, FW 1373.

Solubilities: Soluble in water; very slightly soluble in ethanol; insoluble in xylene. Aqueous solutions of the dye form violet precipitates on addition of concentrated hydrochloric acid.

This is an extremely hydrophilic, very large tetrakisazo hexasulfonated anionic dye. It has a large, planar aromatic system.

Applications: Coloration of collagen, when substituted for acid fuchsine in the Van Gieson stain, is currently the major application of sirius red F3B. One widely used procedure uses polarization microscopy to increase sensitivity (Junqueira *et al.*, 1979; Puchtler and Sweat, 1964; Puchtler *et al.*, 1973; Williams *et al.*, 1996). Sirius red–Van Gieson methods are also used for quantitation of collagen using image analysis methods (Malkusch *et al.*, 1995). The method has also been adopted for staining collagen in meat products (Flint and Pickering, 1984) and for tissues embedded in glycol methacrylate (Cannon *et al.*, 1992). Other staining applications include use as an eosin substitute for counterstaining alum hematoxylins (Clark, 1981); and as a Congo red substitute for staining amyloid (Llewellyn, 1970).

Industrial uses have included textile dyeing, and the coloration of leather and paper.

Purity of commercial batches: These can be chromatographically homogeneous; with dye contents up to 30%. There is a paper chromatographic analytic method available (Rosenthal *et al.*, 1965).

Stability: Light fastness is good.

Luxol fast blue G

CI Solvent blue 34
The tetradiarylguanidine salt is: $C_{102}H_{97}N_{19}O_{13}S_4$, FW 1926

Absorption maximum: 598 nm in ethanol.
Solubilities: Very soluble in ethanol; also soluble in methanol; insoluble in water.

This is a diphenylguanidine salt of sirius light blue G (CI 34200, Direct blue 78). The anionic moiety of luxol fast blue G is thus a tetrasulfonated trisazo dye with a large, not entirely planar, aromatic system.

Unfortunately the biomedical literature commonly refers to 'luxol fast blue' with no further identifying terms. The dye referred to may of course have been luxol fast blue G. However, it may instead have been luxol fast blue MB or luxol fast blue ARN; qv. These latter two dyes, although chemically distinct, are also diarylguanidine salts of anionic dyes with large conjugated systems. In view of these ambiguities of nomenclature it is fortunate that the three dyes appear to behave similarly as staining reagents. The staining mechanism, which results in the deposition of the dye anion, has been studied by Clasen *et al.* (1973). These authors also made several 'luxol' dyes in the laboratory, by precipitation of diazo dyes with diphenylguanidine.

Applications: The principal application of luxol fast blue is for staining myelin, as indicated by current staining manuals (Bancroft and Gamble, 2002). Luxol fast blue is in addition occasionally used to provide selective staining of dense protein-rich granules such as those in gastric parietal cells (Shah and Miller, 1991) or cultured eosinophils (Khan *et al.*, 1997). By exploiting the marked solvent effects on staining selectivity seen with this dye, luxol blue G has also been used for selective staining of collagen and elastic fibers (Salthouse, 1965).

Stability: Myelin stained sections stored conventionally do not fade over a period of many years.

Luxol fast blue ARN (CI Solvent blue 37) is the diarylguanidine salt of an azo dye of undisclosed structure, but probably an azo dye analogous to sirius light blue. This dye has been used to stain myelin (Salthouse, 1962) and various other phospholipid-containing structures (Lycette *et al.*, 1970).

12

Diazonium salts and their reaction products with coupling agents

R.W. Horobin

Nature and nomenclature of diazonium salts

Aromatic diazonium salts contain the cationic substituent $-N_2^+$. Compounds containing two or three such groups are sometimes referred to in the histochemical literature as tetrazonium and hexazonium derivatives respectively; see below for further comments on the nomenclature of diazonium salts. The dyestuff literature often refers to stabilized diazonium salts as azoic diazo components or fast salts.

Synthesis of diazonium salts

They are usually obtained by the diazotization of homo- or hetero-aromatic primary amines, the reaction mechanism is shown in Figure 12.1. Diazonium salts are reactive electrophiles, and hence are unstable in aqueous solution, especially at extremes of pH. Occasionally only the amine is commercially available,

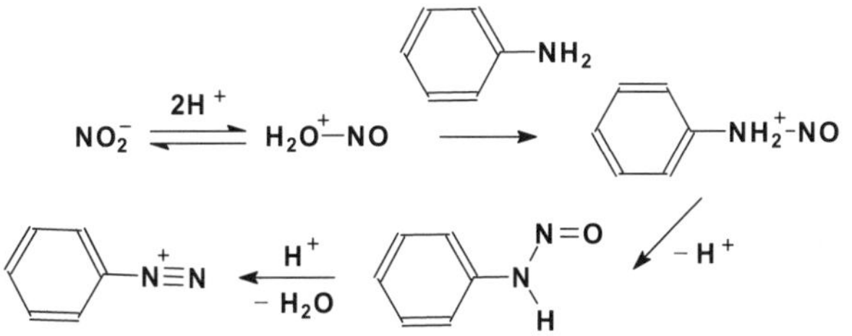

Figure 12.1 Mechanism of diazotization: formation of a diazonium cation (in the example, benzene diazonium) from a primary aromatic amine.

145

the diazonium salt being prepared in the laboratory immediately prior to use. Preparation of hexazonium pararosaniline is a case in point. In many cases, however, stabilized diazonium salts are commercially available, which contain various large anions, such as naphthalene sulfonates, tetrafluorborate, or the $ZnCl_4^{2-}$ species present in zinc chloride double salts.

Diazonium salts in histochemistry

They are used to form azo dye derivatives by azo coupling; for mechanisms of which see Chapter 10, Figure 10.2. This process typically involves reaction of diazonium salts with aromatic amines or naphthols. These are usually released from synthetic substrates by the action of tissue enzymes; for example, from naphthyl phosphates by phosphatases, or naphthylamines from naphthyl peptides by peptidases. Although Figure 10.2 illustrates synthesis of azo dyes for subsequent use in staining solutions, the reaction mechanism is the same as in other biomedical color-generating applications, such as:

(i)   Formation of azo pigments in enzyme histochemistry and immunocyto-chemistry, and azo-protein derivatives when staining protein side-groups; see Chapter 7.
(ii)   Demonstration of enzymes and other proteins in electrophoresis gels, and on thin-layer chromatograms.
(iii)   Identification and assay of drugs and biochemicals.

Diazotizable dyestuffs

Many of the diazonium salts used histochemically, and described below, are colorless compounds. However, dyes may carry diazotizable amino groups, and some coloured diazonium salts are used histochemically:

(i)   Azo dyes, such as fast black K, fast dark blue R, and fast garnet GBC, described in this chapter.
(ii)   Aminotriarylmethane dyes, notably new fuchsine and pararosaniline, described in Chapter 14, see also Figure 12.2.
(iii)   Azine dyes such as safranine O that have diazotizable amine groups.

Rate of azo coupling

Depending on stereochemistry and patterns of substitution, reaction rates of diazonium salts with coupling agents vary markedly; as has been briefly reviewed by Zollinger (1991, p. 110). When azo coupling is used to trap naphtholic reaction intermediates in enzyme histochemistry, fast reactions favor retention of more soluble and nonsubstantive intermediates near enzymic sites (Defendi and Pearse, 1955). Consequently some diazonium salts (e.g. fast garnet GBC and fast red TR) recommended for routine enzyme histochemical staining are rapid couplers. Other recommended diazonium salts (e.g. fast blue B and

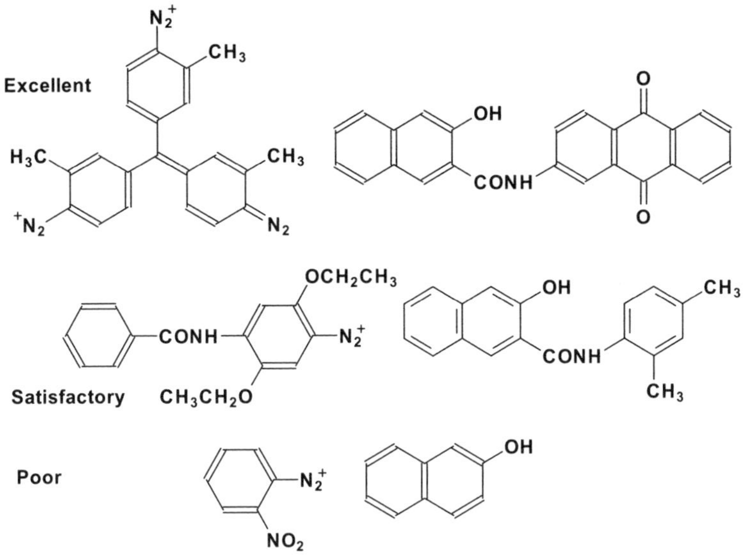

Figure 12.2 Structures of diazonium cations and naphthols giving rise to high (hexa-zonium new fuchsine and naphthylaminoanthraquinone), satisfactory (fast blue BB and naphthol AS-MX), or low (fast orange GR and β-naphthol affinity) histochemical end-products.

fast blue BB) are, however, slow couplers. This reflects the fact that localization of the reaction products near the enzymic site is also strongly favored by binding of the azo dye reaction product to tissue proteins.

Influence of diazonium moiety on azo dye affinity

All final reaction products in azo dye histochemistry are lipophilic, and precip-itate from aqueous incubation solutions. Whilst affinity of azo dyes for tissue proteins will be favored by dye lipophilicity, affinity is also enhanced in dyes with large aromatic systems (Horobin and Bennion, 1973). The latter is clearly the dominant factor, since there is no marked correlation between diazonium salt lipophilicity and azo dye localization, but the influence of the size of the aromatic system of the diazonium salt is marked. Thus the slowly coupling diazonium salts have larger aromatic systems than rapidly coupling salts.

More dramatic examples are provided by the salts derived from new fuchsine and pararosaniline, described in Chapter 14. Azo dye reaction products arising with these reagents are of sufficiently high affinity to resist extraction into dehy-drating and clearing solvents. This is dramatically superior to azo dyes resulting from the diazonium salts discussed in the present chapter. In keeping with this, contributions of new fuchsine and pararosaniline moieties to the sizes of the aromatic systems of azo dye reaction products are considerably greater than for the fast salts discussed in this chapter. For a graphic illustration of such differ-ences, see Figure 12.2.

Influence of naphthoic moiety on azo dye affinity

Naphthoic moieties also influence affinity and localization of azo dye final reaction products in enzyme histochemistry and enzyme-labeled immunocytochemistry and *in situ* hybridization. Staining localization improves on changing from a simple naphthol derivative to amidophenyl substituted AS naphthol derivatives. This parallels increases in sizes of aromatic systems; again lipophilicity does not vary in any systematic way. Note that a naphthol investigated by Burstone (1960), which gave rise to azo dye reaction product with 'very sharp localization', had an aromatic system half as large again as the AS naphthols. The relative sizes of the aromatic systems of such reagents are illustrated in Figure 12.2.

Finding the diazonium salt of interest

In this chapter, diazonium salts are arranged in increasing formula weight of the parent amines. Formula weights of stabilized diazonium salts fail to indicate relative sizes of the reactive species, due to use of large and varied counter ions such as are present in the zinc chloride double salts. Many of the stabilized diazonium salts have been sold under a number of synonyms, which originally gave rise to some confusion, with different authors using the same diazonium salts under different names. To clarify this, a listing of recommended names was set out (Lillie, 1959), and contemporary authors and vendors largely use the names recommended in 1959, as does this chapter.

Fast red B

CI 37125, CI Azoic diazo 5
Synonyms: 2-methoxy-4-nitroaniline
Amine: $C_7H_8N_2O_3$
FW amine 168, FW naphthalene-1,5-disulfonate diazonium salt 469, FW tetra-fluoroborate diazonium salt 268

Absorption maxima: for the diazonium salt 375 nm and 266 nm in water.
Solubilities: the naphthalene-disulfonate-stabilized diazonium salt 8% in water, and the tetrafluoroborate 6%; solubilities in ethanol are 2% and 0.1% respectively.

The amine, illustrated, is lipophilic with a small aromatic system; whilst the diazonium cation is hydrophilic. The diazonium salt has been sold stabilized as the 1,5-naphthalene disulfonate, and also by the tetrafluoroborate anion. The diazonium salt is a fast-reacting coupling agent with naphthols (Nachlas *et al.*, 1959).

Applications: Fast red B has been stated to be one of the best diazonium salt stains for enterochromaffin granules (Lillie *et al.*, 1961). The diazonium salt was also investigated for use in enzyme histochemical procedures for esterase and phosphatases, but was not highly thought of. Other laboratory applications include the identification of exo-glucanases in polyacrylamide gels following electrophoretic separation (Vargic and Mrsa, 1994). This fast salt is used industrially as a diazo coupling agent in textile dyeing.

Purity of commercial batches: The amine has been sold diluted to 20%. Samples of the naphthalene-1,5-disulfonate diazonium salt have been available containing 50% reagent, and of the tetrafluoroborate, 95%. Paper chromatographic methods are available for both amine and diazonium salt (Gasparic, 1966).

Stability: Stabilized diazonium salts may be stored satisfactorily in dark, sealed containers.

Hazards: The amine is harmful if swallowed, inhaled or absorbed through the skin.

Sulfanilic acid

Synonyms: 4-aminobenzensulfonic acid
$C_6H_7NO_3S$,FW 173

Absorption maxima: 249 nm plus a shoulder at 290 nm in water; 249 nm in 1M aqueous sodium hydroxide; and 250 nm in ethanol.
pK value: 3.2
Solubilities: ca. 1% in water; practically insoluble in ethanol.

The structural formula corresponds to the free acid and, following common usage, shows a nonionic species. In reality the species with this composition will be a zwitterion, in which both amine and sulfonic acid are ionized. A sodium salt is also commercially available, in which the acid is present as a sulfonate group but the amine is not ionized. In aqueous solution all likely species are hydrophilic, as is the derived diazonium salt, which is made in the laboratory as needed and is known as Griess' reagent.

Applications: Ehrlich (1883) was the first to use the diazonium salt of sulfanilic acid as a biological stain, namely for bilirubin. There are a number of recent staining applications of sulfanilic acid. For instance, to localize cell wall phenolics in maize (Bergvinson *et al.*, 1994), and to prepare a substrate for the enzyme histochemical demonstration of leucine aminopeptidase (Hwang *et al.*, 1995).

There are many current nonstaining biomedical applications of diazotized sulfanilic acid, for instance to determine ascorbic acid in pharmaceutical products and fruit juices (Srividya and Balasubramanian, 1996), and to assess nitrite utilization by *Nitrobacter* in hydrocarbon fuels (Okpokwasili and Odokuma, 1996).

Sulfanilic acid is used industrially as an intermediate in the manufacture of various dyes and pharmaceuticals.

Purity of commercial batches: Analytical grade material is 99% pure. A thin-layer chromatographic method is available (Jones, 1973).

Fast red TR

CI 37085, CI Azoic diazo 11
C_7H_8NCl
FW amine 142, FW zinc chloride diazonium double salt 257

Absorption maximum: of the diazonium salt 285 nm in water.
Solubilities: the zinc chloride double salt, 5% in water and 1% in ethanol.
pK value: 3.9

The amine, illustrated, is lipophilic with a small aromatic system; the diazonium cation is small and hydrophilic. The diazonium salt has been available commercially both as the zinc chloride double salt, and the 1,5-naphthalene disulfonate. The diazonium salt reacts rapidly with naphthols.

Applications: Fast red TR has been recommended for routine enzyme histochemical demonstration of alkaline phosphatase (Bancroft and Cook, 1994) in histopathological material. Such demonstrations can also be achieved in living cells (Espada *et al.*, 1997).

The fast salt is also widely applied to demonstrate alkaline phosphatase used as an enzyme label. In immunocytochemistry a recent application is a microwave immunoenzymatic study of leukemic marrow cells (Ebener and Wehner, 1993). Another frequent application of this type is for visualizing alkaline phosphatase labels used for fluorescent *in situ* hybridization procedures; including detection of human papilloma virus DNA in genital lesions (Lizard *et al.*, 1997).

The diazonium salt has also been used for *in vivo* demonstration of acid phosphatase in the rhizosphere of soil-grown plants, using root prints (Dinkelaker and Marschner, 1992); and for detecting esterase activities of enterococcal bacteria in culture gels (Tsakalidou *et al.*, 1993). This fast salt is used industrially as a diazo coupling agent in textile dyeing.

Purity of commercial batches: A paper chromatographic analytical method is available for both amine and diazonium salt (Gasparic, 1966).

Stability: The stabilized diazonium salt may be stored in sealed, dark bottles.

Hazards: The amine is carcinogenic and acutely harmful if absorbed.

Fast garnet GBC

CI 37210, CI Azoic diazo 4
Synonyms: the amine is CI 11160, Solvent yellow 3.
Amine: $C_{14}H_{15}N_3$
FW amine 225, FW hydrogen sulfate diazonium salt 334

Absorption maxima: of the diazonium salt 360 nm in water, of the amine 328 nm in 50% aqueous ethanol containing 1% 1M hydrochloric acid.
Solubilities: of the hydrogen sulfate diazonium salt 8% in water, and 0.4% in ethanol; of the amine 0.1% in water, and 3% in ethanol.

The amine, illustrated, is lipophilic with an aromatic system of moderate size; and the diazonium salt is weakly hydrophilic. This latter has been sold stabilized as the hydrogen sulfate. The coupling rate of the diazonium salt with naphthols is rapid (Nachlas *et al.*, 1959).
Applications: Fast garnet GBC salt is recommended for routine enzyme histochemical staining of beta-glucuronidase and N-acetyl-beta-glucuronidase (Chayen and Bitensky, 1991) and acid phosphatase (Vacca, 1985). The diazonium salt has also been used by Gossrau (1991) for the demonstration of lysosomal dipeptidyl peptidase I. Fast garnet GBC salt has also be used for staining enterochromaffin (Gomori, 1954).
   Other applications in biology and medicine include various enzyme assays, for example, for the determination of total esterase in yeast cells (Bardi *et al.*, 1993), and for the staining of proteins with esterase A activity following their separation by electrophoresis on polyacrylamide gels (Pond *et al.*, 1996). This fast salt is used industrially as a diazo coupling agent in textile dyeing.
Purity of commercial batches: Dye content of the diazonium salt can be 75%, for which species a paper chromatographic method is available (Gasparic, 1966).
Stability: The hydrogen sulfate diazonium salt is stable stored in sealed, dark containers.
Hazards: The amine may be carcinogenic and is acutely harmful if absorbed.

Fast blue B

CI 37235, CI Azoic diazo 48
Synonyms: diazonium blue, 3,3′-dimethoxy-4,4′-biphenylbisdiazonium chloride, tetraazotized-*o*-dianisidine. Fast blue B salt should not be confused with fast blue BB or with the polyene dye fast blue.
Amine: $C_{14}H_{16}N_2O_2$
FW amine 244, FW zinc chloride diazonium double salt 475

Absorption maxima: of diazonium salt 371 nm and 307 nm in water
Solubilities: of zinc chloride double salt 7% in water, and 0.3% in ethanol.

The lipophilic amine, with an aromatic system of moderate size, is illustrated; the diazonium cation is hydrophilic. The diazonium salt, which carries two reactive groups, is sold stabilized as a zinc double salt. The coupling rate with naphthols is slow (Nachlas *et al.*, 1959). To obtain highly substantive final stains, a copper complex of the azo dye reaction product is sometimes produced, by treatment with copper sulfate.

Applications: The diazonium salt been used to stain enterochromaffin cells directly (Lillie *et al.*, 1961). Aromatic side chains of tissue proteins also couple with fast blue B salt; a naphthol can then be coupled to the unbound diazo group to generate stronger color. A thorough discussion of this latter 'coupled tetrazonium' method was provided by Danielli (1953). More recently, a fluorescent modification of this latter procedure has been devised (Espada and Stockert, 1994).

Fast blue B is also used for the indirect staining of tissue substrates, for instance of carbonyls following their reaction with naphthoic acid hydrazide. This approach is widely applied to demonstrate and assess damage due to oxygen free radicals and related species (Anderson *et al.*, 1997). Such indirect stains involving fast blue B have also been used to stain DNA, and this method can be used for cytochemical quantitation (Nohammer, 1990). The strategy has also been applied to demonstrating protein disulfides following their reaction with dihydroxy dinaphthyl disulfide, for a recent example of which see Nohammer and Desoye (1997).

Staining manuals recommend fast blue B for the demonstration of esterase (Bancroft and Cook, 1994; Chayen and Bitensky, 1991) and aminopeptidase (Chayen and Bitensky, 1991). A recent application of this type is the demonstration of proteases in smears of cells from the gingival cervicular fluid (Kennett *et al.*, 1997). Fast blue B has also been used as a visualizing agent of alkaline phosphatase in flow cytometry (Ross *et al.*, 1990) and for enzyme-labeled *in situ* hybridization probes, for example, for *Drosophila* embryo whole mounts (Hauptmann and Gerster, 1996).

Other applications in biology and medicine include use as a color reagent for demonstrating diethylstilbestrol on thin-layer chromatograms (Medina and Nagdy, 1993), and of esterases in polyacrylamide gels after separation by electrophoresis (Inoue *et al.*, 1997). The diazonium salt has also been widely used, under the synonym diazonium blue, as an aid to the identification of fungal species (Hutchison, 1991). This fast salt is used industrially as a diazo coupling agent in textile dyeing.

Purity of commercial batches: The amine is available at over 90% reagent content, the stabilized diazonium salt at 30%. A paper chromatographic analytical method is available for the diazonium salt (Gasparic, 1966).
Stability: The zinc chloride double salt is stable stored in dark, sealed containers.
Hazards: The amine is carcinogenic and acutely harmful if swallowed, inhaled or absorbed through the skin.

Fast red violet LB

Synonyms: 4'-amino-2'-chloro-5'-methylbenzanilide
Amine: $C_{14}H_{13}ClN_2O$
FW amine 261, FW zinc chloride diazonium double salt 376

Absorption maxima: of the diazonium salt 342 nm in water, of the amine 293 nm in methanol.
Solubilities: of the zinc chloride double salt 2% in water, and 0.2% in ethanol; of the amine 0.1% in water and 0.3% in ethanol.

The amine, illustrated, is lipophilic with an aromatic system of moderate size; the diazonium cation is hydrophilic. The diazonium salt is available commercially, stabilized as the zinc chloride double salt. The coupling rate of the diazonium salt with naphthols is rapid (Burstone, 1957b).
Applications: Fast red violet LB salt was recommended for use in enzyme histochemical methods for acid and alkaline phosphatase (Burstone 1957a, 1958a,b). Such procedures have subsequently been proposed for routine hematological (Tanaka *et al.*, 1984) and histological (Vacca, 1985) use. A flow cytometric procedure for quantitating alkaline phosphatase in bone marrow cells is also described (Kamalia *et al.*, 1992).
Purity of commercial batches: Both amine and diazonium salt are available with reagent contents of about 95%.
Stability: The zinc chloride double salt is stable stored in dark, sealed containers.

Fast blue RR

CI 37155, CI Azoic diazo 24
Amine: $C_{15}H_{16}N_2O_3$
FW amine 272, FW zinc chloride diazonium double salt 388

Absorption maxima: diazonium salt 296 nm in water, amine 319 nm in methanol.
Solubilities: zinc chloride double salt 4% in water and 0.4% in ethanol; amine 0.7% in water and 0.3% in ethanol.

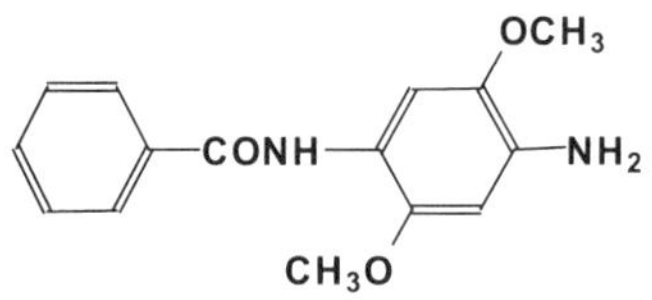

The amine, illustrated, is lipophilic with an aromatic system of moderate size; the diazonium cation is hydrophilic. A zinc chloride stabilized diazonium salt is commercially available.

Applications: The diazonium salt has been used to stain enterochromaffin granules (Lillie *et al.*, 1961). It is recommended for routine enzyme histochemical demonstration of alkaline phosphatase, beta-glucuronidase, and esterase (Bancroft and Cook, 1994; Chayen and Bitensky, 1991). Recent research applications using fast blue RR salt have included investigations of the stereoselectivity of esterases in normal and pathological situations (Yamazaki *et al.*, 1998).

Purity of commercial batches: The diazonium salt is available at a purity of 85%. A paper chromatographic analytic procedure is available for this salt (Gasparic, 1966).

Stability: The dry zinc chloride double salt is stable stored in dark, sealed containers.

Fast blue BB

CI 37175, CI Azoic diazo 20

Synonyms: fast blue 2B. Do not confuse with fast blue B (CI 37235).

Amine: $C_{17}H_{20}N_2O_3$

FW amine 300, FW zinc chloride diazonium double salt 416

Absorption maxima: the diazonium salt at 395 nm in water, the amine at 318 nm in ethanol.

Solubilities: of the zinc chloride double salt 4% in water, and 0.6% in ethanol; the amine 0.6% in water and 2% in ethanol.

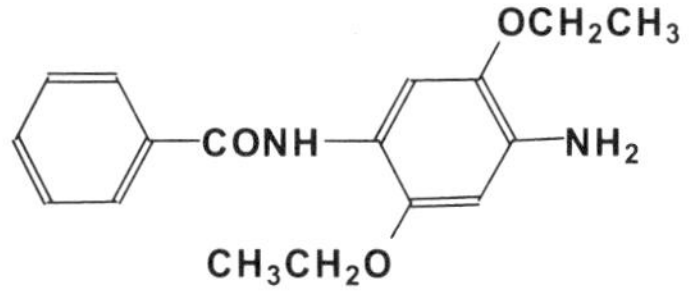

The amine, illustrated, is lipophilic with an aromatic system of moderate size; whereas the diazonium cation is weakly hydrophilic. A diazonium salt is available commercially stabilized as the zinc chloride double salt. Its coupling rate with naphthols is slow.

Applications: Fast blue BB salt has been used to stain enterochromaffin (Lillie *et al.*, 1961). It was used in enzyme histochemistry by Gomori (1954), and remains a reagent of choice for histochemical demonstration of beta-glucuronidase activity (Chayen and Bitensky, 1991) and various peptidases and proteases (Van Noorden and Frederiks, 1992). Other recent applications of fast blue BB salt in enzyme histochemistry include demonstration of serine esterase in blood smears and histological sections (Wagner *et al.*, 1991), and of alkaline phosphatase in glycol methacrylate resin sections (Gerrits and Horobin, 1996).

Fast blue BB salt is used routinely to demonstrate alkaline phosphatase-labeled antibodies (Bancroft and Cook, 1994). Recent applications of this type include its use in a microwave-accelerated procedure for identifying white blood cells (Ebener and Wehner, 1993).

Analytical applications include identification of compounds following their separation by chromatography or electrophoresis, for example, of exo-beta-glucanase on acrylamide gels (Vargic and Mrsa, 1994). The diazonium salt of fast blue BB is used in textile dyeing.

Purity of commercial batches: Suppliers provide the amine with 98% purity and its diazonium salt with zinc chloride at 72%. A paper chromatographic analytical method is available for the diazonium salt (Gasparic, 1966).

Stability: The dry zinc chloride double salt is stable stored in dark, sealed containers. Gossner (1958) noted decomposition of this diazonium salt in aqueous solutions when used in enzyme histochemical procedures.

Fast black K

CI 37190, CI Azoic diazo 38
Amine: $C_{14}H_{14}N_4O_4$
FW amine 302, FW diazonium zinc chloride double salt 418

Absorption maximum: diazonium salt 457 nm in water.
Solubilities: of the diazonium salt 6% in water, and 0.3% in ethanol.

The aromatic system of this lipophilic amino-substituted monoazo dye, illustrated, is of moderate size. The cationic diazonium derivative is hydrophilic. A diazonium salt stabilized as the zinc chloride double salt is available commercially. The rate of coupling of the salt to naphthols is slow.

Applications: Fast black K salt has been used in enzyme histochemistry (Burstone, 1958b), to stain enterochromaffin cells (Lillie *et al.*, 1961), and as an indirect stain for periodate-generated tissue aldehyde groups (Lillie *et al.*, 1961). The latter method was subsequently integrated with lectin histochemistry for investigating neutral complex carbohydrates (Yamada, 1978).

Biochemical applications of fast black K salt include its use as a visualization agent for amphetamines (Munro and White, 1995) and cyanobacterial toxins following their separation by thin-layer chromatography (Pelander *et al.*, 1996). The diazonium salt of fast black K is used industrially in textile dyeing.

Purity of commercial batches: The zinc chloride double salt of the diazotized amine is available at up to 30% purity. A paper chromatographic analytical method is available (Kiel and Kuyper, 1964).

Stability: The dry zinc chloride double salt is stable stored in dark, sealed containers.

Fast dark blue R

CI 37195, CI Azoic diazo 51
Amine: $C_{14}H_{12}N_4O_4Cl_2$
FW amine 371, FW zinc chloride diazonium double salt 487

This is a monoazo dye. The aromatic system of the lipophilic amino derivative, illustrated, is of moderate size; the diazonium cation is weakly lipophilic. A diazonium salt stabilized as the zinc chloride double salt is available. The coupling rate of the diazonium salt to naphthols is slow.

Applications: Fast dark blue R salt has been used to stain enterochromaffin granules (Lillie *et al.*, 1961) and is recommended for routine use in the enzyme histochemical demonstration of acid phosphatase, beta-glucuronidase and N-acetyl-beta-glucuronidase (Bancroft and Cook, 1994; Chayen and Bitensky, 1991).

Purity of commercial batches: A paper chromatographic procedure is available for the diazonium salt (Kiel and Kuyper, 1964).

Stability: Of dry stabilized diazonium salt is good stored in dark, sealed containers; aqueous solutions are stable.

Hazards: The amine is carcinogenic and acutely harmful if swallowed, inhaled or absorbed through the skin.

13

Tetrazolium salts and formazans

R.W. Horobin

BACKGROUND INFORMATION

For a more extensive account of this field, which provides an introduction to a wider literature, Seidler's (1991) monograph is recommended. Note that in the entries which follow these introductory sections, monotetrazolium compounds are placed prior to bistetrazolium salts. The only formazan applied as such is discussed following the entries for tetrazolium salts.

Structures of tetrazolium salts

Tetrazolium salts discussed here are of the 2-H type, and contain one or more five-membered heterocyclic rings, each of which carries a delocalized positive charge. Such a tetrazolium compound may be represented by a pair of Kekulé forms, as seen below.

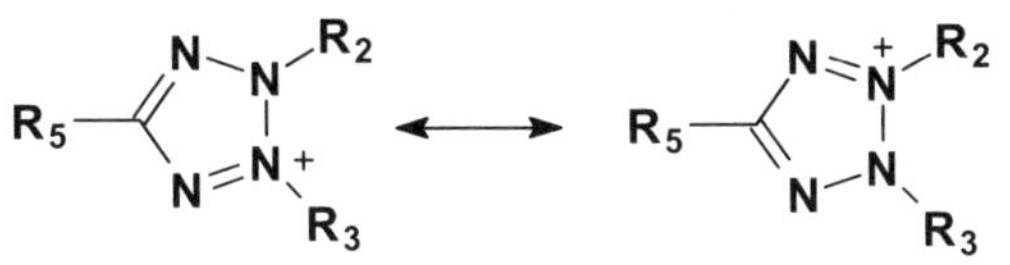

The substituents at N_2 and N_3 of the tetrazolium ring are typically aromatic or heterocyclic; the C_5 substituent may be aromatic or, in the case of cyanotolyl tetrazolium, cyano. Substituents R_2, R_3 and R_5 are usually nonionic, so the overall charge of the tetrazolium compound is positive. Consequently most tetrazolium salts are sold as chlorides or bromides.

Bistetrazolium salts, containing two tetrazolium rings, are also widely applied in the biomedical sciences. All bistetrazolium salts in routine use have the two tetrazolium groups linked via the N_2 and N_3 ring nitrogens, the bridging group usually being a biphenyl (e.g. neotetrazolium).

Reduction of tetrazolium salts

Tetrazolium salts can be reduced to formazans by a variety of mild reductants, including those of enzyme-catalyzed biological systems. The reaction involves uptake of two electrons and a proton by the tetrazolium cation. This proceeds via an unstable tetrazolinyl intermediate, as illustrated below.

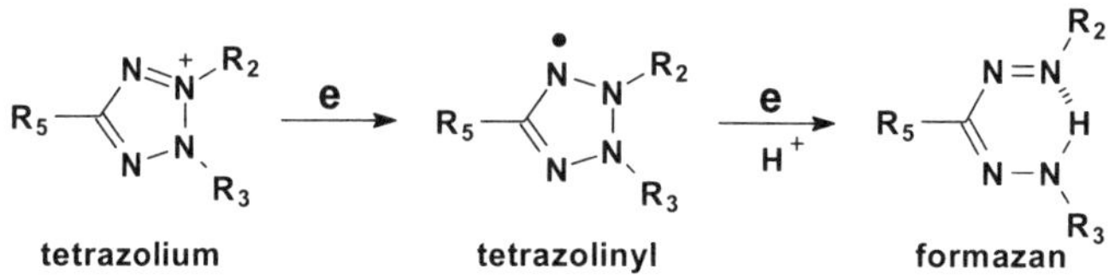

tetrazolium   tetrazolinyl   formazan

In the case of bistetrazolium salts, it is possible for one tetrazolium group to be reduced to a formazan leaving the other in the tetrazolium form. Such stable half-formazans can arise histochemically (Altman, 1974; Altman and Butcher, 1973), resulting in a single, pure tetrazolium salt producing two colors (mono-formazan and diformazan) in a tissue section.

Structures of formazans and of their metal complexes

Formazans generated and used in biology and medicine usually have the *trans–syn* structure shown above. However, one histochemically produced formazan, that derived from yellow tetrazolium (Jones, 1969), has a *trans–anti* structure.

Formazans with suitable chelating groups can give rise to metal coordination complexes under histochemical conditions. An example is the MTT formazan, which gives insoluble complexes with cobalt, copper and nickel ions. The structure of the 2:1 cobalt complex proposed by Lojda (1965) is shown below. As pointed out by Seidler (1991), it seems highly likely that nitrogen and/or sulfur heteroatoms in the thiazole rings are involved as additional ligands. In any event, such metal complexes involve a doubling of the number of aromatic rings compared to the uncomplexed formazan.

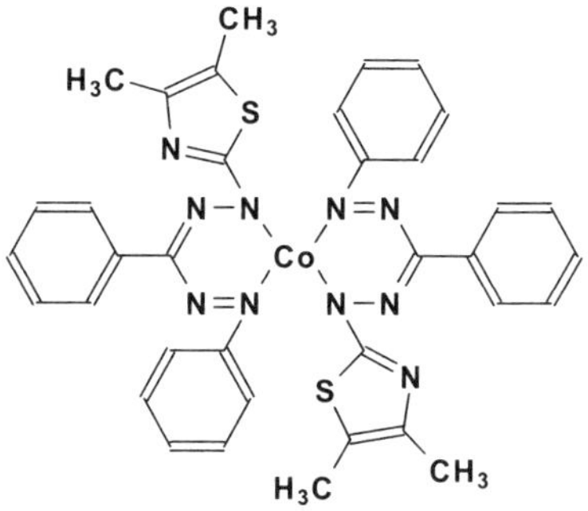

Physicochemical properties significant for staining

To facilitate histochemical application, and to localize enzyme activities, tetrazolium salts should be water soluble, and derived formazans should be insoluble in water and routine processing solvents. The fact that reduction of tetrazolium salts to formazans involves loss of a positive charge, and hence a diminution of hydrophilicity, is therefore advantageous. Localization of staining to the vicinity of the enzymic sites is however largely controlled by generation of formazans with large conjugated systems, because this property enhances binding to cellular proteins (Horobin and Bennion, 1973). Such formazans are obtained histochemically in several ways. Obviously the bistetrazolium salts give an immediate doubling of the conjugated system, albeit at the cost of increased lipophilicity. However, use of dinitro- or trinitro-substituted mono-tetrazolium salts increases conjugation without increasing lipophilicity, and has a similar beneficial effect on localization (Seidler, 1980). Finally, as noted above, the use of suitable metal-chelating monoformazans also doubles the amount of aromaticity, and hence increases substantivity and accuracy of localization.

Tetrazolium salts used histochemically or biochemically must be reducible by the mild biological systems as far as, but no further than, the formazan. Well localized histochemical enzyme demonstrations also need reduction to occur rapidly. In practice, reducibility is a complex property, depending on such variables as pH and the presence of electron transfer mediators, and not merely on the redox potential of the tetrazolium salt. In terms of reducibility, the histochemically applied tetrazolium salts fall into two groups. In the first are the less easily reduced compounds, namely blue tetrazolium, neotetrazolium, tetrazolium violet and triphenyl tetrazolium. The second group, which reflects the beneficial effects of nitro substituents, are the more easily reduced salts: BSPT, MTT, nitro blue tetrazolium and tetranitro blue tetrazolium. Differences within the two groups are small.

Purity of tetrazolium salts

Due to the chemistry of their syntheses, bistetrazolium salts are routinely contaminated with monotetrazolium salts. Moreover the bistetrazolium salts can give rise to half-formazans, as well as formazans. Both types of impurity can result in distinct zones of differing color on tissue sections.

MONOTETRAZOLIUM SALTS

Cyanotolyl tetrazolium

Synonyms: cyanotolyltetrazolium chloride, CTC
$C_{16}H_{14}ClN_5$, FW 312

Absorption maximum: of the CTC formazan is 450 nm in ethanol; with the CTC formazan's emission maximum being 602 nm (i.e. in the red) when excited at 380–420 nm.

Solubilities: The tetrazolium chloride is soluble in aqueous media at pH 7.5.

This monotetrazolium salt, illustrated, has a small, weakly lipophilic cation. Cyanotolyl tetrazolium is routinely sold is the chloride, although use of the bromide has been reported. The nonionic CTC formazan derivative, resulting from reduction of the tetrazolium, is more strongly lipophilic.

Applications: Cyanotolyl tetrazolium is used as an imaging agent in microbiology, for instance, to visualize actively respiring bacteria using fluorescent microscopy, both in cultures (Rodriguez *et al.*, 1992) and for bacteria within solid matrices. The latter include soil (Yu *et al.*, 1995), biofilms from bioreactors (Villaverde and Fernandez, 1997), and peat (Lloyd *et al.*, 1998). Cyanotolyl tetrazolium is also routinely used to distinguish viable from nonviable bacteria (Andreasen and Nielsen, 1997) and to demonstrate viable bacteria in both culturable and nonculturable states (Cappelier *et al.*, 1997). This tetrazolium salt has also been used for the quantitative histochemical estimation of mitochrondrial activity in cultured tumor cells (Stellmach and Severin, 1987).

Essentially the same chemical and biochemical reactions are exploited in the use of cyanotolyl tetrazolium to assess microbial respiratory activity by flow cytometry (del Giorgio *et al.*, 1997). Earlier flow cytometric applications of cyanotolyl tetrazolium have been reviewed (Davey and Kell, 1996).

Purity of commercial batches: A thin-layer chromatographic method is available for CTC formazan (Stellmach, 1984).

Triphenyl tetrazolium

Synonyms: TT, TTC
$C_{19}H_{15}ClN_4$, FW 335

Absorption maxima: of the tetrazolium salt is 247 nm in water; of the TTC formazan is 483 nm in ethanol.

Solubilities: The tetrazolium salt is soluble in both water and ethanol; the formazan is insoluble in water.

This monotetrazolium salt, illustrated, possesses a small, lipophilic cation. Reduction of triphenyl tetrazolium produces the nonionic TTC formazan which is more strongly lipophilic.

Due to the large crystal size of the TTC formazan, it is unsuitable for histological work. However, triphenyl tetrazolium has a wide range of applications in modern biomedical research, as sketched below.

Applications: Triphenyl tetrazolium is widely used for viability assays, for example of cotton cultivars in drought tolerance tests (De Ronde and Van der Mescht, 1997); of bacteria (Muller *et al.*, 1997); and of cardiac muscle following ischemia (Kimura *et al.*, 1998). Other uses include enumeration of bacteria (Parrington and Sharpe, 1998); and to aid identification of microbial species such as mycoplasma (Bradbury *et al.*, 1993).

The salt is also used to stain enzymes separated on electrophoretic gels, for example, beta-1,3-glucanase (Shimoni *et al.*, 1996). In toxicology, triphenyl tetrazolium has been applied to assess the reducing capacity of roots of manganese deficient trees (Hiltbrunner and Fluckiger, 1996).

Purity of commercial batches: Dye content of commercial samples can be as high as 98%; a thin-layer chromatographic method is available (Altman, 1974).

Stability: The dry tetrazolium salt is light sensitive.

Tetrazolium violet

Synonyms: TV, TZV

$C_{23}H_{17}ClN_4$, FW 385

Absorption maxima: of the tetrazolium salt is 244 nm in 10% aqueous methanol; of TV formazan is 512 nm in ethanol-chloroform, 510 nm in dimethylformamide.

Solubilities: 3% in water; 2% in ethanol.

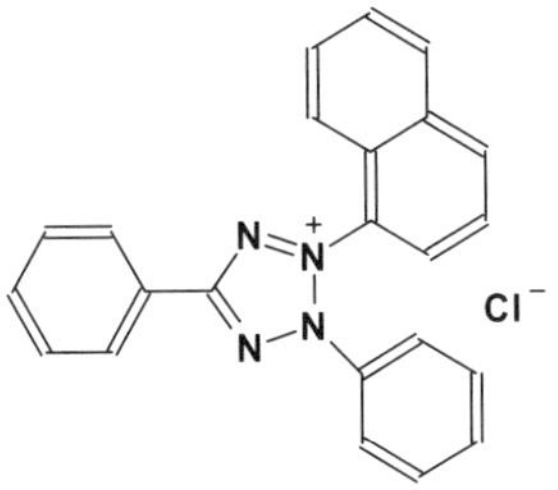

This monotetrazolium salt, illustrated, has a lipophilic cation of small size. Reduction of tetrazolium violet produces a violet, nonionic formazan which is more strongly lipophilic.

Applications: Due to a slow reduction rate and to the large crystal size of its formazan, tetrazolium violet is not used histochemically. However, the salt does have a variety of other laboratory applications (e.g. in microbiology, tetrazolium violet is used as a redox indicator, and to differentiate microbial communities in soil) (Gorlenko and Kozheniv, 1994), and for identify metabolically active microbial populations in cosmetic formulations (Lenczewski *et al.*, 1996).

Purity of commercial batches: Some samples contain only one major component. There is a thin-layer chromatographic method for the TV formazan (Altman, 1974)

Methylthiazolyldiphenyl tetrazolium (MTT)

Synonyms: MTT, thiazolyl blue tetrazolium
$C_{18}H_{16}BrN_5S$, FW 414

Absorption maxima: of the tetrazolium salt is 378 nm in water, and 375 nm in methanol; of the MTT formazan is 560 nm in hexane, and 505 nm in dimethylformamide.

Metal complexation behavior: The MTT formazan forms low-solubility chelates with a variety of transition metals, notably cobalt and nickel.

Solubilities: of the tetrazolium salt is 1% in water, 2% in ethanol; MTT is also soluble in dimethylsulfoxide.

   MTT is a monotetrazolium salt, illustrated, comprising a lipophilic cation of moderate size, sold commercially as the bromide. Reduction produces the lipophilic MTT formazan, convertible to insoluble metal chelates, such as the black, finely granular cobalt derivative.

Applications: MTT is recommended in staining manuals for routine histopathological demonstration of dehydrogenases (Bancroft and Cook, 1994; Bancroft and Gamble, 2002), and has current histochemical research application (Omar and Raoof, 1996).

   This tetrazolium salt has a wide range of other laboratory applications, and a few examples follow. In microbiology MTT is has been applied to assess bacterial activity and viability (Thom *et al.*, 1993), and to detect mycobacteria (Mshana *et al.*, 1998). Toxicologists have applied MTT to assess viability, such as for measurement of the numbers of viable cells in well-cultures (Higa *et al.*, 1998).

Purity of commercial batches: Samples can contain 98% of the major component. There is a thin-layer chromatographic method for MTT formazan (Altman, 1974).

Hazards: The tetrazolium salt is acutely harmful if swallowed, inhaled or absorbed through the skin.

BSPT

Synonyms: BPST, 2(2'-benzothiazolyl)-5-styryl-3-(4'-phthalhydrazidyl) tetrazolium chloride
$C_{24}H_{16}ClN_7O_2S$, FW 502

Absorption maxima: of the tetrazolium salt is 300 nm in 19% aqueous methanol; of the BSPT formazan is 485 nm in acidic dimethylformamide.
Metal complexation behavior: the BSPT formazan chelates various metal ions, such as cobalt, copper and nickel.
Solubilities: The tetrazolium salt is soluble in ethanol, and also in dimethylformamide to at least 1%.

BSPT is a lipophilic monotetrazolium salt, illustrated, of moderate size, available commercially as the chloride. The BSPT formazan, resulting from reduction, is more strongly lipophilic. The BSPT formazan, unlike the tetrazolium salt, rapidly reduces osmium tetroxide.
Applications: BSPT was originally prepared for the ultrastructural demonstration of dehydrogenases by Seligman's group (Kalina *et al.*, 1972), and remains the tetrazolium reagent most widely used for this purpose. Recently BSPT has been applied in both light (Wang *et al.*, 1995) and electron microscopy (Rothe *et al.*, 1998) for the demonstration of nitric oxide synthase (NADPH:tetrazolium oxidoreductase or NADPH diaphorase) in neurons.
Stability: As a dry powder the tetrazolium salt is stable and nonhydroscopic.

Iodonitro tetrazolium

Synonyms: INT, 4-iodonitro tetrazolium violet, 2-(4-iodophenyl)-3-(4-nitrophenyl)-5-phenyltetrazolium chloride
$C_{19}H_{13}ClIN_5O_2$, FW 506

Absorption maxima: of the tetrazolium salt is 248 nm in 10% aqueous methanol; of the INT formazan is 495 nm in tetrahydrofuran or ethanol-chloroform, and 460 nm in dimethylformamide.
Solubilities: The tetrazolium salt is soluble in dimethylsulfoxide and dimethylsulfoxide–water mixtures.

Iodonitro tetrazolium is a monotetrazolium salt, illustrated, with a lipophilic cation of moderate size, usually sold as the chloride. The INT formazan is more strongly lipophilic, and fat soluble.

Applications: Due to the INT formazan's large crystal size and lack of substantivity, histological application of iodonitro tetrazolium has been limited. However, it is used histochemically in microbiology for *in situ* microscopic observation of soil biota (Lussenhop and Fogel, 1993), demonstration of viable but nonculturable bacteria (Gribbon and Barer, 1995), and as a bacterial viability stain (Heidelberg *et al.*, 1997).

Iodonitro tetrazolium also has nonstaining applications, for example, enumeration of hydrocarbon degrading bacteria (Wrenn and Venosa, 1996), and biochemical assays of enzymes such as superoxide dismutase in erythrocytes and whole blood (Haghighi and Wei, 1998).

Purity of commercial batches: these can contain up to 95% tetrazolium salt, and up to 5% methanol. Some commercial samples contain a deiodinated contaminant. A thin-layer chromatographic procedure is available (Altman, 1974).

BISTETRAZOLIUM SALTS

Neotetrazolium

Synonyms: NT, ditetrazolium chloride
$C_{38}H_{28}Cl_2N_8$, FW 668

Absorption maxima: of the tetrazolium salt is 248 nm in 10% aqueous methanol; of the NT diformazan is 560 nm in ethanol and 564 nm in dimethylformamide; and of the half formazan is 500 nm in ethanol and 510–520 in tetrahydrofuran.

Solubilities: The tetrazolium salt is soluble in ethanol and dimethylformamide, which solutions can then be mixed into aqueous media. The NT formazan is insoluble in water, but slightly soluble in lipids.

Neotetrazolium is a bistetrazolium, illustrated, whose large, lipophilic cation has a substantial aromatic system. Neotetrazolium is available commercially as the chloride.

Applications: Neotetrazolium was the first bistetrazolium salt used histochemically for demonstration of dehydrogenases (Antopol *et al.*, 1948). Staining manuals still recommend neotetrazolium for use in routine enzyme histochemistry (Bancroft and Cook, 1994). Such methods have been applied to bacteria (Thom *et al.*, 1993), and to plant material (Skubatz *et al.*, 1993), and for the quantitative histochemical assay of glucose-6-phosphate dehydrogenase activity (Van Driel *et al.*, 1997).

Purity of commercial batches: can contain over 10% of a monotetrazolium salt impurity, resulting in formation of both blue and red formazans. Overall tetrazolium salt content is up to 79%. A thin-layer chromatographic method is available for the formazan (Jones, 1968).

Stability: This tetrazolium salt is light sensitive.

Blue tetrazolium

Synonyms: BT, bimethoxyneotetrazolium
$C_{40}H_{32}Cl_2N_8O_2$, FW 728

Absorption maxima: of the tetrazolium salt is 253 nm in methanol; of the BT diformazan 590 nm in dimethylformamide or tetrahydrofuran.

Solubilities: of the tetrazolium salt is 0.9% in water; and 0.7% in ethanol.

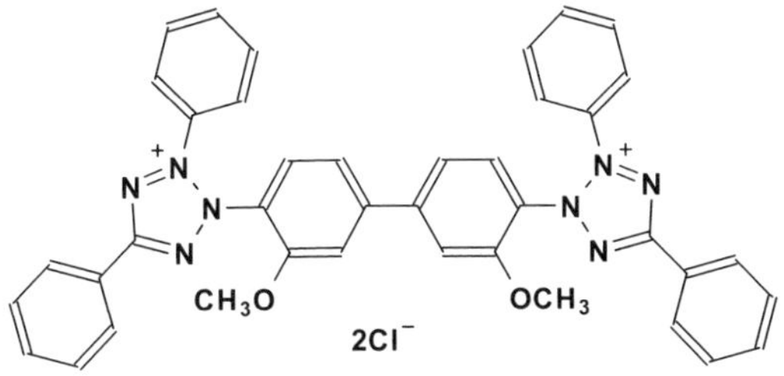

CH_3O OCH_3

$2Cl^-$

Blue tetrazolium is a bistetrazolium, illustrated, whose large, lipophilic cation has a substantial aromatic system. It is typically sold as the chloride.

Applications: This tetrazolium salt was used in histochemical demonstrations of succinate dehydrogenase (Rutenberg *et al.*, 1950) and NAD⁺-linked enzymes (Farber *et al.*, 1956). Such usage has been limited due to the salt's slow reduction rate. However, blue tetrazolium has found use in the biochemical assay of glucose oxidase (Liochev and Fridovich, 1995), as well as in the analysis of various corticosteroids (Ayllon *et al.*, 1994; Kleeman and Bailey, 1985).

Purity of commercial batches: A major monotetrazolium impurity is usually present, and reduction of such samples produces both violet–red monoformazan and blue bisformazan. A thin-layer chromatographic method is available for the BT formazan (Altman, 1974).

Nitro blue tetrazolium

Synonyms: NBT, Nitro BT, ditetrazolium chloride
$C_{40}H_{30}Cl_2N_{10}O_6$, FW 818

Absorption maxima: of the tetrazolium salt is 259 nm in 10% aqueous methanol; of the NBT formazan is 605 nm in dimethylformamide and 540 nm in tetrahydrofuran.
Solubilities: of the tetrazolium salt is 1% in water; and 0.5% in ethanol.

Nitro blue tetrazolium is a bistetrazolium, whose large lipophilic cation has a large aromatic system. Nitro blue tetrazolium is usually sold as the chloride.

Applications: Nitro blue tetrazolium is one of the most widely applied tetrazolium reagents, being recommended by staining manuals for most histochemically demonstrable oxidative enzyme systems, including enzymes used as labels in immunocytochemistry (Bancroft and Gamble, 2002; Van Noorden and Frederiks, 1992).

Applications include several *in vivo* procedures, for instance, demonstration of superoxide production in tracheal explants following exposure to cigarette smoke (Hobson *et al.*, 1991), the assay of respiratory activity of intact plant roots (Connolly and Berlyn, 1996), and the demonstration of cellular damage in livers exposed to carbon tetrachloride (Montosi *et al.*, 1998). Nitro blue tetrazolium has also been used to demonstrate nitric oxide synthase (Matsumoto *et al.*, 1998), and as a visualization reagent for enzyme labels used for *in situ* hybridization (Itoh *et al.*, 1997).

Nonenzymic staining applications of nitro blue tetrazolium include use in a myelin sheath stain (Kaatz *et al.*, 1992), and a procedure for intensifying diaminobenzidine staining, both in immunocytochemistry and in peroxidase tracing (Vaney, 1992). Nitro blue tetrazolium has also found use in a quantitative demonstration of lectins (Hoyer and Kirkeby, 1996).

Other laboratory applications include use as a staining reagent in flow cytometry (Fattorossi *et al.*, 1990), enumeration of bacterial colonies (Barnes *et al.*, 1996), and staining of enzymes separated on electrophoretic gels (Ye *et al.*, 1997).

Purity of commercial batches: Substantial amounts of a monotetrazolium contaminant are routinely present, which yields a red formazan. Overall dye content is up to 92%. Thin-layer chromatographic methods are available for the NBT formazan (Altman, 1974).

Tetranitro blue tetrazolium

Synonyms: TNBT
$C_{40}H_{28}Cl_2N_{12}O_{10}$, FW 908

Absorption maxima: of the tetrazolium salt is 279 nm in methanol; of TNBT formazan is 515 nm in tetrahydrofuran
Solubilities: of the tetrazolium salt 0.3% in water and ethanol; rather more soluble in dimethylformamide.

Tetranitro blue tetrazolium is a bistetrazolium whose large, lipophilic dication has a substantial aromatic system. The salt is sold as the chloride.
Applications: Tetranitro blue tetrazolium is widely recommended for routine dehydrogenase histochemistry (Bancroft and Gamble, 2002; Van Noorden and Frederiks, 1992). Specialized staining applications include use as a stain intensification reagent of diaminobenzidine reaction product in immunoperoxidase and horseradish peroxidase tracing procedures (Vaney, 1992), and for *in situ* hybridization (Asan and Kugler, 1995) and affinity histochemistry (Nichols *et al.*, 1998). The salt has also been used for the vital staining of superoxide radicals (Swei *et al.*, 1997). Nonmicroscopic applications include staining of dehydrogenases separated on electrophoretic gels (Berks *et al.*, 1993).
Purity of commercial batches: A monotetrazolium salt contaminant is routinely present; overall dye contents are up to 89%. There is a thin-layer chromatographic method available for the formazan (Altman, 1974).
Hazards: Acutely harmful if swallowed, inhaled or absorbed through the skin; an animal mutagen on chronic exposure.

FORMAZAN

Zincon

Synonyms: the monosodium salt is 2-[1-(2-hydroxy-5-sulfonatophenyl)-3-phenyl-5-formazano] benzoic acid, monosodium salt
Monosodium salt: $C_{20}H_{15}N_4NaO_6S$, FW 462

167

Absorption maxima: of the monosodium salt is 490 nm in methanol, with a shoulder at 523 nm.

Metal complexation behavior: Forms coordination complexes with a number of metal ions, notably copper and zinc.

Solubilities: of the monosodium salt 1% in water and 0.2% in ethanol.

Zincon is a sulfonated monoformazan, available commercially both as the free acid and its monosodium salt (illustrated). Electric charge, and hence lipophilicity, varies with pH but will be anionic except in a strongly acidic medium, on account of the sulfonate substituent. Multiple metal-binding sites are present. Resulting metal complexes vary in character, but can have zero overall charge and consequent low solubility.

Applications: Zincon has been used for histochemical demonstration of cobalt, copper, magnesium and zinc (McNary, 1960). Following its introduction as an indicator for chelatometric analysis (Kinnunen and Merikanto, 1955), its principal use has been as an analytical reagent in a variety of methodologies, particularly for copper and zinc. Recent applications of this type in biology and medicine include analysis of copper and zinc in children's hair (Liu *et al.*, 1995) and in plants (Oliveira *et al.*, 1996).

Purity of commercial batches: zincon content can be up to 95%.

14

Amino di- and triarylmethane dyes

R.W. Horobin

BACKGROUND INFORMATION

Issues of nomenclature

This and the following chapter use the traditional terms di- and triaryl*methane* dyes. More chemically precise wordings such as di- and triaryl*methine* dyes have been proposed, as noted by Zollinger (1991, p. 71). Other issues of nomenclature, which give rise to false expectations and confusion, are however of more everyday importance.

False expectations may arise with dyes such as acid and basic fuchsines; and alkali, aniline and methyl blues. These terms do not describe single compounds, but standard reference works, including previous editions of this book and the *Colour Index*, illustrate such dyes with single structural formulae. In fact their manufacture inevitably yields heterogeneous and ill-defined dye mixtures, as discussed below and in the relevant dye entries.

Confusion can arise for more trivial reasons. Vendors or end-users are sometimes imprecise with dye names. Thus the terms coomassie blue and Victoria blue are often used in the literature without specifying suffixes. Unfortunately two chemically different coomassie blues and four different Victoria blues have been used. Confusion may also involve misnaming. For instance, methyl green has been unavailable commercially for many years. The dye sold under this name is in fact ethyl green; this is sometimes noted on labels or in catalogs, but rarely if ever in the literature.

Structures of di- and triarylmethanes

Two of the mesomers of the simplest dye of this type, Michler's hydrol, can be drawn as shown below. The linking of the electron donor and acceptor groups via a conjugated system is apparent; as is the similarity with the polymethine dyes discussed in Chapter 23. In Michler's hydrol the donor groups are tertiary amino groups; in others amino di- and triarylmethanes, including many dyes discussed in the present chapter, the donor group may be a primary or secondary amino moiety. Dyes with hydroxy groups, or their anionic conjugated bases, as donor groups are described in Chapter 15.

In triarylmethane dyes, the hydrogen atom on the central methine group (–CH=) of Michler's hydrol is replaced by an aryl grouping. This may be an unsubstituted phenyl ring, as in malachite green; or the phenyl ring may carry an electron donor, as in crystal violet (–N(CH$_3$)$_2$). In other biological stains, such as the various Victoria blues, the third aryl group is naphthyl.

In diphenylmethane dyes used as biological stains, the two aryl rings lie in the same plane. With triarylmethanes steric hindrance makes this impossible. It is considered that symmetrical triphenylmethanes such as crystal violet or pararosaniline resemble three-blade propellers. It is uncertain whether they are symmetrical propellers with all rings equally tilted, or whether one aryl ring is twisted out of the molecular plane more than the others (cf. Gordon and Gregory, 1983, p. 247 with Zollinger, 1991, p. 73). It is clear however that *ortho*-substitution of phenyl rings (e.g. by methyl in coomassie blue G250, or by sulfonate in the patent blues) or presence of naphthyl rings (as in the Victoria blues) results in the bulky aryl systems being rotated well out of the plane.

Physicochemical properties of significance for staining

Electric charges: chromophoric and overall

When the electron donors of triarylmethane dyes are amino groups, as with dyes in this chapter, the chromophore is of course cationic. However, the overall electric charge of a di- or tri-arylmethane dye depends on the substituents present. When these are cationic, for example, the –N(CH$_3$)$_2$C$_2$H$_5$$^+$ of ethyl green, the positive charge of the dye is increased. Monoanionic substitution, as in mono-sulfonated alkali blue, gives rise to a zwitterionic species; whilst di- or polysulfonation, for example, in the trisulfonated fast green FCF, result in anionic dyes.

Bases, pseudobases, and leucobases

Biologists follow tradition and call cationic di- and triarylmethanes basic dyes. This is misleading because the dyes are salts, and moreover the cations do not usually result from the protonation of bases. Indeed dyes such as crystal violet or malachite green have no ionizable hydrogen atoms that could be removed to form free bases. The terminology derives from the 19th-century industrial use for coloration of wool fabrics, using alkaline ('basic') solutions of the dyes.

A proton can, however, be removed from a dye such as pararosaniline that carries primary amino substituents, to form an imine. This ionic equilibrium is established extremely rapidly (ca 10^{-7} s), and has a pK > 13, the imine, shown in Figure 14.1, is a strong base; it is nonionic and more lipophilic than the cation.

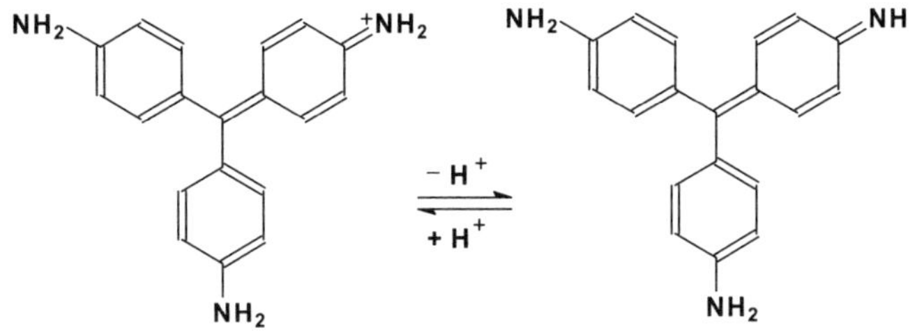

Figure 14.1 Pararosaniline and its imine.

Many of the amino di- and triarylmethane cations, whether or not they can lose a proton in this way, can slowly combine covalently with hydroxide ions. The nonionic products, known as **pseudobases** or **carbinols**, are more lipophilic than their ionic parent compounds. The formation of pararosaniline carbinol is illustrated in Figure 14.2. As a result of the reduction in electron delocalization following attachment of the hydroxide ion, the carbinol is colorless. The rate of formation of pseudobases is much slower than the loss or gain of a proton. For instance the half times of formation of the carbinols of crystal violet and pararosaniline are 2.7 x 10^4 and 3.2 x 10^4 s respectively (several hours). Pseudobases are tertiary alcohols, and so can react with ethanol yielding colorless ethers; one cause of dye fading. Similar reactions occur with any nucleophile, as with sulfite in the formation of Schiff's reagent from basic fuchsine.

There is another type of colorless derivative of triarylmethane dyes; a **leucobase** converts into a colored species following oxidation (Figure 14.2). A variety of oxidants cause this conversion, allowing the use of leucobases of this type as analytical reagents. Leucobase to dye conversions often occur when leucobases are exposed to daylight in the atmosphere, so such reagents are

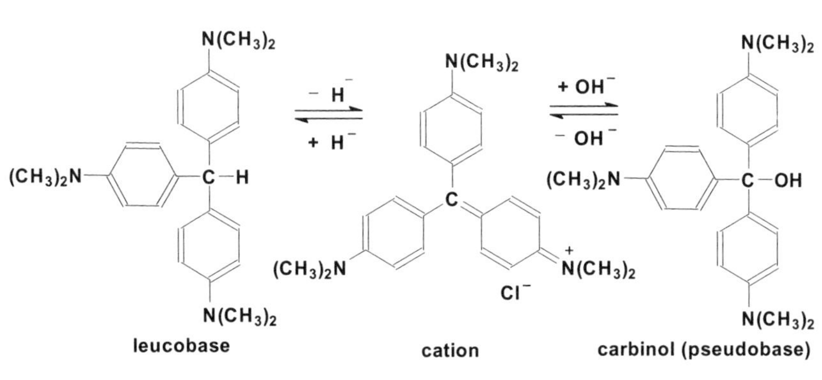

Figure 14.2 Relation of the leucobase and carbinol (pseudobase) with the cationic species, for the dye crystal violet.

routinely stored and used in dark containers. Leucobases are less ionized than their corresponding dye cations, and hence more lipophilic.

Dye size

As discussed below most *N*-phenylated and sulfonated amino triarylmethane dyes are mixtures, varying in degree of phenylation and sulfonation. Fortunately, as long as the dyes are anionic, size variation is irrelevant for many staining applications. This is the case when such acid dyes are used as counterstains for nuclear stains, in mixtures with picric acid such as Van Gieson's stain, or for sequence polychrome stains such as Mallory's or Masson's.

Commercial dye syntheses and dye purity

Commercial syntheses of di- and triarylmethane dyes vary considerably in chemical detail, see for instance Zollinger (1991, pp. 74–78). However when producing dyes whose molecular structures are asymmetric, formation of dye mixtures is inevitable whatever the chemistry. Such mixtures can be separated only by expensive procedures such as chromatography. Consequently many commercial batches of many dyes are heterogeneous. The magnitude of this problem may be illustrated by the case of the basic fuchsines and their sulfonated and *N*-phenylated derivatives, see Figure 14.3.

Modern analytical methods indicate that batches of the symmetrical new fuchsine and pararosaniline dyes usually contain a single major colored

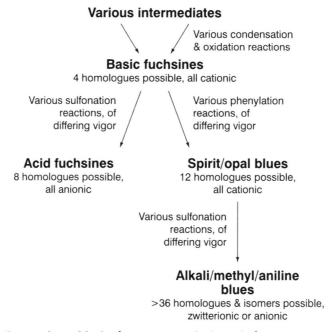

Figure 14.3 Extent of possible dye heterogeneity, for basic fuchsine and its derivatives.

product; as do occasional batches of the fully sulfonated symmetrical acid fuchsine and methyl blue. However all other basic fuchsines and their phenylated and sulfonated derivatives are multicomponent mixtures. The same chemical and economic factors apply to production of all asymmetric dyes of this class, and all are often heterogeneous.

How to find the dye that interests you

Diarylmethanes precede triarylmethanes. There are generic entries for basic fuchsine, methyl green, and Victoria blue. There are also separate entries for certain traditional stains derived from basic fuchsine, namely aldehyde fuchsine, resorcin fuchsine, and Schiff's reagent, because the chemical identities of such materials are now known.

DIARYLMETHANE DYE

Auramine O

CI 41000, CI Basic yellow 2
$C_{17}H_{22}ClN_3$, FW 304

Absorption maxima: 370 nm and 432 nm in water; when excited at 460 nm, its emission maximum is in the yellow at 550 nm.
pK values: 10.7 for the free base–cation equilibrium; 9.8 for the cation–pseudo base equilibrium.
Solubilities: 1% in water; 4% in ethanol; very slightly soluble (0.05%) in xylene.

Auramine O is a weakly lipophilic, cationic diamino diphenylmethane, with a small conjugated system. This dye forms both free base and pseudobase species; the latter being produced from the cation only slowly. The dye is sold as the chloride, illustrated.
Applications: Auramine O is recommended by staining manuals for fluorescent staining of mycobacteria, using rhodamine B as counterstain. Both the leprosy and tubercle bacilli are detectable in this way (Bancroft and Cook, 1994; Bancroft and Gamble, 2002), and a brief review of such methods is available (Churukian, 1991). Auramine O is recommended for detection of nontuberculous mycobacterioses in tissues (Kommareddi *et al.*, 1984). Manuals cite auramine for fluorescent staining of acid-fast bacteria in smears and sections, and for bacterial spores (Clark, 1981). Variant methods are still being developed, such as a phenol-free stain (Cserni, 1998).

Auramine staining has been applied for detecting bacteria in tissue homogenates (Hetland *et al.*, 1998), for demonstration of cryptosporidia in stool samples (Goyena *et al.*, 1997), and for enumeration of reticulocytes, both in smears (Paterakis, 1996) and by using flow cytometry (Rapi *et al.*, 1998).

Nonmicrobiological staining application include preparation of an excellent fluorescent Schiff's reagent (Kasten, 1959), whose use may be quantified (Cihalikova *et al.*, 1985). Elastic fibers (Shelly, 1969) and erythroblasts (Kass, 1980) can also be stained selectively using auramine. Auramine O is used also in vital staining, for example, as a nuclear stain of phytoplankton (Klut *et al.*, 1989), and to aid isolation of viable mucosal mast cells (Stankiewicz *et al.*, 1994).

Industrially auramine O has been used for the coloration of paper, leather, coconut and jute fiber; and for the manufacture of pigments.

Purity of commercial batches: A major component plus one minor species are routinely present; dye contents are up to 85%; the minimum dye content for Commission certification is 80%. A thin-layer chromatographic method is available (Chiang and Lin, 1969). Regarding the Commission's assay procedures see Chapter 28.

Stability: The dry powder is stable when stored in sealed, dark containers; aqueous staining solutions decompose on heating to 60°C; light fastness is poor.

ANIONIC TRIARYLMETHANE DYES

Patent blue V

Synonyms: patent blue VF, patent blue violet, sulfan blue
CI 42045, CI Acid blue 1
$C_{27}H_{31}N_2NaO_6S_2$, FW 567

Absorption maxima: 410 nm and 635 nm in water.
This dye is orange under acid conditions, blue in alkali.
Solubilities: 3% or more in water; 0.7% or more in ethanol; insoluble in xylene.

Various dyes have been sold and used under the name patent blue. The one described here, CI 42045 (Acid blue 1), is widely available commercially at this time. This disulfonated diaminotriphenylmethane has a hydrophilic anion with an aromatic system of moderate size. Due to the *ortho*-sulfonate substituent, the molecule will be markedly nonplanar. Both sodium and calcium salts are sold, the former being illustrated. Confusion over which patent blue is actually supplied perhaps explains discrepancies in published solubility data.

Applications: As a biological stain, patent blue has been used in a long-established method to demonstrate hemoglobin (Lillie and Fullmer, 1976). More recently it has found application in a stain for fungi (Jahnke, 1984).

Other applications in biology and medicine include use as a physiological tracer, to study the movement of fluid within the kidney (MacPhee and Michel, 1995). In oncological surgery, patent blue has been used as a marker dye to aid accurate and complete excision of lymph nodes. To this end, the dye is applied both within liposomes (Hirnle, 1991) and by intradermal injection (Van der Veen *et al.*, 1994), sometimes together with carbon (Lucci *et al.*, 1999). Such procedures amount to vital staining of patients.

Purity of commercial batches: These can contain a single major component; dye contents can be as high as 90%; there are HPLC (Anklam *et al.*, 1995) and thin-layer chromatographic (Hoodless *et al.*, 1971) analytical procedures available.

Stability: Light fastness is poor.

Related dyes, for instance **CI 42051 (Acid blue 3)** and **CI 42080 (CI Acid blue 7)** whose structural formulae are given below, have also been used as biological stains; sometimes intentionally but often doubtless inadvertently under the nonspecific label of 'patent blue'. **Patent blue A (CI 42080)** has been used, as alphazurine A, to prepare microbiological test papers (Ayres and Duda, 1993).

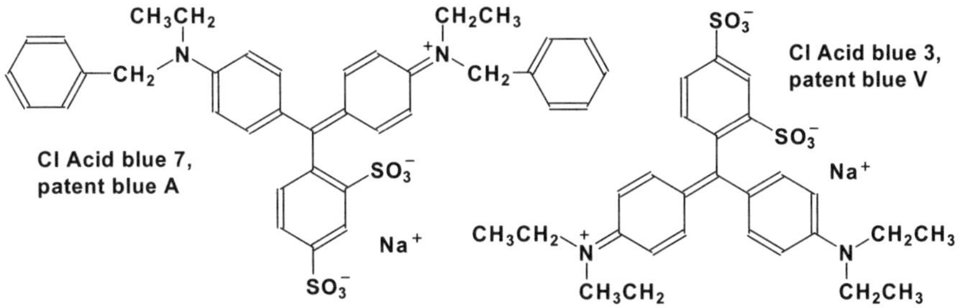

All the patent blues mentioned above are 2,4-disulfonates, but one vendor has identified a product as the 2,5-disulfonate isomer of patent blue V, and has named this material **isosulfan blue**. This dye has been used to identify lymphatics and label sentinel lymph nodes in oncological surgery (Little *et al.*, 1999; Pitman *et al.*, 1998).

Acid fuchsine

CI 42685, CI Acid violet 19

Synonym: acid magenta

$C_{20}H_{17}N_3Na_2O_9S_3$, FW 586 – this corresponds to the sodium salt of a trisulfonated pararosaniline, see below.

Absorption maximum: 546 nm in water acidified with 2.5% 0.1M HCl by volume. When excited in the blue or green, between 530–540 nm, acid fuchsine emits in the red at 630 nm.

Acid fuchsine changes from red to colorless at pH 12–14.

Solubilities: 10–12.5% in water and 0.1–0.3% in ethanol, with significantly higher values having been reported; insoluble in xylene.

Acid fuchsine is prepared by sulfonation of basic fuchsine, or of one of its homologs. As basic fuchsine may contain up to four homologs, and since both di- and trisulfonates will be soluble in water, commercial samples vary markedly in purity and composition. These variations explain the gross heterogeneity of some batches, and probably the wide range of solubilities reported. Only the trisulfonated derivatives of pararosaniline or new fuchsine can be manufactured in substantially pure form.

At this time trisulfonated pararosaniline, illustrated, is often supplied for biological staining. This dye is commercially available as sodium and calcium salts, the ammonium salt having also been sold. Acid fuchsine is a sulfonated triaminotriphenylmethane, containing a very hydrophilic anion whose overall size, like that of its aromatic system, is moderate. Other acid fuchsines have much the same properties, but are less hydrophilic.

Applications: Following its discovery by Caro in 1877, acid fuchsine was soon used as a biological stain, such as in the Ehrlich–Biondi triacid procedure for blood smears reported in 1889 (discussed in Krause, 1926–27), in the Altmann (1890) stain for mitochondria, and in the histopathological polychrome of Pianese (1896).

Polychrome stains originating in the 19th century remain in routine use. Best known are the Van Gieson and Mallory connective tissue stains, first reported in 1889 and 1890 respectively; and the Masson polychrome variant involving acid fuchsine. Versions of all of these are described in current staining manuals, such as that of Bancroft and Gamble (2002). Less common procedures include oversight stains such as a fluorescent acid fuchsine–rhodamine B method for staining Lewy bodies (Issidorides *et al.*, 1991), a celestine blue–acid fuchsine stain for brain histopathology (Zhou *et al.*, 1994), and a methylene blue–acid fuchsine stain for studying assembly of cartilage matrix (Savarese *et al.*, 1996).

Examples of other methods applied to human and animal tissues include a forensic histopathological procedure to stain constriction marks in skin (Wenyou *et al.*, 1991), and various stains for selective identification of neurons damaged by injury or disease (Kiernan *et al.*, 1998). Amongst numerous botanical staining procedures are the fluorescent staining of proteins in starch granules (Wasiluk *et al.*, 1994), and detection of fly eggs within leaves (Nuessly *et al.*, 1995).

A number of staining techniques applied to whole organisms, or at least to large parts of them, use acid fuchsine. Examples include counterstaining of an enzyme histochemical viability test of fungi within plant roots (Schaffer and Peterson, 1993), and staining sections through whole embryos, using a fluorescent Masson trichrome procedure (Apgar *et al.*, 1998). The dye has been used with epoxy-resin embedded tissues (Reempts and Borgers, 1975).

Nonmicroscopic applications of acid fuchsine in biology and medicine include incorporation into an agar growth medium for yeasts (Hagler and Mendonca-Hagler, 1991), fluorescent staining of latent fingerprints (Chesher *et al.*, 1992), and the tracing of water flow within plants (Price *et al.*, 1996).

Industrial uses include coloration of paper, leather, and soap.

Purity of commercial batches: Homogeneity varies from a single major component to mixtures of several major and minor constituents. Overall dye content is up to 70%, with minimum dye content for Commission certification being 55% of anhydrous dye. There is a thin-layer chromatographic procedure (Hals, 1977). Regarding the Commission's assay procedures see Chapter 28.

Stability: Stored as a dry powder in sealed, dark containers the dye is fairly stable; light fastness is poor.

Aniline blue

CI 42780, CI Acid blue 93
Synonyms: cotton blue, ink blue. Related or component dyes include methyl blue, soluble blue and water blue, see below.
$C_{37}H_{27}N_3Na_2O_9S_3$, FW 800

Absorption maximum: 600 nm in water
Aniline blue changes color from blue to orange at pH 10–13.
Solubilities: 7% in water; 0.4% in ethanol; insoluble in xylene

Aniline blue is the traditional name applied to anionic dyes resulting from successive phenylation and sulfonation of basic fuchsine. The degree of methylation and heterogeneity of the basic fuchsine, the degree of phenylation of this material, and of sulfonation of the phenylated intermediates can all vary between batches. Dyes marketed as aniline blue are consequently mixtures. The previous edition of *Conn* suggested that material marketed as aniline blue WS was predominantly a mixture of **methyl blue** (CI 42780, Acid blue 93) and **water blue** (CI 42755, Acid blue 22). Analytical work using modern methods suggests this was over-optimistic. 'Aniline blue', 'methyl blue' and 'water blue' are usually mixtures of much the same set of components.

Occasionally material sold under one of these names contains a single major component. As discussed by Green (1990, pp. 96–97) such material will correspond to methyl blue, and for this reason the physico-chemical information given above relates to this compound. The staining applications, below, were probably carried out with grossly heterogeneous stains, whatever the labeling of the bottles. Batches of aniline blue certified by the Biological Stain Commission perform correctly in the specified stains, but will usually be heterogeneous.

Aniline blue – in the sense of methyl blue – is a triaminotriphenylmethane, each amino group carrying a monosulfonated phenyl substituent, as illustrated. The anion is hydrophilic, of large overall size and with a large aromatic system. Aniline blue is sold both as the sodium and ammonium salts.

The structure of the strongly fluorescent, nontriarylmethane component of commercial aniline blue samples responsible for callose staining is now known. This compound is termed **sirofluor** (Rost, 1995, pp. 373–374).

Applications: Aniline blue, either as such or nominally as methyl blue or water blue, is a constituent of widely used polychrome stains. The AZAN stain of Heidenhain (1916) and the polychrome stain of Mallory (1900) both involve staining steps using aniline blue–orange G mixtures. Mann's stain (1894) involves a methyl blue–eosin mixture. Masson's stain (1929) and Lendrum's MSB technique (Lendrum *et al.*, 1962) both involve complex sequence staining, the final coloration step using aniline blue. These methods are described in various manuals (Bancroft and Cook, 1994; Kiernan, 1999).

Many less widely used stains involve aniline/methyl blue. A recent examples include staining of insect endocuticle (Haas, 1992) and of fungi within plant cells (Williamson *et al.*, 1995). Histones may be stained, this being a standard procedure for assessing nuclear maturity (Morel *et al.*, 1998). Sperm viability can be assessed using the eosin–aniline blue stain (Borg *et al.*, 1997). Selective staining of callose in plant specimens also uses aniline blue, although it is the impurity sirofluor which is the actual staining compound (Yim and Bradford, 1998).

The dye is used to stain tissues embedded in resin, both epoxy (Stockert, 1975) and water-miscible (Thomsen *et al.*, 1998). Aniline blue has also been applied to large, intact specimens such as insect larvae, prior to observation using confocal microscopy (Dahlan and Gordh, 1996), and pollen tubes in flowers (Higashiyama *et al.*, 1997). Vital staining methods using aniline blue include demonstration of septae of dividing yeast cells (Kippert and Lloyd, 1995), and measurement of lymphatic diameters following injection of dye (Nathanson *et al.*, 1997).

Industrial uses – blue inks and, in the form of heavy metal salts, printing ink pigments.

Purity of commercial batches: Material sold as aniline blue, methyl blue or water blue usually contains up to half a dozen major components plus several minor constituents; occasional batches (probably of methyl blue, CI Acid blue 93, see above) contain a single major component. Overall dye contents are up to 60%. There is a thin-layer chromatographic method for aniline/methyl blue (Hals, 1978). Regarding the Commission's assay procedures see Chapter 28.

Stability: The dry powder is moderately stable when stored in sealed, dark containers; typical staining solutions are reasonably photostable to sunlight, though in general light fastness is rather poor.

Related dyes: Structures of dyes differing in degree of phenylation and/or sulfonation to methyl blue – the sulfonated **reflex blue** and **water blue**, and the nonsulfonated **opal blue** and **spirit blue** – were given in the previous edition of Conn. However these are probably also mixtures of homologs and isomers. Spirit blue has been used to stain invertebrate elastic fibers (Elder and Owen, 1967), a procedure requiring an extended conjugated system such as arises from *N*-phenylation.

Light green SF

CI 42095, CI Acid green 5
$C_{37}H_{34}N_2Na_2O_9S_3$, FW 793

Absorption maxima: 422 nm and 630 nm in water.
Solubilities: 10–20% in water; 0.2–4.0% in ethanol; insoluble in xylene.

Light green SF is a diaminotriphenylmethane, with the amino groups both being benzylated. The very hydrophilic anion is large, with quite a large aromatic system. The dye is nominally trisulfonated, but the range of solubilities reported (see above) suggest that some batches may be contaminated by disulfonated compounds, such as **light green L extra**, and perhaps also by a tetrasulfonate.

Applications: Routine staining includes collagen fiber stains in histopathology, especially Masson's trichrome; the Papanicolaou cytological polychrome stains; and the Twort stain for microorganisms in tissue sections. Light green is also widely used as a cytoplasmic counterstain, with such nuclear stains as the

Feulgen method and safranine, for both animal and plant tissues; in the latter case light green also stains cellulose cell walls. These methods are described in standard manuals (Bancroft and Gamble, 2002; Clark, 1981; Kiernan, 1999).

Other staining using Light green SF include procedures for tissues embedded in both epoxy (Burgauer and Stockert, 1975) and water-miscible (Scala *et al.*, 1993) resins. Quantitative protein staining may be carried out with light green (Oud *et al.*, 1984). Hard-tissue applications include detection of resorptive dental lesions (Berger *et al.*, 1996). Light green is a component in a polychrome staining method for distinguishing the secretory cells of the pituitary (Sarasquete *et al.*, 1997).

Industrial uses – the dye has been used to color chromed leather. The barium salt has been used in paper coatings and printing inks; it is also a food dye.

Purity of commercial batches: Typical samples contain a single major component, with several minor contaminants; dye contents are up to 80%, the minimum dye content for Commission certification being 65%. There is a thin-layer chromatographic analytical method (Hals, 1978). Regarding the Commission's assay procedures see Chapter 28. Note: commercial samples labeled light green SF can actually be fast green FCF or some other green acid dye, as described in a Commission report (Penney and Powers, 1995).

Stability: of dry powder is good, when stored in sealed, dark containers; light fastness is poor.

A lower sulfonation product with similar staining properties, **light green L extra**, has been sold. Since sulfonation is difficult to control it is likely that this material will contain a mixture of products.

Fast green FCF

CI 42053, CI Food green 3
$C_{37}H_{34}N_2Na_2O_{10}S_3$, FW 809

Absorption maximum: 624 nm in 50% aqueous ethanol.
The dye is orange in acids, green under neutral conditions, and deep blue in alkali.
Solubilities: 6% in water; 0.5% in ethanol; insoluble in xylene. Much higher aqueous solubilities (up to 16%), and a higher solubility in ethanol (9%), have been reported.

Fast green FCF is a diaminotriphenylmethane, both amino groups being benzylated. The very hydrophilic anion is large, with a large aromatic system; which is markedly nonplanar, due to an *ortho*-sulfonate substituent. The dye is trisulfonated, but the wide range of solubilities reported (see above) suggest batch variation in the degree of sulfonation. It is also likely that other green dyes have been confused with fast green FCF in the biological staining literature.

Applications: Fast green FCF was investigated early in the 20th century as a fade-resistant alternative to light green SF. Such substitutions are widespread, and staining manuals such as Bancroft and Gamble (2002) describe variants of Masson's trichrome, the MSB technique for fibrin, the Twort stain for microorganisms in tissue sections, and so on. The Gomori trichrome is an example of a polychrome stain originally devised to use fast green.

Fast green FCF is also used as a cytoplasmic counterstain for nuclear stains such as the Feulgen method and safranine, for both animal and plant tissues. In the latter case cellulose cell walls are also stained, as described in 1928 by Haynes. Procedures of this type, together with such methods as the Guard stain for sex chromatin, are described in routine manuals and handbooks, see for instance those of Boon and Drijver (1986) and Clark (1981).

Another routine method is the selective staining of basic proteins by alkaline solutions of fast green, as discussed in Chayen and Bitensky's manual (1991). Other recent examples of the substitution of fast green for light green include modified Papanicolaou (Akura and Takenaka, 1991) and Halmi aldehyde-fuchsine (Trindade *et al.*, 1998) stains. In botanical microtechnique fast green FCF is preferred to light green SF because it is less likely to fade with storage (Berlyn and Miksche, 1976; Ruzin, 1999). For the same reason, fast green FCF is also the preferred dye for staining collagen (Gabe, 1976).

Fast green FCF has been used in a surprising range of other staining methods, and a few examples follow. Animal histology procedures such as oversight staining of joints, using fast green-toluidine blue (Burkhardt *et al.*, 1992), and correlation of magnetic resonance imaging with histological findings (Makowski *et al.*, 1998). In botanical histology, fast green selectively stains gelatin fibers in woody plants (Baba *et al.*, 1995).

Fast green FCF is also used in quantitative histochemistry, for instance to quantitate collagen content (Murakami *et al.*, 1997a). Resin sections can be stained, to show chromosomes and histones (Dodge, 1964) and nucleoli (Burgauer and Stockert, 1975) in epoxy resins; and connective tissue fibers in water-miscible resins (Cannon *et al.*, 1992). With living specimens, fast green has been used to mark individual neurons after stimulation with microelectrodes (Ohta *et al.*, 1991), and to assess mechanical damage to seeds (Peterson *et al.*, 1995).

Fast green FCF has been extensively used as a stain of proteins separated by gel electrophoresis, as critically reviewed by Wilson (1983). This can be achieved directly, or following blotting procedures, a recent example of the latter being to stain immunogenic peptides on immunoblots (Kurien *et al.*, 1998).

The dye has been used as a marker, for instance to indicate diffusion of injected GABA antagonists within the brain (Yoshida *et al.*, 1991), and as a tracer, for instance to assess patency of blood vessels within colonic tumors (Leith and Michelson, 1995).

Purity of commercial batches: A single major component plus several minor constituents are routinely present. Note: chemically distinct blue-green dyes have been supplied under this name. Dye contents up to 90% are available, the minimum dye content for Commission certification being 85%. There is a thin-layer chromatographic method (Hals, 1978). See Chapter 28 regarding the Commission's assay procedures.

Stability: The dry powder is reasonably stable stored in sealed, dark containers; light fastness is fair.

Coomassie brilliant blue R250

CI 42660, CI Acid blue 83
Synonyms: brilliant blue R, brilliant indocyanine 6B, coomassie blue R250
$C_{45}H_{44}N_3NaO_7S_2$, FW 826

Absorption maximum: 585 nm in ethanol.
Blue in neutral solutions, violet in alkali.
Solubilities: 7% in water; 1% in ethanol.

Coomassie brilliant blue R250 is a triaminotriarylmethane, with two amino groups being benzylated and one phenylated. The hydrophilic anion is large, as is the aromatic system. The dye is disulfonated, and available commercially as the sodium salt.

Dyes named in the biological literature as coomassie blue, with no further detail, will most commonly be this dye. However the greener (and less water soluble) coomassie brilliant blue G250, as well as the monoazo dye coomassie brilliant blue RL (CI Acid blue 92) have sometimes been referred to as coomassie blue.

Applications: Coomassie brilliant blue R250 has been applied as a protein stain on epoxy embedded tissue sections (Fisher, 1968), and for quantitative micro-densitometry of protein in sections (Cawood *et al.*, 1978). Biological entities stained by coomassie brilliant blue include chromatids (Jan, 1981), actinin microfilaments (Paddock, 1982), developing neurites (Carri and Ebendal, 1989), and sperm acrosomes (Larson and Miller, 1999).

The dye is however widely used for staining of proteins separated on electrophoretic gels, as critically reviewed by Wilson (1983). A more recent high sensitivity variant gel stain has been described (Choi *et al.*, 1996).

Industrially uses – textile dyeing; coloration of feathers and leather; and surface coloration of paper.

Purity of commercial batches: Typically mixtures of dyes, including partially phenylated compounds of redder hues. The degree of sulfonation varies markedly, from less to more than the nominal disulfonation. Overall dye contents are up to 90%. There is a thin layer chromatographic method available (Kundu *et al.*, 1996).

Stability: Light fastness is moderate.

The rather similar dye **coomassie brilliant blue G250 (CI 42655, CI Acid blue 90)** shown below is sometimes used to stain proteins on electrophoretic gels. Recent applications of this type include negative staining proteins separated on casein-polyacrylamide gels (Raser *et al.*, 1995), and staining of serum proteins separated by isoelectric focusing on cellulose acetate membranes (Iijima *et al.*, 1997).

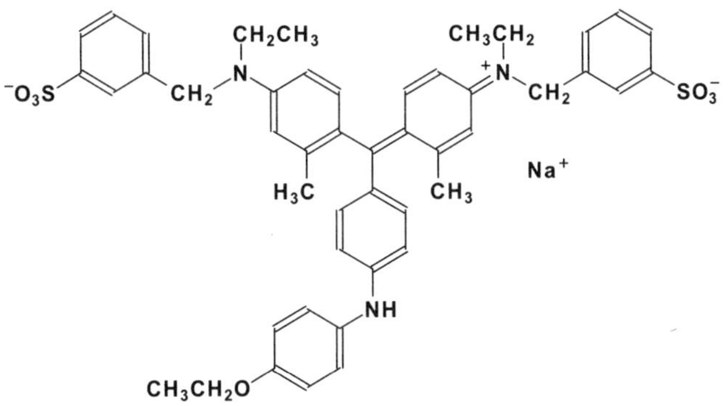

TRIARYLMETHANE ZWITTERIONIC DYE

Alkali blue

CI 42750, CI Acid blue 110
Synonyms: alkali blue 4B or 5B
$C_{32}H_{27}N_3O_3S$, FW 534

Absorption maximum: 603 nm in methanol containing 1% 1M HCl.

The dye changes color from blue to colorless under strongly alkaline conditions.

Solubilities: Insoluble in cold water, slightly or moderately soluble in hot water; moderately soluble in ethanol.

The structures given here are of a diphenylated material, as given in the *Colour Index* for CI 42750. The ionized form is slightly hydrophilic, but due to its zwitterionic nature is of low solubility in water. The pseudobase or carbinol form, formed under strongly alkaline conditions, is colorless.

The previous edition of this book suggested that the term alkali blue referred either to CI 42750, or to a triphenylated material CI 42765. The *Colour Index* suggests single structures for both these dyes. However since these dyes are manufactured by mild sulfonation of spirit blues, both alkali blues are likely to be mixtures of isomers and homologs; a view supported by application of modern analytical methods (Keto, 1984).

Applications: Alkali blue has been used as a selective stain for elastic fibers (Cook and Lamb, 1969). More recent applications include incorporation in a wide-pH range indicator system for pharmaceutical testing (Asikoglu *et al.*, 1995), and as an immobilized chelator for removal of metal ions such as iron (Denizli *et al.*, 1998a) or biochemicals such as bilirubin (Denizli *et al.*, 1998b) from plasma.

Industrially these dyes have been applied as pigments in paints, and for the coloration of paper.

Purity of commercial batches: Typically one or two major constituents plus several minor ones are present. Dye contents can be up to 50%. There is an HPLC method available (Keto, 1984).

Stability: Light fastness is poor.

TRIARYLMETHANE CATIONIC DYES

Basic fuchsine

Basic fuchsines may contain several homologous compounds, including CI 42500, 42501 and CI 42520; see below.

Absorption maximum for Commission certified batches must be in the range 547–552 nm, in 50% aqueous ethanol; when excited around the absorption maximum, basic fuchsine emits around 625 nm.

Basic fuchsine changes from violet to red between pH 1.2–3.0, and from red to colorless between pH 11.6–14.0

Solubilities: A wide range of solubilities has been reported, 0.3–2.4% in water, and 3.7–9.1% in alcohol. This probably reflects variation in homologs and counter ions present in different batches.

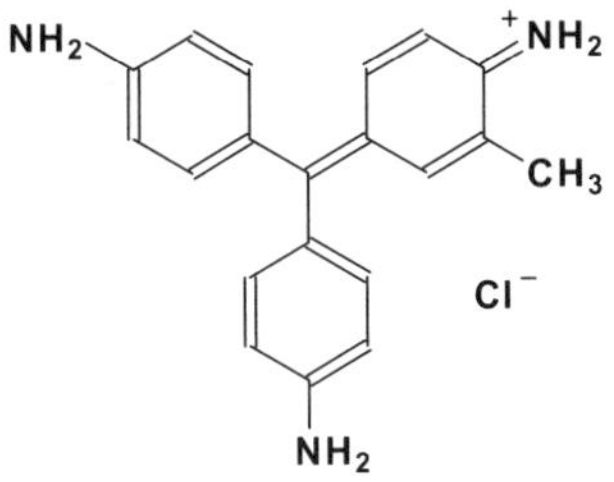

rosaniline 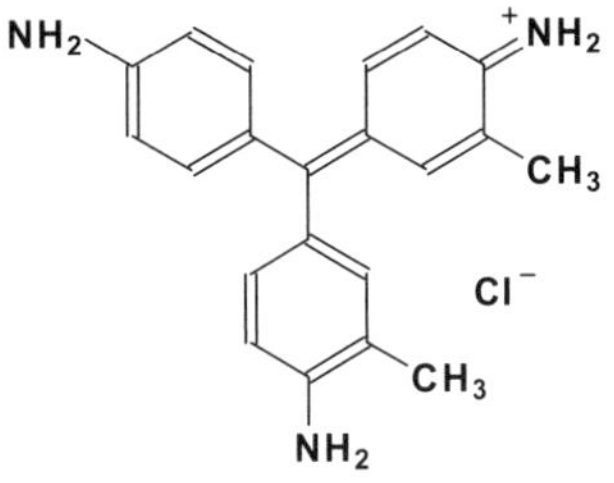   magenta II

Basic fuchsine is the traditional name for various mixtures of four triamino-triphenylmethane homologs, with increasing degrees of methylation; namely **pararosaniline** (CI 42500), **rosaniline** (CI 42510), **magenta II**, and **new fuchsine** (CI 42520). The structures of rosaniline and magenta II are shown here, those for the other two homologs are given in their own entries. All four homologs have rather small cations, and slightly nonplanar conjugated systems. All give rise to free bases when a proton is removed from an amino group, and to pseudobases when a hydroxide or other nucleophilic anion attaches to the central carbon atom. Basic fuchsine has been sold as the acetate, the chloride and, in the past, the sulfate.

Separate entries are provided for commercially available basic fuchsine homologs, namely pararosaniline and new fuchsine. These entries indicate that some staining methods require particular homologs. However, many do not, and numerous procedures are stated to require merely 'basic fuchsine'. Such procedures are decribed in this entry. Certain staining reagents prepared from basic fuchsine – namely aldehyde fuchsine, resorcin fuchsine, and Schiff's reagent – are also described in separate entries.

Applications: Current staining manuals (Bancroft and Gamble, 2002; Kiernan, 1999) describe the direct use of basic fuchsine as a nuclear and mucin stain in histology, and in microbiology in various carbol-fuchsine methods such as Ziehl–Neelsen's stain. Use of other traditional procedures is noted in recent literature, such as, staining of Negri bodies using Seller's basic fuchsine-methylene blue procedure (Ni *et al.*, 1996).

Of course basic fuchsine is the precursor of Schiff's reagent, which itself has many applications, most notably the periodic acid–Schiff's and Feulgen stains. Basic fuchsine is also the precursor of a hexazonium salt, used as a visualizing agent in enzyme histochemistry and for demonstrating enzyme labels in immunostaining.

Illustrative examples of additional staining applications follow. Basic fuchsine is used as an alternative to Schiff's reagent in the periodic acid–Schiff procedure (Perera, 1991), and for coloration of hydrophobic substances in plants (Kraus *et al.*, 1998). The dye is also used for polychrome staining of water-miscible (Abreu *et al.*, 1993) and epoxy (Flores and Hoffmann, 1997) resin sections; the staining of pollen grains trapped on sampling tape (Takahashi *et al.*, 1995); and the coloration of bone prior to investigation by confocal microscopy (Bentolila *et al.*, 1998). Among numerous other applications in biology and medicine are the following: bulk-staining of flower buds, to check for freezing damage (Flinn and Ashworth, 1994), identification of *Brucella* species in microbiology (Corbel, 1991), and as a marker for assessing rate of passage of stained food through the gut of insectivorous bats (Stalinski, 1994).

Clinical applications include assessment of damage to human corneal surfaces (Goffin *et al.*, 1997), and bulk-staining of bones for identifying surface damage (Boyce *et al.*, 1998). In forensic investigations the dye has been used to aid detection of fingerprints (Howard, 1993).

Purity of commercial batches: Material labeled basic fuchsine may be new fuchsine or pararosaniline, *qv.* Other batches contain several, or all four, of the possible homologs. Minimum dye content for Commission certification is 88%. There is a thin-layer chromatographic method (Lyon *et al.*, 1992a,b). See Chapter 28 regarding the Commission assay procedures.

Stability: The dry powder is stable, if stored in dark, sealed containers; light fastness of solutions is poor, of stained sections is varied but can be good.

Pararosaniline

CI 42500, CI Basic red 9
Synonyms: magenta O, parafuchsin
As the chloride: $C_{19}H_{18}ClN_3$, FW 324
As the acetate: $C_{21}H_{21}N_3O_2$, FW 347

Absorption maximum: 545 nm in 50% aqueous ethanol. When excited at 560–570 nm it fluoresces in the red at 625 nm.

Decolorizes between 11.6–14.0; *pK value* for the cation–base equilibrium is >13, for the cation–pseudobase (carbinol) is 7.6, the latter reaction being slow.

Solubilities: The acetate is soluble to 3% in water and 2% in ethanol; the chloride to 0.3% in water and 2.5% in ethanol. Neither salt is soluble in xylene.

Pararosaniline is a triaminotriphenylmethane; with a rather small, hydrophilic cation and a slightly nonplanar conjugated system. The dye is sold both as the acetate and the chloride, which differ in solubility. Both the free base produced when a proton is removed to form an imine, and the pseudobases produced when hydroxide or other nucleophilic anion attaches to the central carbon atom, occur. The free base is commercially available, though it has been mistakenly stated by some vendors to be the pseudobase. The cationic and imino species are illustrated in Figure 14.2 in the **Background information** section, above.

Pararosaniline may be diazotized to give **hexazonium pararosaniline**, used as a visualizing agent in enzyme histochemistry. Other reagents prepared from pararosaniline, namely aldehyde fuchsine and Schiff's reagent, are described in separate entries; as is the trimethylated homolog new fuchsine. Pararosaniline is the major, or only, constituent of many dye samples labeled basic fuchsine. Only staining procedures explicitly stated to use pararosaniline are cited in this entry. Methods stated to use 'basic fuchsine' are described in the entry for that dye.

Applications: Pararosaniline is widely used for preparation of Schiff's reagent. Recent applications include periodic acid–Schiff staining of histopathological specimens (Garvey *et al.*, 1992), and quantitative estimation of DNA by Feulgen staining in thick sections using confocal microscopy (Crist *et al.*, 1996). Pararosaniline has also been used unmodified as a Schiff's reagent substitute, from acidic alcoholic solutions, for staining glycogen following periodate oxidation and for staining nuclei following acid hydrolysis (Horobin and Kevill-Davies, 1971; Cavalcante *et al.*, 1996).

The **hexazonium pararosaniline** derivative is used as a coupling agent in enzyme histochemistry, especially with naphthols in the demonstration of acid phosphatase, as described in staining manuals (Van Noorden and Fredericks, 1992); and with naphthylamine in the demonstration of proteases for trans-mission electron microscopy (Smith *et al.*, 1992). This diazonium salt has also been applied as a fixative agent of cryosections prior to immunostaining (De Jong *et al.*, 1991).

Stains for epoxy resins include oversight staining (Tolivia *et al.*, 1994), and staining of cartilage matrix (Grill *et al.*, 1995). Schiff's reagent prepared from this dye has been used in periodic acid–Schiff staining of glycogen in epoxy embedded material (Ferey *et al.*, 1986), and Feulgen staining of DNA in specimens embedded in water-miscible resins (Braselton *et al.*, 1996). The hexazonium pararosaniline salt has been used for enzyme histochemical demonstration of acid phosphatase and nonspecific esterase in tissues embedded in water-miscible resin (Gerrits and Horobin, 1996).

Industrially pararosaniline has been used for dyeing and printing acrylic textile fibers, and for coloration of leather and paper.

Purity of commercial batches: Most contain only a single major constituent. Dye contents can be up to 95%, with the minimum content of batches certified by the Biological Stain Commission being 88%. A thin-layer chromatographic method is available (Lyon *et al.*, 1992b). Regarding the Commission assay procedures see Chapter 28.

Stability: of dry powder is good if stored in a sealed, dark container; light fastness is poor.

Aldehyde fuchsine

Synonyms: Gomori's aldehyde fuchsine, paraldehyde–fuchsine
Mono-Schiff base form: $C_{21}H_{19}ClN_3$, FW 349
Di-Schiff base form: $C_{23}H_{22}ClN_3$, FW 376
Tri-Schiff base form: $C_{25}H_{25}ClN_3$, FW 403

Absorption maximum: 541 nm in 70% acid ethanol, when all three components noted below are present.
Solubilities: insoluble in water; soluble in 70% acid alcohol.

There are several aldehyde fuchsine components, three being mono-, di- and tri-Schiff base derivatives of pararosaniline, as illustrated. Structures of this type were originally suggested by Bangle (1954), and supporting evidence was subsequently provided by Sumner (1965), Lichtenstein and Nettleton (1980) and Proctor and Horobin (1983). The Schiff base compounds are small lipophilic cations, with more extended conjugated systems than pararosaniline, generated by reacting pararosaniline with acetaldehyde. On standing these Schiff bases are reduced yielding *N*-alkylated products. These have smaller conjugated systems, and no longer act as high affinity stains for elastic fibers. Aldehyde fuchsine can only be prepared from pararosaniline (Lichenstein and Nettleton, 1980; Mowry and Emmel, 1977), not from other components of basic fuchsine.

Applications: Aldehyde fuchsine is used to stain elastic fibers and insulin-containing pancreatic beta cell granules, as described in staining manuals (Bancroft and Cook, 1994; Bancroft and Gamble, 2002). Following oxidation of tissue sections aldehyde fuchsine stains oxytalan fibers and cystine-rich proteins, including keratin and classical neurosecretory material. In a recent investigation, such staining facilitated use of confocal microscopy for evaluation of the three dimensional distribution of the oxytalan fibers within thick sections of the periodontal ligament (Chantawiboonchai *et al.*, 1998). Examples of other applications of aldehyde fuchsine staining from the recent research literature include demonstration of sulfated proteoglycans in cartilage matrix (Yoo *et al.*, 1998), and staining of neurosecretory granules (Honma *et al.*, 1998).

Purity of stain solutions: After preparation by routine room-temperature methods, aldehyde fuchsine initially contains the purple staining components plus residual pararosaniline. After a few days, the three Schiff bases are the principle constituents. After several weeks they slowly transform into nonstaining dyes such as ethyl violet and malachite green. Unfortunately the amount of active staining compounds present is not a simple function of time. Precipitation of stain often occurs, which displays marked batch variation. Thin-layer chromatographic methods are available (Denffer and Heidbrink, 1974).

Stability: of staining solutions is typically up to several weeks. However, the dry powders of precipitated aldehyde fuchsines – such as those of Rosa (1953), Gabe (1976) and Kiernan (1999) – are stable for many years if stored in dark, sealed containers; and even their reconstituted staining solution may be used for more than a year.

Malachite green

CI 42000, CI Basic green 4
Synonym: Victoria green
The basic dye oxalate, containing oxalic acid of crystallization:
$C_{46}H_{50}N_4.2HC_2O_4.C_2H_2O_4$, FW 927
The carbinol base hydrochloride: $C_{23}H_{27}ClN_2O$, FW 383
The carbinol free base: $C_{23}H_{26}N_2O$, FW 347
The leucobase: $C_{23}H_{26}N_2$, FW 330

Absorption maxima: of the cationic species are at 614 and 425 nm in water. In 1N hydrochloric acid the carbinol hydrochloride has, before conversion to the cationic species, a maximum absorbance at 446 nm. The colourless leucomalachite green has an absorption maximum at 262 nm in chloroform.

The color of an aqueous solution of malachite green changes from yellow to green between pH 0.0–2.0, and from green to colorless between pH 11.6–14.0. Both color shifts occur rapidly, in keeping with equilibria involving loss of protons (Baumann, 1995). However the *pK value* for the addition of a hydroxide ion (i.e. the cation–carbinol equilibrium) is 6.9, and this equilibrium is not achieved rapidly.

Solubilities: The oxalate is soluble to 4% in water and 5% in ethanol, and is insoluble in xylene. The carbinol hydrochloride has solubilities of 6% and 2% respectively in water and ethanol. The carbinol base is almost insoluble in water but soluble to 1% in ethanol. Leucomalachite green is also virtually insoluble in water, but slightly soluble (0.4%) in ethanol.

leucomalachite green

malachite green carbinol chloride

Malachite green is a cationic diaminotriphenylmethane. The cationic species typically present in aqueous solutions is small and lipophilic. This cation is present in salts such as the oxalate, and is rapidly generated when the colorless, hydrophilic carbinol hydrochloride is dissolved in water. The carbinol, which is formed slowly from the cation under alkaline conditions, also exists in the lipophilic, nonionic free base form. All these malachite green species are derived, via appropriate oxidations and hydrolyzes, from a leucomalachite green intermediate.

Although the basic dye is usually sold as the oxalate, some batches labeled malachite green chloride contain a zinc chloride double salt. Usually it is the hydrochloride of the malachite green carbinol which is so labeled, although this material is also sold as malachite green carbinol hydrochloride. The colorless free base form of the carbinol (CI 42000:1, Solvent green 1) and leucomalachite green are both available commercially.

Applications: Routine staining applications are described in manuals, for instance staining of bacterial spores (Clark, 1981), and the Gimenez stain for *Rickettsia* and *Helicobacter pylori* (Bancroft and Gamble, 2002). The dye has also been applied in plant pathology as a fluorescent stain for bacterial infections (Zhou and Paulitz, 1993). Malachite green has been widely used in combination with glutaraldehyde as a fixative-stain for phospholipids; as summarized by Hayat (1993, p. 251).

Nonstaining applications of malachite green in microbiology include use for resistogram typing of *Escherichia coli*-isolates from poultry (David *et al.*, 1992), and as a constituent of bacterial culture media (Blivet *et al.*, 1998). Malachite green has also been used as a color-marker of incubating herring gulls (Belant and Seamans, 1993).

Industrial uses – as a fungicide and parasiticide in fish farming, when it is applied as the carbinol hydrochloride to avoid contamination by oxalate or zinc. The phosphomolybdate salt is used as a pigment (CI 42000:2, Pigment green 4). The carbinol base has been used to color fats, oils and waxes.

Purity of commercial batches: A single major component is present in some samples. Dye contents of the oxalate are up to 96%, the carbinol chloride up to 85%. The minimum dye content for Commission certification is not less than 90% as oxalate or 75% as chloride. The carbinol free-base, CI Solvent green 1, is available commercially at up to 90% dye. Thin-layer chromatographic (Rettie and Haynes, 1964) and HPLC (Rushing and Thompson, 1997) methods are available. Regarding the Commission assay procedure see Chapter 28.

Stability: The dry powder form of the basic dye is moderately stable, if stored in dark, sealed containers; light fastness is poor.

New fuchsine

CI 42520, CI Basic violet 2
Synonyms: magenta III, new magenta
$C_{22}H_{24}ClN_3$, FW 366

Absorption maximum: 553 nm in 50% aqueous ethanol.

Changes from purple to red between 1.2–3.0 and from red to colorless between 11.6–14.0.

Solubilities: 1.1% in water; 3.2% in ethanol.

New fuchsine is a triaminotriphenylmethane, with a rather small cation and a slightly nonplanar conjugated system. It is weakly hydrophobic, with a methyl substituent on each aromatic ring. The dye is typically sold as the chloride.

Applications: This entry discusses those staining procedures explicitly stated to use new fuchsine or its derivatives. Certain staining reagents that may be prepared from new fuchsine (Schiff's reagent and resorcin fuchsine) are described more fully in separate entries. Methods stated to use 'basic fuchsine', with no homolog specified, are described in the generic entry for that dye.

New fuchsine is routinely used to prepare Schiff's reagent for use in the Periodic acid–Schiff procedure; in Taylor's stain of Gram-positive and Gram-negative bacteria; and in the Ziehl–Neelsen acid-fast bacterial stain (Churukian and Schenk, 1983). Schiff's reagent prepared from this dye is also suitable for the Feulgen staining of DNA (Schulte and Wittekind, 1989a) despite earlier doubts. New fuchsine is unsuitable for preparing aldehyde fuchsine.

New fuchsine is the precursor of **hexazotized new fuchsine**, a diazonium salt routinely used for enzyme histochemical demonstration of alkaline phosphatase and peptidases. Recent applications include quantitation of alkaline phosphatase in cultured cells (Mozes *et al.*, 1998), and demonstration of alkaline phosphatase-labeled immunostains (Kennett *et al.*, 1997) and *in situ* hybridization systems (Hopman *et al.*, 1997). For peptidase methods see Smid *et al.* (1992) and Smith *et al.* (1992). New fuchsine has also been used to generate a fluorescent Schiff's reagent (Kasten, 1959). It is the best basic fuchsine homolog for preparation of resorcin fuchsine (Proctor and Horobin, 1987).

Purity of commercial batches: Most contain a single major constituent, plus small amounts of other homologs; dye contents of typical samples are up to 80%. A thin-layer chromatographic method is available (Schulte and Wittekind, 1989a).

Stability: The light fastness is poor.

Methyl violet

CI 42535, CI Basic violet 1
Synonyms: methyl violet is sold with various suffixes, e.g. B, 2B, and 6B, see below. Essentially the same dye mixture is also sold as dahlia B, gentian violet (a synonym used only in Europe; in the USA this name usually implies crystal violet), and Paris violet.

Absorption maximum: in the range 583–587 nm in 50% aqueous ethanol; increasing *N*-methylation results in shifts to longer wavelengths. If methyl violet is excited at 540 nm, it emits at 590 nm.
Solubilities: cited values fall in the ranges 2–5% in water, and 0.9–9.5% in ethanol. The variability probably reflects batch variation; see below. The dye is insoluble in xylene.

A mixture of *N*-methylated pararosanilines, predominantly the *N*-penta- (illustrated) and hexamethyl homologs. Less methylated, and hence redder, products have generally been labeled methyl violet R or 2R; the progressively more methylated and bluer materials B, 2B, and so on; with methyl violet 6B having been considered as pentamethylpararosaniline. Application of modern analytical methods to commercial batches suggests that all methyl violets are mixtures of homologs.

The dye components are slightly lipophilic cations of moderate size. The product sold as an indicator or for biological staining is the chloride, CI 42535. The free base derivative of methyl violet is available commercially. This material is CI 42535:1 (CI Solvent violet 8). The names **Hofmann's violet** and **Spiller's purple** have in the past been applied to various dyes containing *N*-methylated pararosanilines and rosanilines, as CI 42530.
Applications: As a biological stain, methyl violet has been used for much the same range of applications as crystal violet, *qv*, excepting the Gram stain. Outcomes are similar, but methyl violet gives redder hues. New applications of methyl violet are still occasionally reported, for instance detection of cryptosporidia in fecal smears (Milacek and Vitovec, 1985), and induction of atrophy in obstructed parotid glands (Wang *et al.*, 1998). The dye is a component of a selective bacteriological culture medium (Ming *et al.*, 1991).

Industrially the basic dye is used as a colorant of inks and lacquers; the free base as a colorant of ball pen and printing inks, copying paper, and typewriter ribbons; and the pigments find application as colorants of paper.

Purity of commercial batches: Several major and numerous minor components are routinely present. Dye contents are up to 85%, with minimum dye content for Commission certification being 75% of anhydrous dye. A thin-layer chromatographic method is available (Marshall, 1976). See Chapter 28 regarding the Commission assay procedures.

Stability: The dry powder is stable if stored in the dark in sealed containers; light fastness is poor.

Crystal violet

CI 42555, CI Basic violet 3
Synonyms: gentian violet (a synonym used only in the USA; in Europe this term usually implies methyl violet), hexamethyl pararosaniline, methyl violet 10B
$C_{25}H_{30}ClN_3$, FW 408

Absorption maximum: 590 nm in water
Changes from yellow at pH 0.0 to blue–violet at pH 2.0.
The *pK* of carbinol formation is 9.4, with the half time of formation being 7.5 hours.
Solubilities: 0.2–1.7% in water; 3–14% in ethanol; also soluble in acetone (0.4%) and chloroform (5.1%); insoluble in xylene. The variation in reported solubilities perhaps reflects presence of lower homologs and variation in counterions in some commercial samples in the past.

Crystal violet is a hexa-*N*-methylated triaminotriphenylmethane with a lipophilic cation of moderate size and a slightly nonplanar conjugated system. The dye is sold as the chloride.

Having no ionizable hydrogen, free base formation does not occur. However, a colorless carbinol (or pseudobase) is formed slowly under alkaline conditions. A colorless leucobase is also formed, on exposure to reducing agents, this reaction being reversible in the presence of suitable oxidants. Crystal violet reacts with resorcinol under oxidizing conditions, yielding a phenylated product used as an elastic fiber stain, see the final section of this entry.

Applications: Crystal violet is routinely used in the Gram stain and its variants; and to stain amyloid, bacterial components and vascular plant tissues; such methods are described in routine staining manuals (Bancroft and Gamble, 2002; Clark, 1981). Applications and mechanism of the Gram stain, which has been in use for more than a century, have been recently reviewed (Popescu and Doyle, 1996).

Other histological procedures include polychrome staining of epoxy resin sections (Gandor and Meyer, 1988), viability staining of cultured neurons (Marin *et al.*, 1994), and to analyze meiotic structures (Chapman and Mulcahy, 1997) using confocal optical sectioning. A nonstaining histotechnical application is the quenching of autofluorescence, for instance in biofilms prior to staining with selective fluorescent reagents (Huan *et al.*, 1996); and in fluorescein-labeled neutrophils prior to investigation by flow cytometry and confocal microscopy (Moffat *et al.*, 1996).

In addition to the Gram stain, crystal violet has various applications in microbiology, for instance in the acridine orange–crystal violet method for intracellular bacteria (Fernandez *et al.*, 1997), and for staining microsporidian spores in tissue sections and cytological smears (Moura *et al.*, 1997).

Crystal violet has numerous nonmicroscopic microbiological applications, for instance to treat nematode infections (Reynolds, 1993, p. 793), for the colorimetric assay of mycelial growth of *Candida albicans* (Abe *et al.*, 1994), and for the detection of bacterial adherence on biomedical polymers (Merrit *et al.*, 1998). It has also found repeated application as a culture broth component, especially to monitor *Salmonella* (Geue and Schluter, 1998). As a solid culture medium component crystal violet is widely applied for detection and identification of bacterial strains by the 'crystal violet reaction', for a critical account of which see Barer *et al.* (1992). An analogous procedure exists for wild yeasts (Kuhle and Jespersen, 1998).

Amongst many other applications in biology and medicine the following are illustrative. Enumeration of cultured cells, including adherent monocytes on inflamed kidney cells (Mene *et al.*, 1995), and counting of gingival fibroblasts (Anderson *et al.*, 1998). The dye is used as a tracer for exploring water flow in plants (Larson *et al.*, 1994), and assessing efficacy of barrier and moisturizing creams (Olivarius *et al.*, 1996).

Industrial uses – preparation of inks; and the coloration and surface coating of paper, for which the phosphomolybdate salt (CI 42555:2, Pigment violet 39) may be used.

Purity of commercial batches: may contain only a single major colored constituent, but more frequently contain moderate amounts of the pentamethyl and lesser amounts of the tetramethyl homologs. Overall dye contents are up to 95%, with the minimum dye content for Commission certification being 88%. Thin-layer chromatographic (Marshall, 1976) and HPLC (Milanova and Sithole, 1997) analytical methods are available. See Chapter 28 regarding Commission assay procedures.

Stability: of dry powder is good in sealed, dark containers; light fastness is poor. The dye is used to prepare **Sheridan's crystal violet**, an elastic stain recommended by routine staining manuals (Bancroft and Cook, 1994). This stain results from reacting crystal violet with resorcinol in the presence of ferric chloride as an oxidant. For a discussion of the chemical background see Proctor and Horobin (1988).

Resorcin fuchsine

Synonyms: resorcinol fuchsine, Weigert's resorcin fuchsine
For the indamine dimer $C_{39}H_{34}ClN_6$, FW 622

Absorption maximum: of dimeric species is 508 nm in 95% acid alcohol.
Solubilities: poorly soluble in water; soluble in 95% acid alcohol.

Resorcin fuchsine prepared from either pararosaniline or new fuchsine (i.e. resorcinol new fuchsine) contains oligomeric cationic compounds, probably including dimeric and trimeric indamine species, as illustrated with pararosaniline as the starting dye (Proctor and Horobin, 1988). All components have large, nonplanar aromatic systems.

Applications: Resorcin fuchsine is recommended in staining manuals (Bancroft and Gamble, 2002) for the demonstration of elastic fibers in routine histological sections. It can also be used to stain elastin in glycol methacrylate-embedded tissues (Cannon and Stuth, 1993), and it has been used to visualize oxytalan fibers (Rosenquist, 1981).

Purity of commercial batches: Even when prepared from single basic fuchsine homologs (pararosaniline or new fuchsine) more than one oligomer is usually present. Thin-layer (Crescenzi *et al.*, 1991) and paper (Rosenthal *et al.*, 1965) chromatographic analytical methods are available.

Stability: of the stain stored as the dry powder is good.

Ethyl green and methyl green

CI 42590 – see below for the complexities of nomenclature

$C_{27}H_{35}BrClN_3.ZnCl_2$, FW for the mixed bromide/chloride zinc double salt; or $C_{27}H_{35}Cl_2N_3.ZnCl_2$, FW 609 for the dichloride zinc double salt – but see below for actuality.

Absorption maxima: 635 nm in water, with a subsidiary peak at 420 nm; emission maximum is in the red.

Changes color from yellow to blue between pH 0.8–1.8

Solubilities: 7% in water; 0.1% in ethanol; insoluble in xylene.

ethyl green, CI 42590   methyl green, CI 42585

The dye sold for many years as methyl green, CI 42590, is more accurately described as ethyl green. The traditional methyl green was a related dye, CI 42585 (CI Basic blue 20), which has probably not been manufactured for decades (Green, 1990, p. 458; Lillie, 1977, p. 280). The structural formulae of these closely related dyes are both shown. The physicochemical information above, and the chemical thumbnail below, describe CI 42585 (ethyl green). This is also probably the case for the applications described, although only one explicitly stated that ethyl green was used, so ambiguity remains. Fortunately there is little or no difference between the staining properties of CI 42590 and CI 42585.

Ethyl green (CI 42590) is a triphenylmethane dye with two *N*-dimethylated substituents fully involved in the delocalized system, and a quaternized *N*-dimethylethyl substituent. Due to the presence of the two positive charges, ethyl green has a hydrophilic cation of moderate size.

The dye is most widely sold as a mixed bromide/chloride, in the form of a zinc chloride double salt. Manufacture usually involves ethylation of crystal violet with ethyl bromide, with dye finally salted out using sodium chloride. So it is unlikely that the bromide–chloride ratio is precisely 1:1.

An important difference between the N-ethylated CI 42590 and traditional methyl green is that the former is more stable and does not readily lose a N-methyl group. Hence, contamination of methyl green with crystal violet does not usually occur with dye of recent manufacture. In any event contaminating dye can be removed from aqueous solutions of old laboratory samples, which may be methyl green, by shaking with repeated changes of chloroform. The correct name, **ethyl green** (CI 42590), should be used in future accounts of applications of this dye, and is assumed in the following paragraphs.

Applications: Ethyl green is widely applied as a nuclear stain, for instance in the *methyl green–pyronine method* for distinguishing structures rich in DNA from those rich in RNA. Ethyl green is also useful as a nuclear counterstain in enzyme histochemistry, immunostaining and *in situ* hybridization, because its bluish-green color contrasts well with red, purple, brown and black products of histochemical reactions. Such methods are found in routine staining manuals (Bancroft and Cook, 1994; Bancroft and Gamble, 2002; Clark, 1981). Other uses noted in reviews and staining manuals include applications of these methods to sections of tissues embedded in water miscible and epoxy resins (Litwin, 1985), and to quantitation of DNA by methyl green staining (Chayen and Bitensky, 1991, p. 87).

There are numerous other staining applications, for instance a modified Poley's acid fuchsine–methyl green stain has been used to demonstrate ligature marks in tissues taken at autopsy (Wenyou *et al.*, 1991), for vital staining of ciliated protozoa (Foissner, 1991), and for localization of double-stranded DNA in epidermal cells examined by confocal microscopy (Ito and Otsuki, 1998). Other laboratory applications include contrast enhancement of latent finger-prints in forensic investigations (Kempton and Rowe, 1992).

Purity of commercial batches: Can be predominantly a single compound. Dye contents up to 90% of zinc-free material are available; the minimum zinc-free dye content for Commission certification being 65%. There is a thin-layer chromatographic analytic method (Lyon *et al.*, 1987). Regarding the Commission's assay procedures see Chapter 28.

Stability: of dry powder is good, in sealed dark containers; acidic staining solutions kept in the dark are stable for weeks or months; light fastness is poor.

Victoria blue dyes

Reports of dyes termed Victoria blue, with no specifying suffix, are common-place. This leads to confusion as there are several chemically distinct Victoria blue dyes. All are triaminodiphenylnaphthylmethanes, with lipophilic cations and large conjugated systems. Some, like Victoria blue B and Victoria blue 4R, have extended aromatic systems due to N-phenylation.

Victoria blue B (CI 44045) and Victoria blue 4R (CI 42563) have traditionally been most widely used as specifically named biological stains. However, at this time, and probably since the 1970s, only Victoria B is being manufactured. The Victoria blue R currently sold is chemically distinct from Victoria blue 4R. Separate entries are provided, below, for the better characterized Victoria blues

B and 4R, and only those applications specifying these particular dyes are cited. Secondary sources have further confused this issue. Either by using the nonspecific term 'Victoria blue' when the primary source specified a particular dye, or by misidentifying dyes from the primary sources (Horobin and Bancroft, 1998, cf. pp. 124 and 126 with the references cited).

Staining applications noted are restricted to those in which the dye used was described merely as 'Victoria blue'. Thus this dye is routinely used as an elastic fiber stain in histopathology, especially in Victoria blue/hematoxylin and eosin stains to visualize blood vessel invasion in tumors (Ogawa *et al.*, 1994). Victoria blue has also been used to demonstrate regions of iron and copper deposition in hepatitis and liver cirrhosis (Ishida *et al.*, 1995) and to show hepatitis B surface antigen within hepatocytes (Lapertosa *et al.*, 1984).

Victoria blue B

CI 44045, CI Basic blue 26
Synonym: fat blue B
$C_{33}H_{22}ClN_3$, FW 506

Absorption maximum: 599 nm in ethanol.
Solubilities: 1% in water; 2% in ethanol; insoluble in xylene.

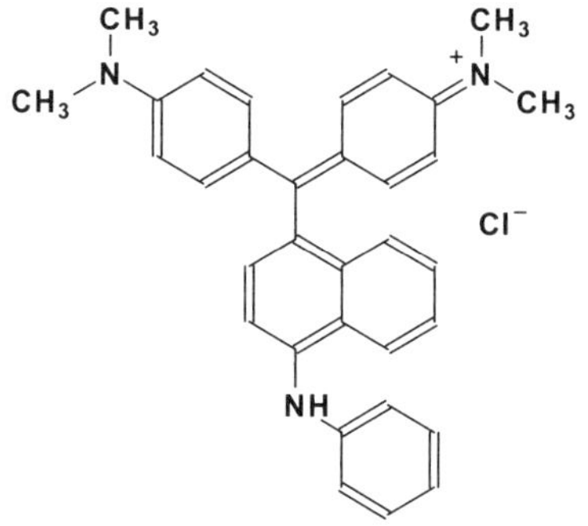

A triaminodiphenylnaphthylmethane dye, whose lipophilic cation is large for a basic dye. It also possesses a large aromatic system, which will be markedly nonplanar due to the bulk of the naphthyl substituent. The commercial material is a chloride.

Applications: Bresslau's 1921 negative stain for cilia and flagella is recommended in staining manuals (Clark, 1981). More recently the dye has been used as a cytological nuclear stain (Schulte *et al.*, 1988), and as a component of an elastic fiber stain (Tsutsumi *et al.*, 1990). Currently this dye is probably the most widely sold Victoria blue; hence it is likely that the recent applications cited in the generic Victoria blue entry in fact used Victoria blue B.

   Industrial applications – the chloride is used to color paper. As CI Pigment blue 2, polyacid salts (notably the phosphotungstate and phosphomolybdate) are used as pigments in copying paper, enamels, paints, printing inks, resins, and typewriter ribbons.

Purity of commercial batches: Two major components and some minor ones are often present. Dye content are up to 85%. There is a thin-layer chromatographic analytical method available (Marshall, 1976).

Stability: When used as an elastic fiber stain, sections do not fade even after several years, and the acidic staining solution is stable for up to 2 years. The *Colour Index* rates light fastness as poor.

Victoria blue 4R

CI 42563, CI Basic blue 8
Synonym: fat blue 4R
$C_{34}H_{34}ClN_3$, FW 520

Absorption maximum: 591 nm in water
Solubilities: 3% in water; 20.5% in ethanol; insoluble in xylene.

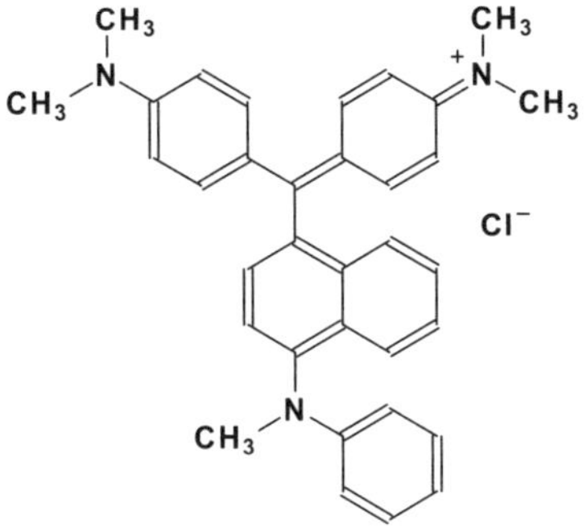

This is a triaminodiphenylnaphthylmethane dye, with a large lipophilic cation. The aromatic system is also moderately large, and will be markedly nonplanar because of the bulky naphthyl substituent.

Applications: Early applications of Victoria blue 4R included staining bacterial flagella (Isada, 1938) and spirochetes (Lipp, 1940). More recently the dye was used to demonstrate neurosecretory material (Dogra and Tandan, 1964) and insulin (Wohlrab and Schwarz, 1988). Victoria blue 4R is recommended in routine staining manuals (Bancroft and Cook, 1994) as a component of the Miller method for demonstration of elastic fibers. However, as the dye is currently not manufactured on any large scale, it is likely that Victoria blue B is actually the dye used in such procedures.

Purity of commercial batches: Dye batches often contain several major components, plus minor ones. A thin-layer chromatographic analytical method is available (Marshall, 1976).

Stability: Light fastness is poor.

TRIARYLMETHANE DYE FORMING COVALENT BONDS WITH TISSUE

Schiff's reagent

Synonyms: Despite Meloan and Puchtler's (1986) accurate comment 'the term leucofuchsin as a synonym for Schiff's reagent is incorrect and should be discontinued' the only synonym in current usage is leucofuchsin.

Absorption maxima: freshly degassed Schiff's reagent absorbs at 260 nm in water. The reaction product with DNA in tissue sections emits around 628 nm.

Solubilities: as routinely prepared, Schiff's reagent is soluble in water and 70% aqueous ethanol. The precipitated sulfate and phosphotungstate salts are soluble in aqueous acid and ethanol respectively.

Modern separation methods, allied to a variety of spectroscopic procedures, have clarified the structure of Schiff's reagent and the mechanism of its reaction with aldehydes (Bedi and Horobin, 1980; Gill and Jotz, 1976; Nauman *et al.*, 1960; Nettleton and Carpenter, 1977; but most unequivocally Robins *et al.*, 1980).

Routinely prepared Schiff's reagent contains organic components derived from pararosaniline or other basic fuchsine homologs, plus oxysulfur anions, in an aqueous solution at about pH 2. The organic components result from reaction of various nucleophilic oxysulfur anions (commonly sulfite and bisulfite, but dithionite and thiosulfate can also be used) with the central carbon atom of a basic fuchsine homolog, such as pararosaniline or new fuchsine. For instance, pararosaniline exposed to sulfite ions will generate a sulfonic acid. Such reactions greatly reduce the extent of the conjugated system, and the sulfonic acid is colorless. This is illustrated in Figure 14.4.

Both sulfonate and amino groups can exist in more than one state of ionization. Consequently, under the acidic conditions of use, more than one Schiff's reagent species can be present even when a single basic fuchsine homolog was used. With other oxysulfur anions, analogous products arise.

After reaction of Schiff reagent with tissue aldehydes, specimens are washed in excess solvent to remove sulfite in solution. Following this the loosely bound sulfonate or analogous substituent on the central carbon atom leaves, as the equivalent anion. The process regenerates an extended conjugated system, giving a colored final reaction product; see Figure 14.4. This stage of the staining process is analogous to the transformation of a colorless carbinol to a dye as hydroxide ion is removed from solution by lowering the pH.

The organic component of Schiff's reagent can be conveniently precipitated as the sulfate (Galassi, 1993). This is stable, but can readily be reconstituted as reactive Schiff's reagent. Schiff's reagent suitable for both the periodic acid–Schiff and Feulgen nucleal (see below) procedures can be prepared from either pararosaniline or new fuchsine, or of course from basic fuchsine (Churukian and Schenk, 1983; Schulte and Wittekind, 1989a).

Applications: The principal applications of Schiff's reagent as a biological stain are the Feulgen nucleal method for DNA and the periodic acid–Schiff procedure for polysaccharides. These are described in routine staining manuals (Bancroft and

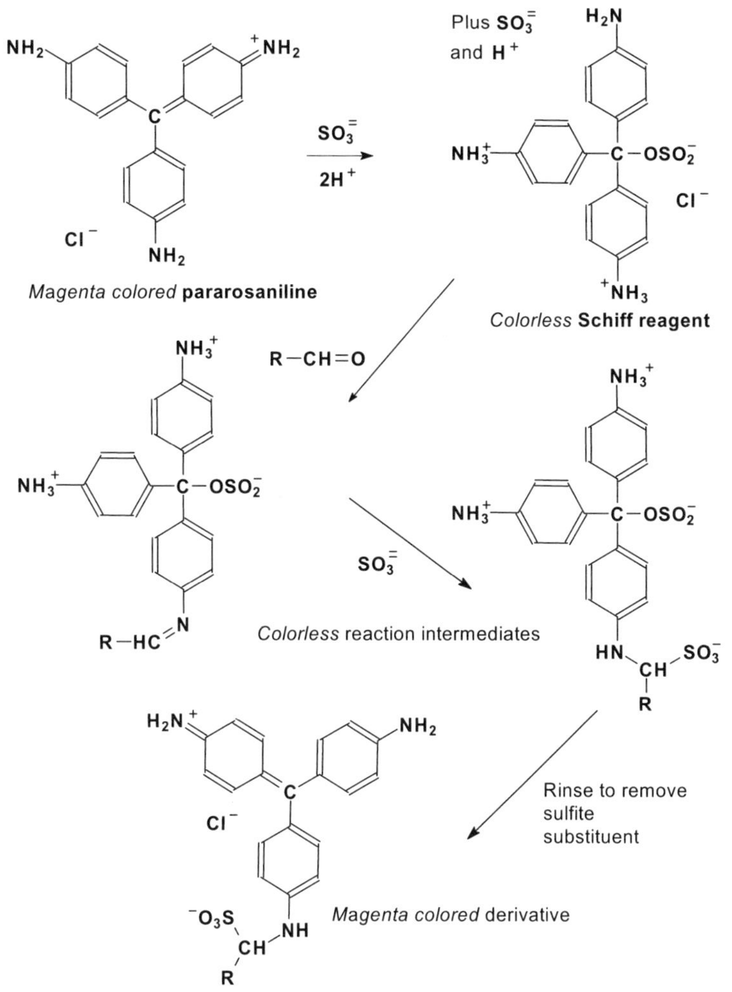

Figure 14.4 Simplified chart illustrating formation of Schiff's reagent from pararosaniline, and subsequent reaction with aldehyde.

Cook, 1994; Bancroft and Gamble, 2002; Kiernan, 1999); as are less commonly used procedures such as the ninhydrin–Schiff's stain for amino groups and the plasmal method for plasmalogen phospholipids. Both the Feulgen nucleal and periodic acid–Schiff procedures can be carried out on epoxy and water-miscible resin sections, as described in staining manuals (Hayat, 1993) and reviews (Litwin, 1985). Less common histochemical applications of Schiff's reagent include direct staining of xylem vessels in wood, with subsequent confocal imaging (Knebel and Schnepf, 1993), and coloration of aldehydes in senile plaques by the Gallyas–Schiff's procedure (Kobayashi *et al.*, 1992).

Other biomedical applications of Schiff's reagent include assessing diffusion of formaldehyde and glutaraldehyde from pulpotomized teeth (Rusmah and

Rahim, 1992), staining rectal mucins smeared onto nitrocellulose membranes (Sakamoto *et al.*, 1993) and glycoproteins on electrophoretic gels (Mori and Masuda, 1993),

Purity of commercial batches: Each basic fuchsine homolog can give rise to more than one colorless reactive derivative. There is a thin-layer chromatographic analytical method available (Bedi and Horobin, 1980)

Stability: Routinely prepared solutions are stable if loss or oxidation of sulfur dioxide is prevented. Specimens sealed in ampules are indefinitely stable; tightly sealed bottles are stable at least for many months. As long as such measures are taken, temperature is irrelevant under usual storage conditions. Sulfate-precipitated Schiff's reagent is also stable indefinitely; even if kept at room temperature in daylight no basic fuchsine has formed after 2 months.

Hazards: As solutions are strongly acidic and contain dissolved sulfur dioxide, inhalation of the vapors or contact with the skin or eyes must be avoided.

15

Hydroxy triarylmethanes

R.W. Horobin

BACKGROUND INFORMATION

Only features peculiar to hydroxytriphenylmethanes are mentioned here. For accounts of phenomena such as triarylmethane stereochemistry and carbinol formation see the introductory section of Chapter 14.

Structures of the hydroxytriphenylmethane chromophores

The electron donor–acceptor pair usually considered characteristic of the hydroxytriphenylmethane chromophore is ionized phenolic hydroxy-keto, so this chromophore is anionic. However, the widespread use of these dyes as acid–base indicators reflects the pH-dependent occurrence of a variety of ionized species with differing spectral features. Since this is a complicated topic, simple hydroxytriphenylmethanes are considered first, and phthaleins and sulfonphthaleins later. Structural descriptions are congruent with Bishop's classic monograph (1972), and with various experimental investigations (Langmyhr and Stumpe, 1965; Tamura *et al.*, 1996; Yamaguchi *et al.*, 1997), but detailed studies of most individual dyes are lacking.

Simple hydroxytriphenylmethanes can also exist as nonionized species, with either one or two of the phenyl rings being *para* substituted by hydroxy groups; the third ring containing a *para* keto group. Such asymmetric species, with restricted electron delocalization, are only weakly colored and typically absorb at the blue end of the visible spectrum. A generic example of such a hydroxy-quinone dye is illustrated by structure **B** in Figure 15.1. Such nonionized species will, in the absence of other electron donating or accepting substituents, be present in aqueous solutions around neutrality.

As the pH rises the phenolic group or groups will ionize. The resulting anion (structure **C** in Figure 15.1), which is electronically more symmetrical, will absorb more intensely and at longer wavelengths. Equilibria between species such as structures **B** and **C** of Figure 15.1, involving proton transfer and consequent spectral shifts, are responsible for most widely exploited color transitions of the hydroxytriphenylmethane indicators.

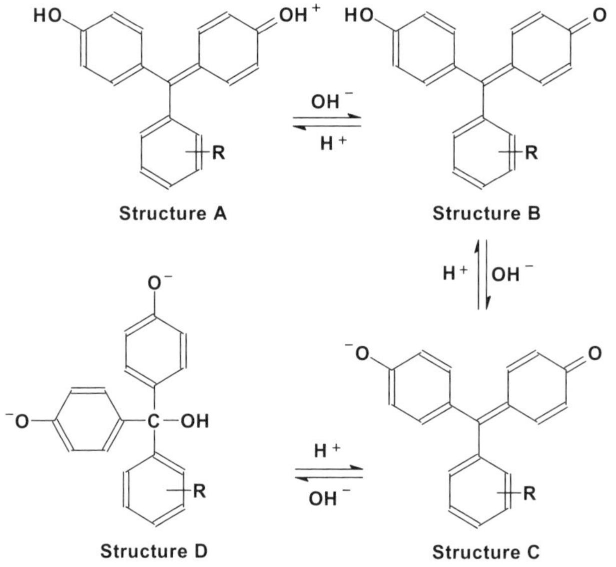

Figure 15.1 Influence of pH on species present in aqueous solutions of a generic hydroxy-triphenylmethane dye.

Under even more alkaline conditions addition of hydroxide ion to the central carbon atom occurs, producing polyanionic carbinols. In the generic example in Figure 15.1 the carbinol is shown as structure **D**. The slow addition of hydroxide ion results in the formation of a colorless species because there is a loss of donor–acceptor pairs, and the polymethine chain-length is reduced by saturation of all the bonds to the central carbon atom.

In strongly acidic solutions, the nonionized hydroxyquinone species may be protonated. This gives rise to the cationic structure **A** of Figure 15.1. Such a species, with extended electronic delocalization, will typically be strongly colored with a red-shifted absorption maximum.

Indicator properties

Phthalein hydroxytriphenylmethane dyes differ from the simple dyes discussed above, as they possess a carboxylic acid substituent *ortho* to the central carbon atom, on the phenyl ring which lacks a hydroxy/keto substituent. In moderately alkaline solutions phthaleins are simple anions corresponding to the generic structure **C** in Figure 15.1, with R being CO_2^-. However as the pH falls the carboxylate reacts intramolecularly to give a lactone ring. This involves saturation of all the bonds to the central carbon atom, and hence a marked reduction in the extent of electron delocalization. Consequently the lactone is colorless. Such an equilibrium is illustrated in Figure 15.2 for the case of phenolphthalein.

In a sulfonphthalein dye an *ortho*-sulfonic acid replaces the carboxylic acid. It has been assumed, in the previous edition of this book and elsewhere, that the colorless spirolactone   colored anion

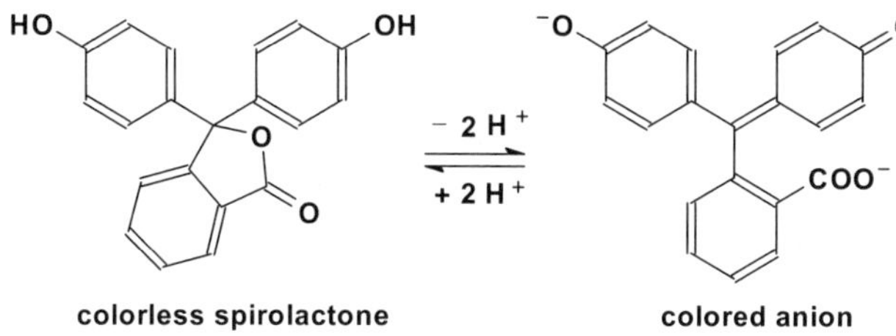

Figure 15.2 The colorless lactone and red anionic forms of phenolphthalein.

pH dependent solution behavior of sulfonphthaleins parallels that of the phthaleins; with transformation of colored anionic species, by protonation, into cyclic sulfonic esters (sultones). This, however, is often not so. The initial product is a noncyclic sulfonated hydroxyquinone; structure in Figure 15.1, with R being $-SO_3^-$. For this reason, sulfonphthaleins typically change from strongly colored species in alkaline conditions to pale yellow on protonation. This is in marked contrast with phthaleins.

How does the absence of sultones in solution relate to the fact that sulfonphthaleins are sold in solid form as sultones? The answer lies in the syntheses employed. Sulfonphthaleins are manufactured by procedures such as fusing sulfobenzoic acid cyclic anhydride with an appropriate phenol in a melt together with anyhydrous zinc chloride. In this unusual fluid a sultone is the stable reaction product. However, when the sultone is dissolved in water, the nature of the species arising largely depends upon solution pH, as described above.

Color transitions of all the hydroxytriphenylmethane dyes occur at higher pH values when the dyes are substituted with electron donating carboxylate or methyl groups *ortho* to the hydroxyl/keto groups. Conversely, the transitions occur at lower pHs when electron accepting bromo and chloro substituents are so situated.

Electric charge on the dyes

As described above, the chromophore of a simple hydroxytriphenylmethane dye in aqueous solution above neutrality will be negatively charged. These dyes become nonionic upon protonation, forming hydroxyquinones. They become polyanionic following hydroxylation at high pH. Of course, the presence of additional ionic or ionogenic substituents, although not affecting chromophoric charge, does alter overall charge. For the dyes in this chapter, the only substituents influencing dye charge are carboxylate and sulfonate. Consequently at low pHs solutions of phthaleins contain nonionic lactones, whereas sulfonphthalein solutions contain zwitterions. At sufficiently low pH all types of hydroxytriphenylmethane dyes probably form cations.

What is present in bottles of dye?

Sulfonphthaleins offered for sale may be described as free acids, as salts, or as sultones; sometimes more than one form of the same dye is sold. The actual identity of commercial materials is not always certain however, as the following examples indicate. Since hydroxyquinone free acids are typically yellow, a batch of phenol red sold as the free acid was unlikely to be as described since it was dark red in color. However, it was probably mostly the zwitterion form, which would be strongly colored. As another example, consider material listed as chlorophenol red, sultone form. This could not have been as described, since the batch was dark green in color, not colorless as the sultone would be. Indeed, few dye batches stated to be sultones are white – although admittedly some are 'beige' or 'light pink'. Probably such products contain a colorless sultone contaminated with a strongly colored zwitterion.

Fortunately the zwitterion, the free acid, and the sultone of a sulfonphthalein dye all have the same formula weight. Moreover, the species present in solution depends only on pH. The complexities described above are therefore of little practical consequence when preparing indicator solutions. The phthaleins, of course, are not problematic since the products sold as lactones are, as expected, white solids.

Some vendors offer the sul**fone** forms of sulfonphthaleins. This confusion – sulfones are chemically distinct from sultones – perhaps arises from a misunderstanding of the nonsystematic name (sulfonphthalein) used for the dye class.

Chelating properties of dyes

Typical hydroxytriphenylmethane dyes giving metal complexes carry carboxylate substituents *ortho* to the hydroxy/keto groups. Such dyes yield a wide variety of complexes – there being considerable pH dependency – with a number of metal ions, including aluminum, chromium and iron. The complexes vary in the mode of binding; both chelation and simple coordination can occur. The metal ion–dye ratios also vary. Consequently both anionic and cationic complexes occur, of various sizes and hydrophilicities.

How to find the dye that interests you

Simple hydroxytriphenylmethanes; dyes with carboxylic acids *ortho* to the central carbon, which can form lactone derivatives (traditional names are phenolphthaleins or phthaleins); and dyes with *ortho*-sulfonic acids (traditional name is phenolsulfonphthaleins) are typically used in different types of application. So each of these three groups is listed separately below, in the order given. Within each group dyes are listed by increasing formula weight of the free acid/lactone/sulfone.

SIMPLE HYDROXYTRIPHENYLMETHANES

Aurin tricarboxylic acid

CI 43810, CI Mordant violet 39

Synonyms: aurin tricarboxylic acid – or ATA – is the term used by those concerned with diabetes and thrombosis, with apoptosis and neuroprotection, and by some analytical chemists. Other chemists, and most histopathologists and microscopists, use the term **aluminon**. The previous edition of Conn preferred the FIAT textile term, chrome violet CG. For the chemical significance of these variations see below.

Free acid: $C_{22}H_{14}O_9$, FW 422

Triammonium salt: $C_{22}H_{23}N_3O_9$, FW 473

Trisodium salt: $C_{22}H_{11}Na_3O_9$, FW 488

Absorption maximum: is 542 nm in aqueous 0.1M sodium hydroxide

Used as a metallochromic indicator, see below

pK values: 8.9 and 9.8

Metal complexation behavior: Strongly colored complexes are formed with metal ions, including aluminum, beryllium, chromium and iron

Solubilities: of the free acid are 0.7% in water and 6% in ethanol; of the triammonium salt are 0.5% in water and 0.1% in ethanol; and of the sodium salt are 3% in water and 0.02% in ethanol

Aurin tricarboxylic acid is a weakly acidic hydroxytriphenylmethane, whose overall size is moderate as is that of its nonplanar aromatic system. The free acid, illustrated, is lipophilic; whereas the trianionic species is strongly hydrophilic. The presence of three carboxylic acid substituents *ortho* to hydroxyl and keto groupings results in metal ion complexation of some complexity. One study found, at various pHs, two aluminum, three chromium(III), and three iron(III) complexes (Bobtelsky and Ben-Bassat, 1956).

The dye is commercially available as the free acid and as its trisodium and triammonium salts. The first of these is often termed **aurin tricarboxylic acid**, the second **aurin tricarboxylic acid trisodium salt** (or some textile dyeing name such as **chrome violet CG**), and the third **aluminon**. However, it is doubtful if literature reports use these terms consistently, and in any event once dissolved in water the chemical species present depend on the pH and buffer salts used.

Applications: Under the name aluminon, this dye has been widely used for histochemical demonstration of aluminum. Aluminon has been shown to be the technically preferred dye to assess bone aluminum levels in kidney dialysis patients (van Landeghen *et al.*, 1998), despite its relatively low sensitivity (Ballanti *et al.*, 1995). Other recent applications include investigation of aluminum-lines in osteomalacia (Boyce *et al.*, 1992). Aluminon has also been used to prepare metal complexes with good nuclear staining characteristics (Berube *et al.*, 1964). Loss of macroscopic staining of agar or acrylamide gels containing aurintricarboxylic acid and aluminum ions is used to detect chemical changes in the root-zones of plants (Dinkelaker *et al.*, 1993).

Purity of commercial batches: These can contain a single major component, with the dye content of analytical grade material up to 97%. A thin-layer chromatographic analytical method is available (Marshall and Horobin, 1975).

Stability: Light fastness of the chromium complex is low.

Chrome azurol B

CI 43830, CI Mordant blue 1

Synonyms: chromoxane pure blue B (FIAT name preferred in the previous edition), eriochrome azurine B or azurol B, solochrome azurine B (name favored by current staining manuals). Chrome azurol B and Mordant blue 1 are the terms used by vendors.

Free acid: $C_{23}H_{16}C_{12}O_6$, FW 459
Disodium salt: $C_{23}H_{14}C_{12}Na_2O_6$, FW 503

Absorption maximum: in water is 424 nm

Metal complexation behavior: Many metal ions are complexed, including aluminum, beryllium, chromium, copper and zinc

Solubilities: soluble in water and ethanol

This dye is of moderate size overall, as is its aromatic system, which is markedly nonplanar due to the chloro substituents. The dye carries several acidic groups, two of which are ionized at neutrality. Hence, under usage conditions chrome azurol S is a hydrophilic anion, but the free acid shown in the structural formula is lipophilic. The carboxylic acid substituents *ortho* to hydroxyl and keto groupings result in effective metal complexation. The commercial product is nominally the disodium salt.

Applications: As solochrome azurine B the dye was used by Pearse (1957) for histochemical demonstration of aluminum and berylium. Current staining manuals recommend it in this role, both for paraffin (Bancroft and Cook, 1994) and glycol methacrylate resin (Bancroft and Gamble, 2002) sections. The dye has also been used to demonstrate aluminum in plant material (Denton and Oughton, 1993).

Industrially – this dye is used for textile dyeing and printing, for coloration of leather and, in the form of the barium salt, as a printing pigment for books and lithography.

Purity of commercial batches: Dye contents are up to 80%. A thin-layer chromatographic method is available (Moukova *et al.*, 1979).

Stability: Light fastness of the chromium complex is moderate.

A structurally related dye, **chromoxane pure blue BLD** (CI 43825, Mordant blue 29), shown below, has also been used for histochemical demonstration of berylium (Wyatt, 1972) and chromium (Suzuki *et al.*, 1978).

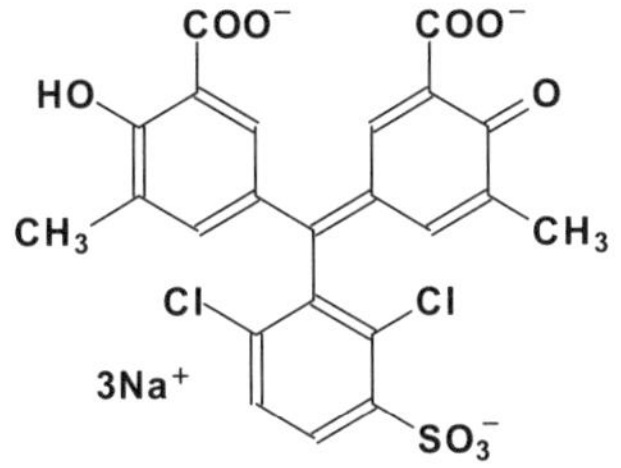

PHTHALEINS – DYES WITH LACTONE SPECIES

Phenolphthalein

Lactone: $C_{20}H_{14}O_4$, FW 318

Absorption maxima: 552 nm, with a subsidiary peak at 374 nm, in water above pH 10. Changes from colorless to violet–red between pH 8.3–9.8, goes red in extremely acidic solutions, and colorless in extremely alkaline solutions.

pK values: pKs of 9.1 and 9.5 correspond to opening of the lactone ring and losses of protons; pK of 12 corresponds to addition of hydroxide to yield a carbinol

Solubilities: the lactone is slightly soluble (0.1%) in water, and soluble in ethanol to 3%.

In aqueous neutral solutions phenolphthalein is present as a nonionic, lipophilic colorless lactone. In alkaline solutions a hydrophilic dianion is formed. These changes, illustrated in Figure 15.2, involve proton losses and hence occur rapidly. The small red anion has a larger conjugated system than the colorless lactone. When the pH is raised further, a pseudo-base (carbinol) is slowly formed by reaction of hydroxide ion with the central carbon. The material available commercially is the lactone.

Applications: Phenolphthalein has long been used as a pH indicator in biology and medicine. Recent applications include assay of total acidity in vinegar (Honorato *et al.*, 1999) and assessment of penetration of sodium hydroxide into potatoes following chemical peeling (Garrote *et al.*, 1998). A traditional clinical application is as a laxative, which role continues to generate research interest (Hopp *et al.*, 1998).

Purity of commercial batches: Dye contents of analytical grade samples are 99%. Thin-layer chromatographic (Brinkman and De Vries, 1972) and HPLC (Gagliardi *et al.*, 1999) analytical methods are available.

Derivatives of phenolphthalein – These have many applications in biology and medicine. For instance **phenolphthalein diphosphate** has been applied as a component of an indicator medium for the histochemical screening of genomic libraries in molecular biology (Riccio *et al.*, 1997). **Phenolphthalein glucuronic acid** (under various synonyms, such as phenolphthalein glucuronide) is widely used to assess liver and enterohepatic circulation (Heppert and Davies, 1999).

SULFONPHTHALEINS

Phenol red

Synonym: phenolsulfonphthalein
Free acid: $C_{19}H_{14}O_5S$, FW 354
Monosodium salt: $C_{19}H_{13}NaO_5S$, FW 376

Absorption maxima: for the sodium salt in methanol is 423 nm; and 557 nm in pH 9 aqueous buffer (i.e. as the dianionic form).

The dye changes from yellow to red between pH 6.8–8.4

pK values: 1.3, 7.9 and 7.6

Solubilities: the free acid is sparingly soluble in water and ethanol, 0.3% and 0.4% respectively; the sodium salt is soluble to 10% in water and 20% in ethanol

In weakly acidic aqueous solutions, phenol red is present as a sulfonated hydroxyquinone. In alkaline solutions the dye forms a hydrophilic dianion. This small, red anion has a larger conjugated system than the yellow hydroxyquinone. At extremes of pH further species arise. Very acid conditions yield a strongly colored zwitterion, whereas very alkaline conditions result in slow formation of a colorless pseudo-base (carbinol) by reaction with hydroxide ion. Phenol red is sold both as the monosodium salt and, nominally, as the free acid. This latter, however, is described as strongly colored not yellow, so is probably a zwitterion. The species illustrated is the sultone, to allow comparison of size and lipophilicity with other dyes.

Applications: Phenol red has been widely used in biology and medicine as a pH indicator. Recent applications of this type include determination of soil acidity (Coscione *et al.*, 1998), and incorporation into microbiological culture media, for example, to identify *Escherichia coli* strains (Kang and Fung, 1999).

The dye is also used as a tracer, for instance to assess the permeability of the ileal mucosal membrane (Grotz *et al.*, 1999), and in dentistry to assess bacterial leakage through root-filling materials (Adamo *et al.*, 1999); and as marker of test meals in gastric emptying studies (Bucaretchi *et al.*, 1999).

Clinical diagnostic applications include tear function assessment in ophthalmology (Patel *et al.*, 1998), and in conjunction with endoscopy for investigation of gastric and duodenal ulcers (Tsuji *et al.*, 1999).

Purity of commercial batches: These often contain a single major component plus a major impurity. Dye contents of analytical grades are up to 95%. A thin-layer chromatographic analytic method is available (Hay *et al.*, 1994).

Cresol red

Synonym: cresolsulfophthalein
Free acid: $C_{21}H_{18}O_5S$, FW 382
Sodium salt: $C_{21}H_{17}NaO_5S$, FW 404

Absorption maxima: of the free acid is 570 nm in water; of the sodium salt, 425 nm in methanol

Cresol red changes from orange to yellow between pH 0.2–1.8, and from yellow to violet between pH 7.0–8.8.

pK values: 1.4, 8.2

Solubilities: of the sodium salt, 3% in water and 2% in ethanol; of the nominal (see below) free acid 0.1% in water and 0.3% in ethanol

Weakly acidic aqueous solutions of cresol red contain a sulfonated hydroxyquinone species. In alkaline solutions the dye is present as a hydrophilic dianion. This small anion is violet in color, having a larger conjugated system than the yellow hydroxyquinone. Under strongly acidic conditions, a zwitterion is formed. The dye is available commercially both as the monosodium salt and, nominally, as the free acid. This latter is strongly colored not yellow, so is probably the zwitterion. The species illustrated is the sultone, to allow comparison of size and lipophilicity with other dyes.

Applications: Cresol red is a widely used indicator for many routine pH assessments, for instance in microbiological culture media (Muniesa Perez *et al.*, 1996). However, the indicator is also used in less commonplace ways, for instance, to assess the effect of calcium hydroxide-containing root canal fillings on the alkalinity of teeth (Staehle *et al.*, 1995), and as a component of instrumentation such as a long-range fiber-optic pH sensor (Gupta and Sharma, 1998). Cresol red has other laboratory applications, as a constituent of polymerase chain reaction reagent systems (Hodges *et al.*, 1997).

Purity of commercial batches: Dye contents of indicator grade are up to 95%. There is a thin-layer chromatographic method for analysis of cresol red (Sheen, 1971).

Eriochrome cyanine R

CI 43820, CI Mordant blue 3
Synonyms: chromoxane cyanine R, solochrome cyanine R
Free acid: $C_{23}H_{18}O_9S$, FW 470
Monosodium salt: $C_{23}H_{15}Na_3O_9S$, FW 492
Trisodium salt: $C_{23}H_{15}Na_3O_9S$, FW 536

Absorption maxima: 435 nm in pH 7 aqueous buffer; 512 nm in methanol.
The color of the dye changes from yellow–orange to red at pH 1.0–2.0; from red to yellow at pH 6.0–7.0; from yellow to blue at pH 11.0–12.0
pK values: ca 0, 1.8, 5.7 and 11.8
Metal complexation behavior: Gives strongly colored products with metal ions such as aluminum, chromium and iron
Solubilities: 5% in water, 2% in ethanol

This dye is of moderate overall size, as is its aromatic system, which is markedly nonplanar due to an *ortho*-sulfonate substituent. The dye carries several acidic groups, and under neutral conditions three groups will be ionized. Under strongly acidic conditions the dye is present as a zwitterion. Dye supplied commercially is stated to be the trisodium salt.

The presence of carboxylic, hydroxyl and keto groupings permits formation of a wide variety of metal complexes. Those significant histochemically are formed under strongly acid conditions with ferric iron. The probable structures of these colored species (Kiernan, 1984a) are shown.

Applications: Iron complexes of eriochrome cyanine R are recommended for rapid staining of both cell nuclei and myelin sheaths in current manuals (Kiernan, 1984b, 1999, p. 114–118). Eriochrome cyanine R also gives colored complexes with aluminum, and this reaction has been used to demonstrate the metal histochemically in barley (Ma *et al.*, 1997).

Purity of commercial batches: usually contain one major and several minor components. Overall dye contents are up to 50%. A thin-layer chromatographic analytical method is available (Dixon *et al.*, 1970).

Stability: Strongly acidic staining solutions of the ferric complexes are stable for many years; light fastness of the chromium complex is moderate.

Bromocresol purple

Synonym: dibromo-*o*-cresolsulfonphthalein
Sultone: $C_{21}H_{16}Br_2O_5S$, FW 540
Monosodium salt: $C_{21}H_{15}Br_2NaO_5S$, FW 562

Absorption maxima: of the 'sultone' is 419 nm in methanol; and of the dianion is 585 nm, with a subsidiary peak at 379 nm, in 0.1 M aqueous sodium hydroxide
The dye changes from yellow to purple between pH 5.2–6.8;
pK values: –0.8, 6.3
Solubilities: of the sultone is 2% in water, and 8% in ethanol; and of the monosodium salt is 8% in water and 2% in ethanol

In acidic aqueous solutions bromocresol purple is a sulfonated hydroxy-quinone. Under alkaline conditions the dye forms a hydrophilic, moderately sized dianion which is purple in color, having a larger conjugated system than the yellow hydroxyquinone. The dye is available commercially both as the sultone, illustrated, and as the monosodium salt.

Applications: Bromocresol purple was used as a vital stain to trace diffusion of fluid out from lymphatics by McMaster and Parsons (1938), and somewhat later as a probe of the blood–brain barrier (Millen and Hess, 1958). Currently the dye is used for live/dead differentiation of yeasts (Kurzweilova and Sigler, 1995).

Bromocresol purple is widely used as a pH indicator, applications include use as a microbiological media constituent, both of nutrient broth (Shelef and Firstenberg-Eden, 1997) and agar (Weenk *et al.*, 1995); demonstration of changes in the rhizosphere of onion roots following fungal infection (Bago and Azcon Aguilar, 1997); and in modern instrumentation systems for ammonia detection (Challener *et al.*, 1999).

Bromocresol purple is also used analytically, for example, to stain carbonic anhydrase isoenzymes on cellulose acetate electrophoresis membranes (Williams and Colman, 1994).

Purity of commercial batches: Dye contents of both 'sultone' and monosodium salt materials are up to 90%. There is a thin-layer chromatographic analytical procedure (Aliotta and Roso, 1971).

Bromothymol blue

Synonyms: bromthymol blue, dibromothymolsulfonphthalein
Sultone: $C_{27}H_{28}Br_2O_5S$, FW 624
Monosodium salt: $C_{27}H_{27}Br_2NaO_5S$, FW 646

Absorption maxima: of the 'sultone' is 420 nm in methanol; and of the dianion is 615 nm, with a subsidiary peak at 392 nm, in 0.1 M aqueous sodium hydroxide
Bromothymol blue changes from yellow to blue between pH 6.0–7.6;
pK values: –1.2, 7.0
Solubilities: the sodium salt is soluble to 6% in water and in ethanol. The sultone is only slightly soluble (0.1%) in water, but soluble to 2% in ethanol.

In acidic aqueous solutions bromothymol blue is a sulfonated hydroxyquinone. In alkaline solutions the dye is present as a moderately sized dianion. This is blue, having a larger conjugated system than the yellow hydroxyquinone. Even the dianion is lipophilic, and the hydroxyquinone is much more so. This dye is sold both as the sultone, illustrated, and the monosodium salt.

Applications: Bromothymol blue has in the past been used as a vital stain to trace movement of fluids from the lymph (McMaster and Parsons, 1938), and also for demonstration of fungal hyphae within plant roots (Garrett, 1937). However, doubtless due to the dye's color transition being so precisely at neutrality, the most important application is as a pH indicator. Uses in biology and medicine include assessment of the soluble acidity within single rice grains (Lii *et al.*, 1999), and various microbiological culture media, for example, in an agar gel medium for enumeration of *Bacillus cereus* (Intveld *et al.*, 1999). The dye is also used in modern instrumentation systems, for example, as a component of a fiber-optic system for chloride and potassium determinations (Werner *et al.*, 1999).

Purity of commercial batches: Typical samples are heterogeneous, with overall dye contents up to 95%. There is a thin-layer chromatographic analytical method available (Saris and Seppala, 1969).

Bromophenol blue

Synonyms: bromphenol blue, tetrabromophenolsulfonphthalein
Sultone: $C_{19}H_{10}Br_4O_5S$, FW 670
Monosodium salt: $C_{19}H_9Br_4NaO_5S$, FW 692

Absorption maxima: of the 'sultone' is 598 nm in methanolic 0.005 M sodium hydroxide; and of the monosodium salt is 589 nm in water, with a subsidiary peak at 383 nm.

The dye changes from yellow to blue between pH 3.0–4.6;
pK values: –1.2, 3.6
Solubilities: the monosodium salt is soluble in water to 3%, but poorly soluble (0.7%) in ethanol; the 'sultone' is poorly soluble in both water (0.3%) and ethanol (0.9%).

In acidic aqueous solutions bromophenol blue is a sulfonated hydroxyquinone. In neutral solutions, and at higher pHs, a weakly hydrophilic dianion is formed. This moderately sized, blue species has a larger conjugated system than the yellow hydroxyquinone. If the pH is raised further, a colorless pseudo-base (carbinol) is slowly formed by reaction of hydroxide ion with the central carbon. Both the sultone, illustrated, and the monosodium salt are available commercially.

215

Applications: Bromophenol blue has been used as a vital stain to probe the blood–brain barrier (Millen and Hess, 1958) and as protein stain (Menzies and Roberts, 1963). The latter application, in conjunction with mercuric chloride, provides semiquantitative protein staining. This technique has been briefly reviewed by Chapman (1975), and used recently by Majumdar *et al.* (1996) and Arellano *et al.* (1999).

More commonly this dye has been used as a pH indicator, for instance, in modern instrumentation such as a long-range fiber-optic pH sensor (Gupta and Sharma, 1998). Other laboratory applications include agglutination of sperm in studies of the reproductive biology of fish (Ota *et al.*, 1996), and as a stain for enhancement of footware marks in forensic science (Glattstein *et al.*, 1996).

Purity of commercial batches: Typical samples contain one major and several minor components, of varied colors; and have dye contents of up to 95%. There is a thin-layer chromatographic method for analysis of the dye (Chapman, 1975).

Stability: The dyes fades in strongly alkaline conditions due to carbinol formation.

Xylenol orange

Free acid: $C_{31}H_{32}N_2O_{13}S$, FW 673
Trisodium salt: $C_{31}H_{29}N_2Na_3O_{13}S$, FW 739

Absorption maximum: 580 nm in 0.1N sodium hydroxide; excited in the ultra violet at 365 nm, it emits at 610 nm when bound to bone. The dye can also be excited in the blue.

The dye changes color from yellow to red between pH 6.4 and 10.4;
pK values: –1.7, –1.1, 2.6, 6.4, 6.5, 10.5 and 12.3

Metal complexation behavior: The dye is used as an indicator for detection and assay of various metal ions, for example, bismuth, cadmium, lead, mercury and zinc.

Solubilities: of the trisodium salt is 20% in water; trivially (0.04%)in ethanol

Xylenol orange carries six potentially anionic substituents, one being a sulfonate. Consequently it is anionic under staining conditions, and during most laboratory applications. The presence of multiple carboxylates, plus keto and phenolic groups, underlie formation of complexes with many metal ions. Xylenol orange is sold both as the free acid and as the sultone. Less ambiguously, it is available as the tri- and tetrasodium salts. A plausible trisodium salt is illustrated.

Applications: Xylenol orange has been used as a vital stain, for marking calcification in developing bone and other hard tissues (Rahn and Perrin, 1971). Such applications are still reported, as in a study of bone injury and repair (Schemitsch *et al.*, 1998). Murray and Ewen (1992) used it in a fluorescent histochemical method for detection of alkaline phosphatase.

Purity of commercial batches: Dye contents are up to 90%.

Bromocresol green

Synonym: bromcresol green
Sultone: $C_{21}H_{14}Br_4O_5S$, FW 698
Monosodium salt: $C_{21}H_{13}Br_4NaO_5S$, FW 720

Absorption maxima: of the 'sultone' is 423 nm in methanol; of the monosodium salt is 617 nm in water, with a subsidiary peak at 400 nm
The dye changes from yellow to blue–green at pH 3.8–5.4;
pK value: 4.6
Solubilities: the monosodium salt is 4% soluble in water, and 6% in ethanol; the 'sultone' is 4% soluble in ethanol, and slightly (0.6%) soluble in water.

In acidic aqueous solutions bromochlorophenol blue is present as the sulfonated hydroxyquinone. At higher pHs the dye forms a large, weakly lipophilic dianion. This is blue–green, having a more extended conjugated system than yellow hydroxyquinone. Bromocresol green is sold both as the sultone, illustrated, and the sodium salt.

Applications: Bromocresol green has been used as a vital stain to investigate permeability of the blood–brain barrier (Millen and Hess, 1958). However, the principal use in biology, medicine and more generally is as a pH indicator. It is a component of microbiological culture media for isolation of lactobacilli from feces (Hartemink *et al.*, 1997); and is used to assess proton fluxes in plant roots by videodensitometry of agarose gel impregnated with the dye (Calba and Jaillard, 1997). Bromocresol green is also used as a selective inhibitor of anion transport in kidney physiology (Masereeuw *et al.*, 1996).

Industrially bromocresol green is used in lithography.

Purity of commercial batches: Dye contents of the sultone are up to 95%, and of the monosodium salt up to 90%. A thin-layer chromatographic analytical method is available (Aliotta and Roso, 1971).

16

Xanthenes

R.W. Horobin

BACKGROUND INFORMATION

Issues of nomenclature

Use of terms such as succinein and rosamine to describe subgroups of xanthene dyes is no longer current, so they are omitted here. Note that naming of xanthene dyes in the biological application literature can be misleading. For instance, ester derivatives of dyes such as BCECF and carboxy-SNARF 1 are often referred to by the names of the free acids; although use of a free acid in place of an ester would usually be unsuccessful.

Even when not casually erroneous, routine nomenclature can still be confusing. For instance, fluorescein derivatives such as carboxyfluorescein and fluorescein isothiocyanate are sold in different isomeric forms. These differ in the location of substituents on the pendant phenyl ring. Unfortunately there are two numbering schemes for the atoms of this ring, with the *Colour Index* using one and *Chemical Abstracts* using the other. The former is most widely used, giving rise to names such as 5- or 6-carboxyfluorescein for the different isomers, and 5(6)- or 5(and 6)-carboxyfluorescein for isomeric mixtures. The *Chemical Abstracts* scheme results in names such as 4(5)-carboxyfluorescein for the same isomeric mixture. The index of this book includes all descriptors in general use.

Structural features

Xanthene dyes can be regarded as derivatives of diaryl- or triarylmethanes in which two aryl rings are bridged by an oxygen atom, as shown in Figure 16.1.

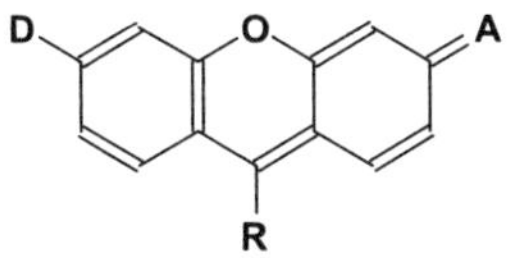

Figure 16.1 Generic structure of the xanthene chromophore. **A** and **D** are respectively electron donor and acceptor groups, **R** is usually a hydrogen atom or a substituted phenyl ring.

There are two different electron donor–acceptor pairs found in the chromophores of commonly used xanthene dyes. The first has an amino group as donor with a cationic quaternary nitrogen as acceptor. An example of such an aminoxanthene dye is rhodamine 123, for which D is $-NH_2$ and A is $-NH_2^+$ in Figure 16.1. In other dyes of this type the donor and acceptor groups can be secondary or tertiary amino groups. In hydroxyxanthenes there is an ionized phenolic hydroxy group as donor with a keto group as acceptor. An example is fluorescein, for which D is $-O^-$ and A is $=O$. In commonly used xanthene dyes the group R may be either a hydrogen atom (e.g. pyronine Y) or a substituted phenyl ring (e.g. fluorescein or rhodamine 6G).

A marked structural modification of the xanthene chromophore can occur in hydroxyxanthene dyes carrying an *ortho* carboxylic acid substituent on the phenyl group R. This permits formation of a cyclic spirolactone (see Chapter 8, Figure 8.1 for the case of fluorescein). Such lactones, due to greatly reduced conjugation, are colorless. They possess two aromatic hydroxy groups, so a variety of ester derivatives may be synthesized.

The xanthene chromophore is a flat heterocyclic aromatic unit. Hence the overall form of dyes in which R = H is planar. However, when R = phenyl, for example, in fluorescein and rhodamine 6G, the pendant aromatic group is forced by steric effects to rotate out of the plane of the chromophore. The spiro-lactone derivatives are also markedly nonplanar, as the central carbon atom carries four approximately tetrahedrally disposed bonds.

Dye properties influencing staining

Under routine staining conditions the chromophore of aminoxanthene dyes is cationic, and that of hydroxyxanthenes anionic. Overall *electric charge* is also influenced by ionizable substituents. For dyes described in this chapter, such substituents are anionic or potentially so. Consequently aminoxanthenes may be cationic (e.g. rhodamine 123), zwitterionic (e.g. rhodamine B) or anionic (e.g. sulforhodamine G). Hydroxyxanthenes are all anionic under neutral or alkaline conditions, with cations only present in strongly acid solutions. Lactone deriva-tives and their esters, and also leucodyes, may be nonionic.

Many xanthene dyes carry carboxylic acid substituents, whose charge is pH dependent. Often, as with carboxyfluorescein, ionization of such substituents merely increases the negative electric charge. Carboxylated aminoxanthenes (e.g. rhodamine B) are more complex; and a number of species of different charges arise across the pH range used in biology.

Changes in *hydrophilicity/lipophilicity* parallel changes in ionization. Examples are provided by the lipophilic lactone esters of hydroxyxanthenes, which derive from hydrophilic acidic hydroxyxanthenes. Such differences are exploited as the basis of viability probes (e.g. fluorescein diacetate). Other dyes, especially the polycarboxylated calcium probes, are available both as very hydrophilic free acid salts (the active chelators) and as acetoxymethyl (AM) esters. The latter are usually lipophilic, or even superlipophilic, and permit internalization of the probes into live cells. The salts on the other hand require maneuvers such as microinjection to achieve cellular internalization.

Some xanthene dyes carry highly *reactive substituents.* For instance isothio-cyanatofluorescein and rhodamine derivatives are used to label molecular probes, see Chapter 7 and below. Other reactive xanthenes include lactone esters, which can be hydrolyzed to the free acids; leucodyes, which can be oxidized to yield the parent dye; mercury derivatives which react with tissue nucleophiles such as thiols; and polycarboxylated dyes, which can chelate calcium ions.

Purity and identity issues

Fluorescein and some of its halogenated derivatives are offered for sale in various forms. However, their identities are not always certain, especially with neutral species. Thus material sold as the free acid form of fluorescein is some-times maroon or red in color, and may be contaminated with a zwitterion.

Some xanthene dyes are sold both as mixtures of isomers and, at higher cost, as pure isomers. However, in the application literature a dye may be described as, say, carboxyfluorescein or fluorescein isothiocyanate, with no indication of which isomer or isomers were used. Confusions concerning naming of isomers have already been mentioned.

A safety issue

Reactive dyes such as isothiocyanates form covalent bonds with biomolecules, and hence have potential risks even at low exposure levels. Such compounds, particularly if supplied as fine powders which are more easily inhaled, could induce allergic responses.

How to find the dye that interests you

Overall electric charges are commonly pH dependent, so where possible placement of entries is in keeping with the charge carried by the technically relevant species. Thus fluorescein diacetate is discussed as a nonionic compound, since the nonionic ester itself is used for detection of intracellular esterase activity.

Free acid and ester forms of some dyes have distinct applications, for example, carboxyfluorescein and carboxyfluorescein diacetate, and each form has a separate entry. However, when a dye derivative has only limited appli-cation, for example, iodoacetamidofluorescein, it is described in an addendum to the principal dye's entry; in this case of fluorescein.

NONIONIC DYES

Fluorescein diacetate

Synonym: FDA
$C_{24}H_{16}O_7$, FW 416

Absorption maximum: 490 nm in methanol, nonfluorescent
Solubilities: slightly soluble in water (0.1%) and ethanol (0.3%), soluble in dimethylformamide.

This is a nonionic, lipophilic lactone of fluorescein in which both phenolic groups are derivitized as acetates, as illustrated. The overall size is moderate, the conjugated system is small. On hydrolysis of the esters, catalyzed by cellular esterases, the lactone ring opens yielding fluorescein, *qv*.

Applications: Fluorescein diacetate (FDA) is a self-indicating esterase substrate, so the dye has been extensively applied to the staining of living cells, for instance as a viability stain of cultured neurons (Katsube *et al.*, 1999) and plant cells (Tandon *et al.*, 1999). Since the ester is lipophilic and many cell types contain esterases, FDA has sometimes been used merely as a way to introduce fluorescein into live cells, for example, the insertion of fluorescein to investigate the cytoplasmic continuity between plant cells (Baron-Epel *et al.*, 1988).

This reagent is noted in flow cytometry manuals for esterase and cholinesterase assays (Macey 1994), and for viability testing and cell sorting of algae and plant protoplasts (Darzynkiewicz and Crissman, 1990). Application in microbiology for viability assessment has been carried out microscopically, for example, on *Giardia* cysts (Hale *et al.*, 1985), and more recently using flow cytometry, on mycobacteria (Moore *et al.*, 1999). Other microbiological applications include quantification of fungal growth in wood (Boyle, 1998) and of microbial activity in soil (Haynes and Williams, 1999).

Purity of commercial batches: Dye contents are up to 98%.
Stability: The dry powder is stable when stored frozen and desiccated to prevent nonenzymic hydrolysis.

Carboxyfluorescein diacetate

Note: information in this entry applies to both isomeric forms.

Synonyms: CFDA, DCF. There are two isomeric forms, usually described as 5- and 6-carboxyfluorescein diacetate, mixtures of the isomers being indicated as 5(6)-carboxyfluorescein diacetate.
Free acid: $C_{28}H_{16}O_9$, FW 460

Absorption maximum: below 300 nm in water, nonfluorescent; free acid similar spectrally similar to fluorescein, *qv*.
Solubilities: almost insoluble in water, soluble in dimethylsulfoxide

This is the lactone of carboxyfluorescein, in which both phenolic groups are derivitized as acetates. The commercially available free acid form, illustrated, is strongly lipophilic. The overall size is moderate, the conjugated system is small. On hydrolysis of the esters, catalyzed by cellular esterases, the lactone ring opens yielding carboxyfluorescein. Both the 5- and 6-isomers, and the isomeric mixture, are available commercially.

Applications: The widespread occurrence of intracellular esterases able to hydrolyze carboxyfluorescein diacetate makes this dye an effective viability stain for cells as varied as human oocytes (Oktay *et al.*, 1997) and bacteria (Miskin *et al.*, 1998). In a related application area the dye is used to investigate membrane integrity, especially in sperm (Gadea *et al.*, 1998). Exposure of cells to the membrane-permeant carboxyfluorescein diacetate – with consequent intracellular generation of carboxyfluorescein – is also used as a procedure for introducing this latter dye into live cells, for instance for measurement of intracellular pH (Breeuwer *et al.*, 1997), and assessment of the role of gap junctions for intercellular connectivity (Carruba *et al.*, 1999).

Purity of commercial batches: Dye content of typical samples is up to 95%. There is a thin-layer chromatographic analytic method (Bruning *et al.*, 1980).

Stability: of dry powder stored desiccated in the freezer is good.

Dihydrorhodamine 123

Synonyms: DHR 123, Rho 123
$C_{21}H_{18}N_2O_3$, FW 346

Absorption maximum: 289 nm in methanol
Solubilities: soluble in dimethylformamide and dimethylsulfoxide

Dihydrorhodamine 123 is a nonionic, lipophilic leuco-dye derived from rhodamine 123. The overall size, and that of the conjugated system, is small. On oxidation, catalyzed by intracellular enzymes such as peroxidases, or by oxygen free radicals, the aminoxanthene chromophore is regenerated, yielding rhodamine 123.

Applications: This chemical transformation is variously exploited in biology, for instance, to assess free radical production in hippocampal slice cultures (Velazquez *et al.*, 1997), and to evaluate intracellular hydrogen peroxide formation in single neutrophils obtained from cytospin preparations (Szucs *et al.*, 1998). Flow cytometric applications include detection of the respiratory burst in cultured neutrophils (Stein *et al.*, 1999), and assessment of increased production of oxidants following phagocyte stimulation (Haugen *et al.*, 1999).
Purity of commercial batches: Dye contents of typical samples are up to 95%.
Stability: of the dry powder, stored desiccated in the freezer, is good; solutions are liable to oxidation and should be stored in the dark under nitrogen.

Dichlorofluorescin diacetate

Synonyms: 2′,7′-dichlorofluorescin diacetate, H_2DCFDA – perhaps originally due to a typographical error, this is often described as the entirely different material 2′,7′-dichlorofluorescein diacetate.
Free acid: $C_{24}H_{16}Cl_2O_7$, FW 487

Absorption maximum: 258 nm in methanol
Solubilities: soluble in dimethylsulfoxide and ethanol

The free acid form illustrated is a nonionic, strongly lipophilic leuco-dye diester derived from 2′,7′-dichlorofluorescein. The overall size, and that of the aromatic system, is moderate. Hydrolysis of the acetate groups, followed by oxidation, regenerates the hydroxyxanthene chromophore, yielding 2′,7′-dichlorofluorescein.
Applications: Intracellular occurrence of these transformations, mediated by esterases and peroxidases, provides the chemical basis of various biological applications of the dye. For instance, 2′,7′-dichlorofluorescin diacetate has been used as a microscopic fluorescent probe of viable mitochondria to assess the role of ceramide in the generation of oxidative stress (Garcia Ruiz *et al.*, 1997); and to demonstrate the production of hydrogen peroxide in living pulmonary endothelia (Minamiya *et al.*, 1998). The dye has been used in an analogous way as a flow cytometric stain, to investigate reactive oxygen generation within endothelial cells (Cominacini *et al.*, 1998).
Purity of commercial batches: A trace of free fluorescent dye is often present.
Stability: The dry powder will hydrolyze and oxidize in moist air, and must be stored desiccated in a sealed container.

Fluorescein

CI 45350 – see below for CI usage names.

Synonyms: the name fluorescein was at one time restricted to the free acid (CI Solvent yellow 94), whereas the disodium salt was termed **uranin** (CI Acid yellow 73). Most application literature now describes both forms as fluorescein, but the salt is still occasionally described as uranin by hydrologists and others. This entry follows the majority usage.

Free acid: $C_{20}H_{12}O_5$, FW 332

Disodium salt: $C_{20}H_{10}Na_2O_5$, FW 376

Absorption maximum: in water this is pH dependent, being 490 nm at pH 9. The emission maximum is also pH dependent, being near 515 nm in neutral and alkaline aqueous solutions.

pK values: 2.2, 4.4, 6.7

Solubilities: in water increase with pH, and the following information relates to commercially available materials. The free acid is slightly soluble (0.2%) in water, but soluble (2.2%) in ethanol and also in dimethylformamide. The solubilities of the disodium salt are 50% in water and 7% in ethanol. The solubilities of the calcium salt are much lower, 1.1% in water and 0.4% in ethanol, which has in the past influenced staining using tapwater in hard water districts.

Fluorescein is a weakly acidic hydroxyxanthene whose overall size is small. Its structure is shown in Chapter 8, Fig. 8.1. The dye has a moderately sized, nonplanar aromatic system. The dianionic form is hydrophilic, the neutral form is lipophilic; in aqueous solutions the hydrophilic anion predominates under alkaline conditions, the lipophilic monoanion around pH 5–6 and the more lipophilic neutral form around pH 4. See Figure 8.1 in Chapter 8 for the structures of the various forms.

In environments other than aqueous solution, forms such as the lactone and zwitterion exist. However, these are of little practical consequence to biologists. Fluorescein is available commercially both as a disodium salt and as a neutral species, the latter nominally being the free acid.

Applications: Being pale yellow, fluorescein was little used for microscopic staining until commercial fluorescence equipment was readily available. A few applications involve staining of dead specimens, for instance, procedures for measuring annual growth rings in hardwood samples and for assessment of water movement into dried timber use fluorescein as a stain (Blais, 1995) and tracer (Matsumura *et al.*, 1998) respectively. Most recent applications however involve living cells and tissues. Examples are the tracing of symplastic pathways in plants (Barnabas, 1994), and organic anion transport into the lumen of the mammalian nephron (Miller *et al.*, 1996). The dye has also been used as a marker for the study of motility of bile canaliculi (Watanabe *et al.*, 1991).

Clinical applications include evaluation of the ocular surface (Tsubota *et al.*, 1999), and delineation of brain regions lacking a blood brain barrier to permit accurate surgical resection of malignant glioma (Kuroiwa *et al.*, 1999). Fluorescein angiography is a routine procedure for investigation of changes in retinal blood supply (Opremcak and Bruce, 1999).

Nonmicroscopic biomedical applications include use as a physiological tracer to characterize the salmon's intestinal permeability (Schep *et al.*, 1998), and for the fluorophotometric measurement of aqueous humor flow (Maus and Brubaker, 1999).

Other scientific and technological uses of fluorescein include checking for contamination of honey bees during pesticide application (Koch and Weisser, 1997), and use as an indicator to assess the pH of paper (Moorthy *et al.*, 1998). Fluorescein has been extensively used as a tracer of environmental water flow and mixing within rivers and lakes (Peeters *et al.*, 1996) and through landfill sites (Johnson *et al.*, 1998). Such applications can require multikilogram amounts of dye per experiment. Industrially fluorescein is used as a colorant in drugs and cosmetics.

Purity of commercial batches: These often contain a single major component plus small amounts of contaminants. Dye contents of typical samples of both the free acid and the disodium salt are up to 95%. A thin-layer chromatographic analytical method is available (Matysik, 1998).

Stability: of dry powder is good; light fastnesses of aqueous solutions are poor.

Carboxyfluorescein

Note: information in this entry applies to both isomeric forms.

Synonyms: CF. The two isomeric forms are usually described as 5- and 6-carboxyfluorescein; with isomeric mixtures being indicated as 5(6)- or occasionally as 4(5)-carboxyfluorescein.

Free acid: $C_{21}H_{12}O_7$, FW 376

Absorption maximum: is pH dependent, for example, 492 nm in aqueous solutions at pH 9; the emission maximum also varies with pH, being 517 nm at pH 9. pK values: 3.3, 4.6, 7.0

Solubilities: soluble in water under neutral and alkaline conditions, also soluble in dimethylformamide.

Carboxyfluorescein is a weakly acidic hydroxyxanthene whose overall size is small. The dye has a moderately sized, nonplanar aromatic system. The di- and trianionic forms are hydrophilic, the monoanion and neutral forms lipophilic. Under physiological conditions the dye is multiply anionic. Carboxyfluorescein is available commercially as the free acid, illustrated.

Applications: Staining applications in living cells include testing for the existence of gap junctions between cells, both in plants (Shepherd and Goodwin, 1992) and animals (Balice-Gordon *et al.*, 1998). Entry of dye into plant cells often utilizes acid-loading, by exposure to a low-pH solution containing the lipophilic free acid form of the dye. With animal cells direct microinjection may be used. For use of a lipophilic ester derivative for this purpose, see the entry for carboxy-fluorescein diacetate. The dye is also used to measure intracellular pH, in situations as varied as mammalian sperm (Jones *et al.*, 1995) and biofilm deposits (Vroom *et al.*, 1999). Other laboratory applications include use as a tracer to probe permeability of cultured cell aggregates (Rojanasakul *et al.*, 1992), and to test for leakage from liposomes (Silvander *et al.*, 1998).

Purity of commercial batches: Samples are available with dye contents of up to 99%.
Stability: Carboxyfluorescein should be stored in the dark, the solutions are susceptible to photofading.

Carboxy-SNARF 1

Information is also provided in this entry concerning the diester derivative, **carboxy SNARF-1 acetoxymethyl ester, acetate**.

Synonyms: carboxy seminaphthorhodafluor-1, c-SNARF 1; the diester is routinely, though imprecisely, referred to as carboxy SNARF-1 AM.

Free acid: $C_{27}H_{19}NO_6$, FW 453
Ester: $C_{32}H_{25}NO_9$, FW 568

Absorption maxima of the acid: this is pH dependent, being 548 nm at pH 6 and 576 nm at pH 10. The emission maximum also varies with pH, being 587 nm at pH 6 and 635 nm at pH 10, when excited in the green.

Fluorescent pH indicator for the range 6.3–8.6; *pK values:* 7.6, 7.8
Solubilities: The acid is soluble in neutral and alkaline aqueous solutions. The diester is soluble in dimethylsulfoxide.

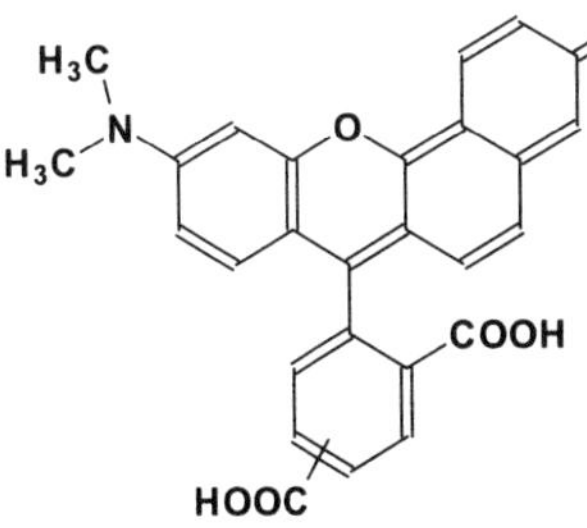

5(6)-carboxy SNARF-1

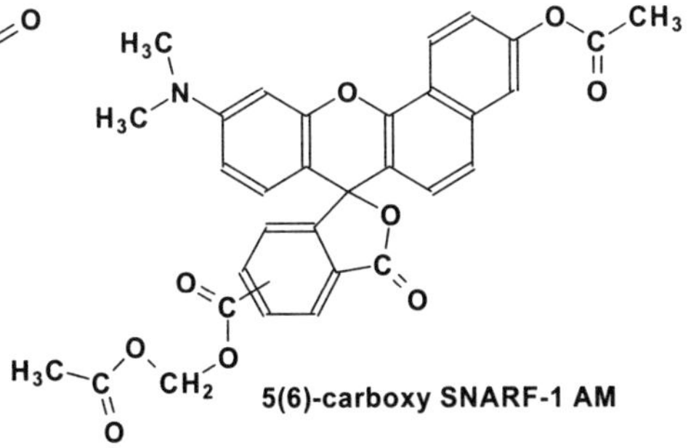

5(6)-carboxy SNARF-1 AM

Carboxy-SNARF 1 is an aminohydroxyxanthene of moderate overall size. The dye has a moderately sized, nonplanar, aromatic system. The dianionic form is hydrophilic, and the zwitterionic form present in weakly acidic solutions is weakly lipophilic, as is the cation. The acetoxylmethyl ester, acetate is also lipophilic. This latter compound is sold as the lactone, the parent dye is sold as the free acid, both are illustrated.

The free acid can be introduced into live cells either by microinjection or by acid loading, the latter using an acidic solution containing the lipophilic forms of the dye. Cellular internalization is also achieved using the lipophilic ester. Below, no distinction is made between the modes of application.

Applications: Carboxy-SNARF 1 is widely used to measure intracellular pH. This can be carried out microscopically, as in the cells of isolated pancreatic ducts (De Ondarza and Hootman, 1997), and cultured aortic endothelial cells (Hu *et al.*, 1998); or in flow cytometric systems, to identify and quantify apoptosis in hybridoma cell cultures (Ishaque and Al Rubeai, 1998). The dye has been recommended in flow cytometry manuals as a probe of intracellular pH (Macey, 1994). Carboxy-SNARF 1 has also been used as an internal fluorescent standard in patch clamp assays of calcium (Spencer and Berlin, 1995).

Purity of commercial batches: There is a thin-layer chromatographic analytical method available (Whitaker *et al.*, 1991).

Stability: The free acid, especially if in solution, should be stored in the dark; the diester can be stored as a solid, desiccated, in a freezer.

Derivatives include **Snarf dextrans**. These polymeric pH indicators are available in various sizes, and remain in the cytosol when microinjected into live cells being unable to enter organelles or to leave the cell. They are slow diffusing, hence their use to assess cytosolic pH gradients (Stewart *et al.*, 1999).

Sulforhodamine B

CI 45100, CI Acid red 52
Synonyms: kiton rhodamine B, lissamine rhodamine B, SRB, sulphorhodamine B
Free acid: $C_{27}H_{30}N_2O_7S_2$, FW 559
Sodium salt: $C_{27}H_{29}N_2NaO_7S_2$, FW 581

Absorption maxima: 566 nm in water; 556 nm in methanol, 554 nm if 0.1N sodium hydroxide is present. The emission maxima are 586 nm in water, 572 nm in methanol.

Solubilities: 2% in water; 0.5% in ethanol.

Sulforhodamine B is an aminoxanthene carrying two sulfonate substituents. Overall size is moderate; as is that of the nonplanar aromatic system. Under neutral conditions this dye is a hydrophilic anion. Both the free acid form and the sodium salt, illustrated, are available commercially. Sulforhodamine B is a precursor in the manufacture of lissamine rhodamine B sulfonyl chloride. This compound, used for the fluorescent labeling of biomolecules, is described later. *Applications:* Sulforhodamine B is used as a microscopic tracer, for example, to assess blood brain barrier integrity by post mortem inspection of frozen sections (Philip *et al.*, 1994); and as a caries detector dye with subsequent inspection of histological material (Boston *et al.*, 1995). However, many applications involve staining and observation of live cells. It was used to visualize rubrospinal neurons by Lustig *et al.* (1998), and as a marker of fluid-phase endocytosis by Vertutdoi *et al.* (1994).

Other laboratory applications include use as a tracer, for example, of water flow between compartments of the eye (Maurice, 1987), and in hydrology, to investigate wastewater and stabilization ponds (Torres *et al.*, 1997).

Industrial uses – Sulforhodamine B is used as a food dye in some countries and has been used in laser technology.

Purity of commercial batches: There is usually a single major component, plus two colored contaminants. The dye content of typical samples of the free acid is 95%, of the sodium salt 75%. There are HPLC (Franke *et al.*, 1997) and thin-layer (Hay *et al.*, 1995) analytical systems available.

Stability: Light fastness of aqueous solutions is poor.

Octadecanoylaminofluorescein

Synonyms: AF, AF18, F18, 5-octadecylaminofluorescein, ODAF
Free acid: $C_{38}H_{47}NO_6$, FW 614

Absorption maximum: is pH dependent, but in methanol containing a trace of potassium hydroxide the value is 497 nm. In the same solvent the emission maximum is 519 nm.
Solubilities: soluble in dimethylsulfoxide, ethanol and methanol.

Octadecanoylaminofluorescein is a weakly acidic hydroxyxanthene whose overall size is moderate. The dye has a moderately sized, nonplanar aromatic system. The long hydrocarbon side chain results in all forms of the dye being strongly lipophilic, and extremely amphiphilic. The dye is sold as the free acid.

Applications: The dye is used as a fluorescent probe of live cells, due to its local-ization within the plasma membrane. Confocal microscopy applications include assessments of membrane fluidity (Ladha *et al.*, 1994) and of the extent of modification of parasitic cell surfaces (Modha *et al.*, 1995). The dye has also been used in spectroscopic procedures to assess cell–virus and cell–cell membrane fusion (respectively by Miller and Hutt-Fletcher, 1992; and Partearroyo *et al.*, 1994).

Stability: Solutions photofade, and should be stored in the dark.

Fluoresceinyl-N-octadecylthiourea, is a related dye in which a similar hydro-carbon side-chain is linked to the chromophore via a thiourea grouping. This has been used as a plasma membrane probe in a flow cytometric application (Radosevic *et al.*, 1990).

BCECF

Information in this entry applies to both isomeric (5- and 6-carboxy) forms. The **ester derivative of BCECF** is also described in this entry.

Synonyms: 2',7'-Bis-(2-carboxyethyl)-5-(and 6)carboxyfluorescein; the ester is usually termed BCECF AM ester, sometimes BCECF acetoxymethoxy or acetoxymethyl ester; see below.

Free acid: $C_{27}H_{20}O_{11}$, FW 520

AM ester: commonly a mixture, the effective formula weight for which is often approximately 615

Absorption maximum: of the acid form is pH dependent, being 503 nm in aqueous pH 9 solutions; emission also varies with pH, and is 528 nm at pH 9.

A fluorescent pH indicator; *pK value* 7.0

Solubilities: The free acid is soluble in neutral and alkaline aqueous solutions. The ester is soluble in dimethylsulfoxide.

BCECF is an acidic hydroxyxanthene whose overall size is moderate. The dye has a moderately sized, nonplanar aromatic system. In aqueous solutions BCECF exists in several ionic forms, and at physiological pH the dye will be multiply anionic. The dianionic and more highly ionized forms are hydrophilic, the monoanionic and neutral forms lipophilic. BCECF is available commercially in the free acid form, illustrated. Although various isomeric BCECF AM esters have been sold, all species are hydrolyzed to the acidic form of BCECF by intra-cellular esterases.

Applications: BCECF is widely used as a fluorescent pH probe of live cells. It is usually introduced as the ester, and converted into the acidic form by cellular esterases; alternatively it can be introduced by use of an acidic solution containing the lipophilic free acid. Examples of such applications include measurement of the internal pH of pollen tubes in plants (Fricker *et al.*, 1997), and of cultured glial cells (Marcaggi *et al.*, 1999). Extracellular pH determinations are also carried out, for example, in lateral intercellular spaces of epithelial cells (Chatton and Spring, 1994). Other microscopic vital staining applications, include investigation of solute transport in plant seedlings (Wright *et al.*, 1996), and the demonstration of hydrogenosomes in protozoa (Scott *et al.*, 1998).

BCECF is also used for nonmicroscopic pH determinations in physiology and related biological fields, including the *in situ* determination of the pH of human tears (Yamada *et al.*, 1998), and intracellular pH of platelets (Zeng *et al.*, 1999). Other nonmicroscopic applications include assessing attachment of cancer cells to bone (Nordstrom *et al.*, 1999), and assessment of cytoplasmic pH using flow cytometry (Darzynkiewicz and Crissman, 1990; Macey 1994).

Stability: The free acid is light sensitive, especially in solution. To prevent hydrolysis, the ester forms are stored frozen and desiccated.

BCECF labeled dextrans are used to measure intracellular pH, for example, in green algae (Plieth *et al.*, 1997). Although they require introduction into cells by a maneuver such as microinjection, once within the cell, leakage is trivial.

Eosin Y

CI 45380, CI Acid red 87 is the sodium salt, the free acid is CI 45380:2, Solvent red 43

Synonyms: the routine synonym is eosin. The sodium salt is manufactured as eosine, with or without suffixes such as A and G, reflecting the older German usage. In biological staining the synonyms eosin yellowish and eosin water soluble are quite widely used. The free acid is termed eosin spirit soluble, not to be confused with ethyl eosin, a different alcohol soluble eosin derivative.

Free acid: $C_{20}H_8Br_4O_5$, FW 648
Disodium salt: $C_{20}H_6Br_4Na_2O_5$, FW 692

Absorption maximum: 516 nm in alkaline or neutral aqueous solutions; emission maximum in neutral aqueous solution is 538 nm, when excited at 516 nm.
Fluorescent pH indicator in the range 0.0–3.0; *pK values:* 2.9, 4.5
Solubilities: of the free acid is 1% in ethanol and trivially (0.08%) in water; of the sodium salt is 44% in water, and 2% in ethanol.

Eosin Y is an acidic hydroxyxanthene, of moderate overall size and with a moderately sized, nonplanar aromatic system. In aqueous solutions eosin Y exists in several ionic forms, being predominantly dianionic in neutral and alkaline aqueous solutions. All species are weakly lipophilic. Eosin Y is available commercially as the sodium salt, illustrated, and as the free acid.

Applications: Eosin Y is widely used to stain cytoplasms and cytoplasmic structures such as eosinophil granules and Negri bodies. Most widely known are the many histological hematoxylin and eosin (H & E) stains, the many Papanicolaou stains in clinical cytology, and the Romanowsky-Giemsa stains for blood smears. Other procedures using eosin Y include oversight polychrome stains such as Mann's methyl blue–eosin, and the counterstaining of various silver staining methods. These procedures are described in standard staining manuals, such as Bancroft and Gamble (2002) and Kiernan (1999).

The chemistry and application of the eosin Y-thiazine Romanowsky-Giemsa stains are complex, for which see Chapter 21. There is a standardized ICSH Romanowsky-Giemsa procedure for blood smears (Lewis, 1984), as well as standardized Romanowsky-Giemsa methods for polychrome staining of cytological smears (Schulte and Wittekind, 1989b) and histological sections (Wittekind *et al.*, 1991).

Many other staining methods make use of eosin Y, including a H & E procedure for staining glycolmethacrylate resin sections (Hayat, 1993), and a method for the selective staining of mineralized bone (Bradbeer *et al.*, 1994). The dye is also used as a diagnostic viability stain of spermatozoa (Verheyen *et al.*, 1997), and of isolated digestive gland cells of the mussel (Mitchelmore *et al.*, 1998).

Nonmicroscopic staining applications of eosin Y include staining of proteins on polyacrylamide electrophoretic gels (Lin *et al.*, 1991), inclusion in a microbiological medium for isolation of *Xanthomonas* species (Gitaitis *et al.*, 1991), and viability staining of *Candida albicans* blastospores in a flow cytometric system (Constantino *et al.*, 1995). Eosin Y has been used as a tracer, for example, for assessing water flow patterns in xylem during ripening of grapes (Creasy *et al.*, 1993).

Industrial uses – eosin Y is used as a colorant of inks, leather and paper; the potassium salt is used to color cosmetics and drugs.

Purity of commercial batches: The tetrabromo compound is always the major species, but varying amount of less brominated compounds are usually present, and sometimes traces of fluorescein. Dye contents of the sodium salt are up to 94%, and of the free acid up to 99%. The minimum dye content for Commission certification is 80%. There are HPLC (Alcantara–Licudine *et al.*,

1997) and thin-layer chromatographic (Marshall and Lewis, 1974a) methods. Regarding the Commission's assay methods see Chapter 28.

Stability: The dry powder is fairly stable stored in sealed, opaque containers. Light fastness is good, although solutions should be kept out of direct sunlight or fluorescent lighting. Aqueous solutions of eosin Y quickly become infected with fungi.

A reactive derivative, **eosin Y isothiocyanate** (also known as **EITC** or **EYNCS**) is commercially available. This has been used as a fluorescent label for immunostaining and in situ hybridization (Deerinck *et al.*, 1994).

Ethyl eosin (CI 45385, Solvent red 45) is an alcohol soluble ethyl ester of eosin Y. This is useful for 'eosin' staining when alcoholic solutions and differentiators are called for, for example, as a counterstain after hematoxylin, and for demonstration of Negri bodies (Clark, 1981).

Phloxin B

CI 45410, CI Acid red 92

Synonym: phloxin; see below for possible confusions arising from this synonym; do not confuse with phloxin G, a cationic cyanine dye.

Disodium salt: $C_{20}H_2Br_4Cl_4Na_2O_5$, FW 829

Absorption maxima: 548 nm in 50% aqueous ethanol, with a subsidiary peak at 510 nm.

A pH indicator, changing from colorless to purple between pH 1.1–3.3; and with fluorescence developing between pH 3.4–5.0.

Solubilities: 11% in water; 5% in ethanol.

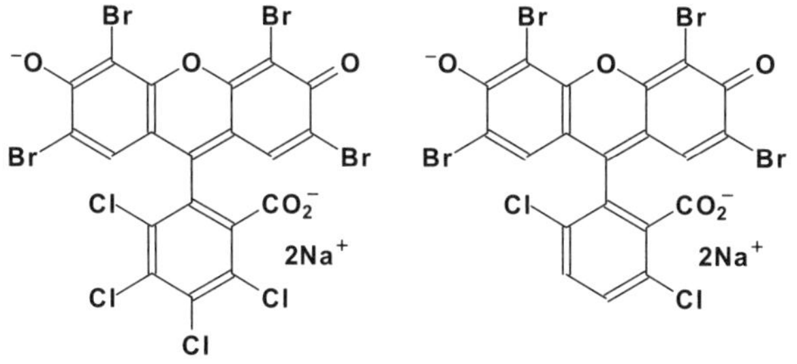

Phloxin B (above left) is an acidic hydroxyxanthene of large overall size, with a moderately sized nonplanar aromatic system. In aqueous solutions phloxin B exists in several ionic forms, all lipophilic. It is predominantly dianionic in neutral and alkaline aqueous solutions. Phloxin B is available commercially as the tetrachloro sodium salt, illustrated.

There is a structurally similar dye, **phloxin** (CI 45405, Acid red 98), the dichloro compound shown above right. While applications papers often state that 'phloxin' was used, it is probable that the dye was phloxin B because CI 45405 has not been produced industrially for many years.

Applications: A routine histopathological stain involving phloxin B is Lendrum's phloxin–tartrazine procedure for selective staining of viral inclusions and other dense structures, such as neurosecretory granules and pancreatic islet secretory products. This is described in staining manuals (Bancroft and Gamble, 2002; Kiernan, 1999). Other routine procedures include hematoxylin–phloxin or –phloxin–eosin oversight stains, also described in manuals (Vacca, 1985). The latter method has also been applied to resin embedded sections, both epoxy (Oldmixon, 1988) and glycolmethacrylate (Zbaeren *et al.*, 1998).

Related methods include stains for nucleoli (Derenzini *et al.*, 1992), and the phloxin–safranine, and hematoxylin–phloxin–safranine, stains for cells of inflammatory exudates (Chartier *et al.*, 1997). Other staining applications include stereomicroscopic evaluation of mucosal surfaces following staining with alcian blue (Mars *et al.*, 1994), and the macroscopic staining of plant–pathogen egg masses adhering to corn roots (Windham and Williams, 1994).

Amongst many other biomedical applications are the use of phloxin-labeled chewing gum to evaluate masticatory function (Matsui *et al.*, 1996), and as a bactericide in plants (Willeford *et al.*, 1998).

Industrial uses – a colorant of inks, lacquers and paper.

Purity of commercial batches: Typically a single major component is present, plus several substantial and many other trace contaminants. Dye contents are up to 85%; minimum dye content for Commission certification being 80%. HPLC and thin-layer chromatographic analytical methods are available (Wright *et al.*, 1997). Regarding Commission assay procedures see Chapter 28.

Stability: of dry powder, stored in opaque sealed containers, is good; staining solutions fade in sunlight, although light fastness of stained sections is satisfactory.

Erythrosin B

The disodium salt is CI 45430, CI Acid red 51; the free acid is CI 45430:2, Solvent red 140.

Synonyms: see below for nomenclature problems.

Free acid: $C_{20}H_8I_4O_5$, FW 834

Disodium salt: $C_{20}H_6O_5I_4Na_2$, FW 880

Absorption maxima: of the free acid is 533 nm in methanol; and of the disodium salt is 525 nm in aqueous alkali. When excited around 530 nm in neutral or alkaline aqueous solutions, erythrosin B emits at 555 nm.

Fluorescent pH indicator, changing from nonfluorescent yellow to fluorescent red over the pH range 0.0–3.6.

pK value: 4.1

Solubilities: of the disodium salt is 11% in water and 5% in ethanol.

Erythrosin B is an acidic hydroxyxanthene of large overall size, with a moderately sized nonplanar aromatic system. The dye can exist in several ionic forms, being predominantly a lipophilic dianion in neutral and alkaline aqueous solutions. The dye is available commercially both as the free acid and the disodium salt.

Various problems of nomenclature and identity arise with erythrosin B. There is a commercially available diiodofluorescein which is sold as **erythrosin Y** (CI 45425, Acid red 95). Another commercial product, **erythrosin yellowish**, is a 9:1 mixture of erythrosin B and eosin Y. Most application literature refers only to 'erythrosin' with no suffix given, so ambiguity exists. However, erythrosin B has always been by far the most widely sold of these products, so investigators mentioning 'erythrosin' probably used the tetraiodo compound. Nevertheless to minimize possible ambiguities recent applications cited below specified use of erythrosin B.

Applications: Erythrosin B has been used as a counterstain to a variety of blue and violet nuclear stains, for example, to stain woody tissue of vascular plants (Clark, 1981). More recently, erythrosin B has been used in cytology to assess stallion spermatozoa (Christensen *et al.*, 1996). The most common current application is as a viability stain, for instance differentiate dead, or membrane-damaged, yeasts (Hutter and Eipel, 1978), or animal cells in xenografts (Abolhoda *et al.*, 1996).

Amongst many nonmicroscopic biomedical applications are its use in dentistry as a disclosing agent to assess plaque formation (Yates *et al.*, 1997), and in experimental neurology to generate localized experimental lesions (Sjolund *et al.*, 1998).

Industrial uses – As CI Food red 14 or FD&C red 3, high purity erythrosin B is used as a colorant of food and pharmaceutical products.

Purity of commercial batches: Typical samples contain the tetraiodo compound together with partly iodinated products and fluorescein. Occasionally rose Bengal B has been supplied as erythrosin B. Dye contents of typical samples of the free acid are up to 98%, and of the sodium salt up to 90%. The minimum dye content for Commission certification is 80%. There are thin-layer (Ganz and Jork, 1994) and HPLC (Alcantara–Licudine *et al.*, 1997) chromatographic analytical methods available. Regarding the Commission assay method see Chapter 28.

Stability: of dry powder is satisfactory, if stored in sealed dark containers. Light fastness is poor.

Hazards: harmful if swallowed.

Amongst various reactive derivatives commercially available **erythrosin isothiocyanate** is most widely used, for instance, to assess mobility of receptors within membranes using fluorescent anisotropy measurements of labeled proteins (Chang *et al.*, 1995), and to estimate intramolecular distances, based on energy transfer between labeled sites (Linnertz *et al.*, 1998).

Rose Bengal B

CI 45440, CI Acid red 94
Synonym: rose Bengale
Disodium salt: $C_{20}H_2Cl_4I_4Na_2O_5$, FW 1018

Absorption maximum: 548 nm in aqueous alkali; excited in the green it emits at 567 nm.
pK values: 3.9, 4.7
Solubilities: 36% in water; 8% in ethanol.

Rose Bengal B is an acidic hydroxyxanthene of large overall size, with a moderately sized nonplanar aromatic system. In aqueous solutions rose Bengal exists in several ionic forms, being present largely as the strongly lipophilic dianion in neutral and alkaline aqueous solutions.

Rose Bengal B is available commercially as the sodium salt. The structurally related **rose Bengal G** (CI 45435, Acid red 93) has apparently not been used in biology or medicine, and is not currently manufactured.

Applications: Rose Bengal B has several routine microscopical staining applications described in staining manuals. These include demonstrating bacteria in soil and foraminifera in marine sediments (Clark, 1981); and counterstaining hematoxylin in the assessment of sperm competence (Bancroft and Cook, 1994). The dye has also been used, together with Bismark brown, in a substitute for the phloxin–tartrazine stain for cellular inclusions (Clark, 1979).

Other laboratory applications include wide use as a constituent of microbiological solid culture media (Mikolon *et al.*, 1998), and as an ophthalmic diagnostic aid to detect corneal abrasion (Tsubota *et al.*, 1999).

Induction of highly reactive singlet oxygen in aerobic environments also finds many biomedical applications. Examples include an experimental treatment of psoriasis (Kumar *et al.*, 1997), an antiviral medication in fish (Maeda *et al.*, 1998), and generation of localized experimental lesions in anatomically specific segments of the cochlea (Wu *et al.*, 1999).

Industrial uses – Colorant of inks, paper, and wood chips; the dye has been used in certain countries as a colorant of cosmetics.

Purity of commercial batches: usually contain two or more major components plus minor contaminants. Dye contents are up to 95%, with the minimum dye content for Commission certification being 80%. There is a thin-layer chromatographic analytical method (Marshall, 1976a). See Chapter 28 regarding Commission assay procedures.

Stability: of dry powder is good when stored in sealed, dark containers; light fastness is poor.

ZWITTERIONIC DYES

Rhodamine B

CI 45170, CI Basic violet 10
Chloride: $C_{28}H_{31}ClN_2O_3$, FW 479

Absorption maxima: in water is at 548 nm, with the emission maximum at 585 nm; in methanol the absorption maximum is at 543 nm. In acidic ethanol the absorption maximum is 554 nm and the emission maximum 627 nm.
pK values: 3.0, 11.8
Metal complexation behavior: Forms insoluble compounds with a wide variety of metal ions, including antimony, gallium, mercury and thorium.
Solubilities: 2% in water; 1.8% in ethanol.

Rhodamine B is an aminoxanthene which carries a carboxylic acid substituent on a pendant phenyl ring. Overall size is moderate, as is that of the nonplanar aromatic system. In neutral aqueous solutions this dye is a hydrophilic zwitterion, illustrated, with both carboxylate and ammonium ion substituents. In acid solutions the dye forms a lipophilic cation, while in strongly alkaline solutions an anion is present. Rhodamine B is available commercially as the chloride. Several modern manuals confuse rhodamine B with a completely different stain, rhodanine, an unfortunate typographical error.

Applications: It seems that when early workers mentioned 'rhodamine' they were using rhodamine B. Thus Griesbach combined the dye with osmium tetroxide to simultaneously fix and stain blood; Ehrlich used rhodamine B as a component of neutral stains, and also for histological work as a contrast stain with methylene blue; it has also been used as a contrast stain with methyl green.

Histological methods using rhodamine B include a fluorescence method for acid fast and mycobacteria, described in manuals (Bancroft and Cook, 1994; Clark, 1981; and briefly reviewed by Churukian, 1991). A selective stain for keratin is also used in histopathology (Shapiro, 1991), and has been recently applied to cultured cells (Ward *et al.*, 1997). Rhodamine B is sometimes used as a fluorescent lipid stain, as it may be applied from aqueous solution (Issidorides and Arvanitis, 1993). Less routine applications include demonstration of fungal hyphae in wood (Pearce, 1984), and as a selective stain for the carious regions of teeth (Berger *et al.*, 1996). There are also applications to whole mounts, including the demonstration of laticifers in plants (Inamdar *et al.*, 1987), and the marking of lipid-rich surfaces of intact, unfixed rat lungs for study by confocal microscopy (Pohl *et al.*, 1998).

Rhodamine B is used as a fluorescent probe of living cells. Routine applications include demonstration of lysosomes and other acidic intracellular compartments (Minier and Moore, 1996) and for retrograde axonal tracing in neuroanatomy (Fu *et al.*, 1996). The dye has also been used to stain fungi in soil (Jensen and Lysek, 1995) and as a cell marker in developmental studies of insects (Van der Reijden *et al.*, 1997).

Other biomedical applications of Rhodamine B include use as a tracer. Macroscopically, it is a constituent of an injection medium for delineating blood vessels in anatomical teaching preparations (Vico *et al.*, 1994), and for assessing permeability of skin patches (Skopp *et al.*, 1996). Microscopically, it reveals the activity of transporters removing xenobiotics from living cells (Smital and Kurelec, 1998) and leakage around dental fillings (Niu *et al.*, 1998). Rhodamine B is also used as a marker, to identify those herring gulls incubating eggs (Belant and Seamans, 1993), and of the bait used in oral rabies immunization of dogs (Matter *et al.*, 1998). Other biomedical applications include incorporation into microbiological solid culture media (Jarvis and Thiele, 1997), and for staining phosphoproteins in polyacrylamide electrophoretic gels (Zhou *et al.*, 1998). Rhodamine B has been used as a hydrological tracer, for example, to investigate the influence of earthworm burrows on water movement in soil (Schrader and Joschko, 1991) and to assess beach contamination from a nearby duck pond (Chen *et al.*, 1991).

Industrial uses – as a colorant of inks, lacquers, leather, paper and wood, and is used in cosmetics in certain countries.

Purity of commercial batches: One major component plus a significant impurity are often present. Dye contents of the chloride are up to 95%. HPLC (Gagliardi *et al.*, 1996) and thin-layer (Naff and Naff, 1963) chromatographic methods are available.

Stability: Light fastness is poor.

Hazards: Harmful if swallowed, inhaled or absorbed through the skin; a possible mutagen and carcinogen if exposure is chronic.

Rhodamine B dextrans have been used to estimate cytoplasmic volume (Ragsdale *et al.*, 1997), and to provide internal standards when using fluorescent intracellular calcium probes (Stricker *et al.*, 1998). **Rhodanile blue**, a complex formed between rhodamine B and Nile blue, is used for polychrome staining, for instance to demonstrate cornification of cultured keratinocytes (Baden *et al.*, 1992). This material appears to be a salt, since on chromatography the component dyes are seen.

CATIONIC DYES

Pyronine Y

CI 45005
Synonym: pyronine G
$C_{17}H_{19}ClN_2O$, FW 303

Absorption maxima: 546 nm in water; 549 nm in 50% aqueous ethanol; emission maximum is 570 nm in water.
pK value: for pseudobase formation is 11.4
Solubilities: of the chloride are 9% in water and 0.6% in ethanol; the tetra-fluorborate is sparingly soluble in water, and even less so in ethanol.

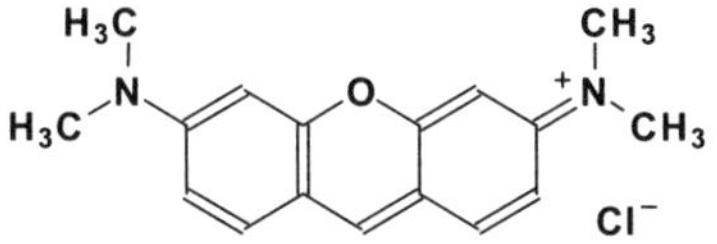

Pyronine Y is an aminoxanthene of small overall size, with a small planar conjugated system. In neutral aqueous solutions this dye is a weakly lipophilic cation, as illustrated. Pyronine Y is available commercially as the chloride.
Applications: The dye is used routinely in histology as a component of the ethyl (methyl) green–pyronine procedure for demonstrating DNA and RNA, as described in routine staining manuals (Bancroft and Gamble, 2002; Kiernan, 1999). This procedure is also discussed in the clinical cytology manual of Boon and Drijver (1986). A standardized methyl green–pyronine procedure, based on pure dyes, has been described (Hoyer *et al.*, 1986). The procedure can be carried out on resin sections, both glycolmethacrylate (Cole and Ellinger, 1981) and epoxy (Jurand and Goel, 1976). The method may also be used quantitatively for DNA and RNA assays (Schulte *et al.*, 1992). Other microscopic applications are simultaneous fixation and staining of phospholipids (Teichman *et al.*, 1972), and demonstration of lipofuscin in neurons (Patro *et al.*, 1996).

Pyronine Y is noted in flow cytometry manuals for RNA staining (Darzynkiewicz and Crissman, 1990). Recent research applications using this methodology include use as a nucleic acid stain both for RNA (Gothot *et al.*, 1997) and for DNA (Schmid *et al.*, 1999).
Purity of commercial batches: Other dyes, such as rhodamine B and rhodamine 6G, have been supplied in bottles labeled pyronine Y. Samples in which the major component is indeed pyronine Y also contain several minor contaminants, and

vary in dye content from 13% to 90%. The minimum dye content for Commission certification is 45%. There is a thin-layer chromatographic analytical method (Lyon *et al.*, 1987). Regarding Commission assay procedures see Chapter 28.

   Stability: of dry powder is good when stored in sealed, dark containers; acidic aqueous solutions are stable.

Rhodamine 123

Chloride: $C_{21}H_{17}ClN_2O_3$, FW 381
Note: commercial samples may be hydrated to varying degrees.

Absorption maxima: 501 nm in water, 507 nm in methanol; emission maximum in methanol 529 nm.
Solubilities: soluble in dimethylformamide and methanol, water solubility at least 0.1%.

Rhodamine 123 is an aminoxanthene, which carries an esterified carboxylic acid substituent on a pendant phenyl ring. Overall size is small, and that of the aromatic system moderate. In neutral aqueous solutions this dye is a lipophilic cation, illustrated. Rhodamine 123 is sold as the chloride, and commercial samples probably contain water of crystallization, although only certain vendors acknowledge this.

Applications: Rhodamine 123 is recommended in staining manuals (Celis, 1998) as a mitochondrial stain in living cells. Such use extends to less routine samples, as in assessment of mitochondrial depolarization in hippocampal slices (Bahar *et al.*, 2000). The dye has been used in experimental dentistry to look for cracks in dentine following laser ablation (Kimura *et al.*, 2000). Rhodamine 123 is recommended by flow cytometry manuals (Darzynkiewicz and Crissman, 1990; Macey 1994) for measurement of membrane potential, and also for assessment of multidrug resistance and of the proliferation of hemopoietic stem cell proliferation.

Purity of commercial batches: Dye content of typical samples is 95%.
Stability: Solutions are light sensitive.

Rhodamine B hexyl ester

Synonyms: hexyl rhodamine B, R6
Chloride: $C_{34}H_{43}ClN_2O_3$, FW 563
Perchlorate: $C_{34}H_{43}ClN_2O_7$, FW 627

Absorption maximum: 556 nm in methanol, in which solvent the emission maximum is 578 nm.

Solubilities: soluble in dimethylformamide, dimethylsulfoxide and methanol.

Rhodamine B hexyl ester is an aminoxanthene carrying an esterified carboxylic acid substituent on a pendant phenyl ring. Overall size is moderate, as is that of the nonplanar aromatic system. In solution a strongly lipophilic cation, illustrated, is present. Rhodamine B hexyl ester is available commercially as the chloride and perchlorate salts.

Applications: As noted by Terasaki (1993, 1998) this dye behaves in the same way as DiOC6(3) and accumulates first in mitochondria and then in the endoplasmic reticulum of living cells. This pattern is exploited to mark particular cell lines, for instance, to distinguish between isolated inner and outer cochlear hair cells (Zajic *et al.*, 1993), and when investigating interspecies protoplast fusion (Pattanavibool *et al.*, 1998). Rhodamine B hexyl ester also stains the endoplasmic reticulum of cells fixed using microwave heating (Barsony *et al.*, 1997) or paraformaldehyde solution (Arregui *et al.*, 1998).

Stability: of dry powder is good if stored in dark desiccated containers in the freezer. Solutions photofade, and should be kept out of daylight.

Octadecyl rhodamine B

Synonyms: ODRB, OR, R18

$C_{46}H_{67}ClN_2O_3$, FW 732

Absorption maximum: 556 nm in methanol, with the emission maximum in that solvent being 578 nm.

Solubilities: soluble in dimethylsulfoxide, ethanol and methanol.

Octadecyl rhodamine B is an aminoxanthene, which carries an octadecyl carboxylate ester substituent on a pendant phenyl ring. Due to the C18 hydrocarbon chain, overall size is quite large, although that of the nonplanar aromatic system is only moderate. In neutral solutions this dye is superlipophilic and highly amphiphilic. The dye is available commercially as the chloride, illustrated. Perhaps due to typographic error, this dye is occasionally described as octadecyl rhodamine 6G.

Applications: The superlipophilic character of the dye results in trapping in biomembranes. This is the basis of the application of octadecyl rhodamine B as a fluorescent probe for plasma membranes imaged using confocal microscopy (Niv *et al.*, 1999); and as a membrane stain in flow cytometric systems (Arienti *et al.*, 1997). An analogous procedure uses the dye to assess the number of membrane layers present in liposomes (Akashi *et al.*, 1996). Other biophysical applications exploit trapping of dye in membranes. Examples are the measurement of lipid mixing by means of changes in self-quenching of the dye (Basanez *et al.*, 1997), and the labeling of liposomes (Ruiz-Arguello and Alonso, 1998).

Stability: of dry powder is good if stored in a desiccated, dark container, preferably in the freezer. Solutions photofade easily.

DYES FORMING COVALENT BONDS WITH TISSUE OR CO-SOLUTES

Fluorescein isothiocyanate

Note: information in this entry applies to both isomeric forms unless noted otherwise.

Synonyms: FITC – There are two isomeric forms, usually described as 5- and 6-fluorescein isothiocyanates, though sometimes termed isomer I and isomer II respectively. Mixtures of the isomers have been described as 5(6)-FITC.
$C_{21}H_{11}NO_5S$, FW 389

Absorption maximum: 494 nm in pH 9 aqueous solution; emission maximum in the same solvent 519 nm.
pK values: 2.2, 4.4, 6.7
Solubilities: soluble in dimethylformamide, ethanol and in neutral and alkaline aqueous solutions; see below for stability in aqueous media.

FITC is a weakly acidic hydroxyxanthene whose overall size is small, with both carboxylic acid and isothiocyanate substituents on the pendant phenyl ring. The dye has a moderately sized, nonplanar aromatic system. FITC is anionic in neutral or alkaline solutions. The isothiocyanate group reacts with nucleophilic substituents (e.g. amino, hydroxyl, thiol) of biomolecules, so providing a way of attaching a fluorescent label; see Chapter 7. FITC also reacts with water and hydroxide ion; hence the dye is unstable in aqueous solutions and the solid dye is moisture sensitive. FITC is sold as the free acid, illustrated. The 5-isomer is the most commonly sold isomerically pure form.

Applications: Many biomolecules have been labeled with FITC, including immunoglobulins, lectins and other proteins and peptides; nucleic acids and polynucleotides; and oligo- and polysaccharides. Such labeled products are used as reagents for affinity, immuno- and *in situ* hybridization staining of sections; for staining living cells; and as stains in flow cytometric procedures. Nonmicroscopic immunoassays and physiological tracing also makes use of FITC-labeled molecules. Whole animal cells and micro-organisms have also been labeled, and observed in the living state. From this wide application range, illustrative examples from the recent literature are provided below.

Tissue sections and cell monolayers may be stained in various ways, for instance, affinity staining of actin, using FITC-phalloidin (Forney *et al.*, 1999); immunostaining, using anti-digoxigenin IgG–FITC in an indirect method to demonstrate nucleotide sequences (Cui *et al.*, 1999); and *in situ* hybridization staining, using FITC–oligonucleotide probes (King *et al.*, 1999).

Staining methods for living cells are also diverse. FITC–dextrans have been used as markers to check whether cells had indeed been transiently permeabilized (Fawcett *et al.*, 1998); and to detect and evaluate diffusion barriers in the gut (Naftalin and Pedley, 1999). The microcirculation of the spinal cord was investigated by introducing extravascularly FITC-labeled erythrocytes (Ishikawa *et al.*, 1999).

FITC is recommended by flow cytometry manuals for labeling and for staining proteins and nuclei and subnuclear particles (Darzynkiewicz and Crissman, 1990; Macey, 1994). Some recent research applications include assessing uptake and distribution of oligonucleotides into cells and tissues following vascular administration of FITC–oligonucleotides (Zhao *et al.*, 1998), and binding of bacteria to buccal epithelium using FITC-labeled micro-organisms (El Ahmer *et al.*, 1999). Strains of yeast have been treated with FITC–concanavalin A, the differentially labeled varieties then being separated using fluorescence activated cell sorting (Buck and Andrews, 1999).

In physiology, FITC-labeled materials are applied as tracers. FITC–albumin is used to assess the blood–nerve barrier (Uncini *et al.*, 1999), and lung permeability for solutes (Lecuona *et al.*, 1999). FITC–inulin is used to investigate single nephron filtration rates (Lorenz and Gruenstein, 1999). White blood cells have been marked with FITC to enable their response to chemoattractants to be measured (Lam, 1998).

Purity of commercial batches: Dye contents are 90–98% for the 5-isomer, and up to 80% for the 6-isomer. There is a thin-layer chromatographic method available

(Rentsch and Wittekind, 1967). For information regarding the Commission assay, see Chapter 28.

Stability: The dry powder is light and moisture sensitive, and should be stored in an opaque and desiccated container, preferably in the freezer. Aqueous solutions are unstable and should not be stored; they also photofade.

Hazards: An irritant which may cause an allergic reaction including asthma with repeated exposure.

Tetramethylrhodamine isothiocyanate

Note: information in this entry applies to both isomeric forms unless noted otherwise.

Synonyms: TMRITC, TRITC – There are two isomeric forms, usually described as 5- and 6-tetramethylrhodamine isothiocyanates, though sometimes termed G and R isomers respectively. Mixtures of the isomers have been described as 5(6)-TRITC.
Zwitterion: $C_{25}H_{21}N_3O_3S$, FW 444
Chloride: $C_{25}H_{22}ClN_3O_3S$, FW 480

Absorption maxima: of the 5-isomer are 543 nm in methanol and 537 nm in pH 8 aqueous solution, with emission maxima at 571 nm in methanol and 564 nm in the aqueous solution; of the 6-isomer are 544 nm in methanol and 555 nm in pH 8 aqueous solution, with emission maxima at 572 nm in methanol and 580 nm in the aqueous solution. The 6-isomer is the more strongly fluorescent compound.
Solubilities: soluble in dimethylformamide, dimethylsulfoxide, ethanol, methanol and water (see below regarding stability in aqueous media).

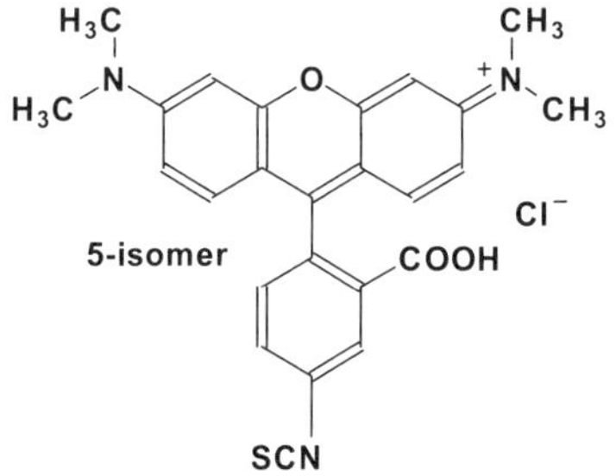

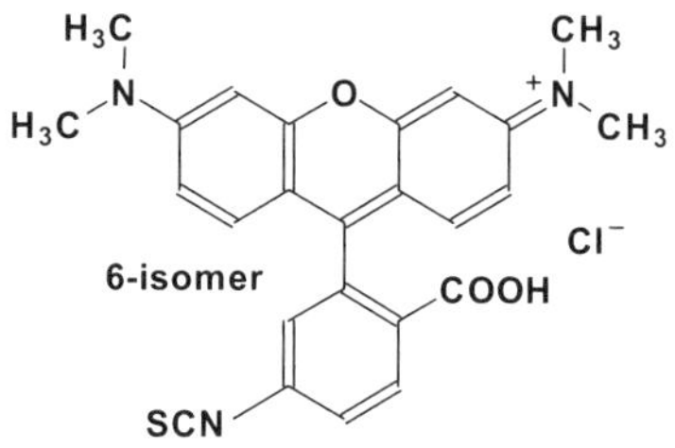

5-isomer   6-isomer

TRITC is an aminoxanthene, whose overall size is moderate, carrying carboxylic acid and isothiocyanate substituents on the pendant phenyl ring. The dye has a moderately sized, nonplanar aromatic system. TRITC forms a lipophilic cation under acid conditions, and a hydrophilic zwitterion around neutrality. The isothiocyanate group reacts with nucleophilic substituents (e.g. amino, hydroxyl, thiol) of biomolecules, providing the means of attaching a fluorescent label; see Chapter 7. TRITC also reacts with water and hydroxide ion, so is unstable in aqueous solutions.

TRITC is most commonly sold as the zwitterion, illustrated; but is also available as the chloride. The 6-isomer (R isomer) is widely sold, but the 5-isomer (G isomer) and isomerically mixed materials are also available commercially.

244

Applications: A variety of biomolecules may be labeled with TRITC, including immunoglobulins, lectins and other proteins; nucleic acids and polynucleotides; and oligo- and polysaccharides. Such products are applied in affinity, immuno- and *in situ* hybridization staining of sections, as probes of live cells, and as stains in flow cytometric procedures. Although Texas red is currently a more popular fluorescent label, the following examples from recent literature demonstrate TRITCs continuing usage.

Amongst the various TRITC-labeled reagents for the affinity staining of sections and cell monolayers are lectins used to distinguish the sialomucins of salivary glands (Menghi *et al.*, 1998). TRITC-antibodies are also used, for instance to identify pathogenic amoebae (Tachibana *et al.*, 1997). TRITC-oligonucleotides are used, an example being the *in situ* hybridization staining of microorganisms in soil (Hahn *et al.*, 1992).

TRITC-labeled reagents have also been used as fluorescent probes of live cells. Labeled latex microspheres have been applied for retrograde tracing of axonal pathways (Vercelli and Innocenti, 1993), and labeled serum albumin as a permeability tracer in cardiac blood vessels (Huxley and Williams, 1996). Flow cytometry of TRITC-labeled slime mold cells has been used to investigate their aggregation behavior (Azhar *et al.*, 1996).

Purity of commercial batches: The dye contents of typical samples of the 6(R)-isomer are up to 80%.

Stability: The dry powder is light and moisture sensitive, and should be stored in a dark, desiccated container, preferably in the freezer. Staining solutions photofade. Aqueous solutions are unstable, and their storage should not be attempted.

Hazards: Harmful if ingested.

Rhodamine B isothiocyanate

Note: information in this entry applies to both isomeric forms.
Synonyms: RBITC, RhIc, RITC, TRITC
$C_{29}H_{30}ClN_3O_3S$, FW 536

Absorption maximum: is 554 nm for the immunoglobulin conjugate in water, with the equivalent emission maximum being 576 nm.

RBITC is an aminoxanthene, whose overall size is moderate, which carries carboxylic acid and isothiocyanate substituents on the pendant phenyl ring. The dye has a moderately sized, nonplanar aromatic system. RBITC forms a lipophilic cation under acid conditions, and a weakly lipophilic zwitterion around neutrality. The isothiocyanate group reacts with nucleophilic substituents (e.g. amino, hydroxyl, thiol) of biomolecules, so permitting attachment of a fluorescent label, see Chapter 7. RBITC also reacts with water and especially hydroxide ion. RBITC is sold as the chloride, and is a mixture of the 5- and 6-isothiocyanates, as illustrated. Note that the use of TRITC as an abbreviation is unfortunate, since there is another dye for which these initials are routinely used.

Applications: This dye has been used as a label of immunoglobulins, as reviewed by Brelje *et al.* (1993), but currently is often applied as a fluorescent probe of living cells. Some applications use RBITC-labeled molecules, especially dextrans. Such dextrans have been applied as probes of a cell's endocytic activity (Lawoko and Tagerud, 1995), and to mark cells to allow their subsequent migration to be detected (Ishii *et al.*, 1996). Bovine albumin heavily labeled with RBITC (Heinicke and Kiernan, 1978) is a valuable tracer of vascular permeability in the nervous system of rodents and fishes, following intravenous, intramuscular or subcutaneous injection; applications of this type have been reviewed by Kiernan (1996).

   RBITC is also used directly as a probe of living systems, for instance, to stain skin grafts so that movements of immune system cells can be traced (Hoefakker *et al.*, 1995), and for retrograde axonal tracing (Deng and Rogers, 1999).

Stability: The dry powder is moisture sensitive and must be stored in a desiccated container, preferably in a freezer. Aqueous solutions are unstable. Solutions photofade readily.

Lissamine rhodamine B sulfonyl chloride

Note: information in this entry applies to both isomeric forms.
Synonyms: lissamine rhodamine B200, LRSC, rhodamine B sulfonyl chloride, RITC
$C_{27}H_{29}ClN_2O_6S_2$, FW 577

Absorption maxima: are 560 nm in water and 568 nm in methanol; the emission maxima in those solvents being 585 nm and 583 nm respectively.
Solubilities: soluble in acetone, dimethylformamide, ethanol and methanol.

Lissamine rhodamine B sulfonyl chloride is an aminoxanthene with one sulfonate plus one sulfonyl chloride substituent on the pendant phenyl ring. Overall size is moderate, as is that of the nonplanar aromatic system. Under neutral conditions this dye is a zwitterion. The sulfonyl chloride group reacts with nucleophilic substituents (especially amino and hydroxyl) of biomolecules, attaching a fluorescent label; see Chapter 7. The dye also reacts with water and hydroxide ion.

The commercial material is zwitterionic; and contains two isomers in which the sulfonate and sulfonyl chloride groups are at the 2,4- and 4,2-positions respectively. The 2,4- isomer is illustrated.

The term **rhodamine B sulfonyl chloride**, sometimes used as a synonym, unfortunately implies the dye is a derivative of rhodamine B, and so is carboxylated. In fact the dye derives from the sulfonated compound sulforhodamine B. Use of the synonym **lissamine rhodamine B200** (with no mention of the sulfonyl chloride) perhaps derives from a typographical error, an unfortunate mistake as an acid dye of that name has itself been used as a protein label. Use of the abbreviation **RITC** is potentially confusing, because this is also used to describe rhodamine B isothiocyanate.

Applications: At one time an important immunofluorescence label, it is still recommended for multicolor work, as reviewed by Brelje *et al.* (1993). Recent microscopic applications of this dye are more often to label other proteins, such as bovine serum albumin used to trace movement of plasma proteins into arterial walls (Weinberg *et al.*, 1994), and the iron transporting protein, transferrin, to assess its endocytic uptake in neurons (Hemar *et al*, 1997). Many current uses involve labeling of phosphoethanolamine lipids, for instance to trace membrane trafficking in the endocytotic–exocytotic pathways (Vidal *et al.*, 1997), and to assess fusion of liposomes into endothelial cells (Cansell *et al.*, 1999).

Stability: The dry powder is light and moisture sensitive, and should be kept in a dark, desiccated container, preferably in the freezer. Solutions photofade, aqueous solutions are unstable. Should not be dissolved in dimethylsulfoxide, with which it reacts.

Oregon green 514 ester

Synonyms: Oregon green 514 carboxylic acid, succinimidyl ester
$C_{26}H_{12}F_5NO_9S$, FW 609

Absorption maximum: is 506 nm at pH 9 in water, in which solvent the emission maximum is 526 nm.
pK value: ca 4.7
Solubilities: soluble in dimethylformamide, dimethylsulfoxide, and in alkaline aqueous solutions.

Oregon green 514 ester is a weakly acidic hydroxyxanthene whose overall size is moderate. The dye has a moderately sized, nonplanar aromatic system. A succinimidyl carboxylate group on the pendant phenyl ring reacts readily with the amino groups of biomolecules. The fluoro substitution of the fluorescein chromophore results in improved photostability, and higher fluorescence with less quenching, compared with fluorescein itself. Oregon green 514 ester is sold as the free acid.

Applications: Oregon green 514 ester is used as an immunoglobulin label for immunofluorescence staining, for example, to localize an acyltransferase in macrophages (Khelef *et al.*, 1998). The dye has been used also in affinity-staining methods, such as Oregon green 514–phalloidin for actin microfilaments in fibroblasts (Sells *et al.*, 1999). There are also nonmicroscopic applications, such as labeling of albumin to study leakage from blood vessels (Rumbaut *et al.*, 1999).

Stability: of dry powder stored in sealed containers, preferably desiccated, is good. Light fastness is good, though storage is best in dark containers.

Texas red

Synonyms: sulphorhodamine 101 acid chloride, TR
$C_{31}H_{29}ClN_2O_6S_2$, FW 625

Absorption maximum: 587 nm in chloroform, emission maximum at 602 nm in that solvent.

Solubilities: soluble in the water-miscible solvents dimethylformamide and acetonitrile; also soluble in chloroform.

Texas red is an aminoxanthene with one sulfonate and one sulfonyl chloride substituent on the pendant phenyl ring. Overall size is moderate, as is that of the nonplanar aromatic system. The cyclic nature of the *N*-substituents results in a relatively rigid, and so strongly fluorescent, molecule. Under neutral conditions this dye is a zwitterion. The sulfonyl chloride group reacts with nucleophilic substituents (especially amino and hydroxyl) of biomolecules, attaching a fluorescent label; see Chapter 7. The dye also reacts with water and hydroxide ion.

The commercial material is zwitterionic; and contains two isomers, in which the sulfonate and sulfonyl chloride groups are at the 2,4- and 4,2-positions respectively, the former is illustrated.

Applications: Texas red is used to label immunoglobulins for staining fixed or frozen sections and cell monolayers, for instance to assess acrosomal status (Kawamoto *et al.*, 1999). In technically related methods other biopolymers are labeled and then used as stains. Examples include plasma membrane protein labeling (Dong *et al.*, 1998), and labeling DNA to assess copy number changes in tumors (Larramendy *et al.*, 1998).

Other staining applications use Texas red-labeled compounds as fluorescent probes of living systems. Thus apotransferrin is labeled with Texas red to permit tracing of its uptake into cultured intestinal epithelial cells (Alvarez Hernandez *et al.*, 1998), and neutrophil elastase inhibitor is labeled to determine its distribution in pulmonary airways (Rees *et al.*, 1999). The endocytic pathway of the antibiotic gentamycin has been explored using a Texas red-labeled derivative (Sandoval *et al.*, 1998), and Texas red-dextrans are used as retrograde axonal tracers (Raybould *et al.*, 1999).

Purity of commercial batches: Dye contents of typical samples are up to 65%, with up to 15% of chloroform also being present.

Stability: The dry powder is moisture sensitive, and should be kept in a dark, desiccated container, preferably in the freezer. Solutions photofade. Aqueous solutions are unstable. Texas red reacts with dimethylsulfoxide, so the dye must not be dissolved in this solvent.

Calcein

Information is also provided in this entry concerning the hexa-ester derivative, calcein acetoxymethyl ester, acetate, usually termed **calcein AM**.

Synonyms: CA; fluorexon, fluorescein complexone, FC (these latter terms being more often used in analytical chemistry). The term 'calcein' is used throughout this entry, even when work cited had used the terms fluorescein complexone or fluorexon.

Free acid: $C_{30}H_{26}N_2O_{13}$, FW 623

Hexa ester: $C_{46}H_{46}N_2O_{23}$, FW 995

Absorption maximum: of the free acid is 494 nm in aqueous solution at pH 9; the emission maximum in that solvent is at 517 nm.

Metal complexation behavior: Forms chelates with calcium and magnesium; as well as with other metal ions such as Cu(II), Fe(II), Fe(III) and Ni(II). Used as a fluorescent indicator for calcium and magnesium.

Solubilities: Calcein is soluble in aqueous solutions above pH 5, being 0.4% soluble in water, and is soluble to 0.6% in ethanol; calcein AM is soluble in dimethylsulfoxide.

Calcein is a hydroxyxanthene whose overall size, due to two methyliminodi-acetic acid groups, is moderate. The dye has a moderately sized, nonplanar aromatic system. Due to the multiple carboxylic acid substituents, the anionic forms present under staining conditions are hydrophilic. The multiple carboxylate groups of this dye provide ligands to chelate calcium ions, yielding fluorescent complexes. Calcein is available commercially as the free acid. At various times products of different isomeric composition have probably been sold, currently the 4′,5′-isomer is available.

In **calcein AM** the four aliphatic carboxyl groups are esterified with acetoxymethyl groups. The compound is a lactone, with two aromatic acetate esters. All ester groups are readily removed by cellular esterases, generating calcein. Calcein AM is membrane permeant and is used to introduce calcein into living cells.

Applications: Calcein is a standard reagent for marking bone growth, permitting subsequent histological study of fixed material; for a recent application see Clement Lacroix *et al.* (1999). The dye has also been used to mark cartilage matrix in fish for age assessment studies (Gelsleichter *et al.*, 1997).

Calcein is applied extensively as a fluorescent probe of live cells. In such procedures entry of dye into cells may be achieved by direct insertion of the hydrophilic calcein, or by uptake of calcein AM with subsequent intracellular formation of calcein. Recent examples include tracing of cytoplasmic continuity between embryonic cells (Gardner and Cockcroft, 1998), and widespread use as a marker of apoptosis, based on entry of the hydrophilic dye into mitochondria following the mitochondrial permeability transition (Lemasters, 1999). Calcein has also been used as a flow cytometric stain to demonstrate the existence of gap junctions joining murine macrophages to intestinal epithelial cells (Martin *et al.*, 1998).

Nonmicroscopic laboratory applications include use as an experimental drug, to assess the value of skin electroporation for facilitating transdermal drug delivery (Chen *et al.*, 1998); and as a stain to determine numbers of cells present in microwell plates (Murphy *et al.*, 1998).

Purity of commercial batches: may contain up to five isomers.

Stability: of dry powder stored in sealed, dark containers is good; light fastness of solutions is poor.

Calcium green-1

Information is also provided here concerning the hexa-ester derivative, calcium green-1 acetoxymethyl ester, acetate; usually termed calcium green-1 AM; and also certain derivatives, such as calcium green-2, calcium green-5N, calcium green-C18 and the calcium green-1 dextrans.

Synonym: CaG

Hexapotassium salt: $C_{43}H_{27}C_{12}K_6N_3O_{16}$, FW 1147

AM ester: $C_{59}H_{53}C_{12}N_3O_{26}$, FW 1291

Absorption maxima: of the hexapotassium salt is 506 nm in water, in which solvent the emission maximum is 531 nm. The AM ester has an absorption maximum of 302 nm in methanol, but is nonfluorescent.

pK value: 6.7

Metal complexation behavior: The dye is used as a fluorescent indicator for calcium ions, which are strongly chelated.

Solubilities: The hexapotassium salt is soluble in aqueous solutions above pH 6; the AM ester is soluble in dimethylsulfoxide and in methanol.

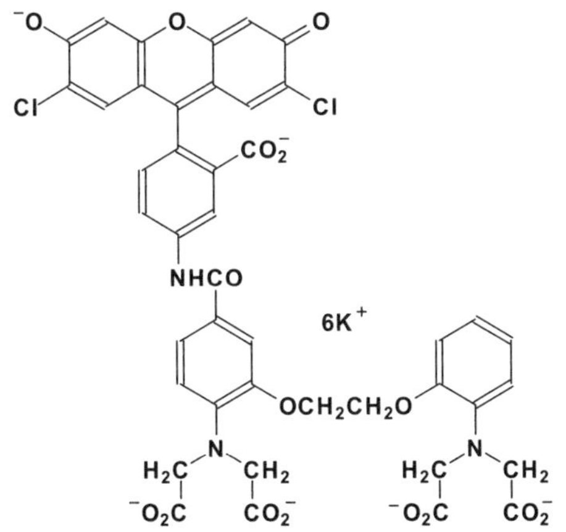

Calcium green-1 has a 2′,7′-dichlorofluorescein fluorophore. Attached to the pendant phenyl ring are aromatic units carrying the multiple iminoacetic acid moieties which are the ligands involved in calcium chelation. The hexaanionic salt, illustrated, is strongly hydrophilic. The calcium green-1 AM ester is a superlipophilic lactone. The four aliphatic carboxyls are esterified as acetoxymethyl groups, and the two hydroxyl groups are derivatized as acetyl groups. This ester is converted to the salt by cellular esterases. Both the hexapotassium salt and the AM ester are available commercially.

Applications: Calcium green-1 finds considerable application as a calcium probe in living cells and tissues. The probe is introduced into cells both directly (e.g. by iontophoresis of the salt) and indirectly by use of the AM ester. Applications cited below involve several modes of dye insertion. Recent examples include confocal microscopy to measure calcium in hippocampal slice cultures (Abdel Hamid and Tymianski, 1997), in insect neurons (Ogawa *et al.*, 1999), and in frog skin epithelia (Brodin and Nielson, 2000).

Stability: of dry powder is good when stored in sealed containers; staining solutions may photofade and are best stored in the dark.

Related compounds: There are several commercially available derivatives. **Calcium green-2**, which carries an additional fluorophore unit, is used in nonimaging systems to record calcium levels. **Calcium green-5N** has only a single fluorophore unit, but is substituted with a nitro group. This dye has been used as a probe of living cells, for example, to estimate calcium levels within photoreceptors using confocal microscopy (Ukhanov *et al.*, 1998). **Calcium green-C18** has a long hydrocarbon sidechain whose resulting amphiphilicity results in accumulation in membranes, facilitating imaging and quantitation of changes in calcium levels at the surfaces of membrane-bound organelles, rather than merely in the cytosol (Tojyo *et al.*, 1998). In terms of relative size, it is easier to consider the various **calcium green-1 labeled dextrans** as dye attached to an oligomer or polymer; all are relatively large and slow diffusing compared to the dye itself. Such products have been used to measure the longitudinal spread of calcium ions within functional rod-segments (Gray Keller *et al.*, 1999), and to assay cytosolic calcium ion in plant root hairs (Lew and Dearnaley, 2000).

17

Acridines and phenanthridines

R.W. Horobin

R.W. Horobin

BACKGROUND INFORMATION

Following the general information, information on specific acridine dyes precedes that on the phenanthridine dyes.

Structural features

The acridine and phenanthridine chromophores both contain a central six-membered aromatic nitrogen heterocycle, with benzo rings fused onto each side. The two types of chromophore are isomeric, differing only in the points of attachment of one of the fused benzo rings; as shown below. Also shown is the IUPAC system for numbering the acridine and phenanthridine ring-substituents, as used below. Some early publications in the biological literature used different numbering systems, as described by Albert (1966).

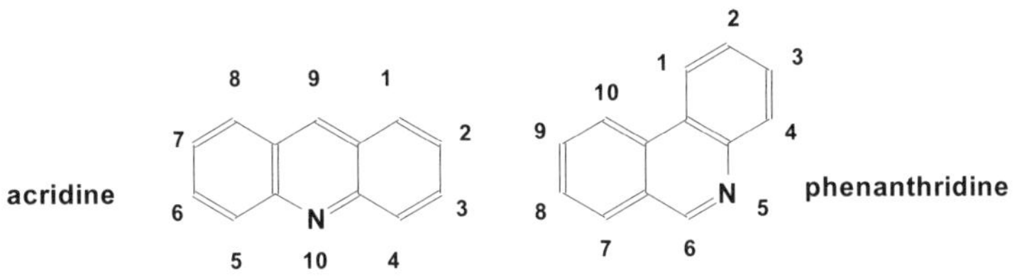

Properties influencing staining

Electric charge – Although sulfonated and hydroxylated acridines have been synthesized, none are used as stains. All the dyes described below contain a basic nitrogen heteroatom; most also carry amino substituent(s) attached to the chromophoric system; and some also have additional amino substituent(s) attached to alkyl side-chains. Consequently when ionized, these acridine and phenanthridine stains are cationic.

Base strength – These stains are mostly moderately strong bases, whose pK values fall in the range 9–11. Some polyamino dyes have an additional pK value around neutrality. In such cases, the first proton is usually attached to the ring nitrogen, although compounds such as quinacrine have both ring and aliphatic nitrogen atoms protonated under physiological conditions. These nitrogen heterocycles can also form pseudobases. This is especially marked with ethidium bromide and propidium iodide, because the carbinol is stabilized by the 6-phenyl substituent. As a result of this range of base strengths, some of the compounds are fluorescent probes for nuclei and some for lysosomes.

The *hydrophilic/lipophilic* properties of the dyes under usage conditions are varied. In part this results from variation in pK; the free bases are commonly lipophilic and the cationic species hydrophilic. This explains why many acridines can enter live cells. The compounds with quaternary nitrogen, such as ethidium bromide, are usually excluded from live cells with intact cell membranes. However, other chemical features, for instance the numbers of amino, alkyl and phenyl substituents, also influence lipophilicity. Thus phosphine is a phenylated and methylated acridine dye containing lipophilic cationic species, and is used as a fluorescent probe for lipids. Nonionized derivatives, such as the pseudobases and the leucodye hydroethidine, are also cell permeant lipophilic species.

Conjugation – Acridine and phenanthridine chromophores are small and planar, and can intercalate into DNA helices. Overall the aromatic units in the common phenyl derivatives of both acridines and phenanthridines (e.g. phosphine and ethidium bromide respectively) are nonplanar, with phenyl units rotated out of the plane of the chromophore. In ethidium homodimer-1 two such moieties are linked by a flexible spacer. In hydroethidine, the formation of a leucodye reduces the size of the conjugated system found in the parent phenanthridine.

Reactive groups – Several reaction classes are exploited in these stains. For instance, the leucodye hydroethidine is used as an oxidation–reduction indicator, to visualize and assay intracellular oxidants. One acridine, quinacrine mustard, yields aziridinium cations in solution, which react rapidly with tissue nucleophiles such as amino and thiol groups. A phenanthridium, ethidium azide, also reacts with nucleophils, but only when photoactivated.

Size – Acridine and phenanthridium dyes used in biology and medicine vary markedly in size, with formula weights between 249 and 857. However, it does not seem that size greatly influences staining patterns.

Safety issues

Although some of the compounds discussed are bacterial mutagens, many are not acutely toxic to eukaryotic cells. Several indeed are used as drugs in humans. The quinacrine mustard, although it has been used as an anticancer agent, is of course a mutagen.

CATIONIC ACRIDINE DYES

Proflavine

CI 790 (1st edition number)
Synonyms: 3,6-diaminoacridine, P_f, PF, PRO
Chloride: $C_{13}H_{12}ClN_3$, FW 246
Sulfate: $C_{26}H_{24}N_6O_4S$, FW 516

Absorption maxima: 444 nm in neutral aqueous solutions, 456 nm in methanol; emission maximum 510 nm in neutral aqueous solution.
pK values: 1.5, 9.6
Solubilities: the sulfate is 1% soluble in water, and 0.2% in ethanol; while the chloride is 0.4% soluble in water and 0.7% in ethanol.

Proflavine is an acridine whose overall size and planar conjugated system are small. The cationic form is hydrophilic, the free base is weakly lipophilic. Proflavine is sold both as the chloride, illustrated, and the sulfate.
Applications: Proflavine has been applied in investigations of chromosome structure by banding studies (Disteche *et al.*, 1980). More recently it has been used for visualizing and quantifying plankton using confocal microscopy (Verity *et al.*, 1996). When injected into laboratory mammals intravenously proflavine does not cross the blood–brain barrier except in regions with permeable vessels. Consequently proflavine can be used as a maker of structures such as the circumventricular organs, which are made conspicuous by their fluorescent cell nuclei (Rodriguez, 1955; Rodriguez-Peralta, 1966).

The dye has been widely applied as a disinfectant, and its activity as a photoactivated bactericide has been investigated (Wainwright *et al.*, 1997).
Purity of commercial batches: Dye contents of chloride and sulfate salts may be over 95%. There is a thin-layer chromatographic method available (Matzke and Thiessen, 1976).
Hazards: bacterial mutagen.

Acriflavine

CI 46000
Synonyms: ACF, trypaflavine
Chloride: $C_{14}H_{14}ClN_3$, FW 260

Absorption maxima: 452 nm in water, and 465 nm in ethanol; emission maximum 510 nm in water.
pK value: 10.1

Solubilities: 15% in water; 1% in ethanol; these values will probably refer to acriflavine contaminated by an uncertain amount of proflavine.

Acriflavine is an acridine of small overall size, with a small planar conjugated system. In aqueous solutions acriflavine is a hydrophilic cation, as illustrated.

This dye is produced by the *N*-methylation of **proflavine**, and commercially this reaction is routinely incomplete. Consequently acriflavine usually contains residual proflavine. Such crude products are sold both as a mixture of chlorides, and as a nominal mixture of acriflavine chloride with **proflavine free base** ('**acriflavine neutral**').

Applications: Acriflavine has been used as a general oversight stain in fluorescence microscopy in entomological specimens (Metcalf and Patton, 1944) and, more recently, visualization of the rat brain using confocal microscopy (Becker *et al.*, 1996) and the identification of plankton from seawater (Naganuma *et al.*, 1998). Acriflavine has also been used to stain sulfated glycosaminoglycans, for both fluorescence (Hollander, 1963) and diachrome (Yamada, 1969) microscopy. A variant, in which an initial acriflavine staining is reacted with dimethylaminobenzaldehyde to give a red reaction product, is recommended in staining manuals for demonstration of sulfatides (Bancroft and Cook, 1994; Bancroft and Gamble, 2002). The acriflavine–sulfur dioxide derivative of the dye was introduced into histochemistry as a Schiff-type reagent some time ago (Kasten *et al.*, 1959; Ornstein *et al.*, 1957). In this application the reagent has a slow fading rate, and is recommended by staining manuals (Kiernan, 1999). In living animals, acriflavine can be injected intravenously to reveal the nuclei of cells in those parts of the nervous system with permeable blood vessels (Kiernan, 1996).

Nonstaining applications are numerous. Historically it was used as a treatment for trypanosome infections, hence its original name 'trypaflavine'. It has since been widely used as an antiseptic agent. Current pharmacological applications include treatment of tick-borne diseases in farm animals (Shen *et al.*, 1997), and use as a virostatic agent for treating HIV-infected human patients (Mathe *et al.*, 1998). In diagnostic microbiology, acriflavine is used in resistogram typing of *Salmonella* serotypes (Purushothaman *et al.*, 1998).

Purity of commercial batches: These usually contain 10–30% of proflavine as contaminant, plus minor amounts of other fluorescent compounds. Occasionally batches are predominantly proflavine. Overall dye content may be up to 90%. There is a thin-layer chromatographic method available (Levinson *et al.*, 1977).

Acridine orange

CI 46005, CI Basic orange 14
Synonyms: acridine orange NO, AO, euchrysine 3R
Chloride hydrate: $C_{17}H_{19}ClN_3.H_2O$, FW 320
Zinc chloride double salt: $C_{34}H_{40}Cl_4N_6Zn$, FW 740

Absorption maxima: 489 nm for the zinc chloride double salt, and 492 nm for the chloride, in water; emission maxima at 535 nm in water, orthochromatic bound to DNA 502 nm, metachromatic bound to RNA 650 nm.
pK *values:* –3.2, 10.5
Solubilities: the hydrochloride is 5% soluble in water and 0.5% in ethanol; the zinc chloride double salt is soluble to 0.6% in water and 0.2% in ethanol.

This dye is an acridine whose overall size and planar conjugated system are small. The cationic form, illustrated, is slightly hydrophilic; the free base is lipophilic. The dye is sold both as a chloride, and as the so-called zinc chloride double salt (in which the anion is the $ZnCl_4^{2-}$ complex ion).
Applications: Several fluorescence histochemical procedures using acridine orange are recommended in staining manuals, including stains of the Bertalanffy type, giving differential staining of DNA and RNA. For instance, in examination of exfoliated material, particularly that of the female reproductive tract (Bancroft and Cook, 1994). Other widely used applications are for staining of acid mucins (Bancroft and Gamble, 2002), and detection of apoptosis (Kiernan, 1999).

Many staining procedures involve demonstration of DNA or DNA-rich structures. For instance C banding of chromosomes in cytogenetics (Cuellar *et al.*, 1999), and the oversight staining of somites in rat embryos (Menegola *et al.*, 1999). A number of procedures involve staining blood smears, to diagnose malaria (Tantular *et al.*, 1999), and to identify and enumerate granulocytes in fish (Morimoto *et al.*, 1999). Acridine orange has also been applied to resin sections for demonstration of plant cell walls and starch granules (Wiatr, 1976), and of nucleic acids (Paul, 1980).

Acridine orange is also used to stain micro-organisms in unfixed specimens. For instance bacteria in marine sediments (Kuwae and Hosokawa, 1999), and in urine (Lorincz *et al.*, 1999). Viral inclusions in living fish cells have also been stained (John and Richards, 1999).

The dye's major vital staining application is, however, with eukaryotic cells, especially to demonstrate low pH structures such as lysosomes (Manyonda and Choy, 1999), and to measure intracellular pH (Rodrigues *et al.*, 1999). Acridine orange is also used to distinguish live from dead cells, for example, in cultures of dinoflagellates (La Barre *et al.*, 1999); and for labeling leukocytes to study their entrapment in the retinal microcirculation (Hiroshiba *et al.*, 1999).

Acridine orange is noted in laboratory manuals as a DNA stain for flow cytometry (Juan and Darzynkiewicz, 1998). Manuals also describe its application for simultaneous RNA and DNA staining (Macey, 1994).

Industrial uses: Acridine orange has been used to color leather.

Purity of commercial batches: One major component and several minor or trace components are routinely present. The dye content of the chloride can be up to 98%, that of the zinc chloride double salt 90%. A thin-layer chromatographic method is available (Marshall, 1976c).

Stability: Light fastness is poor.

Hazards: An animal mutagen.

Phosphine

CI 46045, CI Basic orange 15

Synonyms: aurophosphine G, brilliant phosphine, chrysaniline, phosphine E & 3R; PR. Note that the term **aurophosphine** has been used to describe **euchrysine** as well as phosphine. Related dyes which may have been used as phosphine include CI 46050, sold as **flavophosphine** and **phosphine 2G**; and CI 46065, sold as **benzoflavine** and **flavophosphine**. Phosphine is also the name of a toxic inorganic gas, phosphorus trihydride.

Chrysaniline nitrate: $C_{19}H_{16}N_4O_3$, FW 348

Absorption maximum: 459 nm in methanol; the precise nature of the sample giving rise to this value is uncertain, see below.

Solubilities: 0.6% in water; 0.9% in ethanol; the nature of the sample giving rise to these values is uncertain, see below.

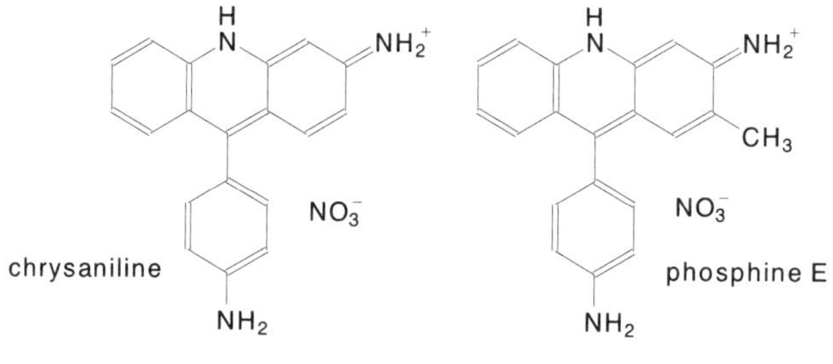

chrysaniline

phosphine E

Descriptions of the nomenclature and nature of phosphine are confusing. The nonmethylated dye (CI 46045) has been termed chrysaniline; various monomethyl homologues have been called **phosphine E**, **phosphine 2G** and **flavophosphine**; and a dimethylated homologue **benzoflavine** or **flavophosphine N**. However since such dyes may be prepared from basic fuchsine-type melts (Colour Index, 1971, pp. 4433–4434), it is probable that these products comprise overlapping mixtures of isomers and homologues. Two of the possible components are illustrated. Phosphine has been sold as the nitrate and as the chloride.

All such phosphines have an additional phenyl ring which is rotated out of the plane of the chromophore. All are moderate in size, with moderately sized aromatic systems. The cationic forms of these dyes will be lipophilic or weakly hydrophilic; the free bases will be lipophilic.

Applications: Phosphine, in the generic sense, has a number of staining applications. It was reported as a fluorescent stain for lipids by Popper (1940), who later (Volk and Popper, 1944) published a detailed account of his procedure. Phosphine is still used for this purpose, for instance as a fluorescent probe of hydroxybutyrate accumulation in bacteria (Bonartseva, 1985), and of lamellar bodies in living alveolar type II cells (Beers, 1996). The dye has also been used as a marker to allow separation of alveolar type II cells by fluorescence-activated flow cytometry (Martin *et al.*, 1993). Phosphine has been used as a histochemical stain for acid mucopolysaccharides in the corneal stroma (Suveges and Modis, 1970).

Industrial uses: Phosphine has been used to color leather.

Purity of commercial batches: Samples may have multiple components and contain up to 70% dye overall. A thin-layer chromatographic method is available (Horobin and Murgatroyd, 1967).

   Stability: of aqueous staining solutions is poor; light fastness is moderate.

Quinacrine

Synonyms: atabrine, atebrine, mepacrine, QFQ
Dihydrochloride, anhydrous: $C_{23}H_{32}Cl_3N_3O$, FW 473
Dihydrochloride, dihydrate: $C_{23}H_{36}Cl_3N_3O_3$, FW 509

Absorption maxima: 445 nm in pH 7 aqueous solution, and 420 nm in pH 9 aqueous solution; emission maximum 500 nm in pH 7 aqueous solution.
pK values: 2.8, 7.5, 10.2
Solubilities: 3% in water, slightly soluble in ethanol.

$$CH_3CH_2 \diagdown$$
$$\quad\quad {}^+NHCH_2CH_2CH_2\overset{\overset{\displaystyle CH_3}{|}}{CH}-NH$$
$$CH_3CH_2 \diagup$$

2Cl⁻   Cl   OCH₃   N⁺H

Applications: Quinacrine is used in routine staining procedures recommended in staining manuals, for example, demonstration of Y chromosomes in buccal smears (Bancroft and Cook, 1994), and fluorescent staining of nucleic acids (Kiernan, 1999). In cytogenetics, quinacrine has been widely used for Q-banding of chromosomes in metaphase spreads, as briefly reviewed by Rost (1995, pp. 153–154).

   In living tissues, quinacrine has been used for staining lysosomes in animal cells (Wittekind and Kretschmer, 1972) and acidic vacuoles in algae (Weiss and Pick, 1991). The dye has also been used to demonstrate peptide-hormone secreting cells such as thyroid C-cells and gastrin secreting cells of the pyloris

(Ekelund *et al.*, 1980); and certain nerve fibres (Crowe and Burstock, 1984), possibly by binding to ATP or inorganic phosphates (Belai and Burnstock, 1994; Zozulya *et al.*, 1997). Quinacrine is used to stain platelets to permit subsequent assessment of their adhesion and aggregation (Joist, Bauman and Sutera, 1998); and to assess the dense granule content of platelets using flow cytometry (Wang *et al.*, 1999).

   Quinacrine has a long history of use as an antimalarial drug, and is still so used where newer drugs cannot be afforded. Other widespread clinical applications of quinacrine are as a treatment of parasitic worm infections (Chung *et al.*, 1991), and as an antiprotozoal agent (Babb, 1995). It has also been used as an antiinflammatory agent in rheumatic diseases (Wallace, 1989), and as a component of a preparation for female sterilization (Kang *et al.*, 1990).

   Purity of commercial batches: dye contents of typical samples are up to 98%. There is a paper chromatographic method available (Albert, 1966, p. 150).

Nonyl acridine orange

Synonyms: AO 10-nonyl bromide, NAO
Bromide: $C_{26}H_{38}BrN_3$, FW 473

Absorption maxima: are at 488 nm in water and 495 nm in methanol; emission maxima are at 525 nm in water and 519 nm in methanol.
Solubilities: soluble in ethanol and dimethylsulfoxide.

$(CH_3)_2N$ N $N(CH_3)_2$

Br^- $CH_2CH_2CH_2CH_2CH_2CH_2CH_2CH_2CH_3$

Nonyl acridine orange is an acridine of moderate overall size, whose planar conjugated system is small. The cationic species is lipophilic. The bromide is available commercially, as illustrated.

   Applications: Nonyl acridine orange is recommended by laboratory manuals as a stain for mitochondria in living cells (Poot, 1998). This dye is applied to investigate mitochondrial distribution, as distinct from mitochondrial activity. Variant applications include assessing mitochondrial volume (Fujii *et al.*, 1997) and visualizing phospholipid domains in bacteria (Mileykovskaya and Dowhan, 2000). Nonyl acridine orange is also used as a stain for mitochondria in flow cytometry (Thomas *et al.*, 1999).
Stability: Solutions are sensitive to light.

Lucigenin

Synonyms: bis-N-methylacridinium nitrate
The nitrate: $C_{28}H_{22}N_4O_6$, FW 511

Absorption maximum: 455 nm in water; emission maximum 505 nm in water. Lucigenin emits chemiluminescence at 470 nm when oxidized in aqueous alkaline solutions.
Indicator properties: fluorescent indicator for reactive oxygen species.
Solubility: soluble in water.

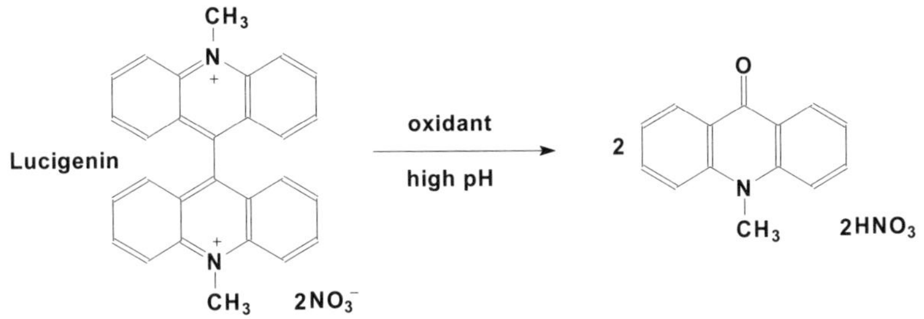

Lucigenin comprises two linked small planar acridine chromophores, the overall molecule is nonplanar. The dicationic form, illustrated, is hydrophilic; the mono-cation resulting from pseudobase formation is very slightly lipophilic. The dye is sold as the nitrate. As shown in the figure, oxidation of lucigenin gives rise to 10-methylacridone, this process being accompanied by chemiluminescence.
Applications: Lucigenin has occasionally been used as a fluorescent probe of living cells. Thus its vesicular uptake has been studied using fluorescence microscopy, in rat hepatocytes (Braakman *et al.*, 1989); and the dye has been used for in vitro assessment of reactive oxygen species of inflammatory cells in asthma patients using a chemiluminescence procedure (Vachier *et al.*, 1994).

Lucigenin chemiluminescence assays of such species are widely applied in nonimaging technologies. For instance to detect superoxide anion in isolated mitochondria (Li *et al.*, 1999), and to investigate the release of reactive oxygen species in cerebral endothelial cells following infection with pneumococci (Koedel and Pfister, 1999).
Purity of commercial batches: There is an HPLC analytical method available (Maskiewicz *et al.*, 1979).
Stability: of dry powder stored in sealed, dark containers is good; solutions are light sensitive.

AN ACRIDINE DYE FORMING COVALENT BONDS WITH TISSUE

Quinacrine mustard

Synonyms: atabrine mustard, QM
Dihydrochloride: $C_{23}H_{30}Cl_5N_3O$, FW 542

Absorption maximum: 450 nm in water.
Solubility: soluble in water.

Quinacrine mustard is an acridine, whose overall size is moderate and whose planar conjugated system is small. The side-chain carrying the nitrogen mustard moiety, with its two chloroethyl groups, will only be present in the solid state. In aqueous solution such chloroethyl substituents rapidly lose chloride ions yielding aziridinium cations. These readily react with the carboxyl, thiol, and heterocyclic ring nitrogen groups of proteins and nucleic acids, or with other nucleophiles such as water; for a brief review see Wilman and Connors (1983). Quinacrine mustard is available commercially as the dihydrochloride, illustrated; as well as a hydrate.

Applications: Quinacrine mustard was the dye first used to band metaphase chromosomes, which procedure transformed cytogenetics (Rost, 1995, pp 153–154). Recent applications include the Q-banding of maize (Song *et al.*, 1994), and karyotyping of fish (Caputo *et al.*, 1996). The dye has also been used as a vital stain of trypanosomes (Schnedl *et al.*, 1982) and of RNA within insect salivary gland cells (Curtis *et al.*, 1987).

Purity of commercial batches: Dye contents of samples labeled as dihydro-chloride are 90% or more, with the hydrate being sold at 98%.

Hazards: An irritant which at high concentration damages mucous membranes and eyes, and blisters the skin; a bacterial mutagen.

NONIONIC PHENANTHRIDINE DYES

Hydroethidine

Synonyms: dihydroethidium, Et^red^, HE, HYD
$C_{21}H_{21}N_3$, FW 315

Absorption maximum: 355 nm in acetonitrile; emission maximum at ca. 420 nm in acetonitrile (unless oxidation to ethidium occurs).
Used as an indicator for reactive oxygen species.
Solubilities: low solubility in water, soluble in dimethylformamide and dimethyl-sulfoxide.

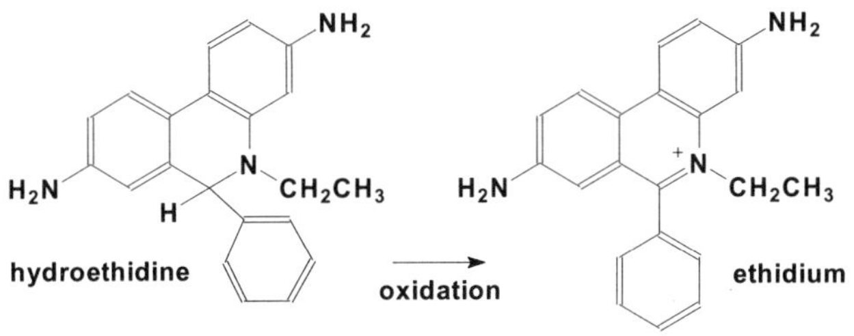

Hydroethidine ⟶ (oxidation) ethidium

Hydroethidine is a nonionic, lipophilic leuco-dye derivative of **ethidium bromide**. The overall size, and that of the conjugated system, is small. Oxidation, for example, by cellular peroxide anions, generates the intensely fluorescent ethidium species; as illustrated in the figure.

Applications: Hydroethidine is most widely used as a fluorescent probe of living cells, usually for the demonstration of reactive oxygen species. Recent applications include demonstration of superoxide levels in blood vessel walls (Nakane *et al.*, 2000), and within the mitochondria of cultured neurons (Castilho *et al.*, 1999). Being membrane permeant, hydroethidine has been used to introduce the DNA probe ethidium bromide into cells, for instance, to estimate the nuclear genome sizes of algal species (Kapraun *et al.*, 1991). The dye is similarly applied in flow cytometry as noted in a laboratory manual (Macey, 1994). Recent applications include use as a marker of eosinophils, prior to fluorescent-activated cell sorting (Mengelers *et al.*, 1995), and to detect reactive oxygen species in irradiated, cultured cells (Vit *et al.*, 2000).

Hydroethidine is also applied as a marker, for example, for the preparation of fluorescent algae used to assess herbivory by microzooplankton (Putt, 1991), and to assess water flow through sandy aquifers (Harvey *et al.*, 1995). Other applications include marking the normal tissue of rat lungs to identify the ischemic regions of ventilated rat lungs (Al Mehdi *et al.*, 1997).

Stability: The dry powder is stable stored frozen, under nitrogen, in the dark. Solutions should be used under nitrogen or argon, and illumination minimized, because they are susceptible to photocatalyzed atmospheric oxidation.

Hazards: Possible bacterial mutagen.

CATIONIC PHENANTHRIDINE DYES

Ethidium bromide

Synonyms: ETDI, ethidium, homidium bromide; the chloride is termed ethidium chloride.

$C_{21}H_{20}BrN_3$, FW 394

Absorption maxima: 480 nm in water, 518 nm in aqueous solutions of DNA, and 525 nm in methanol; emission maxima 605 nm in water, and 620 nm in aqueous DNA solutions.

The pK of pseudobase formation is ca. 10.5

Solubilities: 5% in water, also soluble in dimethylsulfoxide.

Ethidium bromide is a phenanthridine of moderate overall size. The aromatic system comprises the small planar diaminophenanthridium unit, with a nonplanar phenyl substituent. The cationic form of the dye, illustrated in the hydroethidine entry above, is hydrophilic; the pseudobase is lipophilic. Although the dye is by definition the bromide salt, ethidium has also been available as the chloride.

Applications: Ethidium bromide is recommended as a fluorescent nuclear stain by routine manuals (Kiernan, 1999). It is also applied as a fluorescent probe, staining the nuclei of cells with damaged plasmalemmal membranes. Recent applications of this type include assessment of the permeability of the sarcolemmal membranes in the cerebral arteries following chronic spasm (Kim *et al.*, 1999), and detection of apoptosis in the central nervous system following excitotoxic insult (Omar *et al.*, 2000). The dye is also used in a similar range of staining applications in flow cytometry, as discussed in manuals (Darzynkiewicz and Crissman, 1990; Macey, 1994; Shapiro, 1988).

   Other laboratory applications in biology and medicine include wide use as a stain of nucleic acids on electrophoretic gels, as described in laboratory handbooks (Mason, 1999, p. 52), and the staining of nucleic acids on blots (Matyas *et al.*, 1999). In veterinary practice, ethidium bromide is widely used as a trypanocidal drug.

Purity of commercial batches: Dye content of typical samples is up to 98%. There is an HPLC analytical method available (Tettey *et al.*, 1999).

Stability: of the dry powder is good if sealed in dark, airtight containers; aqueous solutions photofade readily.

Hazards: Mutagenic to cultured human cell lines.

   Ethidium monoazide carries an azide substituent in place of the amino group found in ethidium bromide, and when photolyzed in the presence of DNA or RNA, yields fluorescently-labeled nucleic acids. Following such photoconjugation, for instance, to plasmids, intracellular location of nucleic acids is sometimes achieved using confocal microscopy (Mohr *et al.*, 1999), and sometimes by flow cytometry (Tseng *et al.*, 1999). Entry into cells with damaged cell membranes is fast, and as the photoproducts are covalently bound, this probe is suitable for detecting dead cells, for instance spermatozoa, by subsequent flow cytometry (Henley *et al.*, 1994).

Propidium iodide

Synonyms: PI, propidium
Iodide: $C_{27}H_{34}I_2N_4$, FW 668

Absorption maxima: 493 nm in water, and 535 nm in aqueous DNA; emission maxima 636 nm in water, and 617 nm in aqueous DNA.
The pK of pseudobase formation is ca. 10.5
Solubilities: soluble in water and dimethylformamide.

Propidium iodide is of moderate overall size, with an aromatic system comprising a small planar diaminophenanthridium unit, with a nonplanar phenyl substituent. The dicationic form of the dye, illustrated, is hydrophilic; the second cationic group being a pendant quaternary salt. The monocationic pseudobase is also hydrophilic. The dye is sold as the iodide.

Applications: Propidium iodide is used as a DNA and nuclear stain, as noted in staining manuals (Kiernan, 1999). Recent examples include DNA ploidy analysis in fixed, touch preparations of epithelial tumors (Kawamura *et al.*, 2000), the detection of bacteria adhering to contact lenses (Gavin *et al.*, 2000), and as a component of a stain for distinguishing apoptotic from nonapoptotic cells (Mallolas *et al.*, 2000).

Propidium iodide is widely applied as a fluorescent viability stain, as its strongly hydrophilic character restricts entry to cells whose cell membranes are damaged. The dye can be used with isolated cells, as in the detection of viable bacteria in wine (Kopke *et al.*, 2000); assessment of the status of bone marrow cells, prior to trans-fusion (Mascotti *et al.*, 2000). Identification of nonviable cells within tissues is also carried out, for example, within brain slice cultures (Adamchik *et al.*, 2000), and experimental infarcts in heart muscle (Wolff *et al.*, 2000). Although the propidium ion does not generally enter living cells, it is taken up by axonal terminals and retrogradely transported to neuronal cell-bodies (Aschoff and Hollander, 1982). Propidium iodide is also widely applied as a stain in flow cytometry, as noted in laboratory manuals (Juan and Darzynkiewicz, 1998; Ormerod, 1998).

Purity of commercial batches: Dye contents of typical samples are up to 95%.

Stability: Solutions fade in light.

Hazards: Bacterial mutagen.

Ethidium homodimer-1

Synonyms: EB2, ETDI, EthD-1, EtDi, ethidium dimer; sometimes unfortunately referred to as 'ethidium homodimer' (there is an ethidium homodimer-2), or as 'ethidium'.

$C_{46}H_{50}Cl_4N_8$, FW 857

Absorption maxima: 493 nm in water, 587 nm in aqueous DNA solution; emission maximum 617 nm in aqueous DNA solution.

Solubilities: very soluble in water, also soluble in dimethylsulfoxide.

Ethidium homodimer-1 comprises two phenyl substituted phenanthridine chromophores, linked by a flexible alkydiamino chain. The tetracationic form of the dye, illustrated, is strongly hydrophilic; the dication is also hydrophilic. The dye is available commercially as the tetrachloride.

Applications: Ethidium homodimer-1 is widely applied as a fluorescent permeability marker, and thus as a viability stain. Recent applications include a study of the effect on sperm of diluents used in artificial insemination procedures (Sirivaidyapong *et al.*, 2000), and an investigation of the three-dimensional distribution of keratinocytes in the cornea (Hahnel *et al.*, 2000). The dye can also be administered to living organisms, with subsequent observations being made on fixed tissues; for example, the lavaging of the epithelia of the airways of intact lungs, previously exposed to ozone, with ethidium homodimer-1 solution (Postlethwait *et al.*, 2000).

Stability: of dry powder is good, stored in sealed, dark containers; solutions are light sensitive.

18

Azines

R.W. Horobin

Structural features

The characteristic chromophore of the azine dyes is phenazine: a planar, tricyclic dibenzopyrazine unit. However, except for the N-phenazine derivatives the compounds discussed in this chapter contain cationic amino or diaminophenazine chromophores, in which electron delocalization is somewhat extended by the electron acceptor/donor groups. These variants are illustrated below, where **R** can be hydrogen or phenyl and so on.

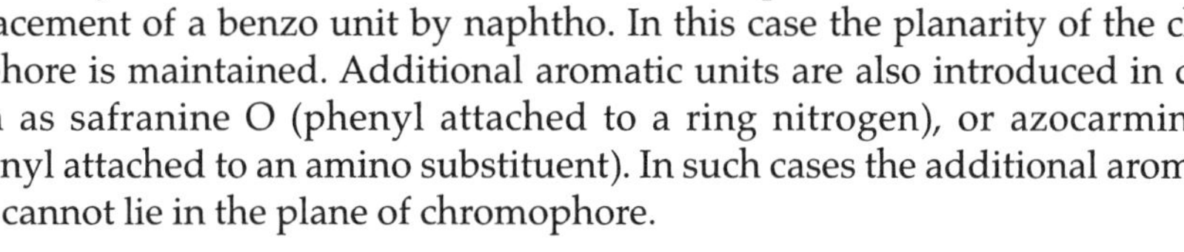

In some dyes, such as azocarmine B, the chromophore is further extended by replacement of a benzo unit by naphtho. In this case the planarity of the chromophore is maintained. Additional aromatic units are also introduced in dyes such as safranine O (phenyl attached to a ring nitrogen), or azocarmine G (phenyl attached to an amino substituent). In such cases the additional aromatic unit cannot lie in the plane of chromophore.

Properties influencing staining

Electric charge – Both cationic and anionic azine dyes are used in histology, for example, acid dyes such as azocarmine B and basic dyes such as safranine O. The anionic substituents are all strongly acidic sulfonates which are fully ionized over the pH range of concern here, but the basicity of the basic dyes varies somewhat. For instance, neutral red is a weak base, permitting its use as a probe of acidic organelles in live cells.

Hydrophilicity/lipophilicity – Amongst the cationic dyes, a compound such as Magdala red is sufficiently lipophilic to have been used as a fluorescent lipid stain; whereas the cation of neutral red is hydrophilic. The sulfonated azines are hydrophilic at physiological pHs. Consequently a dye such as nigrosine WS is used as a viability stain, because it is membrane impermeant and only enters cells if their plasmalemmal membranes are damaged.

Extent of conjugation – This varies from small (neutral red) to large (nigrosine WS). The larger dyes can be used in procedures involving strong van der Waals forces, such as elastic fiber stains.

Redox effects – Many members of the azine dye class undergo oxidation-reduction equilibria under usage conditions. However, phenazine methosulfate and methoxyphenazine methosulfate are specifically used in biology because of their electron transfer properties.

Dye purity

Some azine dyes discussed here are typically very impure because the methods of synthesis inevitably generate mixtures of isomers and homologs. Examples are the azocarmines, nigrosine WS, and safranine O. Other dyes, such as neutral red, are often produced with a single major constituent.

ANIONIC DYES

Azocarmine

Although azocarmine B and azocarmine G are nominally distinct dyes, this difference is not clear cut, see below. Consequently, as biological end-users commonly apply these dyes interchangeably, a single entry is provided. However, specific information relevant to the 'B' or 'G' dyes is provided when available.

Azocarmine G
CI 50085, CI Acid red 101
$C_{28}H_{18}N_3O_6S_2Na$, FW 580

Azocarmine B
CI 50090, CI Acid red 103
$C_{28}H_{17}N_3O_9S_3Na_2$, FW 682

Absorption maxima

511 nm in water

516 nm in water

Solubilities

1% in hot water; 0.1% in ethanol.

2% in water; 0.1% in ethanol.

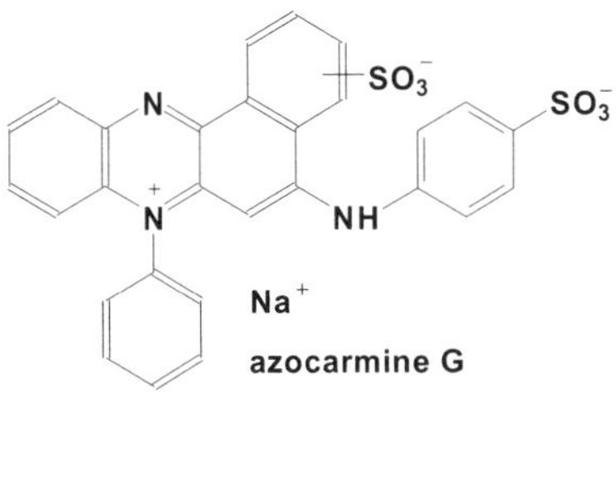

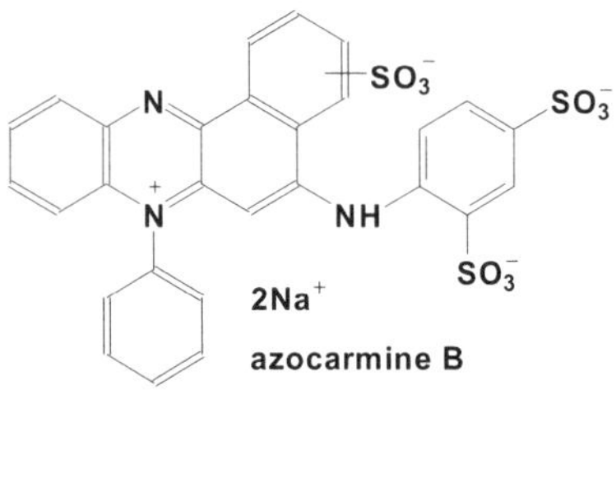

azocarmine G

azocarmine B

In both azocarmine B and G the planar cationic aminophenazine chromophore is extended by an additional benzo unit; a further phenyl substituent is attached to a ring nitrogen. Consequently the amount of aromaticity present in these dyes is substantial. Nominally azocarmines B and G are tri- and disulfonated compounds respectively; being derived from rhodindine by treatment with oleum. Since sulfonation reactions are difficult to control, both dyes will contain anionic products of varying degrees of sulfonation. They are sold as the sodium salts. The overall size of both compounds is moderate, the trisulfonated azocarmine B being, of course, the larger.

Applications: Both azocarmine B and G are recommended by staining manuals for use in Heidenhain's AZAN stain and its variations (Bancroft and Cook, 1994; Churukian, 2000; Gabe, 1976). Both dyes have also been recommended in staining manuals for other histological polychromes, such as Molliers quad stain (Clark, 1981). Other histochemical applications include quenching of background fluorescence in fluorescent Schiff staining (Takamatsu *et al.*, 1981), and as a stain for distinguishing light and dark neurons (T. Murakami *et al.*, 1997). Azocarmine, without further specification, has also been used in neuropathology (Minamitani *et al.*, 1994).

Industrial uses – both azocarmine B and G have been used for the coloration of leather.

Purity of commercial batches of azocarmine B and azocarmine G: Several components are always present. There is a thin-layer chromatographic analytical method (Rao *et al.*, 1985). Dye contents of typical samples of azocarmine B and G are up to 80% and 85% respectively. Regarding Commission testing procedures, see Chapter 28.

Stability: of azocarmine G stored as a dry powder is good in dark, sealed containers. Aqueous stock solutions of both azocarmine B and G are stable for up to a year, although they may need shaking and reheating prior to use. Light fastness of both dyes is moderate.

Rhodindine (CI 50375a, Basic red 6), the precursor of the azocarmines mentioned above, has no recorded use as a biological stain. However, its minor water soluble component **Magdala red** (CI 50375b) has found occasional use as a fluorescent fat stain, and for the demonstration of elastic tissue. The name Magdala red has also been applied to the acid dyes erythrosin and phloxin.

Nigrosine WS

CI 50420, CI Acid black 2
Synonyms: nigrosine water soluble; also sold as nigrosine B, MS, N, W, with various prefixes. In the biological literature this dye is often referred to merely as nigrosine.

No empirical formula or FW are given, as this dye is innately a mixture.

Absorption maxima: 570 nm in 50% aqueous ethanol, 575–580 nm in water.
Solubilities: 5–10% in water; 0.0–0.4% in ethanol; variability probably due to the variable chemical nature of different batches of this material, see below.

Nigrosine WS results from heating nitrobenzene with aniline, to produce a complex mixture of basic azine dyes, which are then sulfonated. An example of an anionic is diaminophenazine thought to be present in nigrosine WS is illustrated. Although precise formulations vary between batches and brands, species with large aromatic systems and of large overall size will be present. The products are sold as sodium salts.

Applications: Microbiologists using nigrosine WS for the negative staining of micro-organisms expect black staining, the traditional stain for this purpose having been India ink. However, nigrosine WS is usually dark blue or violet. Consequently products prepared specifically as biological stains sometimes contain 10–20% of a yellow or orange dye such as orange G, to provide blacker staining.

A routine procedure recommended in microbiological and histological staining manuals is the use of nigrosine WS as a background ('negative') stain for study of unstained bacteria, fungi and spirochetes (Clark, 1981; Lechevalier and Roisen, 1973) and for cilia of protists. Such procedures are also found in the research literature, for example, for determining size and shape of micro-organisms using image analysis (Makarov *et al.*, 1998), and for identifying new ciliate species (Song *et al.*, 1998).

Another widely used staining application described in cell biology manuals (Celis, 1998) is assessment of cellular viability. Other vital staining procedures include identification of protist species (Hill and Borror, 1992), and the diagnostic detection of helminth eggs in human feces (Ramsan *et al.*, 1999).

Numerous nonmicroscopic applications in biology and medicine include staining of proteins on electropherograms (Allchin and Evans, 1986; Kohn, 1987). Nigrosine WS is used as a marker of bait used in studies of termites' foraging behavior (Lelis, 1992), and as a tracer of water flows through soils (Kobr and Linhart, 1994) and biofilms (Sharp *et al.*, 1999).

Industrial uses – as a colorant of anodized aluminum, inks (including ink-jet printer products) and paper; as a leather dye; and in some countries for blackening olives.

Purity of commercial batches: Multiple components are routinely present. Regarding Commission testing procedures, see Chapter 28.

Stability: Light fastness is moderate.

Neutral red

CI 50040, CI Basic red 5
Synonyms: the nonprotonated form is termed toluylene red
$C_{15}H_{17}ClN_4$, FW 289

Absorption maxima: 454 nm in aqueous solution at pH 8.1, 529 nm in aqueous solution at pH 5.8; 541 nm in 1:1 ethanol-1% aqueous acetic acid, emission maximum in the latter solvent is 640 nm.
Used as a pH indicator, color transition from red to yellow is between pH 6.8 and 8.0; *pK value* 6.7.
Solubilities: 4.0% in water; 1.8% in ethanol

Neutral red is a protonated diaminophenazine of small overall size, with a small planar conjugated system. In acidic aqueous solutions this dye is present as a hydrophilic cation, as illustrated. Under alkaline conditions the lipophilic base **toluylene red** is present; not to be confused with the azo dye toluidine red (CI 12120). Neutral red is available commercially as the chloride.
Applications: Neutral red has several staining applications noted in manuals, including use as a nuclear stain (Kiernan, 1999), a counterstain in the Gram procedure, either alone (Boon and Drijver, 1986; Kiernan, 1999), or as a component of the Twort stain, a fast green-neutral red mixture (Bancroft and Cook, 1994; Bancroft and Gamble, 2002). The Twort stain is also recommended as a histological polychrome (Kiernan, 1999). Other histological applications include fluorescent staining of nucleic acids in nervous tissue (Allen and Kiernan, 1994); of lipids and hydrophobic substances such as suberin (Lulai and Morgan, 1992); and of phenolic compounds in algal cell walls (Schoenwaelder and Clayton, 1999).
  Neutral red, because of its selective uptake into lysosomes and other acidic organelles of living cells, is also recommended by manuals for viability assessment (Boon and Drijver, 1986; Clark, 1981).
  Recent applications of this type include detection of viable yeast cells in dandruff samples (Pierard Franchimont *et al.*, 1998), and of mussel larvae (Horvath and Lamberti, 1999). Staining of living structures is also carried out for other purposes, for example, to identify 5-HT-containing cells in whole-cell patch clamp preparations (Ito *et al.*, 1999), and to determine the spatial patterns of neuronal activation in the cerebellar cortex observed in the intact animal (Hansen CL *et al.*, 2000). Neutral red has been used as a stain in flow cytometry,

for example, to permit the separation of pancreatic islets by cell sorting (Jiao *et al.*, 1991), and to detect activated neutrophils and monocytes (Sipka *et al.*, 2000).

Other nonmicroscopic applications in biology and medicine include tracing movements of termites (Sajap, 1999) and marine invertebrates (Holmquist, 1998) after uptake of stained foods (paper and algae respectively). The dye is also used as a marker to assess root growth in plants (Schumacher *et al.*, 1983), The use of neutral red to visualize bacterial plaques in recommended in handbooks of cell biology (Celis, 1998).

Purity of commercial batches: All contain a major component plus minor contaminants; occasional batches contain a second major component. Dye contents of typical samples are up to 90%. Minimum dye content for Commission certification is 50%. A thin-layer chromatographic method is available (Marshall, 1976c). Regarding Commission testing procedures see Chapter 28.

Stability: of dry powder is good if stored in sealed, dark containers; solutions can photofade, although acidic solutions are stable for long periods. The factors influencing stability of neutral red solutions have recently been investigated (Hall *et al.*, 1998).

Phenazine methosulfate

Synonyms: N-methylphenazinium methosulfate, PMS; a synonym for methosulfate is methylsulfate.

$C_{14}H_{14}N_2O_4S$, FW 306

Absorption maximum: 386 nm in water
E_0' *value:* + 0.08 V at pH 7.0 and 30°C
Solubilities: 5% in water; 1% in ethanol

Phenazine methosulfate is an N-methylated phenazine. This hydrophilic cation has a small planar conjugated system. Available commercially both as the commonly used methylsulfate and as the ethylsulfate.

Applications: Phenazine methosulfate is recommended by staining manuals (Kiernan, 1999; Stoward and Pearse, 1991; Van Noorden and Frederiks, 1992) as an intermediate electron carrier in dehydrogenase histochemistry. Such methods may be used for quantitative enzyme determinations (Lofgren and Soderberg, 1998). The reagent has also been used for macroscopic dehydrogenase demonstration when investigating cardiac infarctions in the autopsy room (Derias and Adams, 1982), and for the staining of enzymes separated on electrophoretic gels (Havemeister *et al.*, 1999).

Purity of commercial batches: Dye contents of typical samples are up to 99%. Thin-layer chromatographic (Engel and Sawick, 1967) and HPLC (Nakamura *et al.*, 1998) methods are available for this reagent.

Stability: The photochemical stability of solutions is sometimes poor, although the precise circumstances and implications of this have been vigorously debated (compare Henderson, 1983 with Van Noorden, 1983).

Hazards: Acute exposure causes sneezing.

Methoxyphenazine methosulfate carries an additional, methoxy, substituent. It is also recommended by staining manuals (Kiernan, 1999; Stoward and Pearse, 1991; Van Noorden and Frederiks, 1992) as an intermediate electron carrier in dehydrogenase histochemistry, especially for mitochondrial enzymes. Although widely used, there has been disagreement as to the relative merits of this reagent and the parent phenazine, see Kugler (1982). The reagent has been applied in quantitative histochemical enzyme assays (Frederiks *et al.*, 1995). Use of methoxyphenazine methosulfate in analytical biochemistry has also been reported (Tanabe *et al.*, 1987).

Phenosafranine

CI 50200
Synonyms: phenosafranine B
$C_{18}H_{15}ClN_4$, FW 323

Absorption maximum: 519 nm in 50% aqueous ethanol; emits in the red.
pK value: 6.4
Solubilities: 6.5% in water; 5.3% in ethanol; insoluble in xylene.

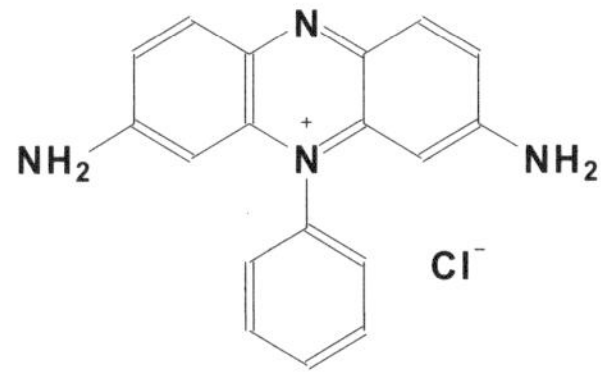

Phenosafranine has a planar, cationic phenazine chromophore. The phenyl substituent attached to a ring nitrogen gives rise to a markedly nonplanar distribution of the aromatic rings. Since the dye has few substituents its overall size is nevertheless fairly small. The cationic form, illustrated, is hydrophilic. The dye is sold as the chloride.

Applications: Phenosafranine can be used to prepare a fluorescent Schiff-type reagent (Kasten *et al.*, 1959) which has been used for quantitative histochemistry (Prenna *et al.*, 1962). The dye is currently applied as a fluorescent viability probe of plant cells (Li S *et al.*, 1999). Phenosafranine has also been used as a tracer to assess the efficacy of cell permeabilization procedures (Joersbo *et al.*, 1990).

Purity of commercial batches: The dye content of a typical sample is 80%.

Related dyes: N-Alkylated derivatives of phenosafranine have occasionally been used in biology. For instance the N-dimethylated **methylene violet RR**

(CI 50205, Basic violet 5) was converted to a diazonium salt for staining hematoidin (Lillie and Pizzolato, 1970); and the bis-*N*-diethyl derivative **amethyst violet** (CI 50225) has been used for differential staining of DNA and RNA in histological material (Roque *et al.*, 1965), and to measure membrane potentials of living cells (Ehrenberg *et al.*, 1988). Note that the biological literature sometimes fails to distinguish methylene violet RR from methylene violet Bernsthen; the latter is a thiazine dye with very different chemical and staining properties (see Chapter 20).

Safranine O

CI 50240, CI Basic red 2
Synonyms: safranine T
$C_{20}H_{19}ClN_4$, FW 351

Absorption maximum: 530 nm in 50% aqueous ethanol; emits in the red.
A nitritometric indicator.
pK value: 6.4
Solubilities: 4.5% in water; 3.5% in ethanol.

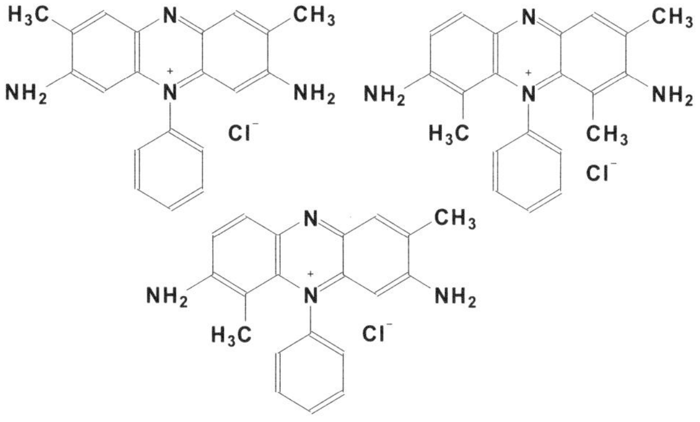

Safranine O contains a planar, diaminophenazine chromophore, with a phenyl substituent attached to a ring nitrogen, giving rise to a markedly nonplanar distribution of the aromatic rings. c-Methylation makes this dye of moderate overall size. Due to the mode of synthesis (see *Colour Index*, 1971, p. 4451) there are inevitably several components present, including those illustrated. Consequently although the literature distinguishes safranine O from safranine T, both refer to some mixture of such isomers and homologs. Below the term **safranine** will be used without further specification to emphasize this. The trimethyl cationic form of safranine is lipophilic, whereas the dimethyl forms are weakly hydrophilic. The dye is sold as the chloride.
Applications: Safranine is recommended in laboratory manuals as a microbiological stain, both as a component of the Gram stain for distinguishing Gram-positive from Gram-negative micro-organisms (Churkian, 2000; Clark, 1981;

Seeley *et al.*, 1991), and for staining bacterial spores (Barrow and Feltham, 1993). Staining manuals note use of the dye with human and animal histological specimens to demonstrate glycosaminoglycans (Clark, 1981) and nuclei (Clark, 1981; Kiernan, 1999), and as a component of Benda's polychrome stain (Gabe, 1976). Botanical staining applications are also recommended in manuals, for instance, as a component of various polychromes and lignin stains (Berlyn and Miksche, 1976; Clark, 1981; Ruzin, 1999). Manuals of resin section staining also describe the use of safranine to demonstrate glycosaminoglycans and proteoglycans in epoxy sections (Hayat, 1993; Litwin, 1985).

Less routine histological staining applications include use in quantitative assays of glycosaminoglycans, both directly using microdensitometry (Haapala *et al.*, 2000) and via use of stereology (He and Roach, 1994). Safranine has also been applied for demonstration of mast cells in cytospin preparations (Harris *et al.*, 1999), and for staining surgical frozen sections (Tran *et al.*, 2000). In botany, the dye has been used to stain lignin in specimens embedded in water-miscible resin (Gutmann, 1995), in a microwave-accelerated version of Johansen's polychrome stain (Schichnes *et al.*, 1998), and as a bulk-stain prior to confocal microscopy (Gray *et al.*, 1999).

Safranine has been applied as a fixative agent for proteoglycans and glycosaminoglycans (Kiraly *et al.*, 1996a), and used to stain organic material in soil specimens (Kooistra, 1991).

Purity of commercial batches: All batches contain substantial amounts of several components, plus traces of others. Dye contents of typical samples are up to 95%. Minimum dye content for Commission certification is 80%. A thin-layer chromatographic method is available (Marshall, 1976c). See Chapter 28 regarding Commission testing procedures.

Stability: of dry powder is good stored in a sealed, dark container; light fastness is poor.

A dye structurally related to safranine, and of some historical interest, is **mauve**. This was the first synthetic dye produced by Perkin in 1856. There is a recent popular account of the history of mauve's invention and manufacture, and of its economic and cultural significance (Garfield, 2000).

19

Oxazines and related dyes

R.W. Horobin

BACKGROUND INFORMATION

Structural features

The dyes described in this chapter contain planar dibenzo heterocyclic chromophores. Several oxazine – more precisely oxazinium – dyes carry amino subsitutents. The related oxazone chromophore is a planar neutral species, however this chromophore is often extended by hydroxy substituent. These structures are illustrated below. In the figure **R** is hydrogen or an alkyl group; **A** is hydrogen or a dialkylamino; and **B** is hydrogen or hydroxyl. Resazurin is an oxazone N-oxide derivative.

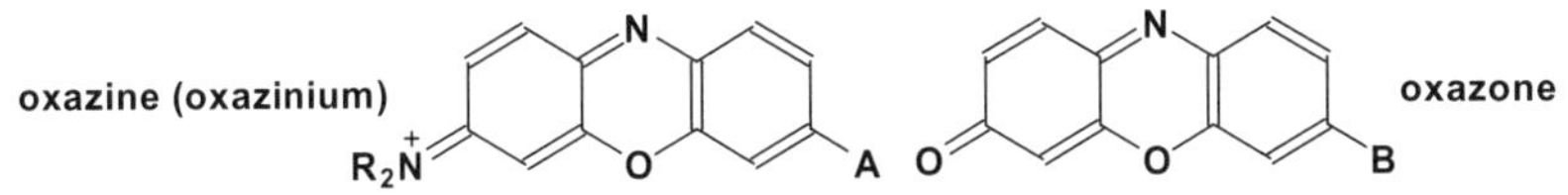

Metal complexes

A number of oxazine dyes give rise to metal complexes which are useful stains, for example, celestine blue, gallamine blue and gallocyanine. Various substituents can act as ligands; in the case of the gallocyanine chromium alum complex, these are carboxy and the ring nitrogen.

Properties influencing staining

Electric charge – Assigning electric charges to these dyes is often complicated. Even a simple oxazone such as Nile red, which is not ionized under usage conditions and whose entry is therefore in the nonionic subsection of this chapter, is cationic in acidic solutions due to its amino substituent. An oxazone such as resorufin carries ionizable hydroxy groups and is anionic under alkaline conditions but uncharged at neutral pH. Oxazines such as cresyl violet being based on a cationic chromophore and carrying amino substituents, are cationic. However, dyes with multiple acidic groups (e.g. celestine blue, gallocyanine) may be zwitterionic or anionic under neutral or slightly alkaline conditions.

277

Acid–base properties – Accurate measurements of pK values for these dyes are sparse. However, pK values will range from ca. 3–4 in the case of aromatic amines, around neutrality for carboxy, to 10–11 for some aryl hydroxy groups.

Hydrophilic/lipophilic properties – Since each ionic sub-species of a given dye will differ, hydrophilicity/lipophilicity is also a complex matter. Probably the most hydrophilic dyes will be metal complexes such as gallocyanine chrome alum, and the most lipophilic is Nile red, a neutral oxazone.

Extent of conjugation – The electron delocalization of the small, fused tricyclic rings is extended in some dyes by oxo and hydroxy substituents (in resorufin), or by amino group(s) (e.g. brilliant cresyl blue). Conjugation is further extended in dyes such as Nile blue by an additional, fused, benzo ring. In gallocyanine chrome alum there are two separate conjugated domains, so the overall amount of aromaticity is doubled.

Reactivity – The dyes celestine blue, gallamine blue, and gallocyanine react with a variety of metal ions. However, whether or not the resulting metal complex dyes are themselves reactive with tissues (i.e. mordanting, as traditionally conceived) is unclear.

Several dyes are useful because they participate in redox reactions with tissues or biological molecules; for example, resazurin, resorufin. No covalent bonds are established to tissue or biomolecule however, so these are not categorized as reactive dyes in this chapter.

Overall dye size – Dyes vary in size from resazurin (formula weight 251) to Nile blue (formula weight 367). Gallocyanine chrome alum contains two gallocyanine moieties and is considerably larger with a formula weight of around 726.

How to find the dye that interests you

Lacmoid (synonym: resorcin blue), litmus and orcein are innately hetero-geneous and contain both acidic and basic compounds. Consequently they are discussed at the end of the chapter under the rubric 'Special cases.'

NONIONIC DYES

Resazurin

Synonyms: alamar blue, 7-hydroxy-3*H*-phenoxazin-3-one-10-oxide
Resazurin, free acid: $C_{12}H_7NO_4$, FW 229
Resazurin, sodium salt: $C_{12}H_6NNaO_4$, FW 251

Absorption maxima of the salt: 598 nm in water, 478 nm in methanol
Resazurin is a redox indicator, changing on reduction from bluish violet through pink to colorless. E_0 +0.324v (pH 2.09, 23°C).
Resazurin is also a pH indicator, changing from orange to bluish violet between pH 3.5 and 6.5

Solubilities of resazurin free acid: insoluble in water except under alkaline conditions, slightly soluble in ethanol
Solubilities of resazurin sodium salt: 2% in water, 0.6% in ethanol

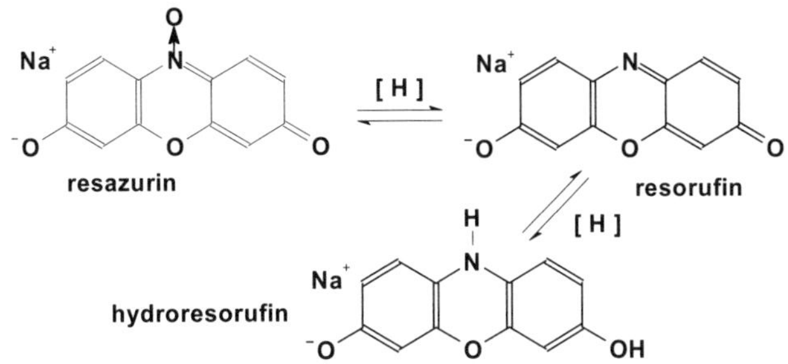

Resazurin is a small *N*-oxide derivative of dihydroxyphenoxazone, with a small and planar chromophore. The slightly lipophilic free acid is orange, the hydrophilic anion is bluish violet. On reduction resazurin first forms the red, fluorescent hydroxyphenoxazone resorufin (see entry below), with further reduction giving rise to colorless hydroresorufin as shown in the figure. Resazurin is commercially available as the sodium salt. Note: alamar blue is a proprietary mixture of resazurin and various stabilizing agents; for discussion of which see Rasmussen (1999) and O'Brien et al. (2000).

Applications: Although not a microscopic stain, the redox indicator character of resazurin is widely exploited in biology and medicine. Originally used for detection of bacterial reduction in milk, resazurin is currently widely applied in viability assays. For instance as a component of culture media for detection of viable bacteria (Smith and Townsend, 1999), and to determine the number of motile sperm in sheep semen (Martin *et al.*, 1999). Due to commercial secrecy, many applications of this type (although none of the above) are described in the literature as using alamar blue, a proprietary resazurin preparation. Resazurin is also recommended by flow cytometry manuals for assay of lactate dehydrogenase (Macey, 1994) and used as an analytical reagent, for example, determination of L-glutamate (Chapman and Zhou, 1999).

Purity of commercial batches: Dye contents of typical samples of resazurin sodium salt are up to 85%. The minimum dye content for Commission certification is 70%. Regarding Commission assay procedures see Chapter 28.

Stability: of dry powder stored in dark, sealed containers is good; solutions are light sensitive.

Resorufin

Synonym: 7-hydroxy-3H-phenoxazin-3-one
Resorufin, free acid: $C_{12}H_7NO_3$, FW 213
Resorufin, sodium salt: $C_{12}H_6NNaO_3$, FW 235

Absorption maximum: 571 nm in water at pH 9, emitting at 585 nm in this solvent. Absorption maxima below pH 7 is ca. 480 nm

Resorufin is a redox indicator, changing on reduction from pink to colorless.
pK value: ca. 6
Solubilities of resorufin sodium salt: soluble in water and ethanol to about 2% and 0.6% respectively.
Solubilities of resorufin free acid: soluble in dimethylformamide, and in water above pH 7.

Resorufin is a small dihydroxyphenoxazone, with a planar chromophore. The red anionic form is hydrophilic, the free acid is lipophilic. Reduction of resorufin gives rise to colorless hydroresorufin; see the previous entry for structures. Resorufin is commercially available as both the sodium salt and the free acid. Several useful derivatives are also sold. Most widely used are esters such as 7-acetylresorufin, and ethers such as 7-ethoxyresorufin and resorufin-beta-D-galactopyranoside.
Applications: Resorufin as such has few applications in biology and medicine. However resorufin esters, ethers and *N*-acetyl derivatives are widely applied, typically as self-indicating substrates for demonstration or assay of enzymes. Most commonly used is **7-ethoxyresorufin**, often used for demonstration of dealkylases such as the CPY1A1 member of the cytochrome P450 family, occasionally applied microscopically, for example, to evaluate the enzyme within cultured multicellular hepatocyte spheroids (Wu F.J. *et al.*, 1999).

Purity of commercial batches: Dye contents of typical samples of resorufin sodium salt are up to 95%.
Stability: Solutions are photolabile.

Nile red

Synonyms: Nile blue A oxazone, Nile pink
$C_{20}H_{18}N_2O_2$, FW 318

Absorption maximum: 553 nm in methanol, with the emission maximum in that solvent at 636 nm: if excited at 485 nm in heptane, it emits at 525 nm
Solubilities: soluble in dimethylformamide, dimethylsulfoxide, heptane and xylene; 1% in acetone, 0.1% in ethanol; 0.02% in water.

Nile red is a small, uncharged, lipophilic benzooxazone with a planar conjugated system of moderate size. Nile red is available commercially, although some workers still prepare their own from Nile blue.

Applications: Nile red is recommended in manuals for the fluorescent staining of lipid in cryosections (Kiernan, 1999). Another common application is as a fluorescent probe of lipids in living cells, for example, the vacuolar membranes in fungi (Hansen *et al.*, 2000b), and lipid droplets in viable glands from the silkmoth (Fonagy *et al.*, 2000). The dye has also been used to stain fat in food emulsions prior to observation using confocal microscopy (Brooker, 1991). The dye is described in flow cytometry manuals as a lipid stain (Ormerod, 1994). A recent application of this type was to identify and separate lung fibroblast subsets (Awonusonu *et al.*, 1999).

Other laboratory applications include staining of bacterial colonies producing poly(3-hydroxybutryic acid), for instance by incorporation of the dye into the growth medium (Spiekermann *et al.*, 1999). The dye is also used as a marker, for example, of the oil used to fill intercellular gas spaces in maize roots (Michael *et al.*, 1999).

Purity of commercial batches: There is a thin-layer chromatographic method available (Dunnigan, 1968).

Stability: Of dry powder is good if stored in sealed, dark containers; solutions photofade readily.

CATIONIC DYES

Cresyl violet

Synonyms: cresyl violet acetate, oxazine 9; related dyes are cresyl echt violet and cresyl fast violet, as discussed below.
Acetate: $C_{18}H_{15}N_3O_3$, FW 321
Perchlorate: $C_{16}H_{12}ClN_3O_5$, FW 362

Absorption maxima: 596 nm in 50% aqueous ethanol; 601 nm in ethanol, with the emission maximum in ethanol being 630 nm
Solubilities of the acetate: 2% in water; 1% in ethanol. The perchlorate is insoluble in water and soluble to 0.01% in ethanol.

As manufactured in the
USA since ca 1950

As manufactured in the
USA between the mid
1930s and 1950

As manufactured in
Germany at least until
the mid 1960s

Cresyl violet is a small diaminobenzooxazine, with a planar conjugated system of moderate size. The cation is somewhat hydrophilic. The dye is available commercially as the acetate, as illustrated, and as the perchlorate. As indicated several related but chemically distinct oxazines, with similar names, have been sold over the years. There is unfortunately no simple correlation between the names given such dyes – see above for synonyms – and the actual component(s) present (Green, 1966).

Applications described below may have used any of the synonyms noted above. However for convenience all are here referred to as cresyl violet. This dye is recommended for the demonstration of nuclei and Nissl substance in neuroanatomical investigations by staining manuals (Bancroft and Cook, 1994; Churukian, 2000; Kiernan, 1999). Similar staining has been used with water miscible resin sections (Litwin, 1985). Some interesting recent applications of this type are use of cresyl violet for fluorescent Nissl staining (Alvarez-Buylla *et al.*, 1990), and for quantitative neuro-imaging using flat-bed scanning (Schmitt and Eggers, 1999).

Cresyl violet is also a standard stain for the identification of *Helicobacter* (Bancroft and Gamble, 2002) and of sex chromatin (Bancroft and Cook, 1994). The dye has been used diagnostically as a vital stain in endoscopic procedures (Jung and Kiesslich, 1999).

Purity of commercial batches: The identities of dyes labelled cresyl echt violet, cresyl fast violet, cresyl violet, cresyl violet acetate depends on date and place of manufacture (Green, 1966). All of these can hydrolyze, giving rise to keto derivatives, often present as impurities. Dye contents of typical samples of cresyl violet acetate are 75%. Minimum dye content for Commission certification is 65%. There are paper chromatographic (Rosenthal *et al.*, 1965) and thin-layer chromatographic (Horobin and Murgatroyd, 1967) analytical methods available. Regarding Commission testing procedures see Chapter 28.

Stability: of dry powder is fairly good if stored in dark, sealed containers; of typical staining solutions is moderate.

Brilliant cresyl blue

CI 51010

Synonyms: several different chemical structures have been referred to by this name, see below.

Hemi zinc chloride $C_{17}H_{20}ClN_3O.1/2\ ZnCl_2$, FW 386

Absorption maximum: 622 nm in 50% aqueous ethanol
pK values: 6, 11
Solubilities: 3% in water and 2% in ethanol. The uncertain identity of brilliant cresyl blue batches makes these solubility data of uncertain value.

Brilliant cresyl blue is a diaminobenzooxazine whose planar conjugated system is small, as is the dyes overall size. The cation is weakly hydrophilic. The dye is sold as the hemi zinc chloride, and has been sold as the chloride (FW 318).

Several chemically distinct, though related, oxazines have been sold under the name brilliant cresyl blue (Green, 1990, 2000, and personal communication). Structure I was the material manufactured in Germany until World War II, and to a limited extent until the 1980s. Structure III is the material manufactured in the USA since the mid 1950s, sometimes termed brilliant cresyl blue ALD. Structure IV was made in the USA after World War II. Structure II, despite being the *Colour Index's* favored formulation, has probably never been available commercially. The cations of these species are illustrated.

Applications: Probably the most widely used application of brilliant cresyl blue has been to provide reticulocyte counts for blood or marrow, as recommended in hematological (Simmons, 1997) and other (Clark, 1981; Lillie and Fullmer, 1976) staining manuals. The dye is still used in this way, as well as for investigation of erythroblast pathology (Beris *et al.*, 1999).

Brilliant cresyl blue has a long history as a vital stain, its introduction as a supravital reticulocyte stain by Levaditi being as long ago as 1901. Current applications of this type include assessing the ability of boar sperm to penetrate oocytes (Roca *et al.*, 1998), and viability assays of *Blastocystis hominis* recovered from human patients following drug treatment (Haresh *et al.*, 1999).

Purity of commercial batches: The major component of different batches of dye sold under this name may have different chemical structures, and in any event batches routinely contain multiple components. Dye contents of typical samples are up to 65%. The minimum dye content for Commission certification is 60%. There is a thin-layer chromatographic method available (Frodyma and Frei, 1969). Regarding the Commission testing scheme see Chapter 28.

Stability: of dry powder is good when stored in a sealed, dark container.

Gallocyanine

CI 51030, CI Mordant blue 10
$C_{15}H_{12}ClN_2O_5$, FW 336

Absorption maxima: 550 nm at pH 1.6 in water, and 564 nm at pH 4.0; 620 nm in ethanol.

In aqueous solution gallocyanine changes from purple to blue between pH 3 and 4, and from blue to purple around pH 7.

pK values: 3.0, 6.1 and 10.1

Metal complexation behavior: Intensely colored compounds are formed with various metal ions, for example, iron, lead and chromium; for details of the chromium complex see below.

Solubilities: 1% in hot water; 0.7% in ethanol

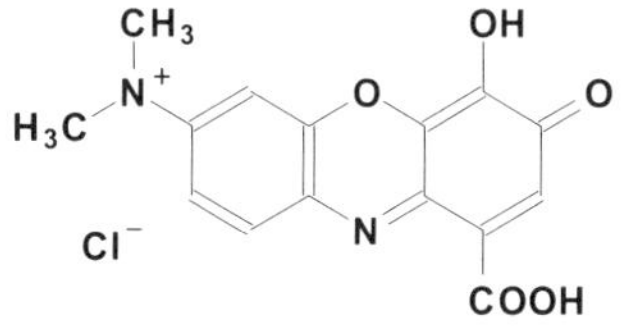

Gallocyanine has a small planar conjugated system, and its overall size is small. The previous edition of this book, in agreement with the *Colour Index* and other sources, suggested the predominant tautomer to be a dihydroxy species. Other workers favor a hydroxyketo structure, as suggested by the pK value of 3.0 and as illustrated here. The dye carries both basic and acidic substituents, so whatever the molecular detail, overall electric charge shows complex pH dependence. Thus, gallocyanine is cationic under acidic conditions, anionic under alkaline conditions, and a poorly soluble zwitterion around neutrality. The dye is sold as the chloride.

Applications: The only routine staining application of gallocyanine as a stain is as its chromium complex, see below. Occasionally the dye has been used to stain eosinophil granules (Lillie and Fullmer, 1976) and, following iron mordanting, nuclei and myelin (Augulis and Sepinwall, 1971).

Purity of commercial batches: Commercial batches contain one major component. The dye contents of typical samples are up to 90%. A thin-layer chromatographic method is available (Horobin and Murgatroyd, 1968).

Related Dyes: **Gallocyanine chrome alum** (Gallocyanin chromalum, GCA) is a widely used stain derived from gallocyanine by reaction with chromium(III) salts. The metal complex common to the various GCA preparations has been shown (Horobin and Murgatroyd, 1968; Marshall and Horobin, 1972a) to be a hydrophilic, cationic 2:1 gallocyanine–chromium complex, as illustrated; the actual counter ion will be sulfate. Physicochemical data given below relate to this complex.

Absorption maximum: 574 nm in pH 4.0 aqueous solution
Solubilities: soluble in water below pH 5.5, insoluble in ethanol or in alkaline aqueous solutions.
Applications: Gallocyanine chrome alum is recommended by staining manuals for staining of DNA and RNA, for instance in botanical (Ruzin, 1999), histological (Bancroft and Gamble, 2002), and neurological (Churukian, 2000) specimens in paraffin sections. Material in resin sections is also demonstrated with this stain (Litwin, 1985). Examples of specialist applications include quantitative determination of DNA (Schulte *et al.*, 1991) in smears and cell layers, and identification of microcracks in human bone (Villaneuva *et al.*, 1994). Schmitt and Eggers (1999) found gallocyanine chrome alum the best of several dyes tested for automated identification of neurons and glial cells by pattern recognition of nuclear texture.
Purity: All gallocyanine chrome alum preparations such as Berube's, Einarson's, Gray's, and Stenram's contain the 2:1 complex described above. Some contain additional complexes, and some unreacted gallocyanine; all contain free chromium(III) salts. There is a thin-layer chromatographic method available (Horobin and Murgatroyd, 1968).
Stability: Berube's precipitated stain is stable stored in sealed, dark containers; staining solutions are stable for a week; light fastness of stained sections is fairly good.

    Gallamine blue (CI 51045, Mordant blue 45) is structurally similar to gallocyanine, but with a carboxyamide substituent not a free acid. Gallamine blue has been used to demonstrate calcium ions in tissues, and its aluminum complex has occasionally been recommended as a nuclear stain (Gray, 1973).

Darrow red

Synonyms: 9-acetylamino-5-aminobenzo[a]phenoxazonium chloride
$C_{18}H_{14}ClN_3O_3$, FW 340

Absorption maximum: 502 nm in 50% aqueous ethanol
Solubilities: 0.1% in water, 0.5% in ethanol

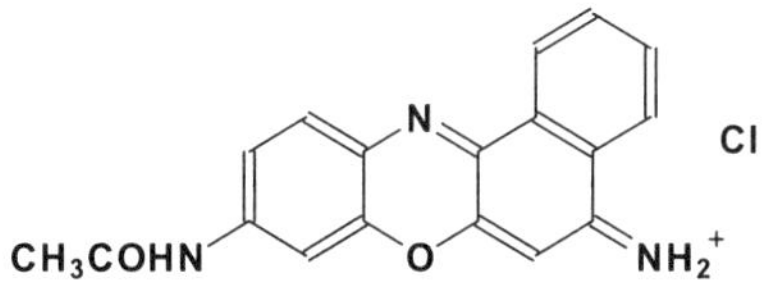

Darrow red is an aminobenzooxazine, whose planar conjugated system is small, as is the dye's overall size. The cation is hydrophilic. The dye is sold as the chloride.

Who was Darrow? This dye, introduced into microtechnique by Powers *et al.* (1960), memorializes Mary Darrow, assistant and laboratory technologist to Dr Harold Conn, founder figure of the Biological Stain Commission.

Applications: Darrow red has been recommended as a Nissl stain for use with blue or black myelin stains, such as luxol fast blue or Weil's iron hematoxylin (Clark, 1981), and is still occasionally used in this way (Ohm *et al.*, 1992). The dye has also been used in biochemistry as an enzyme inhibitor (Wang *et al.*, 1997).

Purity of commercial batches: Dye contents of typical samples are up to 75%. Minimum dye content for Commission certification is 65%. Regarding Commission testing methods see Chapter 28.

Celestine blue

CI 51050, CI Mordant blue 14
$C_{17}H_{18}ClN_3O_4$, FW 364

Absorption maximum: 642 nm in water
Metal complexation behavior: forms complexes with aluminum, chromium and iron
Solubilities: 2% in water; 1.5% in ethanol

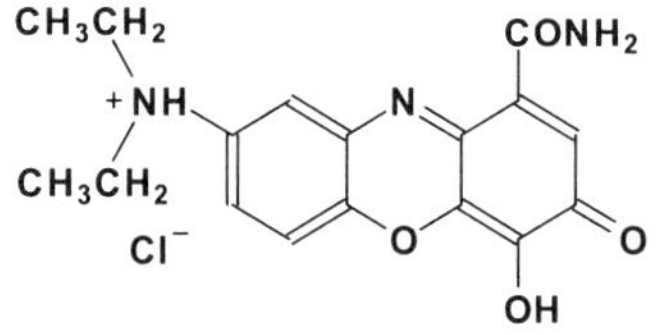

Celestine blue, with a small planar conjugated system and of small overall size, carries *N*-diethylamino and arylhydroxy substituents. It is unclear which tautomer predominates. The previous edition of this book, in agreement with the *Colour Index*, suggested a dihyroxy tautomer, whereas other workers favor the hydroxyketo species illustrated here.

Applications: Iron-mordanted celestine blue is recommended by staining manuals as an acid-resistant nuclear stain applied in combination with an alum hematoxylin and eosin (Bancroft and Gamble, 2002; Clark, 1981). The iron celestine blue complex is also combined with an acid fuchsine cytoplasmic stain (Jarolim *et al.*, 2000), which can be applied to cryosections (Garvey *et al.*, 1996) and resin embedded materials (Lopez and Kornegay, 1991).

Purity of commercial batches: Typical batches contain a single major component. The dye contents of typical samples are up to 85%. There is a thin-layer chromatographic analytical method (Rao *et al.*, 1985).

Stability: Depending on the mode of preparation, solutions of the iron complex may be stable for a few days or several months.

Nile blue

CI 51180, CI Basic blue 12
Synonyms: Nile blue A, Nile blue sulfate

Chloride: $C_{20}H_{20}ClN_3O$, FW 354
Perchlorate: $C_{20}H_{20}ClN_3O_5$, FW 418
Sulfate: $C_{40}H_{40}N_6O_6S$, FW 733

Absorption maxima: 633 nm in 50% aqueous ethanol, 660 nm in methanol; emits at 660 nm in methanol
Changes from blue to purple red between pH 10–11; *pK value:* 9.7
Solubilities of the sulfate: 2% in water and 3% in ethanol. The chloride is less soluble, and the perchlorate hardly soluble at all, in water.

Nile blue is a small aminobenzooxazine, with a planar conjugated system of moderate size. The cation is lipophilic. Available commercially as the sulfate, the perchlorate, and also as the chloride, illustrated.
Applications: Nile blue serves as a single stain which differentiates neutral fats and cholesteryl esters from free fatty acids and from phospholipids, as described in routine animal and plant staining manuals (Bancroft and Gamble, 2002; Gahan, 1984). Nile blue has also been recommended in manuals for detection of fetal cells, and to stain lipids in resin sections (Bancroft and Cook, 1994 and Litwin, 1985 respectively). Nile blue will stain other hydrophobic materials, for example, degenerating myelin in the human brain (Miklossy and Vander Loos, 1991) and polyhydroxyalkanoic acids in bacterial cells (Spiekermann *et al.*, 1999). *Note:* for these procedures to be effective the dye must contain significant amounts of Nile red, either as a contaminant or an additive (see separate entry for Nile red).

Nile blue, being a basic dye, may also be used to demonstrate apoptotic nuclei, for instance in the mouse fetus (Kimura and Shiota, 1996). Other staining applications make use of the dye's fluorescent properties, for instance to stain starch granules in industrial products viewed using confocal microscopy (Lynn and Cochrane, 1997), and to assay DNA (Chen *et al.*, 1999).

Living cells are also investigated using Nile blue fluorescence. For instance esophageal glands in parasitic worms are stained (Humphries and Fried, 1996), and embryonic cells can be marked to follow their subsequent movements

(Kominami and Masui, 1996). The dye is also used nonmicroscopically to mark termites for assessment of their foraging patterns (Marini and Ferrari, 1998).

Purity of commercial batches: One major and two minor blue components are often present, plus variable amounts of the red oxazone derivative. Dye contents of typical samples of the chloride or sulfate are 90%, with the laser grade perchlorate being available at 95%. The minimum dye content for Commission certification is 75%. A thin-layer chromatographic method is available (Dunnigan, 1968). For information concerning Commission testing procedures see Chapter 28.

Stability: of acetone stock solution is good when stored in a sealed dark container in the refrigerator; light fastness is good.

Related dye: **Rhodanile blue** is a salt prepared from Nile blue and rhodamine B. The suggestion that the two dyes were linked covalently via an amide moiety (Gurr, 1965) is incorrect, and they can be separated chromatography (Floyd Green, personal communication). Use of this salt as a histological polychrome is nevertheless occasionally found useful (Baden *et al.*, 1992).

SPECIAL CASES – LACMOID, *LITMUS* AND **ORCEIN**

Lacmoid, litmus and orcein are variable and poorly defined oxazine and oxazone mixtures, with some components being present in more than one dye. A general introduction to their chemical nature is followed by entries in alphabetical order.

Historically the raw materials for litmus and orcein derived from various lichens. For instance cudbear was obtained from Scotland. These were converted to phenols, predominantly orcinol, a name derived from a lichen known as orchil. On oxidation of such materials in the presence of ammonia, condensation reactions occurred giving rise to orcein. In the presence of alkali carbonates condensation proceeds further yielding litmus.

Perhaps because the original sources of raw materials were nonindustrial, litmus and orcein have been termed natural dyes. However, the orcinol and resorcinol used for dye manufacture have been produced industrially since the 19th century, and neither dye is itself a natural product. The terms 'natural' and 'synthetic' still applied to orcein batches by some vendors thus seem inappropriate. A related dye manufactured in much the same way – but starting from synthetic resorcinol – was termed lacmoid, or resorcin blue. Variants of this material were also made.

The products are mixtures of related compounds, and elucidation of their chemistry required modern separation and spectroscopic methods. Such work was carried out by Musso's group, and was summarized by Beecken *et al.* (1961). These investigators found that all three dyes contained both acidic and basic compounds. Many of these compounds had aromatic systems of considerable size, and litmus contained substantial amounts of polymeric oxazone species of over 3000 daltons in size. Lacmoid and litmus share some constituents, as do litmus and orcein; the principal constituents are illustrated in Figure 19.1. The crowding of substituents seen in these structures for the case of orcein is relieved in chemical reality by rotation of groups in space.

In the entries below physicochemical information is given for illustration. However, it should be appreciated that such data derive from mixtures of compounds that show marked batch variation.

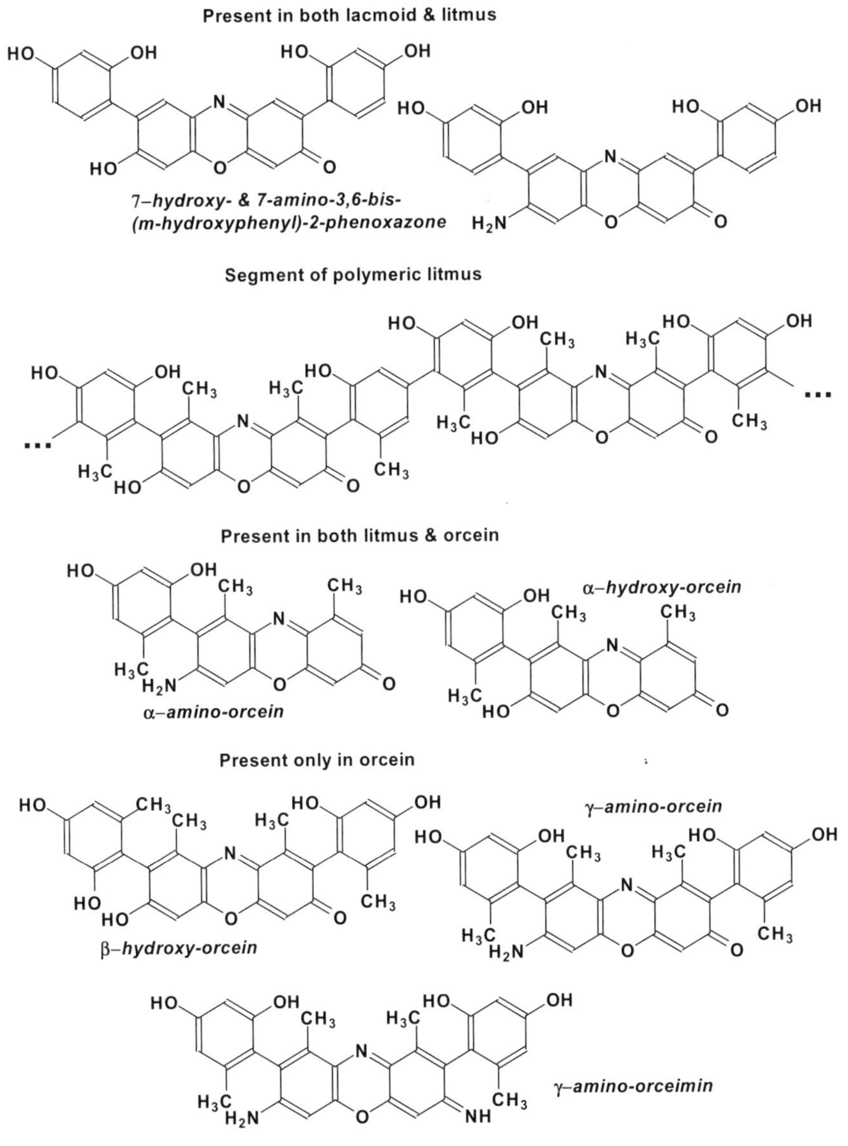

Present in both lacmoid & litmus

*7–hydroxy- & 7-amino-3,6-bis-
(m-hydroxyphenyl)-2-phenoxazone*

Segment of polymeric litmus

Present in both litmus & orcein

α–amino-orcein

α–hydroxy-orcein

Present only in orcein

β–hydroxy-orcein

γ–amino-orcein

γ–amino-orceimin

Figure 19.1 Some major components of lacmoid, litmus and orcein.

Lacmoid

Synonyms: resorcein, resorcin blue

Absorption maximum: 611 nm in methanol
Lacmoid is a pH indicator, and changes from red at pH 4.4 to blue at pH 6.4
Solubilities: reported aqueous solubility values vary from "sparingly" to 10%; and solubility in ethanol from 2% to 10%; probably due to batch variation.

Applications: Recommended in staining manuals (Jensen, 1962; Ruzin, 1999) as a stain for xylem vessels and callose in botanical material. Lacmoid is also used as a stain for cell nuclei, for example, of oocytes to check developmental stage (Willis *et al.*, 1994) and of occurrence of sperm penetration (Palomo *et al.*, 1999). The dye is also used as a pH indicator in physiology (Terra and Regel, 1995).
Purity of commercial batches: When prepared as sketched above, two major components, and a number of minor ones, are commonplace. However, according to the *Colour Index*, quite dramatically varied products have been sold under this name in the past. The dye content of current samples sold for biomedical use is up to 60%. There is a thin-layer chromatographic method available (Musso *et al.*, 1961).

Litmus

CI (ed.1) 1242, CI Natural red 28
Synonyms: lacmus

Absorption maximum: marked batch and pH dependency, 575 nm in water.
Litmus is a pH indicator, and changes from red at pH 4.5 to blue at pH 8.3
Solubilities: 5% in water, 0.3% in ethanol

Applications: Though litmus is not used as a stain, it has clinical and laboratory applications. These include pH monitoring, such as assessment of leakage following pancreatectomy (Yamaguchi *et al.*, 1998), and of gastro-esophageal reflux (James and Ewer, 1999). Litmus has also been used in microbiology for milk testing (Chaves *et al.*, 1999).
Purity of commercial batches: complex mixture, containing oligomeric material with up to 10 or more oxazine or oxazone units as well as small amounts of low molecular weight dyes. There is a thin-layer chromatographic analytical procedure for the low molecular weight constituents (Musso *et al.*, 1961).

Orcein

CI (ed.1) 1242, CI Natural red 28

Absorption maximum: 575 nm in aqueous pH 9.2 solution; emission maximum at pH 9.2 is 585–590 nm

The various compounds present in orcein have widely scattered *pK values*, from β- and γ-amino-orceins at 4.0, to α-hydroxy-orcein at 6.9, and β- and γ-amino-orceimins at 13.4.

Solubilities: dependent on batch and on pH, reported values vary from 'practically insoluble' to 2% soluble in water; and from 0.1% to 4% in ethanol.

Applications: Orcein is recommended in staining manuals as a stain for elastic fibers and lamellae (Bancroft and Cook, 1994; Kiernan, 1999). Variants include use of orcein for staining whole mounts (Vandaele *et al.*, 1995) and cryosections (Wilhelm *et al.*, 1999), and for the quantification of elastin (Ortiz *et al.*, 2000). Fluorescent staining of elastic fibers in epoxy resin sections has also been carried out (Molero *et al.*, 1985). The dye is used to demonstrate related connective tissue elements, such as elaunin (Augsburger, 1997) and oxytalan fibers (Bachmaier and Graf, 1999).

Other routine applications of orcein described in manuals involve staining of chromosomes in plants (Berlyn and Miksche, 1976; Ruzin, 1999) and human sex chromatin (Bancroft and Cook, 1994). Orcein is also used as a nuclear stain, for example, for wholemount staining of embryos (Wang W.H. *et al.*, 1999), and to assess maturation of oocytes (Bruck *et al.*, 2000). Manuals of *in-situ* hybridization recommend aceto-orcein to determine the stages of meiosis in plant meiocytes (Schwarzacher and Heslop-Harrison, 2000). Further methods in general use include demonstration of hepatic inclusions, especially of hepatitis B antigen (Bancroft and Gamble, 2002; Churukian, 2000) but also of copper–protein deposits (Bancroft and Cook, 1994).

Purity of commercial batches: up to eight major components, and six minor ones, however there is a great deal of batch variation. A paper chromatographic method is available (Darrow, 1952). Regarding Commission testing procedures see Chapter 28.

Stability: of acid alcoholic solutions is up to 6 months

20

Thiazines

R.W. Horobin

BACKGROUND INFORMATION

Use of thiazines as components of Romanowsky–Giemsa stains is described in Chapter 21. The present account is more general, with the same layout as other dye-class chapters.

The structure of the thiazine chromophore

Dyes described in this chapter contain a planar dibenzothiazine heterocyclic chromophore. Most dyes with the cationic thiazine, more precisely thiazinium, chromophore carry conjugated amino substituents. A related chromophore is the nonionic ketothiazine/thiazone heterocycle, as seen in methylene violet Bernthsen and the related thionol and thionoline dyes. All these structures are illustrated below.

Dye properties influencing staining

All the thiazine dyes used in biology and medicine are small, with small conjugated systems. They are cationic, or in one case neutral, under usage conditions.

Acid–base properties – Most thiazines are strong bases, with pKs well above neutrality. Presence of nitro or carbonyl substituents results in less basic dyes.

Hydrophilic/lipophilic properties – The cationic thiazines are hydrophilic, albeit some only weakly; the neutral ketothiazine is lipophilic. This group of dyes illustrates rather neatly the distinction between hydrophilicity and water solubility, the latter having no simple relationship to the former. Also illustrated is the marked dependency of solubility of ionic dyes on the counter ion present.

Redox properties – Thiazine dyes can often be reduced by various components of biological systems. This involves disruption of the electronic structure of the chromophore, with reduced species being colorless.

Purity – The homologs and derivatives of methylene blue require comment. Azure A, azure B, azure C and methylene violet Bernthsen – and polychrome methylene blues made for incorporation into Romanowsky stains such as

Giemsa's, Leishman's, Wright's and that ilk – are routinely manufactured by oxidation of methylene blue. As has been repeatedly demonstrated – most thoroughly by Marshall and Wittekind (see Chapter 21) – this process gives rise to many dyes, in particular those illustrated in Figure 20.1. Consequently polychromed methylene blues and dyes labeled azure A, azure B or azure C usually contain numerous compounds, not merely the one on the label. Indeed only in the case of azure B is the named component likely to be the most abundant dye in the mixture (Marshall, 1976). Certain dyes – routinely methylene blue and thionine, occasionally azure B – are manufactured by direct synthesis, and hence are much purer. The traditional terms **azure I** and **azure II**, still found in catalogs and the research literature, describe polychrome methylene blue and polychrome methylene blue with added methylene blue respectively.

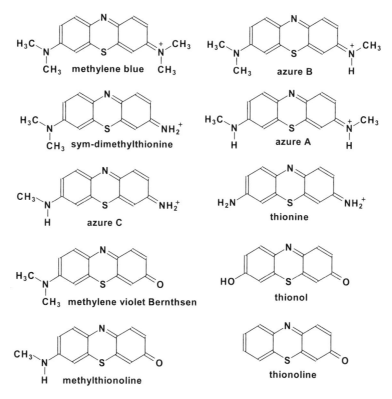

Figure 20.1 Dyes produced by oxidation (polychroming) of methylene blue. Counter ions are not shown.

NONIONIC DYE

Methylene violet Bernthsen

CI 52041
Synonym: unfortunately the term methylene violet is widely used without further qualification; do not confuse with the azine dye methylene violet 3R
Free base: $C_{14}H_{12}N_2OS$, FW 256

Absorption maxima: 580 nm in 50% aqueous ethanol containing HCl (predominantly the cation) and 601 nm in methanol (predominantly the free base)
The dye changes color from grayish violet to blue at pH 4
Solubilities: trivial (<0.1%) in water unless acidified; 0.4% in ethanol. These values probably relate to dyes highly contaminated with thionol and thionolines.

Methylene violet Bernthsen is an aminoketothiazine whose planar conjugated system is small, as is the dye's overall size. The commercially available free base, illustrated in Figure 20.1, is lipophilic.

Who was Bernthsen? He prepared this component of polychrome methylene blue by direct synthesis, in 1885, to establish its structure.
Applications: Addition of methylene violet Bernthsen to various Romanowsky polychromes used to be recommended (Gray, 1954). Such protocols must be regarded as dubious, and the only current staining application of this dye is as a component of MacNeil's tetrachrome stain for blood cells (see Chapter 21). Methylene violet Bernthsen has, however, been found to nick DNA when photoactivated (Morrison *et al.*, 1997) and to be effective for the photoinactivation of both intra- and extracellular viruses in red cell suspensions (Skripchenko *et al.*, 1997).
Purity of commercial batches: Typically heterogeneous, with overall dye contents up to 80%. The minimum dye content for Commission certification is 65%. Thin-layer (Marshall, 1976) and HPLC (Mohammad and Morrison, 1997) chromatographic methods are available. Regarding the Commission's testing procedure see Chapter 28.
Stability: of dry powder is poor.

CATIONIC DYES

Thionine

CI 52000
Synonyms: Lauth's violet, thionine Ehrlich, thionine O.
Acetate: $C_{14}H_{13}N_3O_2S$, FW 287
Chloride: $C_{12}H_{10}ClN_3S$, FW 264
Do not confuse thionine with thionine blue GO (thionine blue; CI 52025; Basic blue 25).

Absorption maxima: 598 nm in water and 602 nm in methanol; the emission maximum in water is at 625 nm when excited at 550 nm
pK values: 2.5, 11.3
Solubilities: of the acetate are 3% in water and 0.3% in ethanol; and of the chloride are 1% in water; 0.1% in ethanol

Thionine is a diaminothiazine whose planar conjugated system is small, as is the dye's overall size. The cation, illustrated in Figure 20.1, is hydrophilic. The dye is sold as the acetate; although the perchlorate is also available. Some current literature refers to the chloride, but this salt is not widely available commercially.

Who were Lauth and Ehrlich? Lauth synthesized thionine, Ehrlich recommended its use as a stain.

Applications: Several staining procedures involving thionine are found in manuals, including metachromatic staining of mast cell granules (Bancroft and Gamble, 2002; Presnell and Schreibman, 1997) and Nissl substance (Clark, 1981; Presnell and Schreibman, 1997) and staining of lacunae and canaliculi of bone (Drury and Wallington, 1980). Demonstration of lipofuchsin pigment (Bancroft and Cook, 1994) and use in the thionine–methyl green RNA/DNA stain (Presnell and Schreibman, 1997) are also recommended. In botany thionine has long been used as a polychrome oversight stain for hand-cut sections (Clark, 1981), and more recently as an oversight stain (Hayat, 1993) and as a component of a polychrome stain (Ruzin, 1999) for epoxy resin sections.

The dye is used to prepare a thionine–Schiff reagent, applied together with routine Schiff reagent for two-color demonstration of sialoglycoproteins (Bancroft and Gamble, 2002; Kiernan, 1999). This derivative is also used in a Feulgen-type nuclear stain, for example, with cytospin derived cells (Gersak *et al.*, 2000). Less routinely, thionine is used to identify cells, mineralizing fronts and cement lines in calcified bone (Watanabe *et al.*, 1998).

Purity of commercial batches: A single major component plus minor impurities are usually present. Occasionally batches labeled 'thionine' are actually methyl-ene blue or polychromed methylene blue. Dye contents of the acetate are up to 90%; the perchlorate is sold at 95% dye. The minimum dye content for Commission certification of the acetate is 85%. There is a thin-layer chromatographic method available (Marshall, 1976). Regarding the Commission testing procedures see Chapter 28. Certified batches of thionine are the only ones suitable for staining bone (Allison, 1995).

Stability: of dry powder stored in sealed, dark containers is good.

Azure B – with comments on other azures

CI 52010
Synonym: azure blue
Chloride: $C_{15}H_{16}ClN_3S$, FW 306
Tetrafluoroborate: $C_{15}H_{16}BF_4N_3S$, FW 357

Absorption maximum: 639 nm in methanol
Solubilities: the chloride is 5% soluble in water, 0.8% in ethanol; the tetrafluoro-borate is only 0.3% soluble in water and 0.1% in ethanol, but is more soluble in dimethylsulfoxide

Azure B is a diaminothiazine whose planar conjugated system is small, as is the dye's overall size. The cation, illustrated in Figure 20.1, is weakly hydrophilic. Dye prepared by direct synthesis is available commercially as the chloride, tetra-fluoroborate and thiocyanate salts. The bromide has also been sold. Many, perhaps most, batches labeled azure B will have been prepared by poly-chroming of methylene blue. Such material may contain as much methylene blue as azure B, plus other azures and derivatives.

Applications: Azure B is recommended in staining manuals for a variety of routine procedures. A number of these are stains for nuclei (Clark, 1981; Presnell and Schriebman, 1997), or nuclear constituents such as DNA (Clark, 1981). Variants of such stains may be used with water miscible (Hayat, 1993; Litwin, 1985) and epoxy (Litwin, 1985; Ruzin, 1999) resin sections. Because of its staining of basophilic materials, the dye is recommended for demonstration of viral inclusions (Clark, 1981); and for the metachromatic staining of mucins (Bancroft and Gamble, 2002) and neuroendocrine cells (Bancroft and Cook, 1994). Azure B of high purity is used in standardized Romanowsky stains, both hematological (Kiernan, 1999; for the original International Committee for Standardization in Haematology recommendation see Lewis, 1984) and histological (Wittekind *et al.*, 1991; for a microwave accelerated variant see Schulte and Wittekind, 1987).

Related applications in the research literature include a standardized method for reticulocyte counts (Wittekind and Schulte, 1987b), and quantitative esti-mation of RNA using microdensitometry (Serino *et al.*, 2000). The dye is applied in botany to demonstrate RNA (Bonner *et al.*, 1991); as a component of a general oversight stain (Graham and Joshi, 1996); and to demonstrate lignified cell walls (Fineran, 1997).

Azure B is used, directly and indirectly, to stain living cells, for instance to visualize enzyme-labeled axonal tracer (Dolleman van der Weel *et al.*, 1994), and to image neurons in the ganglion cell layer of the living retina using an infrared sensitive camera (Hu *et al.*, 2000).

Purity of commercial batches: Dye prepared by direct synthesis comprises a single major component plus trace impurities; dye prepared by polychroming meth-ylene blue is a complex mixture of which azure B is a component. The chloride, fluoroborate and thiocyanate salts of the direct synthesis material have dye contents of 88%, 95% and 98% respectively. The minimum dye content for Commission certification is 89%. Both HPLC (Raffaelli *et al.*, 1999) and thin-layer (Marshall, 1976) chromatographic methods are available. Regarding Commission testing procedures see Chapter 28.

Stability: of dry powder stored in dark, sealed containers is good; stock solutions are more stable in dimethylsulfoxide than in methanol, and the tetra-fluoroborate salt is more stable than the chloride.

Other azure dyes – applications and implications: Commercially available batches labeled **azure A** and **azure C** are best regarded as polychrome methylene blues. Though azure C has no modern staining applications, this is not the case for azure A. However in most staining procedures requiring this latter dye – for example, various nuclear and nucleic acid stains, the histological and hematological neutral stains, and metachromatic staining of mucins – azure B or toluidine blue may be substituted. Perhaps the only exception is the use of azure A to prepare a Schiff-type reagent, for which purpose thionine may be substituted.

Toluidine blue

CI 52040, CI Basic blue 17
Synonyms: TBO, tolonium chloride, toluidine blue O.
Chloride: $C_{15}H_{16}ClN_3S$, FW 306
Do not confuse toluidine blue with toluylene blue, (CI 49410), which is an indamine dye.

Absorption maxima: 626 nm in water, 630 nm in methanol
pK values: 2.4, 11.6
Solubilities: 3.3% in water; 1.8% in ethanol

Toluidine blue is a diaminothiazine whose planar conjugated system is small, as is the dye's overall size. The cation, illustrated, is hydrophilic. Currently available commercially as the chloride, the dye has been sold as the zinc chloride double salt.
Applications: Toluidine blue is recommended for a number of routine methods by staining manuals. For instance to demonstrate amyloid (Bancroft and Gamble, 2002), DNA in plant material (Gahan, 1984), mast cells (Carson, 1997), Nissl substance (Churukian, 2000), nuclei (Kiernan, 1999), and sulfatides (Bancroft and Gamble, 2002). Toluidine blue is also used for the metachromatic staining of sulfated mucins and related materials (Kiernan, 1999) and, after acid hydrolysis, of argyrophil cells (Bancroft and Gamble, 2002). The dye is used as an oversight stain in frozen sections (Presnell and Schreibman, 1997), and for animal (Hayat, 2000) and plant (Ruzin, 1999) material in epoxy resin sections. Other routine applications to resin sections include staining of mucins in water miscible resins and of nucleic acids in epoxy resins (Litwin, 1985), and it is used together with basic fuchsine in polychrome staining of epoxy-embedded material (Hayat, 1993). Toluidine blue dissolved in aqueous glycerol is used as a stain-mountant for food products (Flint, 1994). The dye is also used for the assay of proteoglycans in sections (Cake *et al.*, 2000) and as an endoscopic vital stain for detection of cervical carcinoma (Nishimura *et al.*, 2000).

Purity of commercial batches: A single major component and several minor ones are usually present. Dye contents of the chloride are up to 85%, and the minimum dye content for Commission certification is 50%. Thin-layer chromatographic methods are available (Marshall, 1976). Regarding the Commission's testing procedures see Chapter 28.

Stability: of dry powder stored in sealed, dark containers is good; acidic aqueous solutions are stable for months or years; light fastness is poor.

Methylene blue

CI 52015, CI Basic blue 9
Chloride hydrate: $C_{16}H_{18}ClN_3S.3H_2O$, FW 374
Zinc chloride double salt hydrate: $C_{16}H_{18}ClN_3S.1/2ZnCl_2.H_2O$, FW 406

Absorption maxima: 661 nm in water, 656 nm in methanol
pK values: 2.6, ca.11.2
Solubilities of the chloride: 9.5% in water; 6% in ethanol
Solubilities of the zinc chloride double salt: 1% in water, <0.1% in ethanol

Methylene blue is a diaminothiazine with a small planar conjugated system, and of small overall size. The cation, illustrated in Figure 20.1, is weakly hydrophilic. It is commercially available as the chloride and the zinc chloride double salt.

Applications: The dye is widely used to manufacture the polychrome methylene blue used in preparation of Romanowsky stains, as described in Chapter 21. However, in addition to these procedures there are a number of routine staining applications of methylene blue recommended in manuals, for instance, as a component of a polychrome for staining bone marrow (Presnell and Schreibman, 1997), and as an oversight stain for epoxy resin sections, both of animal (Hayat, 1993) and plant (Ruzin, 1999) material. The cationic character of methylene blue underlies its routine use as a general bacterial stain (Barrow and Feltham, 1993; Seeley *et al.*, 1991) as well as a counterstain for Ziehl-Neelsen and other stains for acid fast bacteria (Carson, 1997; Churukian, 2000). The dye is also used to stain Rickettsiae and viral inclusions in animal cells (Clark, 1981; Presnell and Schreibman, 1997). Among less widely applied staining applications of methylene blue are the differentiation of leukocytes in a hemocytometer (Brattig *et al.*, 1993), detection of leukocytes in fecal specimens (Saha *et al.*, 2000), and coloration of gut mucosal fragments to evaluate surface topology (Le Leu *et al.*, 2000).

Methylene blue has also been recommended as a supravital stain for nerves and nerve terminals in muscle (Bancroft and Gamble, 2002; Kiernan, 1999) and for neuronal processes in the central nervous system (Muller, 1998). Occasionally the dye is used to stain nonneuronal live cells, for example, of esophageal tumors (Jung and Kiesslich, 1999), and for viability testing of *Schistosoma mansoni* (Gold, 1997).

Nonmicroscopic applications of methylene blue are diverse. Methylene blue is used as a surgical marker in urology (Crew and Fellows, 1996). It has also been applied as a surgical tracer for checking tubal patency following sterilization procedures (Sporri *et al.*, 2000) and to detect sentinel lymph nodes (Landra *et al.*, 2000) – and has been used to check the permeability of surgical gloves (Hentz *et al.*, 2000). As a physiological tracer methylene blue has been used to assess red cell permeability after membrane damage (Gavrilov *et al.*, 2000), and to visualize the cardiac region at risk following experimental anoxia (Hasegawa *et al.*, 2000).

Industrial uses – coloration of leather, paper and wood; manufacture of pigments. *Purity of commercial batches:* One major component with traces of azure B, and sometimes other demethylated derivatives. Dye contents of chloride samples are up to 85%. The minimum dye content for Commission certification is 82% of anhydrous dye chloride. There are thin-layer chromatographic methods for methylene blue (Marshall, 1976) and for leuco methylene blue (Yatome and Ogawa, 1995). Regarding the Commission's testing procedures see Chapter 28.

Stability: of dry powder stored in sealed, dark containers is good; acidic aqueous staining solutions are stable for months; light fastness is poor

Related dyes: **Leucomethylene blue** is a colorless compound formed when methylene blue is reduced, as shown in the **Redox indicators** sections of Chapter 8. This can be used to demonstrate and assay oxidation/reduction phenomena such as the oxidizing activity of rice roots growing in water-logged soil (Sadana and Claassen, 1996).

Taylor's blue, or **9-Dimethylmethylene blue**, illustrated below, was synthesized by the eponymous Taylor and finds occasional application as a stain, for example, to assay proteoglycans (Muller and Hanschke, 1996).

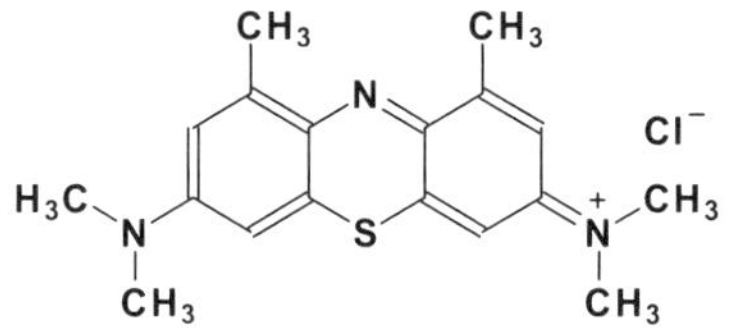

New methylene blue

CI 52030, CI Basic blue 24
Synonym: new methylene blue N
Chloride: $C_{18}H_{22}ClN_3S$, FW 348
Zinc chloride double salt: $C_{18}H_{22}ClN_3S \cdot \frac{1}{2}ZnCl_2$, FW 416

Absorption maxima: 591 and 630 nm in water
Solubilities: 4% in water; 1% in ethanol

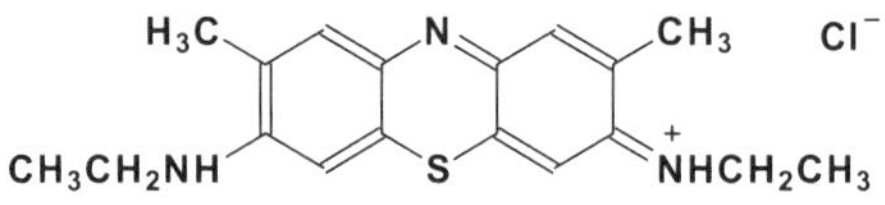

New methylene blue is a diaminothiazine with a small planar conjugated system, and of small overall size. The cation, illustrated, is weakly lipophilic. Currently sold as the zinc double chloride, it was earlier sold as the chloride.

Applications: Originally used as a metachromatic mucin stain (Highman, 1945), new methylene blue is currently mentioned in staining manuals for demonstration and enumeration of reticulocytes (Presnell and Schreibman, 1997; Simmons, 1997). The dye is also used as a microscopic stain for demonstration of Heinz bodies (Fallin and Christofer, 1996) and for the micronucleus test in cytogenetics (Sugihara *et al.*, 2000). It has also been used as a tracer to assess intracranial migration of epidural injections (Hendrickson *et al.*, 1998), and as an experimental antibacterial photosensitizer (Wainwright *et al.*, 1998).

Purity of commercial batches: Some 'new methylene blue' samples are brilliant cresyl blue. This problem aside, commercial samples contain one major component, with several minor and trace compounds. The dye contents of typical samples of the zinc chloride double salt are up to 90%. A thin-layer chromatographic method is available (Marshall, 1978b).

Stability: Light fastness is moderate

21

Romanowsky–Giemsa Stains

D.H. Wittekind

HISTORY, NOMENCLATURE AND OTHER BACKGROUND INFORMATION

The term *Neutralfarbstoff* (neutral stain) was coined by Paul Ehrlich. His plan was to allow a basic dye to react with an acid one, making a new substance with new properties: a salt whose anion and cation were of different colors. The Biological Stain Commission also defines a neutral stain as a compound of an acid dye and a basic dye, recognizing that both the cation and the anion contain chromophoric groups (see Lillie, 1977, Chapter 18, and earlier editions of *Conn's Biological Stains*). The term *neutral dye* is ambiguous and should not be used. In the past it has been applied to combinations of oppositely charged different dyes, to individual zwitterionic dyes (which may exist as uncharged molecules at a particular pH), and to colored compounds with uncharged molecules, such as solvent dyes.

Ehrlich and Lazarus (1909) published a detailed account of their joint work on this topic, describing various mixtures, among them the once famous *triacid stain*, which Ehrlich had described in 1882. This, which contained the anionic dyes orange G and acid fuchsine and the cationic dye methyl green, can colorfully distinguish different types of leukocyte (Baker, 1958). It involves, however, a critical, subjective differentiation (partial destaining) step, and the colors fade in some mounting media (Gabe, 1976). Since the beginning of the 20th century, the *thiazine eosinates* have been the most important neutral stains; in these, blue cationic thiazine dyes are combined with red anionic xanthenes. Even when they were poorly understood, these pairs of dyes became pre-eminent in hematology. This chapter is therefore devoted to the thiazine eosinates, which are used principally for identifying cells in blood and hemopoietic tissues.

Three excellent texts on biological staining contain chapters written by experts in the application of cationic/anionic dye-pairs to staining in cytology. Baker (1958), with his characteristic dislike of the word 'stain', entitled his appropriate chapter 'The Blood Dyes'. Harms (1965) discussed the subject in relation to the dyes used, and Lillie (1977), in Chapter 18 of the 9th Edition of *Conn's Biological Stains*, provided a critical historical review of the preparation and usage of many combinations of eosins with thiazine dyes. Lillie himself contributed much to the development of thiazine–eosinate techniques in hematology and histology. He preferred to use, in place of the prepared eosinates, a suitably buffered aqueous solution of eosin Y or eosin B combined at the time of using with an individual thiazine dye such as azure A or azure B (Lillie, 1944b; see also Lillie and Fullmer, 1976). This was done to simplify the application of dye-pairs to smears of cells and sections of solid specimens. Marshall (1978b, 1979a,b) reviewed the classical and more recent literature in the context of modern knowledge of thiazine dye chemistry, and he also evaluated several commercially available thiazine–eosinate stains.

Practical instructions for the application of oppositely charged thiazine–eosin dye-pairs can be found in many books, including those edited or written by Luna (1968), Cook (1974), Gabe (1976), Lillie and Fullmer (1976), Clark (1981) and Kiernan (1999). Anyone wanting to try Ehrlich's triacid stain or other early methods should consult an older encyclopedic work such as Jones (1950), Gatenby and Beams (1950), Gray (1954) or Gabe (1976).

Many other mixtures of anionic with cationic dyes have been used in histology, but most have failed in the competition with thiazine–eosinate techniques (for a brief review, see Lillie, 1977). **Twort's stain** is a neutral stain that is still used in many routine histopathology laboratories. It has neutral red (an azine dye) as the cation and either light green or fast green FCF (triarylmethane dyes) as the anion. This method and its variants can impart a variety of colors to bacteria and animal tissues (Monroe and Frommer, 1967; Ollett, 1951), but they have not yet been critically studied. For technical details and discussion see Clark (1981) and Kiernan (1999).

THIAZINE EOSINATES AND THE ROMANOWSKY–GIEMSA EFFECT

Preparation of a thiazine eosinate stain

In all the commonly used neutral stains for blood cells, the anionic component is a red xanthene dye, eosin Y, and the cationic components are one or more blue thiazines. The thiazine dyes have quite small planar molecules. Those present in neutral stains are formed during prolonged boiling of aqueous solutions of methylene blue, usually with alkali and sometimes with an oxidizing agent. Such *polychroming* generates a mixture of several blue cationic thiazine dyes (notably azures A, B and C, together with some unchanged methylene blue) and at least one uncharged dye known as methylene violet Bernthsen, which is not related to other dyes that have 'methylene violet' in their names (Lillie, 1977; Marshall, 1978b). This crude mixture of thiazine dyes, known as **polychrome methylene blue** can be bought or made in the laboratory.

On mixing aqueous solutions of eosin Y and polychrome methylene blue (or pure thiazine dyes), a thiazine eosinate precipitate is formed. This is insoluble in water but soluble in methanol, ethanol and many other organic solvents. The alcoholic solution has no staining power; it must be diluted with water immediately before it is used. The period of efficacy of an alcohol–water solution of a thiazine eosinate is limited. It lasts from the beginning of dissociation of the cationic–anionic dye-pair molecules in the water until the unavoidable reprecipitation of the water-insoluble neutral stain.

The length of the period of high tinctorial power depends on the chemical natures of the two dyes involved and on their concentrations, first in alcohol, then in the water–alcohol mixture. The higher the ultimate concentration of the dyes in the water–alcohol mixture the earlier will precipitation occur. Empirical experiments have determined optimal concentrations of the two dye ions for producing the desired staining pattern before the unstable aqueous solution deteriorates. A more concentrated solution has a shorter lifetime than a dilute one, but it can provide increased stability of dye-binding to tissues (Gabbett, 1887), probably because nonionic as well as ionic forces can be mobilized (Goldstein, 1963a; Muller-Walz and Zimmermann, 1987).

In the generation of a neutral stain from two oppositely charged and water-soluble dyes, binding forces additional to the electrostatic ones must come into play. Consider a crystal of an inorganic salt such as sodium chloride, in which coulombic forces operate exclusively. When NaCl is dissolved in water the dissociation into Na^+ and Cl^- is instantaneous and complete. On the other hand, the main characteristic of a neutral stain is that it is virtually insoluble in water. Somewhat regrettably the modes of cooperation of nonionic adjuvant binding forces in dye precipitates have not yet been examined by physico-chemical methods. The most likely candidate seems to be hydrophobic bonding (Horobin, 1982; Zollinger 1991), which can be expected to be effective in water but of minor importance in an alcoholic environment.

Ehrlich discovered that a neutral stain precipitated from water could be brought into solution by adding an excess of either the anionic or the cationic parent dye. In practice, an excess of thiazine dye over eosin is used for this purpose. A search of the literature leaves the impression that the factors contributing to this effect are not yet fully understood. This is regrettable because the significance of Ehrlich's discovery can hardly be overemphasized: it means that the presence of one dye-ion in excess of the other dye-ion prevents the formation of an insoluble dye-ion-pair. It also means also that a small excess of one dye allows the insoluble neutral stain to change back to the two dye-ion state in an aqueous solution. Consequently, Ehrlich's authority cannot be claimed as a witness for the correctness of the postulate that 'neutral stains as such' (see below) exhibit tinctorial power.

The behavior of a thiazine eosinate solution is affected by the ratio of alcohol:water in the solvent (Frank and Ives, 1966; Levine, 1939). A stained section will hardly release dyes into pure ethanol or pure water, but cationic dyes in particular are quickly removed from tissues by 30–70% ethanol. This observation seems pertinent in the present context because solvent mixtures with high water:alcohol ratios mediate the transition from combined neutral dye ion-pairs in alcohol to separate dye ions, which bind to oppositely charged tissue sites. The water–ethanol mixture that provides the optimal conditions for the ionization of thiazine eosinates, while keeping all the ingredients in solution, has still to be found.

STAINING OF BLOOD CELLS AND PARASITES

The great value of thiazine–eosinate neutral stains resides in the purple colors imparted to the nuclear chromatin of all cells, and to the cytoplasmic granules of those polymorphonuclear leukocytes that proliferate in response to acute bacterial infection. Leukocytes with purple-staining granules were named *neutrophils* from the early belief that they were stained by an electrically neutral ion-pair formed from eosin and a thiazine dye. Leukocytes with granules assumed to be stained by the thiazine dye alone (dark violet or purple, varying with the technique) were called *basophils,* and cells with granules colored orange–red by eosin were called *eosinophils.* These names are still applied to the principal types of granular leukocyte, but the reasons for the colors are not as simple as was first thought.

Another asset of thiazine–eosinate staining is the different color imparted to the nuclei of parasitic protozoa, notably malaria parasites. Romanowsky (1891) and Malachowski (1891) independently discovered that an aqueous mixture of partly oxidized methylene blue eosin Y allowed the differential coloration of malarial parasites in blood films: the trophozoite was clearly blue, the nucleus contrasting strongly by its carmine–violet color. Of equal importance was the finding in both studies that leukocyte nuclear staining was also very good. They showed purplish colors in heat- or alcohol-fixed blood films. These observations were confirmed and extended by Giemsa (1902a,b, 1904, 1910), who also introduced variations in the composition of the staining solution. Malachowski's paper remained unnoticed for many decades. Consequently, the special properties of neutral stains, especially the purple coloration of certain blood cells, are now generally known as the *Romanowsky effect* (Marshall, 1978b, 1979b) or the *Romanowsky–Giemsa effect* (Wittekind, 1983, 1991; Wittekind and Gehring, 1985; Wittekind *et al.,* 1976). Clinicians, hematologists in particular, enthusiastically developed Romanowsky's discovery as an invaluable aid in morphological diagnosis.

Through the following decades until about the mid-1970s, most investigations of the Romanowsky–Giemsa staining were technical or technological. The capriciousness of the method was the first obstacle that the early workers (Giemsa, 1902a,b, 1904, 1924, 1934; Leishman, 1901; Lillie, 1942, 1943a,b, 1944b; MacNeal, 1925; Nocht, 1898; Roe *et al.,* 1940; Unna, 1891; Wright, 1902) had to overcome on the way to a standard stain formula. In the various methods attention was paid to the identity and sometimes also purity of dyes used, and to the influence of fixation, solvents and of the pH of staining solutions (Pischinger, 1926; Tolstoouhov, 1928). All these variables had to be taken into account in the efforts to render Romanowsky staining reproducible, but the mechanisms involved in staining were not seriously investigated until the 1970s. The components of some thiazine eosinate stains are listed in Table 23.1.

It would be easy but unfair to criticize those early, and also later, workers for the empirical methods they adopted to approach the Romanowsky–Giemsa phenomenon, the color purple. Methods applied were indeed sometimes amateurish, but in the early decades of this century where was there an

Table 21.1 Composition of some thiazine–eosinate staining solutions.
The table shows a trend with time from complex mixtures of uncertain composition (polychrome methylene blue) through mixtures of known dyes to a combination of two pure dyes.

Name stock solution(s)	Dyes	Solvent(s) for
Romanowsky (1891)	aged methylene blue, excess eosin	water
Nocht (1898)	polychrome methylene blue, methylene blue, eosin	17% acetone
Jenner (1899); also May and Grünwald (1902)	methylene blue, eosin Y	methanol
Reuter (1901)	methylene blue, eosin Y	ethanol
Leishman (1901)	polychrome methylene blue, eosin B	methanol
Wright (1902)	polychrome methylene blue, eosin B	80% methanol
Giemsa (1902a)	excess azure I (impure azure B), eosin Y	water
Giemsa (1902b, 1924)	excess azure II (impure azure B with methylene blue), eosin Y	50% methanol, 50% glycerol
MacNeal (1925) (tetrachrome stain)	methylene blue, azure A, methylene violet Bernthsen, eosin Y	methanol
Roe *et al.* (1940) also Lillie (1943a)	methylene blue, azure B, azure A, eosin Y	50% methanol 50% glycerol
Marshall *et al.* (1975)	methylene blue, azure B (pure), eosin Y	50% methanol 50% glycerol
Wittekind *et al.* (1976)	azure B (pure), eosin Y	50% methanol 50% glycerol
Wittekind and Kretchmer, 1987	azure B (pure), eosin Y	40% dimethylsulfoxide 60% methanol

advanced theory of staining to which biological dyers (not stainers!) could refer? We should recall that Vickerstaff's book *The Physical Chemistry of Dyeing* did not appear until 1954. It marked the introduction of expert scientific explanations of dye–substrate relations applicable not only to textile dyeing, but also to biological staining. From then on, the complicated theories of staining gradually expanded. New insights into dye–substrate relations have been greatly aided by applying computer-guided instrumentation, which permits quantitative measurements that were impossible for the pioneers of Romanowsky–Giemsa research. It must also be stated, however, that practitioners of dyeing did not likewise benefit from the upsurge of theoretical research, mainly because the majority of dye consumers were unable or unwilling to mobilize the basic theoretical knowledge (organic and inorganic chemistry, physics) necessary to achieve the transposition, even on a modest scale, of theory into everyday staining practice. On the other hand, it must be admitted that hardly any of the theoreticians undertook to explain this background information to pathologists, who have to interpret the staining results rather then the staining mechanisms.

As a side remark, it should be added that this drifting apart of theory and practice of staining is not a phenomenon specific to the biological field. Turner (1955) clearly pointed out the steadily broadening gap between those who developed new textile dyes and those who used them in practice. The remarks in this and the preceding paragraph were not interpolated here just for the sake of adding some 'staining philosophy' to the text. Rather, it seemed desirable to contribute to an understanding of the conditions under which the long, irksome and colorful (in the true sense) history of staining was written, with the development of thiazine–eosinate neutral stains as an interesting subchapter.

Dyes responsible for the Romanowsky–Giemsa effect

Earlier work with neutral stains was based on the assumption that a mixture of thiazine dyes, combined with an eosin, was necessary in order to obtain the expected combination of colors in a stained blood film (Lillie, 1977; Marshall, 1978b). It is now known that these colors are attributable only to azure B and eosin Y (Wittekind *et al.*, 1991). The leading role of azure B, long ago assumed by Giemsa and more recent investigators (Marshall and Lewis, 1974b; Roe *et al.*, 1940) could only be finally established by the application of the highly purified dye, separated from the initial thiazine mixture (polychrome methylene blue) by column chromatography (Lohr *et al.*, 1974, 1975). Since the early 1980s, azure B of high purity has been available commercially as its perchlorate, thiocyanate and tetrafluoroborate salts, with the chloride available at 90% dye content. Azure B made by direct synthesis is more stable in solution than that prepared by purification from polychrome methylene blue (Green, 1990).

The combination of pure azure B with eosin Y was introduced into cytology by Wittekind *et al.* (1976). They showed that the full Romanowsky–Giemsa pattern was achieved, not only the nuclear purple color but also the cytoplasmic clear blue color, and dark blue in RNA-rich cytoplasms. No addition of methylene blue was required to achieve those colors. In complementary experiments azure A was deliberately added to mixtures of azure B with eosin Y. Unwanted grayish and grayish-violet cytoplasmic hues were then seen. In the early Romanowsky–Giemsa stain formulae, methylene blue was added to compensate for grayish and violet hues due to the presence of Azure A, which could not be selectively removed from polychrome methylene blue. Giemsa's 'azure II' a mixture of methylene blue with impure azure B, served this purpose of cytoplasmic color correction.

STAINING MECHANISMS

Nature of the purple product

When azure B was recognized as the sole effective cationic dye in the Romanowsky–Giemsa stain (Wittekind and Gehring, 1985; Wittekind *et al.*; Zipfel Grèzes *et al.*, 1984), these authors assumed a sort of molecular interaction between the two dyes as the substrate of the purple color. The fundamental physicochemical studies of Zimmermann and his team (Friedrich *et al.*, 1990; Huglin *et al.*, 1986; Muller-Walz and Zimmermann, 1987; Zimmermann, 1983; Zipfel *et al.*, 1984) established that a DNA–azure B–eosin Y complex was formed. The complex is purple, with wave number 18100 and absorption maximum 552 nm.

In the simple eosinate, 2:1 is the most likely molar ratio of azure B to eosin Y (Dean *et al.*, 1976). This gives the nuclear purple color but it does not provide a perfect Romanowski–Giemsa staining pattern, mainly because of deficiencies in cytoplasmic coloration. The most effective molar ratio of azure B to eosin Y is about 10 to 1. The reason for the better performance of the higher molar ratio of the cationic to the anionic dye is to be sought in the staining mechanism. Azure B must first bind to DNA and form molecular aggregates. Eosin Y then binds to the aggregated (and therefore metachromatic) azure B. The high concentration of azure B facilitates formation of the DNA–azure–eosin complex, the geometry of which is not yet understood (Friedrich *et al.*, 1990).

The purple product of Romanowsky–Giemsa staining does not form instantaneously. When watching the process under the microscope one will see cell nuclei first turning bright blue, then purple after about 1–3 min. These observations show that azure B and eosin Y act as individual dye ions until they form a new, externally neutral, complex in conjunction with nuclear DNA and certain other substrates. After that short time of exposure the colors are unstable and will disappear even from an air-dried smear of cells. Some additional time in the stain is required to obtain a permanent preparation: 5–10 min to survive brief passage through water and alcohol, or about 90 min to enable the purple color to withstand all the conventional histological post-staining procedures. Some mechanism of stabilization must occur while the cells are in the dye solution, but its nature is still unknown. One might speculate that the azure B–eosin Y polymer grows on suitable substrates over the course of some minutes, thereby mobilizing additional binding forces, likely of the van der Waals type. That might obtain for extremely thin but extended surfaces. But on most substrates, including cell nuclei, the density of the purple color increases with prolongation of staining time, suggesting that the dye ions may be deposited in multiple layers.

The formation of the purple product is pH-dependent. Below pH 5.25, azure B aggregates, which are indispensable for formation of the DNA–dye complex, cannot exist (Zipfel *et al.*, 1984). Stains for blood are usually applied at pH 6.8, and the water used to rinse the stained preparation should also have this pH. The dye complex is easily removed by alcohols, in the order methanol > ethanol > isopropanol > *t*-butanol.

Formation of purple material at surfaces

Generation of the purple color occurs only at interfaces. It has never been observed in aqueous solution, but the water–air interface can serve as a suitable surface. A bronze-colored insoluble film or scum forms on the surfaces of the unstable aqueous staining solutions. It was first observed by Romanowsky (1891). The scum, which contains aggregated azure B associated with eosin Y, possesses the most significant criterion of a neutral stain; it is insoluble in water, otherwise it could not exist on top of an aqueous solution. The scum can be transferred to a deparaffinized section by careful handling, but there is no evidence to indicate that a purple color is imparted by the insoluble material to the tissue.

Other surfaces on which the azure B–eosin Y complex can be formed most spectacularly, and least taken notice of in cytological practice, are those of glass: when an aqueous azure eosinate solution is discarded after long standing the walls of the vessel show a purple covering. No other dye combination does this, certainly not Jenner's (1899) or May and Grunwald's (1902) methylene blue–eosin Y combinations in water–alcoholic mixtures (unless polychrome methylene blue is used, providing a source of azure B). Of course, no colored covering is formed from contact of glass with azure B or eosin Y alone.

Substrate thickness and azure B–eosin Y complex formation

Another significant variable, mostly in histology, is the thickness of the object being stained. This presents no difficulty with thin monolayer cell films, but satisfactory staining cannot be obtained in sections that are more than 6 or 7 μm thick. A thickness of 5 μm is preferable. In sections more than 10 μm thick both dye ions penetrate separately and bind separately. Nuclei are then colored dark blue and erythrocytes brilliant red. The situation is worst in areas of dense cell populations (e.g. lymphatic tissues). There, uniform blue coloration generally prevails. The same phenomenon may be seen in cytological preparations and is well illustrated in several of the micrographs in Boon and Drijver (1986).

TISSUE COMPONENTS EXHIBITING THE ROMANOWSKY–GIEMSA EFFECT

The nuclear staining is attributed to DNA, but certain other components of tissues may also be stained purple by a solution containing azure B and eosin Y. Preponderantly hyaluronic acid gives the purple color on collagen fibers, provided the specimen was fixed immediately *post mortem* in neutralized formaldehyde; autopsy material fails to show this effect. Other Romanowsky–Giemsa-positive objects include secretory granules in the adrenal medullary cells, Auer rods in leukocytes in myeloid leukemia, thrombocyte granules, certain cells of the anterior lobe of the hypophysis, Howell–Jolly bodies (fragments of chromatin in erythrocytes, after splenectomy), and neutrophil granules.

Neutrophil granules

Ehrlich believed that the granules of polymorphonuclear leukocytes were colored by both anionic and cationic components of his 'triacid' stain, and that is why he called them neutrophils. He regarded the specific dyeing of these granules as an important property of neutral stains, not to be obtained without their use. However, this theory was based on the assumption that neutral stains gave better indications of acidophilia and basophilia than any arbitrary mixture or succession of dyes. More is known about the composition of the neutrophil granules today than in the classical period of biological staining. The granules are considered to be primary lysosomes, containing several enzymes and basic proteins (Hayhoe and Quaglino, 1988; Zucker-Franklin, 1968). The components that most affect staining, however, are polyanions (Bainton *et al.*, 1971; Fedorko and Morse, 1965). It had been surmised that the polyanionic materials neutralize the basic proteins also present in primary lysosomes. Extensive attempts at establishing correlations between the components of 'neutral' granules and their reactions to traditional stains have apparently not been undertaken so far. The colors imparted to DNA and to neutrophil granules are not completely identical. Microspectrophotometric analysis of the various stained substrates might clarify the nature of the polymer that accounts for the characteristic staining properties of neutrophil granules.

Eosinophil and basophil granules

Traditionally it was maintained that only the neutrophil granules took up both the cationic and anionic components of a neutral stain. Coloration of the eosinophil and basophil granules, as their names imply, is attributed to one dye only: anionic dyes bind mainly to cationic arginine side-chains in the eosinophils, whereas cationic dyes attach to the strongly anionic heparin of basophil and mast cell granules. The staining of these cytoplasmic granules may, however, be a more complex process.

Pappenheim (1912) recommended the sequence of the Jenner or May–Grunwald technique (methylene blue–eosin Y), followed by the Giemsa method (using an eosinate of polychrome methylene blue containing about 35% of azure B). He pointed out that the eosinophil granules were colored more brilliantly red after that sequence than after Giemsa's method alone, which imparted somewhat dull red–violet hues to the same granules. This mixed color can be considered due to the simultaneous uptake of azure B and eosin Y. A combination of pure azure B with eosin Y, which has no colored contaminants, also gives a mixed color to eosinophil granules; it is not seen after any other combination of eosin Y with a cationic dye. The existence of granule–eosin Y–azure B aggregates is inferred, but has yet to be determined. The heparin-containing granules are stained blue–violet by azure B alone, and purple by azure B–eosin Y. This appears to be a true Romanowsky–Giemsa effect, and it cannot be copied by eosin Y combined with other cationic dyes.

22

Polyene dyes and fluorochromes

R.W. Horobin

Structural features

Polyene dyes and fluorochromes contain linked methine groups (–CH=) forming a chain of *trans*-oriented conjugated carbon–carbon double bonds, as shown below in structure **a**. The number of methine groups varies, but in the absence of additional nonpolyene conjugated units 22 methine groups must be present for absorption to occur in the visible range. The end-groups of the polyene chains vary. In the naturally occurring carotenoids these are lipophilic alicyclic or aliphatic groups (as in lycopene) or hydrophilic groups such as hydroxy or carboxyl (as in crocetin). In synthetic polyene dyes, aromatic end groups are usual. The stilbenes comprise a well-known example, in which a two-methine chain has terminal phenyl substitutents, as illustrated below in structure **b**.

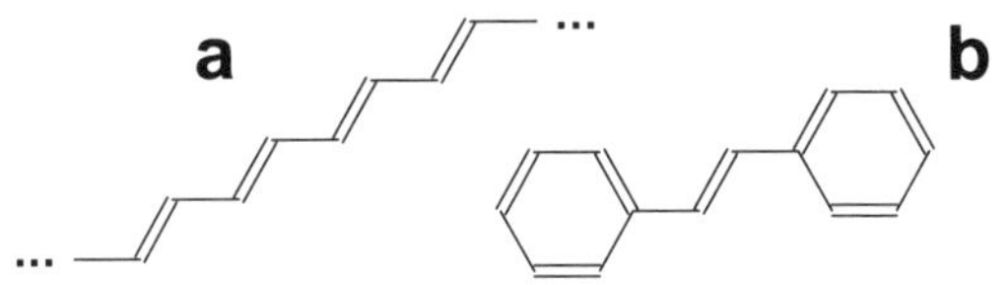

The polyene category so specified subsumes various traditional dye classes. These include carotenoids, the stilbenes, and two-methine polyenes with end-groups other than phenyl; the latter have been discussed elsewhere as ethylene derivatives.

Confusing usage of commercial names

Certain staining reagents are sometimes referred to as blankophore, calcofluor, tinopal or uvitex. Unfortunately these terms do not specify individual stilbene compounds, as they are names of product *ranges* of fluorescent brightening agents. Thus eight chemically distinct stilbenes have been sold in the calcofluor

313

range, whilst other members of this range have azole, benzidine, coumarin, naphthalimide or triazole fluorophores. As a further example of variability within these dye ranges consider tinopals CBS and AN, both described below. Papers cited in this chapter used fully specified compounds, whose structures have been disclosed.

Properties influencing staining

Electric charge – Included in this chapter are anionic, cationic and uncharged dyes. Some cationic dyes are quaternary salts (e.g. TMA–DPH), others are protonated salts of strong bases (e.g. amidines, such as fast blue). Some anionic dyes are salts of strong acids (sulfonates, e.g. calcofluor white M2R), some of weak acids (carboxylic, e.g. crocetin). Crocetins are unionized esters.

Hydrophilic/lipophilic properties – Some dyes are lipophilic (e.g. DPH), some very hydrophilic (e.g. uvitex 2B).

Extent of conjugation – from small (e.g. crocetin) to large (e.g. calcofluor white M2R). Lengths of methine chains vary between 2 (the stilbenes) and 22 (lycopene) methine units.

Reactivity – Two of the dyes are considered to form covalent bonds with proteins, namely SITS and DIDS.

Dye size – varies from small (e.g. DPH) to large (e.g. calcofluor white M2R).

Synthesis

The carotenoids are natural products, and saffron and lycopene are still extracted from plant material. Industrial syntheses of stilbenes are, however, of interest. Some of these compounds, including SITS, are prepared by condensation reactions starting from 4-nitrotoluene-2-sulfonic acid. Such processes give rise to mixtures including stilbene and azostilbene oligomers and polymers, and removal of such impurities is time consuming and expensive.

Purity and safety of reactive dyes

Some commercial batches of SITS and related materials are impure, and biologists do not seem aware of this (however see Payne *et al.*, 1983). Neither is it generally appreciated that DIDS and SITS decompose when dissolved in water, especially under alkaline conditions. Reactive dyes may modify human proteins, resulting in hypersensitivity responses.

How to find the dye that interests you

Carotenoids and stilbenes are traditional dye categories, and dyes of these types are collected together in the first two sections of this chapter; with all other polyene dyes being placed in a third section. Within each section the usual sequence of nonionic, anionic, cationic and reactive dyes is followed; within these subsections dyes are listed in order of increasing formula weight.

CAROTENOIDS

Saffron, crocins and crocetin

The term saffron describes the naturally occurring yellow material present in the flowers and fruit of various plants. The *Colour Index* lists material obtained from Saffron flowers and Wongshy fruit as CI 75100, Natural yellow 6; and notes that the same compounds occur amongst the colored material obtained from Toon flowers, which latter material is termed CI 75100, Natural red 1.

The yellow compounds are crocins; mono- and diglycosyl esters of the dicarboxylic acid crocetin. The solubility of crocins in water and ethanol permit their extraction from plant material. Crocins are easily hydrolyzed, for example, by dilute alkali, to crocetin. Since it is uncertain which substances are present in the solutions used as stains, data are given for both saffron and crocetin.

Synonyms: the term crotin given in one standard text is a typographic error. Do not confuse saffron with the cationic azine dye safranine.

Saffron

Crocetin

$C_{20}H_{24}O_4$, FW 328

Absorption maxima:

427 and 452 nm in ethanol. When excited at 435 nm, emits at 543 nm.

411, 436 and 464 nm in pyridine.

Solubilities:

3% in water; 0.1% in ethanol; soluble in aqueous alkali, but see below.

Very sparingly soluble in water; soluble in aqueous alkali and in pyridine.

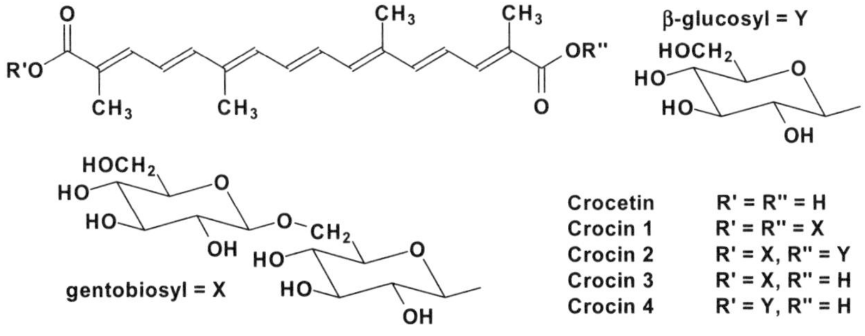

Crocetin	R' = R" = H
Crocin 1	R' = R" = X
Crocin 2	R' = X, R" = Y
Crocin 3	R' = X, R" = H
Crocin 4	R' = Y, R" = H

The composition of saffron and its constituents is outlined above. Crocetin is an all-*trans* polyene dicarboxylic acid, whose overall size and conjugated system are small. The dianionic form of crocetin is weakly hydrophilic, the monoanion and free acid species are lipophilic. Crocin 3, a gentiobiose monoester, is a considerably larger (FW 668), hydrophilic anion. The nonionic crocin 1 gentiobiose diester is large (FW 1008) and hydrophilic.

The question arises, what species are present in a staining solution of saffron? Boiling in water for an hour, one procedure noted in the staining literature for extracting colored material from saffron, will probably result in partial hydrolysis. Staining solutions produced in this way will consequently contain crocetin and crocin 3. Another procedure uses ethanolic extraction, with water being carefully excluded both during extraction and whilst staining. The major components extracted will be the crocins. Such solutions are sometimes used with sections previously treated with phosphotungstic acid, so *in situ* hydrolysis may occur in collagenous connective tissues containing the polyacid. Yet other staining procedures use alkaline solutions of saffron, and these will largely comprise the acid dye crocetin.

Applications: Staining animal material for microscopy using saffron was reported by Leeuwenhoek as early as 1719 (Clark and Kasten, 1983). Saffron has been recommended in staining manuals as a selective collagen and bone stain (Gray, 1973), and as a general counterstain following the rubeanic acid method for the demonstration of copper (Vacca, 1985). Current uses of saffron are as an oversight fluorochrome (Trigoso and Stockert, 1995), and in hematoxylin–eosin– or hematoxylin–phloxin–saffron polychromes for histopathological staining (Rostoker *et al.*, 2001).

Saffron and its components have other laboratory applications. In experimental oncology crocin is used to induce apoptosis (Thatte *et al.*, 2000). Crocetin is used both as an inhibitor of induced genotoxicity and neoplastic transformation (Chang *et al.*, 1996), and as a selective cytotoxic agent for cultured tumor cell lines (Jagadeeswaran *et al.*, 2000). Crocin has also been used as a reagent to assay radical-scavenging activity of human plasma (Lussignoli *et al.*, 1999).

Industrial uses – saffron is used as a food colorant and spice

Purity of commercial batches: The amount of colored material in saffron batches varies widely (Diaz-Marta *et al.*, 1998). There are HPLC (Li *et al.*, 1999) and thin-layer (Corti *et al.*, 1996) chromatographic methods available for crocetin and the crocins.

Stability: Light fastness of saffron batches varies but is usually poor.

STILBENES

Tinopal CBS

The tinopal dyes used in biology and medicine are not always further specified. As tinopals are chemically diverse this is unfortunate. The present entry describes the compound most often precisely identified; whilst tinopal AN is described at the end of the entry.

CI Fluorescent brightener 351

Synonyms: tinopal CBS-X, uvitex NFW

$C_{28}H_{20}Na_2O_6S_2$, FW 569

When excited in the violet the dye emits in the blue.

Solubility: soluble in water

Tinopal CBS comprises two stilbene moieties linked via a bridging diphenyl. The molecule's overall size, and that of the conjugated system, is moderate. The dianion, illustrated, is weakly hydrophilic. The dye is sold as the disodium salt

Applications: Tinopal CBS is used to detect parasitic worms in histopathological material (Green *et al.*, 1994) and has also been used with Papanincolaou stained cytological smears (Meistrich and Green, 1989). Other applications include use as an apoptotic tracer in plants (Peterson *et al.*, 1981), and as a stain for microorganisms in flow cytometry (Davey and Kell, 1997).

Purity of commercial batches: The fluorochrome content is up to 90%

Stability: Light fastness is good

Other tinopals have been used in biology and medicine. Most often reported are **tinopal LPW**, a synonym for calcofluor white M2R, *qv.*; and **tinopal AN**. The latter is not a stilbene but a cationic benzoxazole of lipophilic character. This fluorochrome has been used to demonstrate bacteria in plant tissues (Eng and Cole, 1976), and to inhibit mitochrondrial enzymes (Anderson and Delinck, 1987).

Calcofluor white M2R

CI 40622, CI Fluorescent brightener 28

Synonyms: calcofluor white ST, cellufluor, CF, CFW, tinopal 4BMA, tinopal LPW and tinopal UNPA-G.

Note 1: only synonyms found in the biomedical literature are listed here

Note 2: calcofluor white MR (CI Fluorescent brightener 9) is a different, albeit structurally related, stilbene

$C_{40}H_{42}N_{12}Na_2O_{10}S_2$, FW 961

Absorption maximum: 350 nm in methanol; when excited in the violet or near ultraviolet the dye fluoresces blue–white

Solubilities: 8% in water, and 0.2% in ethanol

The dye is a stilbene with additional triazinyl and phenyl aromatic groups. The overall size is large, as are the number of aromatic rings. The dianionic sulfonate is hydrophilic. Calcofluor white M2R is sold as the disodium salt, illustrated.

Applications: Calcofluor white M2R is recommended in staining manuals for the demonstration of cellulose plant cell walls (Kiernan, 1999; Ruzin, 1999) and as a counterstain for *in situ* hybridization staining of plant material (Schwarzacher and Heslop-Harrison, 2000). It is used as a viability stain of plant cells (Gahan, 1984). The dye has also been recommended as a flow cytometry stain for assessing viability of animal cells (Shapiro, 1988).

Other microscopic staining includes demonstration of chitin. These applications range from detection of microsporidian spores in feces (Schottelius *et al.*, 2000) to the staining of fungi grown on solid agar (Thrane *et al.*, 1999). Other methods involving live cells include staining cellulose microfibrils in plant protoplasts (van Amstel and Kengen, 1996) and viability testing of spores of *Plasmodiophora brassicae* (Narisawa *et al.*, 1996). The dye is used to assess platelet binding to yeasts, using a hemocytometer (Robert *et al.*, 2000). Other applications in histopathology and cytology are detection of *Dirofilariasis* worms in paraffin sections (Green *et al.*, 1994), and staining of micro-organisms in bronchoalveolar washes (Maymind *et al.*, 1996).

Nonmicroscopical laboratory applications are varied, for instance, as a component of a chitin gel-overlay for detection of chitinase activity on electro-pherograms (McBride *et al.*, 1993). As a marker the dye finds application in assessing leaf growth (Soros and Dengler, 1996) and in tracing foraging patterns of ants (Vega and Rust, 2001). Calcofluor white M2R is an inhibitor of cell wall production in yeasts (Rodriguez-Pena *et al.*, 2000). As tinopal LPW the dye is widely used an enhancer of viral insecticides (Arakawa *et al.*, 2000).

Industrial uses – as a fluorescent brightening agent for cellulosic and polyamide fabrics, paper, and in detergents and soaps

Purity of commercial batches: Dye contents of typical samples are up to 90%

Stability: Alkaline solutions tend to precipitate and fade; light fastness of stained specimens is good

Uvitex 2B

CI Fluorescent brightener 362
$C_{40}H_{40}N_{12}Na_4S_4O_{16}$, FW 1164

Solubility: soluble in water

The dye is a stilbene with additional triazinyl and phenyl aromatic substituents. The overall size is large, as are the number of aromatic rings. The tetraanionic sulfonate is extremely hydrophilic, and is sold as the tetra sodium salt, illustrated. *Applications:* Uvitex 2B has been used diagnostically as a microscopic stain to demonstrate intestinal protozoal parasites both in tissue sections (Franzen *et al.*, 1995) and stool samples (Franzen *et al.*, 1996). Other applications include staining of microsporidians in nonhuman material such as smears and sections of fish muscle (Yokoyama *et al.*, 1996). Parasitic fungi of cereals, such as rusts, may also be detected (Bender *et al.*, 2000).
Stability: Light fastness of stained specimens is good

Fluoro-gold

Synonyms: fluorogold, hydroxystilbamidine, 2-hydroxy-4,4′-diamidinostilbene, OHSA
Dimethanosulfonate: $C_{18}H_{29}N_4O_7S_2$, FW 473
Diisethionate: $C_{20}H_{28}N_4O_9S_2$, FW 532

Absorption maxima: 344 nm at pH 5.0 and 318 nm at pH 9.0 in aqueous solutions; if excited at 360 nm there are emission maxima at 448 and 582 nm
pK value: ca 11.6
Solubilities: the dimethanosulfonate is moderately soluble in water; the diisethionate is freely soluble in water and is 1% soluble in ethanol

Fluoro-gold is a stilbene with two amidino substituents. The overall size of the cation is small, as is the size of the conjugated system. The cation is hydrophilic, the free base lipophilic. The dye is sold as the dimethanosulfonate, illustrated, for biological staining; and as the diisethionate for pharmaceutical applications. *Applications:* Fluoro-gold has been applied as a fluorescent Schiff reagent-substitute for demonstration of DNA and mucopolysaccharides (Murgatroyd, 1982), and as a direct DNA stain (Arndt-Jovin and Jovin, 1989).

Most staining applications of fluoro-gold involve living cells or creatures, although microscopic observation may not be of live specimens. The most common application is retrograde axonal tracing, as discussed in a recent review (Kobbert *et al.*, 2000). Labeled neurons may be identified *in situ* by microscopy (Zhang *et al.*, 2000) or, following tissue dissociation, counted by flow cytometry (Yang *et al.*, 2000). A related method is the long-term marking and identification of microglia. Thus if, after retrograde filling with dye, neurons are killed, then microglia will phagocytose fragments of labeled cells, so becoming labeled themselves. This procedure has been reviewed (Thanos *et al.*, 1994). Fluoro-gold has also been used as a viability stain in flow cytometry (Barber *et al.*, 1999).

The diisethionate is used as an antiprotozoal drug, especially for leishmaniasis.
Purity of commercial batches: Stilbamidine is routinely present as an impurity.
There is a HPLC chromatographic method available (Wessendorf, 1991).
Stability: Both the solid and staining solutions should be stored in the dark; light
fastness of stained sections is satisfactory.

SITS

Synonyms: 4-acetamido-4'-isothiocyanostilbene-2,2'-disulfonic acid, stilbene
isothiocyanate sulfonic acid
Disodium salt: $C_{17}H_{12}N_2Na_2O_7S_2$, FW 498

Absorption maxima: 336 nm in pH 7 aqueous solution, 320 nm in 90% aqueous
sulfuric acid; emits at 436 nm in pH 7 aqueous solutions
Solubilities: soluble in methanol and dimethylsulfoxide; slightly soluble in water,
in which solvent SITS is unstable except at low pH or with NaCl present.

SITS is a disulfonated stilbene also carrying reactive isothiocyanate and
acetamido substituents. The overall size is moderate, the size of the conjugated
system is small. The anion is hydrophilic. SITS is considered to attach covalently
to biomolecules (Horobin *et al.*, 1987). The material commercially available is the
disodium salt of the *trans* isomer.
Applications: The dye has been used as a fluorescent stain of fixed material, for
instance of neutrophil and eosinophil granules (Rothbarth *et al.*, 1976). It has also
been applied as a fluorescent antibody label, with an unusual emission color
(blue–white). Its lack of popularity in this latter role has been attributed by
Gilbert and co-workers (1982) to the fact that "it seems more difficult to
conjugate to proteins and is reported to fade rapidly." It has been suggested this
is because SITS may not bind covalently to proteins (Horobin *et al.*, 1987). A flow
cytometry manual recommends this fluorochrome as a stain (Macey, 1994).
   SITS has more often been applied to living cells and organisms, for instance,
as a viability stain (Benjaminson and Katz, 1970), to surface-label cultured cells
(Juliano, 1974), and to give cytoplasmic staining of algal cell cytoplasm (Klut *et
al.*, 1989). SITS is sometimes used as a retrograde axonal tracer (Qu *et al.*, 1996).
However, the active compound in this application is not SITS but an impurity
only present in certain batches of the fluorochrome. This may underlie reproducibility problems arising with this tracer (Horobin *et al.*, 1987).
   Other laboratory applications of SITS include incorporation as a ligand in an
affinity chromatography system for enzyme purification (Okamura *et al.*, 1991),
imaging the distribution of protein in erythrocyte membranes (Rodgers and
Glaser, 1993), and use as a chloride blocker and inhibitor of anion-exchange
proteins (Mahieu *et al.*, 1994; Wilson *et al.*, 2000).

Purity of commercial batches: Some are predominantly SITS whilst others contain only a trace of the nominal dye. Some batches are red–orange due to presence of azo or azoxystilbene contaminants. Dye contents are up 80%. There is a thin-layer chromatographic method available (Horobin *et al.*, 1987). Note: chromatographic methods using alkaline solutions cause decomposition of SITS.

Stability: The dry powder reacts with, and must be protected from, water vapor; aqueous staining solutions are unstable. Benchworkers who dissolve SITS in aqueous alkali, and then adjust to physiological pH, are actually using decomposition product(s) of SITS.

Hazards: Probably reacts with proteins, giving rise to hypersensitivity responses.

DIDS is a related more symmetric stilbene, with two isothiocyano substituents. DIDS is also used in physiology as a chloride channel inhibitor (Dietz *et al.*, 2000).

OTHER POLYENES

Fast blue

*Synonyms: trans-*1-(5-amidino-2-benzofuranyl)-2-(6-amidino-2-indolyl)ethylene dihydrochloride, diamidino 453/50, FB. Do not confuse with the stabilized diazonium salts described in Chapter 12, namely fast blues B, BB and RR

$C_{18}H_{19}C_{12}N_5O$, FW 392

Absorption maxima: 382 nm in methanol, 372 nm in water; emits at ca 400 nm in water

Solubilities: soluble in dimethylsulfoxide and water

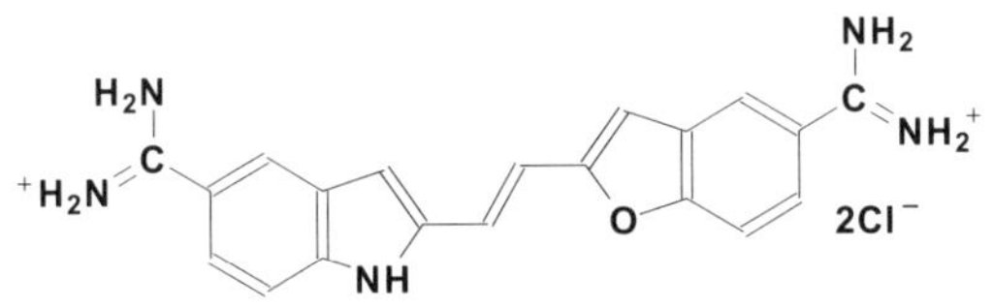

Fast blue has a two methine polyene chain, with benzofuranyl and indolyl terminal aromatic rings; each ring carries an amidino substituent. The sizes of the molecule and of its conjugated system are moderate. The dication is hydrophilic, the monocation and free base are lipophilic. The fluorochrome is sold as the chloride, illustrated.

Applications: Fast blue is widely used as a neuroanatomical tracer, and this application has recently been reviewed (Kobbert *et al.*, 2000). The dye has also been used to label glial cells grown in culture, prior to their injection into lesioned spinal cords (Prieto *et al.*, 2000).

In a related compound, **true blue**, both end-units are benzofuranyl, as illustrated below. This fluorochrome is also used as a retrograde axonal tracer (King and Bradley, 2000).

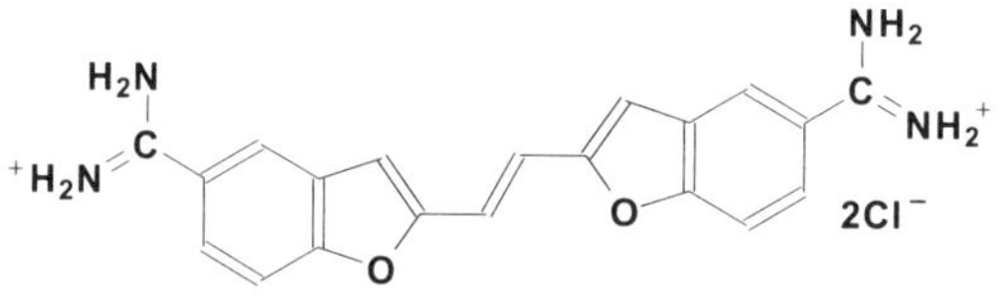

TMA-DPH

Synonyms: trimethylammonium diphenylhexatriene, 1-(4-trimethylammoni-umphenyl)-6-phenyl-1,3,5-hexatriene p-toluenesulfonate
p-Toluenesulfonate: $C_{28}H_{31}NO_3S$, FW 462

Absorption maximum: 355 nm in methanol, emission maximum in that solvent is 430 nm
Solubilities: soluble in dimethylformamide, dimethylsulfoxide and methanol; slightly soluble in water

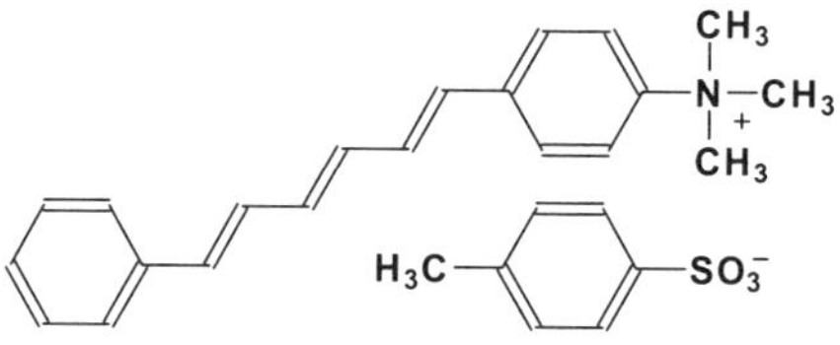

This polyene has a central chain of six methine units, with terminal phenyl groups, one of which carries a cationic trimethylammonium substituent. Overall size is small, as is that of the conjugated system. The cation is lipophilic and surface active. The probe is sold as the p-toluenesulfonate, illustrated.
Applications: TMA-DPH has been used to stain cell membranes for investigation of endocytosis using fluorescence microscopy. Such studies have been carried out with cultured animal cells (Illinger and Kuhry, 1994) and fungal hyphae (Fischer-Parton *et al.*, 2000). The probe is also widely used to study membrane fluidity. This usually involves fluorimetry (Kantar *et al.*, 1999) or flow cytometry (Benderitter *et al.*, 2000), though occasionally microscopy is used (Illinger and Kuhry, 1994).
Purity of commercial batches: Dye contents of typical samples are up to 95%
Stability: Photostability poor, especially of solutions
   A related reagent is **DPH**, or **diphenylhexatriene**, which has no hydrophilic cationic substituent, as illustrated below.

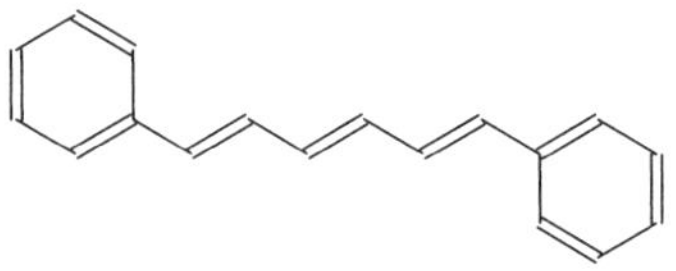

This has been used as a lipid stain in living cells (Collard and De Wildt, 1978) but is currently most commonly used for investigation of membrane fluidity (Kremer *et al.*, 2000).

23

Polymethine dyes – 1. Cyanines, oxonols, benzimidazoles, indolenines and azamethines

R.W. Horobin

BACKGROUND INFORMATION

Structural features of polymethine dyes

The core feature of dye classes described in Chapters 23 and 24 is a polymethine chain with an electron donor at one end and an electron acceptor at the other. This model is diagrammed in Figure 23.1.

   In dye classes whose core structure is cationic, electron acceptor and donor groups are respectively cationic and uncharged nitrogen. In other dye classes acceptor/donor pairs are carbonyl/ionized hydroxyl, in which case the core charge is anionic; or carbonyl/hydroxyl, in which case the core charge is zero. The bridging methine units can be replaced by aza (indamine) nitrogens (–N=). The methine units of the chain are probably *trans* orientated in most cases (Sturmer, 1977). Structural formulae in these chapters are drawn as all-*trans* forms, in the absence of hard data for individual compounds. For an illuminating overview, which also describes the complexities of nomenclature, see Zollinger (1991).

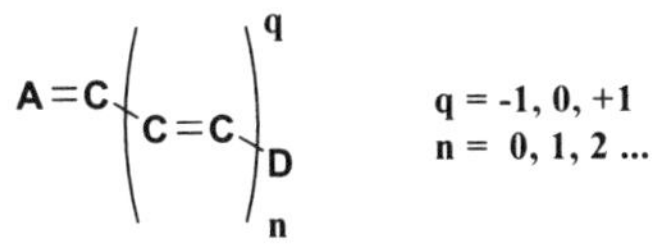

$$A=C\left(\begin{matrix} \\ C=C \\ \end{matrix}\right)^{q}_{n} D \qquad \begin{matrix} q = -1, 0, +1 \\ n = 0, 1, 2 \ldots \end{matrix}$$

Figure 23.1 Core structural features of polymethine dyes. **A** and **D** are electron acceptor and donor groups respectively.

Issues of nomenclature

Systematics

The nomenclature of these dyes is confusing, as several different schemes have been used in the chemical literature. To aid understanding of other texts, some terms are noted below although few are subsequently used.

Cationic polymethines (q = +1 in Figure 23.1) have been subdivided on the basis of the nature of the acceptor and donor groups. In the dyes most used in biology and medicine, acceptor and donor groups are nitrogen. According to whether both, one, or neither nitrogens are components of rings, the dyes are respectively called *cyanines, hemicyanines* or *streptocyanines*. If end-groups are identical the dye is termed *symmetrical*, if not then *asymmetrical*. In cyanines and hemicyanines terminal heterocycles may contain additional heteroatoms, such as oxygen in the *oxacyanines* and sulfur in the *thiacyanines*.

Polymethines whose core charge is zero (q = 0 in Figure 23.1) are termed *neutrocyanines*, and have a variety of acceptor/donor pairs. Polymethines with an anionic core charge (q = –1) are *oxonols*, whose acceptor/donor pairs are carbonyl/ionized hydroxyl.

The polymethine chain is also used for dye classification, for instance, chain lengths may be specified as *mono-, tri-* and *penta-methine*, or alternatively as *mono-, tri-* or *pentacarbo* (i.e. n = 0, 1, 2 in Figure 23.1). The term *carbocyanine* was traditionally used for trimethine dyes, and this word survives in some current commercial names. One or more methine units may be replaced by –N= to give an *azamethine* dye.

Mundane realities

Due to these complexities, polymethine dyes are usually given simpler and less informative trivial names. Some schemes are semisystematic, such that used by Waggoner's group for cyanines (Sims *et al.*, 1974). In this scheme dye names take the general form di$Y$–C$_n$–(2m + 1), where Y designates the non-nitrogen hetero- or homo-atom of the terminal ring (e.g. –O– or –C(CH$_3$)$_2$), n indicates the length of the alkyl substituents attached to the acceptor/donor nitrogens, and m has the same significance as does n in Figure 23.1.

Some names of dyes are derived from the name of the heterocycle containing an acceptor/donor nitrogen, as with benzimidazole and indolenine dyes. Other naming schemes have more distant relationships to the chemistry, for example, YOYO-1 is a structural mnemonic for a 'dimeric' product in which two yellow oxazole dyes are linked via a spacer. Some names are personalized and lack any chemical information, RH 155, for instance, was the 155th dye synthesized in Rina Hildesheim's laboratory, while Hoechst 33342 was the 33342nd compound investigated by the Hoechst company.

Properties influencing staining

Electric charge – As described above the chromophores themselves can be anionic (oxonols such as DiSBAC$_2$(3)), cationic (cyanines such as DiO) or uncharged (neutrocyanines, such as merocyanine 450). Overall charge also depends on any ionic substituents. A dye such as indocyanine green has a cationic cyanine chromophore but carries two sulfonate substituents, so is an anion under usage conditions.

Acid–base properties – dyes such as the oxonols are anionic only under alkaline condition, whereas the sulfonated dyes are ionized under all conditions of use. The pH at which cyanines become protonated, and hence colorless, varies markedly; as does perhaps their tendency to form pseudobases under alkaline conditions. No critical compilation of data is available, although Sturmer (1977) provides an extensive bibliography addressing both protonation and pseudobase formation.

Hydrophilic/lipophilic properties – These are influenced both by dye ionization and by the alkyl side-chains and quaternary ammonium groups present in some compounds. Dyes range from extremely hydrophilic, for example, TO-PRO-3 which is dicationic; through moderately lipophilic, for example, the oxonol V anion; to superlipophilic, for example, DiI which carries two octadecyl side-chains. Dyes such as YOYO-1 carry multiple quaternary ammonium substituents.

The extent of conjugation – Varies from small (e.g. DAPI) to moderate (e.g. Cy5). The 'dimeric' cyanine dyes such as TOTO-1, in which two chromophores are separated by an alkyl spacer, have two quite separate conjugated regions. However, structures which can be drawn on paper with a single conjugated region may contain several conjugated segments due to steric crowding forcing rotation of aromatic units; an example is RH 155.

Reactivity – The Cy dyes carry one or more succinimidyl ester groups, CellTracker CM-DiI has a chloromethyl substituent. All react with nucleophilic groups found on nucleic acids and proteins. Indo-1 carries multiple carboxylates, which form complexes with calcium ions.

Dye size – This varies from small (e.g. pinacyanol) to large (e.g. CellTracker CM-DiI), with certain of the 'dimeric' dyes being even larger (e.g. YOYO-1).

Safety issues

Although no published work has reported such effects, reactive dyes such as Cy3, Cy5 and CellTracker CM-DiI have the potential to modify human proteins resulting in hypersensitivity responses.

How to find the dye that interests you

Since it would be perverse and misleading to ignore the widely used traditional categories, this chapter describes dyes using the headings cyanines, oxonols, benzimidazoles and indolenines, and azamethines. Within each of these groups the usual sequence of anions, cations, and reactive dyes is followed when appropriate. Within each subgroup dyes are listed in order of increasing formula weight. The remaining groups of polymethine dyes are discussed in Chapter 24.

Merocyanine 540

Synonym: MC540
$C_{26}H_{32}N_3NaO_6S_2$, FW 570

Absorption maxima: 500 nm and 534 nm in neutral aqueous solution; 590 nm in dioxane. When excited at 534 nm in neutral aqueous solution, it emits at 577 nm. Excited at 530 or 570 nm in 95% aqueous dioxane it emits at ca 585 nm. Above pH 7.6 in aqueous solutions a derivative is formed which absorbs at 390 nm and emits at 500 nm.
pK value: 1.6
Solubilities: soluble in dimethylsulfoxide, dioxane, ethanol and water

Merocyanine 540 is an asymmetric neutrocyanine in which the electron donor, a nitrogen atom in a benzoxazole heterocycle, is joined to the thiobarbiturate heterocycle containing the carbonyl electron acceptor by a four methine chain. The compound carries three short alkyl chains, one with a terminal sulfonate substituent. The overall size, and that of the conjugated system, is moderate. The anion is lipophilic and amphiphilic. Merocyanine 540 is sold as the sodium salt.
Applications: Merocyanine 540 has been used to stain normal and leukemic granulocytes in fixed blood smears (Kass, 1986), but is usually used as a fluorochrome of live cells. It is used as a probe of plasma membrane reorganization during maturation of sperm (Sivashanmugam and Rajalakshmi, 1997), to detect apoptosis during photodamage of myeloid leukemic cells (Chen *et al.*, 2000), and to demonstrate cell proliferation in culture (Siboni *et al.*, 2001). Merocyanine 540 is applied as a stain in flow cytometry to identify apoptotic cells (Reid *et al.*, 1996), and to study alterations in membrane properties (Liang and Huang, 2001).

Merocyanine 540 has a variety of nonstaining laboratory applications, often exploiting its induction of photodamage in live cells. Examples are removal of microbial and leukemic contaminants from blood products (reviewed by Corash, 1999), and killing transformed cells, as in photopurging bone marrow of leukemic patients (Danilatou *et al.*, 2000). The dye has been incorporated into an optical sensor for potassium (Krause *et al.*, 1999).
Purity of commercial batches: Dye content of typical samples is up to 97%. A thin-layer chromatographic method is available (Hirpara *et al.*, 2000).
Stability: The dry powder, and even more so solutions, photofade. The dye precipitates from aqueous solutions below pH 1.7. Above pH 7.6 irreversible formation of a hydroxyl derivative occurs.

Indocyanine green

Synonym: cardio green
$C_{43}H_{47}N_2NaO_6S_2$, FW 775

Absorption maximum: 775 nm in water; the emission maximum is in the near infra-red at 835 nm.
Solubilities: soluble in water, slightly soluble in ethanol

Indocyanine green is a symmetrical cyanine; the heterocyclic rings containing the donor/acceptor nitrogens are bridged by a seven methine chain. The compound carries two alkyl chains each with terminal sulfonate substituents. The overall size, and that of the conjugated system, is moderate. The dye is sold as the sodium salt, illustrated, of the hydrophilic anion.
Applications: The first applications in medicine were for blood volume measurement (Picker *et al.*, 2001) and hepatic function testing, and it now also has several uses as a stain and marker. Its use in ophthalmology as an angiography tracer has been reviewed by Zarfati *et al.* (2000). Indocyanine green has been investigated as a reagent for estimating burn depth by infra-red fluorescence following intravenous administration (Green *et al.*, 1992), as a marker for sentinel lymph nodes following gastric cancer surgery (Hiratsuka *et al.*, 2001), and for the photochemotherapy of Kaposi sarcoma (Szeimies et al., 2001).
Purity of commercial batches: Dye contents up to 90% are available. There is a HPLC analytical method (Niemann *et al.*, 2000).
Stability: Routine aqueous solutions are stable for only a few hours.

CATIONIC CYANINES

Pinacyanol

CI Ed 1 808
Synonyms: 1,1'-diethyl-4,4'-carbocyanine; quinaldine blue
Chloride: $C_{25}H_{25}ClN_2$, FW 389

Absorption maxima: 608 nm in ethanol; 600 nm, 550 nm and ca 520 nm in water
An oxidation indicator; *pK value:* 3.1
Solubilities: 0.7% in water and 2% in ethanol

Pinacyanol is a symmetrical cyanine, the heterocyclic rings containing the donor/acceptor nitrogens are bridged by a three methine chain. The overall size, and that of the conjugated system, is moderate. The cation is lipophilic, and the dye is sold as the chloride (illustrated), bromide and iodide.

Applications: Pinacyanol has been applied as a component of a polychrome stain used for frozen sections (Proescher, 1933), and Bensley (1952) used its erythrosin salt as a sensitive stain for mast cells. The latter application is recommended in staining manuals (Bancroft and Cook, 1994), and is occasionally reported in the research literature (Florenzano and Bentivoglio, 2000). Pinacyanol has been used as a vital stain for mitochondria in leukocytes (Hetherington, 1936), doubtless due to the lipophilic character of its cation (Rashid and Horobin, 1990).

As an oxidation indicator pinacyanol has been applied during investigations of the catalytic activity of cytochrome C (Vazquezduhalt *et al.*, 1993), and of the oxidation of copper complexes with hydrogen peroxide (Robbins and Drago, 1997). The dye has also been used as an experimental photosensitizer of yeasts (Iwamoto *et al.*, 1990).

Purity of commercial batches: Dye contents are up to 95%.

Hazards: Harmful if adsorbed.

DiOC$_5$(3)

Synonyms: dioc-5 and DiOC(5); 3,3′-dipentyloxacarbocyanine iodide
C$_{27}$H$_{33}$IN$_2$O$_2$, FW 544

Absorption maximum: 484 nm in methanol, in which solvent it emits at 500 nm
Solubilities: soluble in dimethylsulfoxide and methanol

DiOC$_5$(3) is a symmetrical cyanine, the benzoxazole heterocycles containing the donor/acceptor nitrogens being bridged by a three methine chain. The overall size, and that of the conjugated system, is moderate. The cation is lipophilic, with two *N*-pentyl chains, and is weakly amphiphilic. This dye is sold as the iodide, illustrated.

Applications: DiOC$_5$(3) is used to demonstrate endoplasmic reticulum in living cells (Kamisaka *et al.*, 1999), and occasionally as a probe for mitochondria (Ricken *et al.*, 1998); these latter organelles are routinely stained prior to uptake of dye by the reticulum. DiOC$_5$(3) has been used as a membrane stain to follow liposome–cell interactions (Miller *et al.*, 1998), and as a probe for detection of membrane efflux pumps (Prudencio *et al.*, 2000). The fluorescence of DiOC$_5$(3) in membranes is potential-sensitive, providing the basis for potential measurements with either imaging methodology or flow cytometry, as reviewed by Shapiro (1994) and Plasek and Sigler (1996) respectively.

CH₂CH₂CH₂CH₂CH₃ CH₂CH₂CH₂CH₂CH₃

DiOC$_5$(3) is recommended in flow cytometry manuals for measuring plasma membrane potential of living cells, and as a general nuclear stain to permit sorting of fixed cells (Darzynkiewicz and Crissman, 1990), and for detection of hyper-polarized cells (Macey, 1994). The potential-sensitivity has been exploited also for assessing susceptibility of microorganisms to drug treatments (Favel *et al.*, 1999).
Purity of commercial batches: Dye contents up to 98% are available.
Stability: Photofades in solution.

DiOC$_6$(3)

Synonyms: 3,3′-dihexyloxacarbocyanine iodide; DiOC6
C$_{29}$H$_{37}$IN$_2$O$_2$, FW 573

Absorption maximum: 484 nm in methanol, its emission maximum in that solvent being at 501 nm
Solubilities: soluble in dimethylsulfoxide and methanol

CH₂CH₂CH₂CH₂CH₂CH₃ CH₂CH₂CH₂CH₂CH₂CH₃

DiOC$_6$(3) is a symmetrical cyanine, the benzoxazole heterocycles containing the donor/acceptor nitrogens being bridged by a three methine chain. The overall size, and that of the conjugated system, is moderate. The cation, carrying two N-hexyl chains, is lipophilic and weakly amphiphilic. The dye is sold as the iodide, illustrated.
Applications: DiOC$_6$(3) is occasionally used as a general stain for biomembranes in fixed (Bassnett, 1997) and living (Jesuthasan, 1998) cells. Routine applications of DiOC$_6$(3) are as a fluorescent probe of living cells. It is recommended in manuals and reviews (Celis, 1998; Sabnis *et al.*, 1997) as a stain for the endo-plasmic and sarcoplasmic reticulum, and to demonstrate the nuclear envelope (Koning *et al.*, 1993) and mitochondria (Bereiter-Hahn and Voth, 1994). The potential-sensitive fluorescence of the dye allows it to be used to assess mito-chondrial membrane potential (Isenberg and Klaunig, 2000). DiOC$_6$(3) has also been used for selective staining of ascomycete fungi within plant material (Duckett and Read, 1991).

The dye is used as a stain in flow cytometry. It is noted in manuals as a reagent for measurement of mitochondrial and plasma membrane potentials (Darzynkiewicz and Crissman, 1990; Macey, 1994), and for the detection of mitochondria (Macey, 1994). $DiOC_6(3)$ has been used in fluorescent activated cell sorting to separate viable from apoptotic leukocytes (Belloc *et al.*, 2000). Nonstaining applications include use as an enzyme inhibitor (Anderson *et al.*, 1993), and as a microtubule-damaging sensitizer in photodynamic therapy research (Lee *et al.*, 1995).

Purity of commercial batches: Dye contents up to 98% are available.

Stability: Photofades in solution

YO-PRO-1

$C_{24}H_{29}I_2N_3O$, FW 629

Absorption maximum: 491 nm in an aqueous DNA solution, in which system the emission maximum is 509 nm

Solubilities: soluble in dimethylsulfoxide and water

YO-PRO-1 is an unsymmetrical cyanine, with the benzoxazole and quinoline heterocycles containing the donor/acceptor nitrogens being linked by a one methine bridge. The di-cation, which carries a *N*-propyl side-chain with a terminal trimethylammonium group, is hydrophilic. The dye is sold as the di-iodide, illustrated.

Applications: In fixed tissues YO-PRO-1 has been used as a nuclear counterstain (Dupuis *et al.*, 2000), and in living cells the hydrophilic character of the dye has been used to give selective staining of apoptotic cells (Castaneda and Kinne, 2000). Other microscopic staining applications include visualization of specific sequences in single stretched DNA molecules (Oana *et al.*, 1999), and counting of viruses in aquatic systems (Bettarel *et al.*, 2000). YO-PRO-1 is used as a stain in flow cytometry, for instance to identify apoptotic and necrotic cells (Hubl *et al.*, 1998), and to detect DNA in permeabilized cells as discussed in a review by King (2000).

Further nonimaging applications directly involving nucleic acid staining include spectrofluorimetric tracking of cell movements into solid matrices (Gohla *et al.*, 1996), and detection of nucleic acid in electropherograms (Nishimura and Tsuhako, 2000). Other applications exploit the dye's hydrophilic character by using it as a permeability tracer. Examples include monitoring entry of phage into bacterial cells (Bonhivers *et al.*, 1998), and investigation of cation channels in fibroblast membranes using a patch-clamp technique (Schilling *et al.*, 1999). YO-PRO-1 has also been used as a component of a miniaturized device for DNA extraction (Tian *et al.*, 2000).

Stability: Photofades in solution

JC-1

Synonym: CBIC$_2$(3)
C$_{25}$H$_{27}$C$_4$IN$_4$, FW 652

Absorption maximum: 514 nm in methanol, in which solvent the emission maximum is 529 nm
Solubilities: soluble in dimethylformamide, dimethylsulfoxide and methanol

JC-1 is a symmetrical cyanine, the benzimidazole heterocycles containing the donor/acceptor nitrogens being connected by a three methine chain. The overall size, and that of the conjugated system, is moderate. The cation is lipophilic. The dye is sold as the iodide, illustrated.
Applications: As a fluorescent probe JC-1 is used to demonstrate mitochondria, as isolated organelles (Tiano *et al.*, 2000) or in living cells (Gravance *et al.*, 2001). Such staining is membrane potential dependent and so can be used to assess mitochondrial functionality (Mathur *et al.*, 2000). The dye has been used to detect apoptotic cells using laser scanning cytometry (Bedner *et al.*, 1999).
  JC-1 is used for assessment of mitochondrial membrane potential by flow cytometry, as described in a review by Shapiro (2000). Spectroscopic measurements of JC-1 have been used to determine ion channel activity (Chanda and Mathew, 1999), and mitochondrial function in brain slices (Sick and Perez-Pinzon, 1999).
Stability: Photofades in solution

TO-PRO-3

Synonym: TP3
C$_{26}$H$_{31}$I$_2$N$_3$S, FW 671

Absorption maxima: 641 nm in 20% aqueous ethylene glycol, and 642 nm in aqueous DNA solution; the emission maxima are 661 nm in the glycol and 661 nm in the DNA solutions respectively.
Solubilities: soluble in water and dimethylsulfoxide

TO-PRO-3 is an unsymmetrical cyanine, with the thiazole and quinoline hetero-cycles containing the donor/acceptor nitrogens being linked by a three methine bridge. Both the overall size and size of the conjugated system are moderate. The dication, which has a *N*-propyl side-chain with a terminal trimethyl-ammonium group, is hydrophilic. The dye is sold as the di-iodide, illustrated.

Applications: TO-PRO-3 is used as a nuclear counterstain for fixed immunos-tained material (Tsai *et al.*, 2000), including whole-mount embryos (de Maziere *et al.*, 1996). It is also used to stain polytene chromosomes in squash preparations (Kirsch *et al.*, 1998). In flow cytometry TO-PRO-3 is used to label bacteria (Nebe-von Caron *et al.*, 2000) and to evaluate their permeability (Shapiro, 2001), and to label DNA of skin cells (Mommers *et al.*, 2000). It is also used to stain DNA on agarose gel electropherograms (Guttman *et al.*, 2000).

Stability: Photofades in solution

DiO

Synonyms: $DiOC_{18}(3)$; 3,3'-dioctadecyloxacarbocyanine perchlorate
$C_{53}H_{85}ClN_2O_6$, FW 882

Absorption maximum: 484 nm in methanol, in which solvent the emission maximum is at 501 nm
Solubilities: soluble in dimethylformamide, dimethylsulfoxide and methanol

DiO is a symmetrical cyanine with the benzoxazole heterocycles containing the donor/acceptor nitrogens being bridged by a three methine chain. The overall size is large due to the two *N*-octadecyl side-chains; that of the conjugated system is moderate. The cation is superlipophilic. The dye is sold as the perchlorate, illustrated.

Applications: DiO has been widely used as a stain of biomembranes in both living and fixed cells. Applications of such procedures to animal cells have been reviewed by Vercelli *et al.* (2000). These authors concluded that DiO was most satisfactory with single cells, cultured or disaggregated, and less effective with fixed material than dyes such as DiI. Recent vital staining applications include labeling of tumor cells to study their penetration into brain tissue (Khoshyomn *et al.*, 1998), selective staining of plasma membrane to label a cell line (Shan *et al.*, 2000), and detection of glial cells following their phagocytosis of DiO-labeled neurons (Thanos *et al.*, 2000).

Whatever the validity of the above critique, DiO is still used for axonal tracing in fixed postmortem specimens, as reviewed by Köbbert *et al.* (2000). DiO in stained neurons has been used to photopolymerize diaminobenzidine for stain intensification and for generation of a permanent reaction product visible in the light or transmission electron microscope (Lubke, 1993).

DiO is applied in flow cytometry, for instance, to stain plasma membranes of cellular targets of cytotoxic T-cells (Mattis *et al.*, 1997), and to label phospholipid emulsions for studies of binding of low density lipoprotein to leukocytes (Boullier *et al.*, 2000). Applications in virology include use as a photosensitizer in a chemical study of viral membrane activity (Pak *et al.*, 1997), and as a tracer to assess spectroscopically viral envelope permeability (Munoz-Barroso *et al.*, 1998). *Purity of commercial batches:* Commercial samples are available with dye contents up to 99%.

Stability: Photofades in solution.

A related dye, **DiOC$_{16}$(3)**, has slightly shorter, hexadecyl, side-chains. This is occasionally used for staining biomembranes in living cells (Flock *et al.*, 1998). Yet another structural variant is **Fast DiO**, with two unsaturated *N*-linoleyl substituents. This is occasionally used in fluorescent staining of the plasma membrane of live cells (Ziv and Smith, 1996) and as a neuronal tracer (Belluscio *et al.*, 1999).

DiI

Synonyms: DiIC18; DiIC$_{18}$(3); 1,1'-dioctadecyl-3,3,3',3'-tetramethylindocarbo-cyanine perchlorate
C$_{59}$H$_{97}$ClN$_2$O$_4$, FW 934

Absorption maximum: 549 nm in methanol, the emission maximum in that solvent being at 565 nm
Solubilities: soluble in dimethylsulfoxide, ethanol and methanol

DiI is a symmetrical cyanine, the heterocycles containing the donor/acceptor nitrogens being joined by a three methine chain. The overall size is large due to the two *N*-octadecyl side-chains. The size of the conjugated system is moderate, and the cation is superlipophilic. The dye is sold as the perchlorate, illustrated.

Applications: The superlipophilic character of DiI results in its irreversible attachment to the first biomembrane it contacts. This phenomenon is exploited in a variety of ways. Thus, it selectively stains the plasma membrane by application of dye solution to the cell (Hostager *et al.*, 2000), or stains the endoplasmic reticulum following microinjection of DiI saturated oil droplets (Terasaki, 2000). After plasma membranes have been stained with DiI, their deposition onto the surface across which the cell is moving may be tracked (Hakansson *et al.*, 1999). DiI labeling is also used to follow cell movements. Examples include investigation of glial cells following their phagocytosis of damaged DiI-labeled neurons (Thanos *et al.*, 2000), and the movements of labeled endothelial progenitor cells after myocardial ischemia (Kawamoto *et al.*, 2001). DiI staining of lipid domains is used to investigate low density lipoprotein uptake into vascular smooth muscle (Llorente-Cortes *et al.*, 2000). The use of DiI to trace neurons in living organisms and in aldehyde-fixed post-mortem specimens has been reviewed by Kobbert *et al.* (2000) and Vercelli *et al.* (2000). Cell lines can be labeled with DiI to permit subsequent identification using flow cytometry (Czyz *et al.*, 2000).

There are some nonstaining applications of DiI in biology and medicine. These include its use as a photosensitizer to destabilize endocytosed liposomes (Miller *et al.*, 2000), and as a component of an experimental microfluidic device for combinatorial fusion of liposomes and cells (Stromberg *et al.*, 2001).

Purity of commercial batches: Material with dye content of up to 97% is available
Stability: Photofades in solution

A structural variant is **Fast DiI**, with two unsaturated *N*-linoleyl side-chains. This dye is used as a neuronal tracer, as discussed in a review by Sparks *et al.* (2000).

DiD

Synonyms: DiC18; DiC$_{18}$(5); DiD oil; indocarbocyanine
C$_{61}$H$_{99}$ClN$_2$O$_4$, FW 960

Absorption maximum: 644 nm in methanol, in which solvent the emission maximum is 665 nm
Solubilities: soluble in dimethylsulfoxide, ethanol and methanol

DiD is a symmetrical cyanine, the heterocycles containing the donor/acceptor nitrogens being bridged by a five methine chain. The overall size is large due to the two *N*-octadecyl side-chains; the size of the conjugated system is moderate; and the cation is superlipophilic. The dye is sold as the perchlorate, illustrated.
Applications: The superlipophilic character of DiD underlies several staining applications. Most direct is its use as a probe for the plasma membranes of living cells (Servant *et al.*, 1999) and, in permeabilized or fixed cells, of biomembranes in general (Jesuthasan, 1998). The permanent nature of cellular staining also makes DiD a useful cell marker, as noted in reviews concerned with migrations of embryonic cells (Fraser, 1996; Gan *et al.*, 2000). A clinical application of cell marking is erythrocyte labeling, for evaluation of retinal and choroidal circulation (Khoobehi and Peyman, 1999). Membrane labeling also underlies use for axonal tracing in fresh post-mortem material (Agmon *et al.*, 1995).
   Marked cells are usually detected microscopically, but they may also be counted using flow cytometry (Celluzzi and Falo, 1998). DiD has been used also as a component of a potassium-detecting fiber-optic sensor (Roe *et al.*, 1990).
Purity of commercial batches: Dye contents of typical samples are up to 95%
Stability: Photofades in solution

YOYO-1

Synonyms: oxazole yellow dimer (or homodimer)
$C_{49}H_{58}I_4N_6O_2$, FW 1271

Absorption maximum: 491 nm in aqueous DNA solution, in which solution the emission maximum is 509 nm
Solubilities: soluble in dimethylsulfoxide and water

YOYO-1 comprises two separate cationic cyanine moieties, separated by a spacer containing two quaternary ammonium groups. The benzoxazole and quinoline acceptor/donor heterocycles in each cyanine are linked by one methine bridges. The overall size of YOYO-1 is large, although the individual conjugated systems are moderately sized. The tetracation is hydrophilic. This dye is sold as the iodide, illustrated.

Applications: The hydrophilic character of YOYO-1 results in a membrane impermeant character, which underlies application as a nucleic acid stain with living cells. Thus the dye has been used to demonstrate ribosomal RNA when microinjected into the cell (Terasaki, 1994); DNA when present extracellularly, adsorbed onto the cell surface (Clamme *et al.*, 2000); and as a viability stain, with access only to cells with damaged cell membranes (Weinhaus *et al.*, 2000). YOYO-1 has also been used for viability assays in a microwell system (Becker *et al.*, 1994).

   With fixed tissues YOYO-1 has been used to identify mitotic figures (Yan *et al.*, 1998), and as a nuclear counterstain for immunostained yeasts (Serpe *et al.*, 1999). The dye is recommended in a review by Crissman and Hirons (1994) for DNA staining in flow cytometry.

   YOYO-1 has been used in analytical biochemistry for assay of mRNA (Miura *et al.*, 1996), in molecular biology for optical mapping of single chromosomes (Jing *et al.*, 1999), and as a component of a picoliter PCR system (Nagai *et al.*, 2001).

Stability: Photofades in solution

Related dyes: In an otherwise structurally similar dye, **TOTO-1**, the hetercycles are thiazoles. This dye is recommended in a review of DNA stains for flow cytometry (Crissman and Hirons, 1994) and has occasionally been used as an imaging probe, for instance, to assess the viability of cultured neuroblasts (Broadus and Doe, 1997). Yet another variant is **TOTO-3**, which differs from TOTO-1 in having a three methine bridge in each cyanine moiety. The most common application of this reagent is as a nuclear counterstain in a variety of immunostaining methods (Knaut *et al.*, 2000).

CYANINES – FORMING COVALENT BONDS WITH TISSUE OR CO-SOLUTES

Cy3

$C_{35}H_{40}KN_3O_{10}S_2$, FW 766

Absorption maximum: 550 nm in water, in which solvent the emission maximum is 570 nm

Solubility: soluble in water

Cy3 is a symmetrical cyanine, in which the heterocycles containing the donor/acceptor nitrogens are connected by a three methine bridge. The dye carries a *N*-pentyl substituent with a terminal succinimidyl ester, reactive with nucleophiles such as the amino groups of proteins and nucleic acids. Overall size is substantial, that of the conjugated system small. The chromophore being a cyanine with two sulfonate substituents, this dye forms a hydrophilic anion in neutral aqueous solutions, and is sold as the potassium salt.

Applications: Cy3 is widely used as a fluorescent label of nucleotide, polynucleotide and nucleic acid probes. The results of the interaction of such probes with cells and cell products are sometimes detected microscopically, for example, following use of Cy3-dUTP to detect regions of DNA replication in cell nuclei (Sadoni *et al.*, 2001); or using flow cytometry, as in an investigation of the penetration of oligosaccharides into bacteria (Fuchs *et al.*, 2001). Most applications of labeled nucleotide probes, however, involve detection with a micro array scanner or with related technologies (Balazs *et al.*, 2001).

Cy3 is also widely applied as a protein label. This includes labeling of antibodies, when binding to biological materials may be demonstrated microscopically, for example, to localize membrane components in sperm (Rattanchaiyanont *et al.*, 2001). Labeled antibodies are also used for protein detection in blotting procedures (Gingrich *et al.*, 2000). Cy3 labeling of other proteins is also widely used, for example, of substance P for investigating its localization within smooth muscle cells (Southwell and Furness, 2001).

Stability: Photofades, slowly decomposes in aqueous solution

Hazards: The dye is reactive with protein and therefore may cause hypersensitivity responses

Cy5

$C_{45}H_{51}KN_4O_{14}S_2$, FW 975

Absorption maximum: 649 nm in water, in which solvent the emission maximum is 670 nm

Solubility: soluble in water

Cy5 is a cyanine, with the heterocycles containing the donor/acceptor nitrogens joined by a five methine bridge. The dye carries two *N*-pentyl substituents with terminal succinimidyl esters, reactive with nucleophilic moieties such as the amino groups of proteins and nucleic acids. Overall size is large, that of the conjugated system is quite small. The chromophore is a cyanine carrying two sulfonate substituents, and the dye therefore forms a hydrophilic anion in neutral aqueous solutions. It is sold as the potassium salt.

Applications: This dye is widely used as a fluorescent label of nucleotide, polynucleotide and nucleic acid probes. The results of the interaction of such probes with cells and cell products are sometimes detected microscopically, as with the *in situ* hybridization staining of chromosomes (Konig *et al.*, 2000), or indeed in a procedure for observing hybridization in living cells (Tsuji *et al.*, 2001). More often, however, some nonimaging methodology is used, as in human leukocyte antigen (HLA) typing using a micro array system (Balazs *et al.*, 2001).

Cy5 is also applied as a label for antibodies and other proteins. Binding of the fluorescent immunoglobulin to biological materials may be imaged either macroscopically, for example, for noninvasive detection of tumors (Ramjiawan *et al.*, 2000); or microscopically, such as to localize serotonin(5A) receptors in brain specimens (Duncan *et al.*, 2000). Cy5-labeled antibodies are also used as stains in flow cytometry, for instance, to label marker antibodies for lymphocyte categorization (Bellido *et al.*, 2001); and in other systems such as flow-based immunoassay (Ohmura *et al.*, 2001).

Labeling of other proteins is also carried out. For instance, Cy5-low density lipoprotein has been used as a marker for endosome trafficking, followed microscopically (Kluve-Beckerman *et al.*, 2001), and Cy5-phycoerythrin is used in a flow cytometric system to detect apopotosis (Takahashi *et al.*, 2001).

Stability: Photofades, slowly decomposes in the presence of water

Hazards: The dye is reactive with protein and therefore may cause hypersensitivity responses

Cell Tracker CM-DiI

Synonym: although sometimes described as DiI, it should not be confused with that dye

$C_{68}H_{105}Cl_2N_3O_7$, FW 1051

Absorption maximum: 553 nm in methanol, in which solvent the emission maximum is 570 nm

Solubilities: soluble in dimethylsulfoxide, ethanol and methanol

Cell Tracker CM-DiI is based on the symmetrical cyanine DiI, in which the heterocycles containing the donor/acceptor nitrogens are connected by a three methine chain. The Cell Tracker derivative has a reactive chlorobenzyl substituent, which can link to nucleophilic groups such as amino or hydroxyl. In addition there are two octadecyl side-chains. Consequently the overall size is large, although that of the conjugated system is moderate. The cation is super-lipophilic. The dye is sold as the perchlorate, illustrated.

Applications: The superlipophilic character of Cell Tracker CM-DiI underlies its use as a selective stain for the plasma membrane in live cells (McConalogue *et al.*, 1999), and indeed as a stain of biomembranes in general when cells are permeabilized before exposure to the dye (McLean PJ *et al.*, 2000). The dye is also applied for the irreversible labeling of liposomes, to study their interactions with endothelia (McLean JW *et al.*, 1997).

The irreversible character of such staining underlies use of the dye as a long-term cell marker. Such applications may involve subsequent microscopic study, as in the movement of transplanted cells (Hebda and Dohar, 1999); or labeled cells may be assayed using flow cytometry (Andrade *et al.*, 1998). Flow cytometry has also been used to trace endocytosis of labeled bacteria (Harf *et al.*, 1997).

Stability: Photofades in solution. Reacts slowly with water, more rapidly in aqueous alkali

Hazards: The dye is reactive with protein and therefore may cause hypersensitivity responses

OXONOLS

Oxonol V

Synonyms: MC-V, OX-V
$C_{23}H_{16}N_2O_4$, FW 384

Absorption maxima: of the anion is 610 nm in methanol, in which solvent the emission maximum is 639 nm. In methanol acidified using hydrogen chloride the free acid is present, whose absorption maximum is 473 nm

A potientiometric indicator

Solubilities: soluble in dimethylsulfoxide, ethanol and methanol; and in alkaline aqueous solutions

Oxonol V is a symmetrical oxonol, the heterocycles carrying the donor/acceptor oxygens being bridged by a five methine chain. The overall size and that of the conjugated system is moderate. The dye is sold as the lipophilic free acid, illustrated. When this dye was first applied in biology its structure was uncertain, this being established later by Smith *et al.* (1976).

Applications: The dye's fluorescence is potential-sensitive. This is exploited to assess membrane potential in a variety of methodologies, and has been discussed in reviews, for instance by Plasek and Sigler (1996) and Wolosker *et al.* (1996).

Purity of commercial batches: Dye contents are up to 95%. A thin-layer chromatographic method is available (Smith *et al.*, 1976).

Stability: Photofades in solution

Oxonol VI is a related dye, with two *N*-propyl substituents rather than the two phenyl groups present in oxonol V. Oxonol VI is sometimes used for measuring membrane potentials in cell physiology (Guffanti *et al.*, 1998).

DiSBAC$_2$(3)

Synonyms: bis-oxonol, trimethine oxonol. In the literature these two terms are also used to refer to related dyes

$C_{19}H_{24}N_4O_6S_2$, FW 437

Absorption maximum: 535 nm in methanol, in which solvent the emission maximum is 560 nm

A potientiometric indicator

Solubilities: Soluble in dimethylsulfoxide, ethanol, methanol and in aqueous alkali

DiSBAC$_2$(3) is a symmetrical oxonol, with the thiobarbiturate heterocycles carrying the donor/acceptor oxygens being connected by a three methine chain. The overall size and that of the conjugated system are moderate. The dye is sold as the lipophilic free acid, illustrated.

Applications: DiSBAC$_2$(3) is used to assess the membrane potential of live cells. This has been carried out using fluorescence and confocal microscopy, for example, to investigate individual nasal respiratory cells (Renier *et al.*, 1995), and fibroblasts (Dall'Asta *et al.*, 1997). More often, however, DiSBAC$_2$(3) is applied with populations of cells, investigated by spectroscopy (Maechler *et al.*, 1999), or flow cytometry (Suller and Lloyd, 1999).

Stability: Photofades in solution

DiBAC$_4$(3)

Synonyms: trimethine oxonol. The dye is sometimes called bis-oxonol, a term also used for the functionally related but chemically distinct DiBAC$_4$(5)
$C_{27}H_{40}N_4O_6$, FW 517

Absorption maximum: 493 nm in methanol, in which solvent the emission maximum is 516 nm
A potientiometric indicator
Solubilities: Soluble in dimethylsulfoxide, ethanol and methanol

DiBAC$_4$(3) is a symmetrical oxonol, the barbiturate heterocycles carrying the donor/acceptor oxygens are bridged by a three methine chain. The overall size and that of the conjugated system is moderate. The dye carries four *N*-butyl side-chains, and is sold as the lipophilic free acid, illustrated.
Applications: This dye is sometimes used for membrane potential-dependent microscopic imaging, to assess viability of bacteria (Lopez-Amoros *et al.*, 1997) and of cultured embryonic cortical cells (Maric *et al.*, 2000). Differential permeability of the cysts and trophozoites of parasitic amoebae to DiBAC$_4$(3) has been exploited as an alternative to traditional hemocytometry or plate counts (Connell *et al.*, 2001).

More commonly however, DiBAC$_4$(3) is used to assess membrane potential using flow cytometry as discussed in a review by Shapiro (2000), or spectroscopy, often in a microwell plate system (Straub *et al.*, 2000).
Purity of commercial batches: Dye contents up to 98% are available.
Stability: Photofades in solution

DiBAC$_4$(5)

Synonym: The dye is sometimes referred to as bis-oxonol, a term also used for the functionally related but chemically distinct DiBAC$_4$(3)
$C_{29}H_{42}N_4O_6$, FW 543

Absorption maximum: 590 nm in methanol, in which solvent the emission maximum is 616 nm
A potientiometric indicator
Solubilities: Soluble in dimethylsulfoxide, ethanol, and methanol

DiBAC$_4$(5) is a symmetrical oxonol, the barbiturate heterocycles carrying the donor/acceptor oxygens being bridged by a five methine chain. The overall size and that of the conjugated system is moderate. The dye carries four *N*-butyl side-chains, and is sold as the lipophilic free acid, illustrated.

Applications: DiBAC$_4$(5) is used to image membrane potential in single cells, as discussed in the review by Zochowski *et al.* (2000). It is also applied microfluorimetrically, for example, with smooth muscle cells (Cornfield *et al.*, 1994). In flow cytometry the dye is used to estimate membrane potential (Ferlini *et al.*, 1999), and the membrane potential dependent staining has also used as a marker of cell type with fluorescence activated cell sorting (Fiszman *et al.*, 1990).

    Stability: Photofades in solution

RH 155

C$_{49}$H$_{71}$N$_7$O$_8$S$_2$, FW 950

Absorption maximum: 650 nm in methanol
Potentiometric indicator
Solubilities: Soluble in dimethylsulfoxide, ethanol and methanol

RH 155 is a symmetrical oxonol, the pyrazole rings carrying the donor/acceptor oxygens being bridged by a five methine chain, which carries a central phenyl substituent. The pyrazole heterocycles are substituted by phenylsulfonic acid groups. Consequently the overall size is large, but steric crowding results in nonplanarity and consequent loss of conjugation despite the continuity of alternating single and double bonds in the two-dimensional structural formula. RH 155 is sold as the hydrophilic triethylamine salt, illustrated.

Applications: RH 155 is widely used to image membrane potential and thus synaptic activity in neural specimens, such as brain slices or cultured cells (Senseman, 1996), and in other electrically excitable cells such as skeletal muscle

(Heiny and Jong, 1990). Applications of these types have been reviewed by Zochowski *et al.* (2000).
Stability: Photofades in solution

CATIONIC BENZIMIDAZOLES AND INDOLENINES

DAPI

Synonym: 4′,6-diamidino-2-phenylindole dichloride
Dichloride: $C_{16}H_{17}Cl_2N_5$, FW 350
Dilactate: $C_{22}H_{27}N_5O_6$, FW 457

Absorption maxima: 344 nm in water and 358 nm in aqueous DNA; the emission maxima in these two solvents are 450 nm and 461 nm respectively; emits at 493 nm when used as a neuronal tracer
Solubilities: soluble in water and dimethylformamide

DAPI is an indolenine dye carrying two amidine substituents, one directly on the indole moiety and the other on a pendant phenyl ring. The overall size, and that of the conjugated system, is small. The free base is weakly lipophilic; the cationic species are hydrophilic. The dye is sold as the dichloride, illustrated, and also as the dilactate.
Applications: DAPI is often used as a DNA and nuclear stain, as discussed in the reviews of Crissman and Hirons (1994) and Kapuscinski (1995). The dye is recommended in manuals as a nuclear stain for botanical material (Ruzin, 1999), general animal histology (Kiernan, 1999), and as a fluorescent nuclear counter-stain following *in situ* hybridization (Schwarzacher and Heslop-Harrison, 2000). DAPI is also used as a counterstain for immunofluorescence procedures (Jones *et al.*, 2000). Other staining applications include bacterial enumeration, reviewed by Kepner and Pratt (1994), and chromosome staining, discussed by Kapuscinski (1995). Bacterial counting can also be carried out in hydrophilic resin sections (Decho and Kawaguchi, 1999).
The dye is used as a fluorescent probe for living cells and tissues, as in investi-gations of the penetrability of gap junctions (Elfgang *et al.*, 1995), and to stain bacterial nucleoids (Fishov and Woldringh, 1999). DAPI has been used as a retrograde axonal tracer, as in a study of the innervation of the developing testis (Hrabovszky *et al.*, 2001).
Perhaps DAPI's most common application is as a flow cytometry stain. The dye is recommended in manuals as a nuclear stain of isolated cells, and for DNA analysis of material from fixed tissue blocks (Darzynkiewicz and Crissman,

1990; Macey, 1994). DAPI has been used for the fluorescent assay of DNA in solution, as discussed in Kapuscinski's review (1995); and for the staining of DNA in electrophoresis gels (Buel and Schwartz, 1993).

Purity of commercial batches: Dye contents up to 98% are available. There is a thin-layer chromatographic method (Gluth *et al.*, 1986)

Stability: Photofades in solution

Hazards: A potential mutagen

Hoechst 33342

Synonyms: bisbenzimidazole 342, Hoechst 342; sometimes termed bisbenzimide, but not to be confused with Hoechst 33258

Trihydrochloride trihydrate: $C_{27}H_{37}Cl_3N_6O_4$, FW 616

Absorption maximum: 350 nm in aqueous DNA, in which solvent the emission maximum is 461 nm

Solubilities: soluble in water and dimethylformamide

Hoechst 33342 comprises two linked benzimidazole units, with an alicyclic amine at one end and a phenyl ring carrying an ethoxy substituent at the other. The overall size, and that of the conjugated system, is moderate. The mono-cation is lipophilic, and the di- and trications are hydrophilic. The dye is sold as the trihydrochloride, illustrated.

Applications: The dye is widely used as a DNA, and thus nuclear, stain as reviewed by Crissman and Hirons (1994). Some particular applications to fixed tissues include chromosome banding in spreads, reviewed by Schweizer and Ambros (1994), as a nuclear counterstain for *in situ* hybridization stains (Breininger and Baskin, 2000), and as a component of BrdU cell cycle investigations (Tang *et al.*, 2000).

With living cells, applications include detection of apoptosis (Doostzaheh-Cizeron *et al.*, 2000) and as a counterstain for *in vivo* immunostaining (Weisbart *et al.*, 2000). Hoechst 33342 is also applied to live cells as a label, for instance, to sperm, so that subsequent adhesion to oviduct epithelial cells can be detected (Thomas and Ball, 1996); and to lymphocytes, enabling developmental migrations to be traced (Aboussaouira *et al.*, 1998).

Hoechst 33342 has been recommended as a DNA stain in flow cytometry. Related applications are cell cycle analysis and as a perfusion probe of cell spheroids (Darzynkiewicz and Crissman, 1990). The dye can also be used as a flow cytometric viability stain (Ellwart and Dormer, 1990).

Purity of commercial batches: Material with dye content up to 98% is sold. There is a HPLC method available (Harapanhalli *et al.*, 1994).
Stability: Photofades in solution
Hazards: Possible mutagen

Nuclear yellow (Hoechst S769121) is a structurally related dye in which the weakly lipophilic ethoxy substituent is replaced by a hydrophilic sulphonamide. This dye was introduced to biology as an axonal tracer, and is still used for this purpose (Katoh *et al.*, 2000). It has also been used for chromosome banding (Pinna-Senn *et al.*, 2000).

Bisbenzimide

Synonym: Hoechst 33258
Trihydrochloride pentahydrate: $C_{25}H_{37}Cl_3N_6O_6$, FW 624

Absorption maxima: 352 nm in aqueous DNA solution, 343 nm in methanol; the emission maximum is 461 nm in aqueous DNA
Solubilities: 2% in water, also soluble in dimethylformamide

Bisbenzimide comprises two linked benzimidazole units, with an alicyclic amine at one end and a phenyl ring carrying a phenolic substituent at the other. The overall size, and that of the conjugated system, is moderate. The monocation is weakly lipophilic, but the di- and trications are hydrophilic. The dye is sold as the trihydrochloride, illustrated.
Applications: Bisbenzimide is widely used for fluorescent nuclear staining, as recommended in a review of DNA staining in cell biology (Crissman and Hirons, 1994). More specifically bisbenzimide is used as a nuclear counterstain after autoradiography (Schnell and Wessendorf, 1995), immunostaining (Lecuit and Wieschaus, 2000), and *in situ* hybridization (Estil *et al.*, 2000). Related techniques include chromosome banding, as recommended in Ruzin's (1999) manual on botanical histochemistry and more generally by Sumner (1994), and for the detection of apoptosis (Landriscina *et al.*, 2000).

Bisbenzimide is recommended as a stain for DNA and nuclei in living cells in general (Crissman and Hirons, 1994), and in particular for yeasts (Pringle *et al.*, 1989). This provides a means to detect contamination of cell cultures by mycoplasma (Battaglia *et al.*, 1994). The dye is also recommended in flow cytometry manuals for cell division and cell cycle analysis, chromosome sorting, and *in situ* hybridization counterstaining (Darzynkiewicz and Crissman, 1990).

Purity of commercial batches: Dye contents up to 98% are available. There is a HPLC method for bisbenzimide (Harapanhalli *et al.*, 1994)
Stability: Photofades in solution
Hazards: Potential mutagen

INDOLENINE DYE FORMING COVALENT BONDS WITH CO-SOLUTES

Indo-1

The acetoxymethyl ester derivative Indo-1 AM is also described in this entry
Synonym: Indo-1 AM is often misleadingly referred to as Indo-1.

Indo-1   **Indo-1 AM ester**

Potassium salt: $C_{32}H_{26}K_5N_3O_{12}$, $C_{47}H_{51}N_3O_{22}$, FW 1010
FW 840

Absorption maxima

346 nm in water, 330 nm in   356 nm in methanol, in which
aqueous calcium salts; in   which solvent the
which solvents the emission   maxima are 475 and 401 nm
maximum is 478 nm

Solubilities

Highly soluble in water   Soluble in dimethylsulfoxide
above pH 6   and methanol

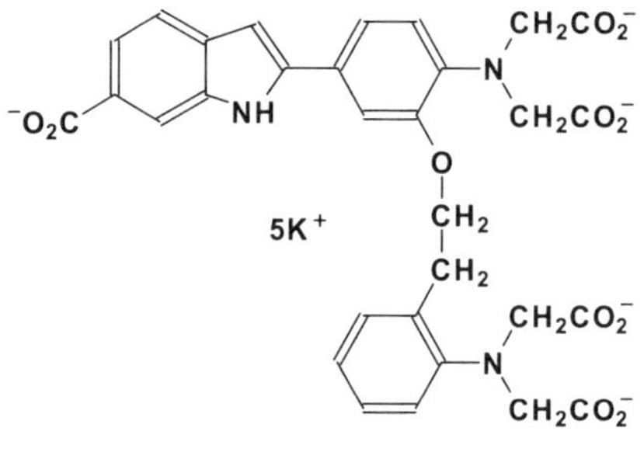

Indo-1 has two aromatic domains, one an indole, the other a phenyl ring, separated by a spacer. Calcium ion chelating carboxylic substituents are attached to both aromatic regions. In indo-1 AM all carboxyl groups have been transformed into acetoxymethyl esters. The extremely hydrophilic, penta-anion of indo-1 is of moderate overall size, with a small conjugated system; the nonionic indo-1 AM ester is superlipophilic and large. The compound is widely sold and used both as the pentapotassium salt, illustrated, and the ester.
Application: Indo-1 finds wide application as a fluorescent probe for calcium in live cells. Imaging calcium in this way has been discussed repeatedly in general

346

reviews (Brownlee, 2000; Takahashi *et al.*, 1999), and also in reviews of more specific topics, such as the investigation of the relationship between cytosolic and nuclear calcium (Himpens *et al.*, 1994). For such applications both the salt and the AM ester are used. Indo-1 is a membrane impermeable species which gains access to cellular interiors only following a permeabilizing maneuver, such as electroporation or microinjection or, in the case of plant cells, acidification of the medium to form the less hydrophilic free acid. Indo-1 AM, or perhaps some partially de-esterified derivative, gains entry by passive diffusion. Indo-1 AM is also used as an intracellular calcium stain in flow cytometry as discussed in a review by June and Rabinovitch (1994) and a manual by Macey (1994).

Purity of commercial batches: Dye contents of typical samples of indo-1 are up to 97%, and of indo-1 AM up to 95%

Stability: Both salt and AM ester photofade in solution

AZAMETHINES

This subgroup includes compounds traditionally known as indamine and indophenol dyes.

Dichloroindophenol

Synonym: sodium 2,6-dichloroindophenolate hydrate
Sodium salt: $C_{12}H_6Cl_2NNaO_2 + XH_2O$, FW 290 + XH_2O

Absorption maximum: 605 nm in aqueous alkali
A pH and redox indicator: when oxidized is red in acid and blue in alkali; colorless when reduced
Solubilities: 3% in water; 2% in ethanol

Dichloroindophenol is an azamethine, with the phenyl rings carrying the donor/acceptor oxygens bridged by a single aza group. On reduction this becomes a colorless diphenylamine derivative, as illustrated. The red free-acid species is lipophilic, the blue anionic species is weakly hydrophilic. The dye is sold as the sodium salt hydrate.

Applications: Dichloroindophenol has various applications as an electron acceptor. Its application to enzyme histochemistry was discussed by Pearse (1972, p. 921) and the dye is occasionally applied in this way, for instance to demonstrate NADPH diaphorase (Wehby and Frank, 1999). Dichloroindophenol is a routine reagent for assay of vitamin C (Paim and Reis, 2000). It has been used to stain glutathione reductase in electrophoretic gels (Ye *et al.*, 1997), and for biochemical enzyme assays (Sakuraba *et al.*, 2001).

Purity of commercial batches: Dye contents of typical samples are up to 88%.

Stability: Dry powder is hygroscopic

Related dye: Other azamethine dyes have occasionally been applied in biology and medicine. An example is **indophenol blue** (CI 49700), illustrated below, which is a small, lipophilic, nonionic dye. It was once used by Herxheimer early in the 20th century as a fat stain (Lillie, 1977), and was the detected product in early biochemical and histochemical studies of cytochrome oxidase (Kiernan, 1999).

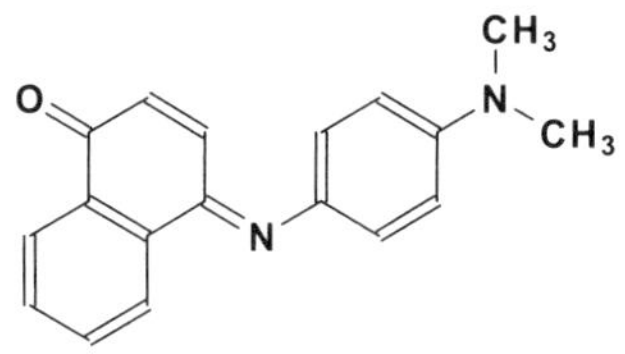

24

Polymethine dyes – 2. Styryls, thiazoles, coumarins and flavonoids

R.W. Horobin

Here are more polymethine dyes, for which the introduction to Chapter 23 provides background notes on chemistry and nomenclature. Comments on the specific dye classes described in this chapter follow.

Structural features

Styryl dyes are rather loosely defined. Some contain an obvious styryl group with two methine units attached to a phenyl ring (e.g. DiA); others, however, have longer methine chains (e.g. FM 4-64) or a naphthyl ring (e.g. Di-4-ANEPPS). All are hemicyanines with one acceptor/donor nitrogen being in a pyridine heterocycle and the other being an amino substituent on the phenyl/naphthyl ring.

    Thiazole dyes contain one or more benzothiazole units conjugated with various acceptor/donor moieties. Thiazole orange has an *N*-methylated thiazole ring with the other nitrogen being in a quinoline system, hence it is a cyanine. The thiazole ring nitrogen is also methylated in thioflavine T, but here the other nitrogen is an aminophenyl, making this a hemicyanine. In primuline and thioflavine S neither nitrogen atom of the acceptor/donor pair carries a positive charge, and these dyes are neutrocyanines.

    Coumarin dyes have benzo-2-pyrone fluorophores containing carbonyl groups as electron acceptors. Most coumarins used histochemically have electron donor 7-amino or 7-hydroxy groups, and so are neutrocyanines.

    Flavonoids are derivatives of benzo-4-pyrone (i.e. flavone), and the acceptor/donor pairs of dyes discussed here are hydroxy/carbonyl; hence they are neutrocyanines. The colorless precursors of two flavonoids, brazilin and hematoxylin, are also described.

Complications of nomenclature

Systematic dye names are not used by biologists, although some terms do contain chemical hints. Thus ANEPPS stands for amino-naphthyl-ethene-pyridiniumsulfonate, and the number in a name such as Di-8-ANNEPS indicates the lengths of the carbon chains attached to the terminal amino group. Other names lack any chemical information. The FM styryl dyes, for instance, correspond to page numbers in the laboratory notebook of Fei Mao, who first synthesized them. Origins also inspired hematoxylin and brazilin, reflecting the names of the trees from which these compounds are extracted.

Properties influencing staining

Electric charges of different chromophores differ, as previously described. Overall charge depends also on substitution patterns. Thus DiA is a cationic hemicyanine but Di-8-ANEPPS, although a hemicyanine, is monosulfonated and hence a zwitterion.

Hydrophilic/lipophilic character varies widely, also largely due to the substituents carried. DiA with two hexydecyl chains is superlipophilic, whereas FM 4-64 is dicationic and hydrophilic. In the biological literature it is often suggested that styryl dyes have distinct lipophilic and hydrophilic domains, and hence are strongly amphiphilic. Although this view oversimplifies, by ignoring charge delocalization, some dyes (e.g. Di-8-ANEPPS) are amphiphilic, if not as markedly as suggested.

The *extent of conjugation* varies from small (e.g. DASPI) to large (e.g. the tribenzothiazole component of primulin).

The *overall sizes* also vary widely from hematoxylin (FW 302) to DiA (FW 787).

Reactivity – Most coumarin dyes are used as reactive labels of peptides and proteins. The reactive groupings vary, and include carboxylic acids, reactive esters, and maleimides.

Purity – Industrial thiazole syntheses typically generate mixtures of dyes.

Safety issues

Some of the coumarins described here react with proteins; they may give rise to hypersensitivity responses, although there are currently no reports of such.

How to find the dye that interests you

To avoid confusion traditional dye class names are used. Within the sections describing styryl, thiazole and coumarin dyes the usual sequence of nonionized dyes, anions, zwitterions, cations, and reactive dyes is followed where appropriate. Other traditional dyes – brazilin, hematoxylin and morin – are grouped as together as flavonoids.

ZWITTERIONIC STYRYL DYES

Di-4-ANEPPS

Synonyms: 4-(2-(6-(dibutylamino)-2-naphthalenyl)ethenyl)-1-(3-sulfopropyl)-pyridinium, inner salt; JPW-211
$C_{28}H_{36}N_2O_3S$, FW 481

Absorption maxima: 480 nm in neutral aqueous solution, 496 nm in methanol, and 465 nm in biomembranes; emission maxima are 750 nm, 705 nm and 665 nm respectively.
This dye is a potentiometric indicator.
Solubilities: soluble in dimethylsulfoxide, ethanol and methanol

Di-4-ANEPPS is a hemicyanine whose pyridine and naphthyl rings, respectively containing and carrying the acceptor/donor nitrogens, are bridged by a two methine chain. The *N*-propyl group on the pyridine has a terminal sulfonate; the naphthylamine carries two *N*-butyl groups. Overall size is moderate and that of the conjugated system small. The weakly lipophilic and slightly amphiphilic zwitterion is available commercially.
Applications: Di-4-ANEPPS is used as a fluorescent probe for microscopic observation of living cells. Due to its amphiphilicity the dye is taken from solution into plasma membranes; following which its potential-dependent fluorescence permits determination of membrane potential. Such applications have been repeatedly discussed in reviews (Loew, 1993; Zochowski *et al.*, 2000). Applications to other types of specimen include assessment of membrane potential in endothelia of intact capillaries (Beach *et al.*, 1996), and detection of neural activity in brain slices (Tominaga *et al.*, 2000). Since di-4-ANEPPS is not highly amphiphilic and is slightly lipophilic errors can arise due to dye internalization (Chaloupka *et al.*, 1997). This does, however, allow the dye to be used to stain biomembranes within cells (Campagnola *et al.*, 1999).
Purity of commercial batches: There is a thin-layer chromatographic method (Hassner *et al.*, 1984)
Stability: Photofades in solution

Di-8-ANEPPS

Synonyms: 4-[2-[6-(dioctylamino)-2-naphthalenyl]ethenyl]-1-(3-sulfopropyl)-pyridinium, inner salt; JPW-1153
$C_{36}H_{52}N_2O_3S$, FW 593

Absorption maxima: 455 nm in neutral aqueous solution, 498 nm in methanol, and 471 nm in biomembranes, in which media the emission maxima are 700 nm, 713 nm and 627 nm respectively

This dye is a potentiometric indicator
Solubilities: soluble in dimethylsulfoxide, ethanol and methanol

Di-8-ANEPPS is a hemicyanine whose pyridine and naphthyl rings, respectively containing and carrying the acceptor/donor nitrogens, are bridged by a two methine chain. The N-propyl group on the pyridine has a terminal sulfonate; the naphthylamine carries two N-octyl groups. Overall size is moderate and that of the conjugated system small. The lipophilic and strongly amphiphilic zwitterion is available commercially.

Applications: The amphiphilic nature of this dye permits accumulation in biomembranes from aqueous solution. Consequently Di-8-ANEPPS is used to stain T-tubule membranes in skeletal muscle (Kim and Vergara, 1998), and the plasma membranes of dissociated adrenal medullary cells (Inoue *et al.*, 2000). The dye has been applied for axonal tracing (Tsau *et al.*, 1996), and general staining of biomembranes (Conklin *et al.*, 2000).

Fluorescence of Di-8-ANEPPS is voltage dependent, and membrane-bound dye can report on membrane potentials. This has been done in a range of cells, for example, cultured neurons (Bedlack *et al.*, 1994), and cardiac muscle cells (Cheng *et al.*, 1999). The procedure has been applied to other types of specimen, for example, endothelia of intact blood capillaries (McGahren *et al.*, 1998), and in patch-clamp preparations (Zhang *et al.*, 1998).

Stability: Photofades in solution

CATIONIC STYRYL

DASPI

Synonyms: DASPMI, 4-Di-1-ASP; 4-(4-(dimethylamino)styryl)-N-methylpyridinium iodide
$C_{16}H_{19}IN_2$, FW 366

Absorption maximum: 475 nm in methanol, in which solvent the emission maximum is 605 nm
Solubilities: Soluble in dimethylformamide and methanol

DASPI is a hemicyanine whose pyridine and phenyl rings, respectively containing and carrying the acceptor/donor nitrogens, are bridged by a two methine chain. The *N*-substituents are all methyl groups (i.e. $R_1 = R_2 = CH_3$ in the structural formula). Overall size, and that of the conjugated system, is small. The weakly hydrophilic cation is available commercially as the iodide.

Applications: The dye is used to demonstrate mitochondria in living cells, such as yeasts (Rinaldi *et al.*, 1998) and cultured neurons (Hoyt and Reynolds, 1996). This is surprising, because the cation is not lipophilic, and indeed DASPI has been found to be inferior in this application to its more lipophilic homologs (Irion *et al.*, 1993). Perhaps this characteristic underlies the use of DASPI for staining metabolically active mitochondria, in which dye uptake is driven by membrane potential, not by lipophilicity alone. DASPI has also been used as a tracer to assess the activity of membrane cation transporters (Hohage *et al.*, 1998).

Purity of commercial batches: There is a thin-layer chromatographic method (Irion *et al.*, 1993)

Stability: Photofades in solution

Related compounds: A structurally similar compound is **4-Di-2-ASP** (R_1 = methyl, R_2 = ethyl). This dye stains mitochondria and mitochondria-rich structures such as motor end plates in living cells (de Paiva *et al.*, 1999). Another dye with the same core structure is **4-Di-10-ASP** (synonym: **DiASP**), which is lipophilic and strongly amphiphilic (R_1 = methyl, R_2 = decyl). The resulting strong staining of plasma membranes makes the dye a useful axonal tracer as discussed in a review by Kobbert *et al.* (2000). Glial cells can be identified following their uptake of stained neuronal fragments, as discussed in a review by Thanos *et al.* (2000). **DASPEI** (synonym: **DMP+**) is structurally similar to DASPI, and has been used as a mito-chondrial probe in live cells (Hiroi *et al.*, 1999). Use of DASPEI for selective staining of cells in the lateral line of fish is probably also driven by locally high membrane potentials (Jones and Corwin, 1996).

RH 414

Synonym: N-(3-triethylammoniumpropyl)-4-(4-(4-(diethylamino)phenyl)buta-dienyl)pyridinium dibromide

$C_{28}H_{43}Br_2N_3$, FW 582

Absorption maximum: 532 nm in methanol, in which solvent the emission maximum is 716 nm

This dye is a potentiometric indicator

Solubilities: Moderately soluble in water; also soluble in dimethylsulfoxide, ethanol and methanol

RH 414 is a hemicyanine whose pyridine and phenyl rings, respectively containing and carrying the acceptor/donor nitrogens, are bridged by a four methine chain. The pyridine ring carries an *N*-propyl chain with a terminal triethylammonium group. Overall size, and that of the conjugated system, is small. The hydrophilic and moderately amphiphilic dication is sold as the bromide.

Applications: The amphiphilic character of RH 414 results in uptake into surface membranes of live cells. It is used to stain the plasmalemma and T-tubule membranes of skeletal muscle (Krolenko *et al.*, 1995). Probably because RH 414 is not highly amphiphilic, dye may be removed from plasma membranes by glycerol extraction. With longer exposures to dye, RH 414 serves as a marker of internalized membrane, as with recycled synaptosomes (Costanzo *et al.*, 1999), endocytosis and endosomes in general (Minshall *et al.*, 2000). Fluorescence of RH 414 is voltage-sensitive and is used to assess membrane potential, especially in neurons as discussed in the review by Cinelli and Kauer (1992). This methodology is also used to visualize neuronal activity across large areas of the brain, including the olfactory bulb (Lam *et al.*, 2000).

Stability: Photofades in solution

FM 4-64

Synonym: N-(3-triethylammoniumpropyl)-4-(6-(4-(diethylamino)phenyl)hexa-trienyl)pyridinium dibromide
$C_{30}H_{45}Br_2N_3$, FW 608

Absorption maxima: 558 nm in chloroform, 560 nm in methanol; in which solvents emission maxima are 734 nm and 767 nm respectively

Solubilities: Soluble in water and dimethylsulfoxide

FM 4-64 is a hemicyanine whose pyridine and phenyl rings, respectively containing and carrying the acceptor/donor nitrogens, are bridged by a six methine chain. The pyridine ring carries an *N*-propyl chain with a terminal triethylammonium group. Overall size is moderate, and that of the conjugated system small. The hydrophilic and amphiphilic dication is sold as the iodide.

Applications: Due to its amphiphilic character FM 4-64 is taken into the surface membranes of live cells. Consequently the dye is used to stain bacterial membranes (Fishov and Woldringh, 1999) and the plasmalemmas of cultured eukaryotic cells (Janecki *et al.*, 2000). Following extended staining times membrane and hence dye may be internalized. Consequently FM 4-64 is used as a marker of endocytosis and endosomes in various cell types, including fungi (Fischer-Parton *et al.*, 2000), and as a marker for reuptake of synaptosomes and hence a stain of such entities as nerve terminals (Herrera *et al.*, 2000) and neuronal processes in brain slices (Castejon and Sims, 1999). The same phenomena underlie use of FM 4-64 in flow cytometric cell-sorting, for example, to separate yeast mutants with membrane internalization defects (Wendland *et al.*, 1996).

Stability: Photofades in solution

FM 1-43

Synonym: N-(3-triethylammoniumpropyl)-4-(4-(dibutylamino)styryl)pyridinium dibromide

$C_{30}H_{49}Br_2N_3$, FW 612

Absorption maxima: 512 nm in methanol and 479 nm when bound to biomembranes; emission maxima are 626 nm and 598 nm respectively

Solubilities: Moderately soluble in water; also soluble in dimethylsulfoxide, ethanol and methanol

FM 1-43 is a hemicyanine whose pyridine and phenyl rings, respectively containing and carrying the acceptor/donor nitrogens, are bridged by a two methine chain. The pyridine ring carries an *N*-propyl chain with a terminal triethylammonium group; the phenylamine carries two *N*-butyl chains. Overall size is moderate, and that of the conjugated system small. The hydrophilic and amphiphilic dication is sold as the bromide.

Applications: The amphiphilic character of FM 1-43 results in uptake from solution into the surface membranes of live cells, hence its application as a plasmalemma probe of cultured cells (Mihai *et al.*, 2000). After extended staining times, membrane and hence dye may be internalized. If the resulting vesicles are subsequently exocytosed, as during secretion or release of neurotransmitters, dye will again be found extracellularly. In this way FM 1-43 is used as a marker of endocytosis and exocytosis in a wide variety of cell types, as has been reviewed many times (Cochilla *et al.*, 1999; Cousin and Robinson, 1999). FM 1-43 is also used to image synaptic activity in brain slices (Kay *et al.*, 1999). Other applications include detection of early events in apoptosis

(Kawasaki *et al.*, 2000), and staining of mitochondria (Fischer-Parton *et al.*, 2000; Nishikawa and Sasaki, 1996). Selective entry of the hydrophilic FM 1-43 into membrane-damaged cells probably underlies the former application. Accumulation in mitochondria is puzzling, however, and may be due to a lipophilic cationic impurity. FM 1-43 has also been used as a plasma membrane stain in flow cytometry (Bajno *et al.*, 2000).

Purity of commercial batches: A lipophilic cationic impurity is sometimes present

Stability: Photofades in solution

Related compound: **FM 2-10** is a structurally similar dye carrying two N-ethyl in place of N-butyl groups. It is used for imaging vesicle trafficking in live cells (Cousin and Nicholls, 1997). The slightly more hydrophilic character is useful when easier release of probe from the plasma membrane is required (Pyle *et al.*, 2000).

DiA

Synonyms: di-16-ASP; 4-(4-(dihexadecylamino)styryl)-N-methylpyridinium iodide $C_{46}H_{79}IN_2$, FW 787

Absorption maximum: 491 nm in methanol, in which solvent the emission maximum is 613 nm

Solubilities: Soluble in dimethylsulfoxide, ethanol and methanol

DiA is a hemicyanine whose pyridine and phenyl rings, respectively containing and carrying the acceptor / donor nitrogens, are bridged by a two methine chain. Overall size is substantial due to two N-hexadecyl chains, although the size of the conjugated system is small. The superlipophilic cation is available commercially as the iodide.

Applications: As a consequence of the superlipophilic character of DiA, living cells accumulate this dye in the plasma membrane. Staining applications include axonal tracing in live and in fixed postmortem specimens (Kobbert *et al.*, 2000; Vercelli *et al.*, 2000), and subsequent detection of glial cells by phagocytic labeling (Thanos *et al.*, 2000). The almost irreversible character of such dye uptake has been exploited in cell labeling for developmental studies (Tuvia *et al.*, 1997).

Stability: Photofades in solution

THIAZOLE DYES

Thiazole orange

Synonym: TO
Tosylate: $C_{25}H_{24}N_2O_3S_2$, FW 477

Absorption maximum: 512 nm in methanol, with an emission maximum around 533 nm
Solubilities: Soluble in methanol

Thiazole orange is a cyanine whose *N*-methylthiazole and quinoline rings, containing the acceptor/donor nitrogens, are bridged by a single methine group. Overall size, and that of the conjugated system, is small. The weakly hydrophilic cation is available commercially as the *p*-tosylate.
Applications: Thiazole orange is used as a nuclear counterstain following *in situ* hybridization of fixed cells (Kahn *et al.*, 1999) and as a fluorescent stain for RNA (Abramsson-Zetterberg *et al.*, 2000), and it is applied to living cells to stain cell nuclei (Bassnett, 1992). This dye is most widely used, however, in flow cytometry. The most common application is to identify reticulocytes, as described in manuals (Mundee *et al.*, 2001; Shapiro, 1988). The dye is also used to stain platelets (Fujii *et al.*, 2000).
Purity of commercial batches: Dye contents are up to 90%

Primuline

CI 49000, CI Direct yellow 59
Nominal major component: $C_{21}H_{14}N_3NaO_3$, FW 476

Absorption maximum: 351–356 nm in water, the emission maximum being 432–450 nm; the ranges are probably due to the heterogeneous nature of commercial primuline, discussed below
Solubilities: 0.25–2% in water, 0.03–1% in ethanol; again see below

The previous edition of this book, following the *Colour Index* (1971, volume 4, p. 4441), emphasized a species containing two conjugated benzothiazole units. Analytical work, however (Kelenyi, 1967), suggests that mono-, di- and tribenzo-thiazole species are all present in commercial products, as illustrated. These compounds are of moderate overall size, with conjugated systems varying from small to large. All are sulfonated anions, the smallest being hydrophilic and largest lipophilic, and the dye is sold as the sodium salt.

Applications: Primuline has been used as a stain of fixed tissues such as lignified cell walls, as described in the botanical staining manual of O'Brien and McCully (1981). More generally it has been applied to living cells, for example, as a viability stain of yeasts (Graham and Caiger, 1964), of starch in phytoplankton (Klut *et al.*, 1989), and to demonstrate yeast cell walls as discussed in the review of Pringle *et al.* (1989). This latter process may involve binding to chitosan (Slaninova *et al.*, 2000). Primuline has been used as a tracer of the movement of organic anions in the teleost kidney (Lavrova and Natochin, 1985), and as an axonal tracer (Kucheryavykh *et al.*, 1999).

Industrial use – primuline is a leather dye

Purity of commercial batches: A mixture of three components with much batch variation, see above. Overall dye contents are up to 75%. There is a paper chromatographic method (Kelenyi, 1967).

Stability: Light fastness is good

Related dye: The structure of **thioflavine S** (CI 49010, Direct yellow 7) is uncertain but commercial batches are probably mixtures of the components found in primulin together with their methylated derivatives. This dye is widely used as a fluorescent stain for amyloid, especially of plaques found in Alzheimer's disease, in paraffin sections (Bancroft and Gamble, 2002; Churukian, 2000; Westermark *et al.*, 1999). There is also a technique for thick frozen sections (Guntern *et al.*, 1992). Thioflavine S is also used to detect and label patent blood vessels by endothelial staining (Wu *et al.*, 1998).

Thioflavine T

CI 49005, Basic yellow 1

Nominal major component: $C_{17}H_{19}ClN_2S$, FW 319

Absorption maximum: 412 nm in water; with a silvery-white fluorescence

Solubilities: 2% in water; 1–3% in ethanol

Thioflavine T was described in the previous edition of this book, following the
Colour Index (1971, volume 4, p. 4441), as a cationic *N*-methyl monobenzothiazole, as illustrated. However, analytical work (Kelenyi, 1967; Stiller *et al.*, 1972)
suggests it is a mixture of that compound with di- and tribenzothiazole derivatives. The overall size, and that of the conjugated system, of the slightly
lipophilic monobenzothiazole cation is small.
Applications: The principal application is to detect amyloid, as described in
staining manuals (Bancroft and Cook, 1994; Bancroft and Gamble, 2002).
Variants include a method for demonstration of amyloid in cytological needle
aspirates (Halliday *et al.*, 1998), and use of flow cytometry to characterize
thioflavine T stained amyloid fibrils (Wall and Solomon, 1999).
   Industrial uses – colorant of ballpoint inks and, as the phosphomolybdate, a
pigment
Purity of commercial batches: One major and several minor components are
usually present. Dye content of typical samples is up to 75%. There is a paper
chromatographic method (Stiller *et al.*, 1972).

COUMARIN DYES

Calcein blue

Synonym: 4-methylumbelliferone-8-methyliminodiacetic acid. The acetoxymethyl
ester, calcein blue AM, is also described in this entry; unless stated otherwise information below refers to the free acid.
Free acid: $C_{15}H_{15}NO_7$, FW 321
AM ester: $C_{21}H_{23}NO_{11}$, FW 465

Absorption maxima: 360 nm in pH 9 aqueous solution, in which solvent the
emission maximum is 449 nm. In methanol the AM ester has an absorption
maximum at 322 nm and emits at 435 nm.
Fluorescent indicator for calcium and other metal ions; *pK value is* ca. 7.8
Solubilities: soluble in water above pH 8, and in methanol. The AM ester is
soluble in dimethyl sulfoxide and methanol.

In calcein blue the coumarin fluorophore carries a calcium-chelating methyl-iminodiacetic acid group. The overall size is small, as is the conjugated system. The free acid is lipophilic; all the anionic species are hydrophilic. The strongly lipophilic AM ester is of moderate size. Calcein blue is sold as the free acid, illustrated, and as the AM ester.

Applications: Due to its calcium-chelating character calcein blue has been used as a marker of bone growth, and to identify microcracks in bone (Lee *et al.*, 2000). Chelation of calcium also occurs in an enzyme histochemical application, to generate a fluorescent reaction product from calcium carbonate deposited in a Gomori-type method for alkaline phosphatase (Murray and Ewen, 1992). Calcein blue AM ester is hydrolyzed to calcein blue by nonspecific intracellular esterases. Hence the ester has been widely applied as a viability stain, both microscopically (Imbert and Cullander, 1999) and in fluorescence activated cell sorting systems (Krasnow *et al.*, 1991). Ionized calcein blue is hydrophilic and lost from cells only slowly, allowing use of the ester to mark living cells. An example is the long term labeling of Schwann cells at neuromuscular junctions (O'Malley *et al.*, 1999).

Purity of commercial batches: Dye content of the free acid is typically 90%

Stability: Photofades, especially in solution

DACM

Synonyms: DACM-3; *N*-(7-dimethylamino-4-methylcoumarin-3-yl)maleimide
$C_{16}H_{14}N_2O_4$, FW 298

Absorption maximum: 383 nm in methanol; in which solvent the emission maximum is 463 nm

Solubilities: Soluble in dimethylsulfoxide and methanol

DACM is a 7-aminocoumarin with a thiol-reactive maleimide substituent. Overall size of this lipophilic compound, like that of its conjugated system, is small.

Applications: DACM has been used as a quantitative stain for thiol groups (Maezawa *et al.*, 1997), and as a label of peptides for use as substrates for the imaging of protein kinase activity in living neural cells (Higashi *et al.*, 1997).

Purity of commercial batches: Dye contents up to 99% are available

Stability: Photofades, especially in solution

Hazards: The dye may react with human proteins, and hypersensitivity responses could occur, although such effects have not been reported

Related compounds: Other coumarin maleimides are occasionally used as stains. **CPM** (synonym: 7-diethylamino-3-(4-maleimidylphenyl)-4-methylcoumarin)

has been used to demonstrate thiol groups in nucleolar organizer region asso-
ciated-proteins (Mehes *et al.*, 1993). **MDCC** (synonym: 7-diethylamino-3-(2-
maleimidylethylaminocarbonyl)coumarin) is used as a reactive protein label, for
example, of phosphate binding protein for microscopic studies of intact muscle
fibers (He *et al.*, 2000).

AMCA

The term 'labeled with AMCA' when applied to labeled antibodies, peptides
and other biomolecules refers to labeling by a reactive coumarin approxi-
mating to aminomethylcoumarin acetic acid. Unfortunately several chemically
distinct coumarin labels are described in the literature as AMCA, and from the
information provided it is often impossible to know which label was used.
Consequently this entry has the nature of a general exposition, not a
prescriptive description.
Related compounds: That have been termed AMCA include *7-amino-4-methyl-3-
coumarin acetic acid N-succinimidyl ester* (illustrated), and various 7-dimethyl and
diethyl derivatives; also 7-amino compounds with spacers of various lengths
between the coumarin nucleus and the carboxyl ester; and 7-amino compounds
with the acetic acid ester at other positions on the coumarin ring. Indeed even
7-hydroxycoumarin-3-carboxylic acid succinimidyl ester has been termed AMCA.

The protein or polynucleotide reaction products of many of these compounds
emit in the blue
Solubilities: These compounds are often soluble in dimethylformamide and
dimethylsulfoxide, some are soluble in water

These coumarins all carry a carboxylic acid moiety, and are usually applied as a
reactive *N*-succinimidyl ester. This group forms covalent links with amino
groups on biomolecules.
Applications: While bearing in mind the ambiguities described above, some
applications of AMCA labeling follow. Labeled peptides are widely used, for
instance to identify macrophages by their uptake of a marker peptide (Otto and
Bauer, 1996), and to visualize oligopeptide transporters in living cells
(Groneberg *et al.*, 2001). Labeling of immunoglobulins is also carried out with
this reagent (Ray *et al.*, 2000). AMCA has also been used to label bromod-
eoxyuridine (Ffrench *et al.*, 1994); and tyramides for *in situ* hybridization
methods (Speel *et al.*, 1997).
Stability: Aqueous solutions of these reactive coumarins slowly decompose,
especially under alkaline conditions

Hazards: These compounds may react with human proteins; hypersensitivity responses could occur, although such effects have not been reported

FLAVONOID DYES

Hematoxylin and hematein

The active compound in staining applications is hematein, usually present as a metal complex. Preparation of these stains starts from hematoxylin, and the traditional names for the stains use that term. Consequently information is provided both for hematoxylin and hematein; notes on molecular structures and preparation of metal complexes follow the body of this entry; as does a note on the related compounds brazilin and brazilein.

CI 75290, CI Natural black 1

Hematoxylin	Hematein

Synonym

Haematoxylin	Haematein, hematine. Do not confuse with the porphyrin derivative hematin

Anhydrous: $C_{16}H_{14}O_6$, FW 302 Trihydrate: $C_{16}H_{14}O_6 \cdot 3H_2O$, FW 356	$C_{16}H_{12}O_6$, FW 300

Absorption maxima

292 nm in methanol	445 nm and 560 nm in acidic and alkaline aqueous solutions respectively

Metal complexation behavior

	colored complexes formed, e.g. with Al(III), Cr(III), Fe(III) and phosphotungstate

Solubilities

3% in water; 3% in ethanol	1.5% in water, 7.5% in ethanol

Acid–base indicator properties claimed for hematoxylin (color changes from red to yellow between pH 0–1, and from yellow to violet between pH 5–6) are due to traces of hematein. The pK value of hematein has been determined to be 6.7

Hematoxylin is a colorless flavone carrying several hydroxyl groups. After oxidation hematein, a colored product, is formed. The nonionized form of hematein is lipophilic, with an acceptor/donor pair comprising carbonyl and hydroxyl groups; its weakly hydrophilic anion being more intensely colored.

The compound forms chelates with various metal ions, see below. Further oxidation can occur, yielding nonchelating, pale yellow oxyhematein (Marshall and Horobin, 1972b). These reactions are illustrated. Hematoxylin is sold as the anhydrous form and as a hydrate; hematein is commercially available in the anhydrous form.

Applications: Hematein has been used as a staining reagent, and is currently recommended in human and animal (Bancroft and Cook, 1994; Kiernan, 1999) and plant (Ruzin, 1999) staining manuals for demonstration of phospholipids. Hematoxylin has also been used to stain glycogen in a Best's-carmine type procedure (Murgatroyd and Horobin, 1969); and to demonstrate iron in resin sections, as noted in Litwin's (1985) monograph.

However, hematein containing stains usually involve metal complexes. The most widely applied are known as aluminum, iron, and phosphotungstic acid hematoxylins; for notes on their chemistry and stain compositions, see below. To avoid confusion, these stains are here referred to in this traditional though chemically inaccurate way.

The name **hemalum** is applied to aluminum hematoxylin staining solutions. Variants are often identified by the inventors' names, e.g. Ehrlich's and Mayer's. These are probably the most widely used stains in biology and medicine. They are applied for the staining of cell nuclei in botanical histo-chemistry (Ruzin, 1999), histopathology (Bancroft and Gamble, 2002), general animal histology (Kiernan, 1999), and clinical cytology (Boon and Drijver, 1986); also for resin sections (Hayat, 1993) and in conjunction with *in situ* hybridization (Schwarzacher and Heslop-Harrison, 2000). Aluminum hema-toxylin is used in botanical histochemistry to stain cellulose cell walls (Berlyn and Miksche, 1976).

As recommended in staining manuals, **iron hematoxylins** are used, under names such as Heidenhain's, to stain such cellular detail as muscle striations and also in methods for myelin (Bancroft and Gamble, 2002; Kiernan, 1999). Other iron hematoxylin myelin staining methods include those of Loyez, Weigert-Pal and Weil (Bancroft and Cook, 1994). Verheoff's iron hematoxylin is used for the demonstration of elastin (Bancroft and Gamble, 2002; Churukian,

2000); and Weigert's for producing an acid-stable stain for nuclei, including use with water-miscible resin sections (Churukian, 2000).

The other hematein–metal complex with routine application is known as **phosphotungstic acid hematoxylin (PTAH)**. This is used to give one-bath poly-chrome staining of histological material, for instance to demonstrate fibrin and glial fibers (Bancroft and Gamble, 2002; Churukian, 2000).

Industrial uses – industrially hematoxylin/hematein was a major natural dyestuff. It is still used in this way for production of 'naturally dyed' textile products, and by craft dyers. Hematein is used for manufacture of colored surgical sutures.

Purity of commercial material: Hematoxylin samples are routinely contaminated with small (<1%) amounts of hematein, while material sold as hematein often contains substantial amounts (5–75%) of hematoxylin. Overall dye content of hematoxylin batches is up to 99%. There is a thin-layer chromatographic method (Marshall and Horobin, 1974b). Regarding Commission testing proce-dures see Chapter 28.

Stability: of dry hematoxylin in sealed, dark containers is good. Alcoholic solu-tions of hematoxylin are stable for months if kept in sealed containers, although aqueous solutions of hematoxylin readily oxidize to hematein, which itself oxidizes further. Light fastnesses of hematein metal complexes, especially iron chelates, are good.

Notes on the structures of hematein–metal complexes

The **aluminum complexes** are derived from hematein by reaction with Al(III) salts. Hematein is routinely produced from hematoxylin by prior, simultaneous or continuous oxidation. In some mixtures the oxidant is air (e.g. Ehrlich's); in others it is a reagent such as iodine or sodium iodate (e.g. Cole's and Mayer's respectively). The absolute concentrations and the ratios of metal ion to hematein vary markedly among different hemalum formulations.

Bettinger and Zimmerman (1991) found several colored complexes in hemalum solutions. These included a red 1:1 cationic hematein–aluminum complex present in strongly acidic conditions, violet 1:1 and 1:2 complexes formed under less acidic conditions, and a blue nonionic 1:1 complex in weakly alkaline solutions. Examples of red and blue isomers are illustrated. The absorption maxima in aqueous solution are 506 nm and 626 nm at pH 2.6 and 8.7 respectively.

low pH isomer, red

high pH isomer, blue

The interconvertibility of these compounds with change in pH probably causes 'blueing' of nuclei when specimens stained in a hemalum are washed in slightly alkaline water. The complexes are soluble in water and ethanol–water mixtures, but insoluble in ethanol. At pH 6–7 the aluminum–hematein complex exists as an insoluble, probably polymeric, dark blue substance.

Less detailed structural work has been carried out on other hematein–metal complexes. A cationic 1:1 structure, possibly an olation polymer, has been suggested for **iron hematein** (Arshid *et al.*, 1954; Shirai and Matsuoka, 1996). Giles' group also suggested an anionic or nonionic 2:1 olation polymer. This multiplicity of charge and size might underlie use of iron hematoxylins to stain both polyanion rich sites and proteins.

In **phosphotungstic acid–hematoxylin (PTAH)**, Terner *et al.* (1964) found a single anionic blue 2:1 hematein–tungsten species and suggested that the contrasting red and blue colors seen with PTAH staining should be attributed to metachromasia. Other analytical investigations found evidence for both blue and red complexes (Puchtler *et al.*, 1980).

Related dyes: **Brazilin** and **brazilein** correspond to hematoxylin and hematein respectively, with one less phenolic hydroxyl. Colored metal complexes may be prepared from brazilein. **Brazalum** (i.e. a brazilein–aluminum complex analogous to a hemalum) is recommended by Gatenby and Beams (1950) and Kiernan (1999) for red staining of nuclei in histological preparations.

Morin

CI 75660, CI Natural yellow 11
Synonym: 3,5,7,2',4'-pentahydroxyflavenol
Hydrate: $C_{15}H_{10}O_7$.aq, FW 302 + aq

Absorption maxima: of the neutral species is in the ultraviolet, giving greenish-white emission in the presence of metal ions; the absorption maximum of the morin–aluminum complex is at 430 nm, and the emission maximum at 495 nm
Metal complex behavior: Reacts with metal ions including Al(III), Be(II) and Ca(II)
Solubilities: Soluble in aqueous alkali and ethanol; barely soluble (0.03%) in acidic aqueous solutions

Morin is a flavonoid carrying several hydroxyl substituents able to chelate metal ions. Overall size, and that of its conjugated system, is small. The lipophilic neutral species, illustrated, is sold as the hydrate.
Application: Morin is recommended in staining manuals for demonstration of metal salts using fluorescence microscopy (Bancroft and Cook, 1994), a recent

research application being detection of aluminum in the brain (Platt *et al.*, 2001). The aluminum complex of morin, both preformed and generated *in situ*, has been used as a fluorescent nuclear stain (Del Castillo *et al.*, 1988).

Industrial use – none currently, though morin was used predindustrially and industrially as a textile dye

Purity of commercial batches: There is an HPLC method (Hsiu *et al.*, 2001)

Stability: Morin decomposes in aqueous solutions exposed to air, but its aluminum complex is stable. Light fastness of the aluminum complex is good.

25

Carbonyl dyes including indigoids, anthraquinones and naphthalimides

R.W. Horobin

BACKGROUND INFORMATION

Carbonyl dyes, notably alizarin, indigo and carmine, were used as colorants for centuries in many countries. For historical and anthropological accounts see the monographs *Indigo* (Balfour-Paul, 1998) and *Madder Red* (Chenciner, 2000); the former is an exceptionally beautifully illustrated book.

Structures and nomenclature

Dyes described in this chapter contain two carbonyl groups linked by a chain of sp^2-hybridized carbon atoms. This generalized conjugated system is shown below. In this figure residues **R** and **R′**, which may or may not be identical, can be linked together, as in the anthraquinones, or be separate as in indigoid dyes.

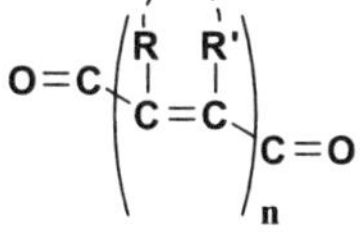

$$O=C\left(\begin{matrix} R & R' \\ | & | \\ C=C \end{matrix}\right)_n C=O \qquad n = 1, 2 ...$$

The *indigoid* chromophore contains a double-bonded carbon–carbon unit substituted by two carbonyl donor groups and two acceptor NH groups. Due to its geometry this system is termed a cross-conjugated or H-chromophore. The *anthraquinone* chromophore contains a central quinone ring, in which the two carbonyl groups are linked by two separate pairs of sp^2-carbons, each pair being part of an aromatic ring. The *naphthalimide* chromophore comprises a naphthalene ring system perisubstituted by a -CO-NH-CO- unit. The carbonyl groups in this are connected both through the nitrogen atom and a chain of three sp^2-hybridized carbons.

367

Properties influencing staining

Electric charge – This depends on the substituents, because the carbonyl dye chromophore is nonionic. Compounds described in this chapter include neutral species such as some indoxyl esters, and anionic dyes such as indigocarmine. Some metal complex derivatives, such as aluminum–carminic acid, are cationic.

Acid–base properties – Many dyes carry multiple acidic substituents, with varying pK values. For instance, alizarin red S has one strongly acidic sulfonate group and two aromatic hydroxyls with pK values of 4.5 and 11.

Hydrophilic/lipophilic properties – vary from extremely hydrophilic, such as Lucifer yellow CH, to the strongly lipophilic indigo dyes produced in enzyme histochemical procedures from indoxyl precursors.

Extent of conjugation – This is rather small in all the dyes.

Reactivity – The polycarboxy moiety of alizarin complexone chelates with calcium ions. Lucifer yellow CH & VS, and the procion and remazol reactive dyes, bind to proteins via various electrophilic groups.

Redox properties – are not relevant to biological or medical uses of carbonyl dyes as such, but in some histochemical methods oxidative coupling of enzymatically generated indoxyls results in the formation of indigoid dyes.

Dye sizes – Range from small (e.g. the indoxyl esters used in enzyme histochemistry) to moderate (e.g. carminic acid). Dye-labeled materials, such as Lucifer yellow–dextrin, are of course much larger.

Safety issues – The protein-reactive dyes may give rise to hypersensitivity responses.

How to find the dye that interests you

Traditional dye classes indigoid, anthraquinone and naphthalimide are used. Within these sections the sequence nonionic, anionic and reactive is followed as appropriate. Within each subgroup dyes are listed in order of increasing formula weight.

INDIGOID DYES AND STAINS

Indigocarmine

CI 73015, CI Acid blue 74, Food blue 1
$C_{16}H_8N_2Na_2O_8S_2$, FW 466

Absorption maximum: 608 nm in water. Reduction yields a fluorescent leuco derivative (Stockert & Trigoso, 1994)
Indigocarmine changes color from blue to yellow between pH 11.5–14
Solubilities: 1.5% in water, very slightly soluble (0.1%) in ethanol

The cross-conjugated indigo chromophore is described above. Indigocarmine contains two benzene rings fused to the central chromophoric unit, each ring carrying a sulfonate group. The resulting hydrophilic dianion is of moderate overall size, with a small conjugated system. Indigocarmine is sold as the sodium salt.

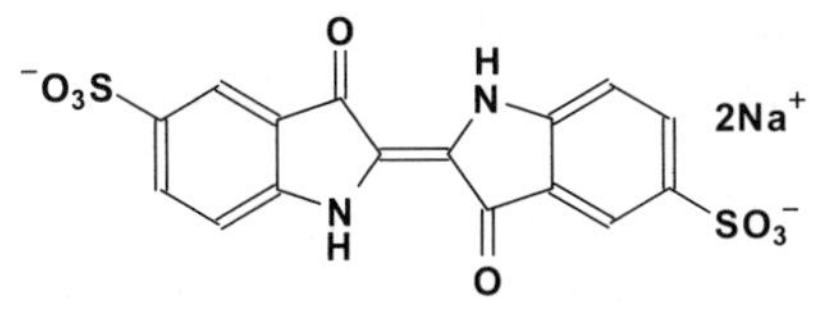

Applications: As a histological stain, this dye is used to stain collagen from a mixture with picric acid (Gabe, 1976; Kiernan, 1999). Other microscopic applications include differential staining of stages of the cell cycle in plant cell nuclei (Swain and De, 1990), and fluorescent staining of eosinophil granules (Stockert and Trigoso, 1994). Clinically, indigocarmine is used as an *in vivo* stain to assist endoscopic diagnosis, for example, of gastric cancer (Fukuzawa *et al.*, 1999); also for detection of sentinel lymph nodes (Imoto and Hasebe, 1999), and as an angiographic tracer (Baba *et al.*, 2000).

Industrial uses – widely used as colorant of foods, some use as a textile dye

Purity of commercial batches: A minor contaminant is usually present. Dye contents of typical samples are up to 85%; the minimum dye content for Commission certification is 80%. There are HPLC and thin-layer chromatographic methods (Marmion, 1991). Regarding Commission testing procedures see Chapter 28.

Stability: of dry powder stored in sealed, dark containers is good; picro-indigocarmine staining solutions keep for 2–3 years in dark bottles; light fastness is poor

Indoxyl esters and glycosides

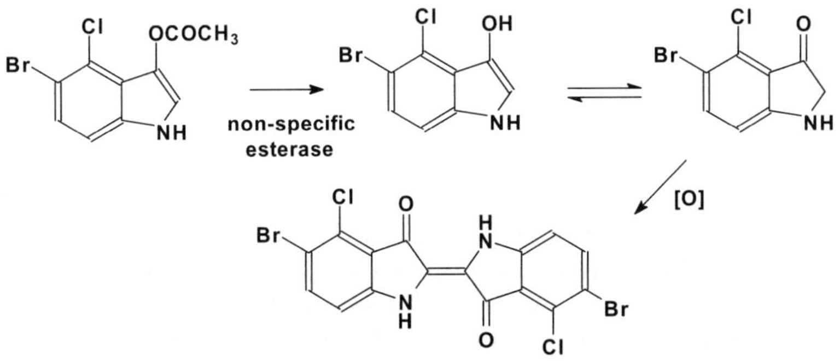

Indigo dyes are generated when colorless indoxyls are exposed to air or an oxidant such as a ferrrocyanide/ferricyanide mixture, as illustrated below. This color reaction is exploited in enzyme histochemistry by incubating specimens in indoxyl esters and amides which yield indoxyls in the presence of esterases and proteases respectively; see the example below. If such reactions are carried out under alkaline conditions oxidative condensation does not occur, but visualization of indoxyls can be achieved by reduction of a tetrazolium salt, yielding two colored species, a formazan and an indigo. *Note:* a widely used synonym for indoxyl is indolyl.

Applications: Histochemical staining manuals recommend such indigogenic methods for nonspecific esterases (Bancroft and Gamble, 2002; Kiernan, 1999), glycosidases (Van Noorden and Frederiks, 1992; also reviewed by Gossrau, 1990), and phosphatases (Van Noorden and Frederiks, 1992). These methods may be used with water-miscible resin sections, as described in reviews (Litwin, 1985) and in the research literature (Frederiks and Bosch, 1993).

Most indoxyl substrates carry substituents such as bromo and chloro groups; representative examples are shown below. Dyes produced from these are particularly insoluble, giving rise to well-localized pigment deposits.

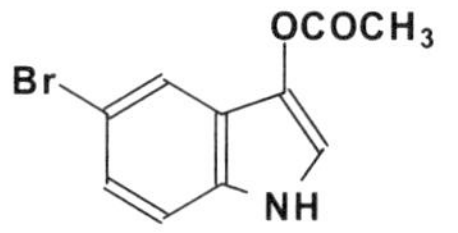

5-bromo-3-indoxyl acetate

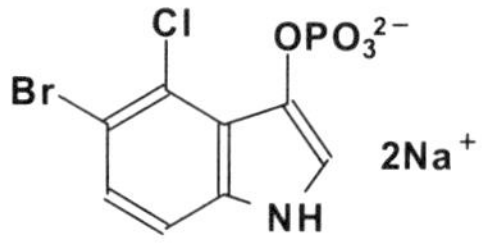

5-bromo-4-chloro-3-indoxyl phosphate

5-bromo-4-chloro-3-indoxyl-β-D-galactopyranoside

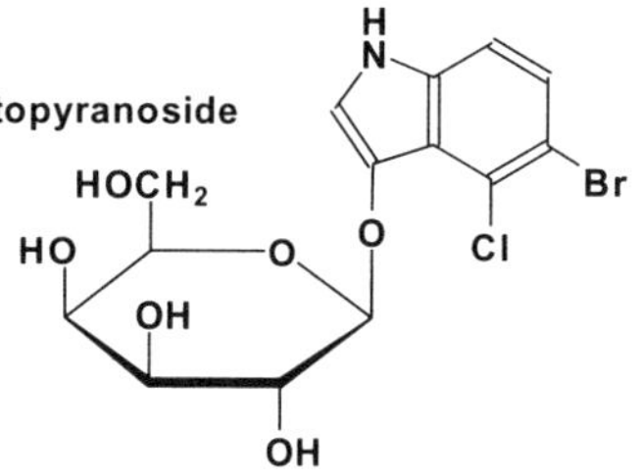

Indigogenic enzyme histochemical methods are also used in microbiology for staining bacterial colonies on solid culture media. Examples include detection of *Listeria monocytogenes* by phosphatase demonstration (Restaino *et al.*, 1999) and of transfected yeasts expressing acetylcholinesterase activity (Villatte *et al.*, 2001). The methods are also applied to visualize enzyme-labeled antibodies (Zeindl-Eberhart *et al.*, 1997) and polynucleotides used for *in situ* hybridization (O'Sullivan *et al.*, 2000).

ANIONIC ANTHRAQUINONE DYES

Alizarin red S

CI 58005, CI Mordant red 3
Monohydrate: $C_{14}H_7NaO_7S.H_2O$, FW 360

Absorption maxima: 556 nm and 596 nm in 0.1M aqueous sodium hydroxide Changes from yellow to red in the range 3.5–6.5, and from orange to violet at pH 9.4–12. Considerable batch variation occurs with respect to colors and pH ranges. The dye has been used as an indicator for aluminum and boron

pK values: 4.5, 11
Metal complexation behavior: forms complexes with a variety of metal ions, including aluminum and calcium
Solubilities: 7.7% in water; 0.2% in ethanol

The hydroxy substituted anthraquinone chromophore can chelate metal ions; a sulfonate substituent is also present. Overall size of the hydrophilic anion is small, as is the conjugated system. The dye is sold as the monohydrated sodium salt.

Applications: The ability to form metal chelates probably underlies use of alizarin red S to demonstrate calcium ions. Manuals and monographs recommend the dye for staining calcified tissue in routine sections (Bancroft and Gamble, 2002; Kiernan, 1999), including application to water miscible resin sections (Litwin, 1985). Such methods may be microwave accelerated (Churukian, 2000). Another widely applied variant is staining of calcified bones in wholemount vertebrate preparations, as reviewed by Klymkowsky and Hanken (1991); which can also be microwave accelerated (Ilgaz *et al.*, 1999).

Other applications of alizarin red S in microscopy include counterstaining donated corneas after vital staining with trypan blue (Mindrup *et al.*,1999), and detection of barium and lead deposits in gunshot wounds (Brown *et al.*, 1999). Macroscopically this dye is used to stain calcium-containing minerals in drill cores (Hitzman, 1999).

Alizarin red S is also used for calcium staining in living creatures. The dye is used as a growth marker in specimens as varied as dogs' teeth (Kagayama *et al.*, 2000) and corals (Rinkevich, 2000); and to mark otoliths in fish using whole-body immersion in staining solutions (Lagardere *et al.*, 2000).

Industrial uses – a textile dye

Purity of commercial batches: A single colored compound is often present, and dye contents are up to 75%. There is a thin-layer chromatographic method (King and Pruden, 1968). Regarding Commission testing methods see Chapter 28.

Stability: Staining solutions may be used for several weeks or longer; light fastness is moderate

Nuclear fast red

CI 60760
Synonym: kernechtrot (strictly: kernechtröt). Do not confuse with neutral red (CI 50040) which is sometimes described as nuclear fast red; or indeed with various histochemically applied diazonium salts named fast red.
$C_{14}H_8NNaO_7S$, FW 357

Absorption maxima: 535 nm and 505 nm in water; if the aluminum complex is excited at 540 nm it emits at 565 nm in water
Metal complexation behavior: chelates various metal ions, including aluminum and calcium
Solubilities: 0.6% in water; scarcely soluble (0.04%) in ethanol

The hydroxy substituted anthraquinone chromophore can chelate metal ions; there are also amino and sulfonate substituents. The overall size of the hydrophilic anion is small, as is that of its conjugated system. The dye is sold as the sodium salt.
Applications: The aluminum complex of nuclear fast red is used to demonstrate cell nuclei, as described in staining manuals (Kiernan, 1999; Churukian, 2000). Recent applications of this type have been as varied as staining sperm and spermatids in semen (Hendin *et al.*, 1998), and identifying areas for PCR analysis in paraffin embedded tissues (Burton *et al.*, 1998). As a nuclear counterstain it has been used following immunostaining (Larouche and Schiffrin, 1999), and silver stains (Churukian *et al.*, 2000). Nuclear staining can also be carried out on epoxy resin sections (Del Castillo *et al.*, 1990).
Purity of commercial batches: Dye contents are up to 96%
Stability: A typical aluminum–nuclear fast red staining solution is stable for several months

Carminic acid, carmine and cochineal

The terms carminic acid, carmine and cochineal have been used interchangeably though inaccurately, see below. Unless otherwise stated, physicochemical data are for carminic acid.

Cochineal describes the dried bodies of female *Coccus cacti*, an insect living on cacti of the prickly pear family. An aqueous extract of cochineal was once used by textile dyers. An alcoholic extract of defatted, dried insects (tincture of cochineal) yields **carminic acid** on evaporation, and this was probably the material intended in early stain formulations.

While the chemical nature of carminic acid is known, see below, **carmine** is a product of variable and ill-defined composition. The major colored species is an aluminum–carminic acid complex, but excess aluminum and calcium salts are often present together with protein and sometimes silica. The nondye substances are added during manufacture, as precipitants and purification agents. Although the chemically defined carminic acid is the active staining species, stains are often prepared from the cheaper carmine. This is perhaps why staining protocols often call for an initial treatment with acid or alkali, because such treatments release carminic acid from carmine.

CI 75470, CI Natural red 4
$C_{22}H_{20}O_{13}$, FW 492

Absorption maxima: 490 nm in 0.02 M aqueous hydrochloric acid, 495 nm in methanol. In the aqueous medium the emission maximum is 590 nm
Changes from yellow to purple at pH 4.8–6.2. *pK values:* 2.9, 5.5 and 8.2
Metal complexation behavior: Reacts with metal ions such as aluminum and iron
Solubilities: 0.4% in water; 0.2% in ethanol

The anthraquinone chromophore of carminic acid carries carboxyl, hydroxyl and *C*-glucopyranosyl substituents; so the overall size is moderate, although the conjugated system is small. The dye is sold as the hydrophilic free acid, probably the hydrate.

The presence of chelating groups on the anthraquinone chromophore permits formation of various metal complexes. Both 1:1 and 2:1 carminic acid–aluminum compounds have been proposed (Meloan *et al.*, 1971; Stapelfeldt *et al.*, 1993). As well as such red complexes, carminic acid forms bluer staining compounds with iron. Although the precise compositions and constitutions of 'carmine' stains are uncertain, such metal complexes may be present in most, except those using strongly alkaline solutions. Indeed this is the case even if no metal ions are added intentionally. Carmine itself contains aluminum ions, and formation of iron–carminic acid complexes may occur even in nominally iron-free aceto-carmine stains if steel needles are used to manipulate specimens (Gatenby and Beams, 1950, p. 129).

Applications: Carminic acid, as such or generated by alkali from carmine, is used in procedures to demonstrate glycogen, for example, the eponymous Best's carmine as described in staining manuals (Bancroft and Gamble, 2002; Churukian, 2000).

A variety of applications of carminic acid–aluminum complexes are described in staining manuals, see below. These methods may use carminic acid plus an aluminum salt, or cochineal plus a salt, or carmine with or without an added aluminum salt. Such **alum-carmines** or **carmalums**, typically containing a large excess of aluminum salt, are red nuclear stains particularly effective with whole-mounts because overstaining does not easily occur (Bancroft and Gamble, 2002; Kiernan, 1999). Red staining of mucins can be obtained with **mucicarmine**, which has a lower aluminum:carminic acid ratio (Bancroft and Cook, 1994; Lillie and Fullmer, 1976). Staining of chromosomes and nuclei is also achieved using acetocarmines, especially in botanical material (Ruzin, 1999), but also in microbiology (Lechevalier and Roisen, 1973;

Schwarzacher and Heslop-Harrison, 2000). In these latter techniques the active constituent is probably an iron–carminic acid complex, see above. For a listing of methods using carminic acid, cochineal and carmine see Gray's (1954, 1973) extensive compilations.

Carminic acid has been used as a marker to allow assessment of gastro-intestinal passage time in various mammals, including human infants (Sievers *et al.*, 1995), and to prepare colored gelatin for injection into blood vessels for anatomical studies (McDougall *et al.*, 2000).

Industrial uses – carmine was used as a textile dye, and currently carminic acid is widely used to color food products and pharmaceuticals

Purity of commercial batches: Single colored components are present in most carminic acid and carmine batches, which are contaminated with protein. The dye content of carminic acid samples is up to 88%. The acid-extractible carminic acid content of carmine batches is up to 43%; with more sometimes released into alkaline solutions. HPLC and TLC methods are available for carminic acid (Berzas-Nevado *et al.*, 1998; Stapelfeldt *et al.*, 1993). A thin-layer chromatographic method is suitable for both carminic acid and carmine (Marshall and Horobin, 1974). There is a Commission testing procedure for carmine, but not for carminic acid; see Chapter 28.

Stability: of stock and staining solutions vary widely; alkaline solutions are unstable; light fastness is good

A<small>NTHRAQUINONE</small> D<small>YES</small> F<small>ORMING</small> C<small>OVALENT</small> B<small>ONDS</small> W<small>ITH</small>
TISSUE OR CO-SOLUTES

Alizarin complexone

Synonym: alizarin-3-methylimino-diacetic acid
Dihydrate: $C_{19}H_{15}NO_8 \cdot 2H_2O$, FW 421

Absorption maximum: 427 nm in methanol; if excited at 580 nm its emission maximum is at 625 nm in bone
Metal complexation behavior: chelates calcium ions
Solubilities: slightly soluble (0.1%) in water, less so (0.02%) in ethanol

The anthraquinone chromophore carries a calcium-chelating methyliminodi-acetic acid group. The overall size is moderate, and the conjugated system is small. The free acid and the monoanion are lipophilic, the dianion is hydrophilic. Commercially available products are the free acid and its dihydrate.

Applications: Alizarin complexone has often been used to label bones (Carls *et al.*, 1997) and calcified structures such as otoliths (Lagardere *et al.*, 2000) in developmental studies in which the structures are later examined microscopically. The dye has also been used as a marker of live cells, such as fish eggs (Nagata and Irvine, 1997).

Purity of commercial batches: Alizarin is usually present as a minor contaminant. Dye contents of typical samples is up to 95%. There is a paper chromatographic chromatographic method (Tusl, 1968).

Procion and remazol reactive dyes

A number of reactive anthraquinone textile dyes have been sold, several of which have been used in biology and medicine. As it is not always clear which dyes are actually being used, illustrative information on CI Reactive blue 2 is provided, with some comparisons to related dyes. Notes on current applications in biology follow.

A representative dye – Reactive blue 2

CI 61211

Synonyms: basilen blue, cibacron blue F3G or 3G, procion blue HB

$C_{29}H_{17}ClN_7Na_3O_{11}S_3$, FW 840

Absorption maximum: 607 nm in water

Solubilities: 0.5% in water – all dyes of this type are water soluble to varying degrees

The anthraquinone chromophore of reactive blue 2 carries amino and sulfo substituents; as well as a lengthy aromatic moiety with sulfo-phenyl groups plus a monochlorotriazine group. This latter can form covalent links with neutrophilic groupings on biomolecules. Other dyes of this general class have different reactive groups, although all react with the amino and hydroxyl groups of biomolecules. Thus **reactive blue 4 (CI 61205**, synonyms: **procion brillant blue MX-R, procion brillant blue M)** is a dichlorotriazine, while **reactive blue 19 (CI 61200, remazol brillant blue R)** carries a vinylsulfone or its precursor.

Purity of commercial batches: The position of the sulfo substituents on the phenyl rings appears to vary between batches, some of which contain more than one isomer. Overall dye contents are up to 60%. Thin-layer chromatographic methods are available for such reactive dyes (Szichy *et al.*, 1974).

Stability: of dry powder is good if kept in anhydrous conditions; aqueous staining solutions decompose due to reaction with water; light fastness is good.

Hazards: Reacts with proteins, and hence can give rise to hypersensitivity responses, as is so with related reactive dyes

Some applications of anthraquinone reactive dyes – Several reactive blue anthraquinone dyes are used below as examplars, with identities indicated by the abbreviations **RB2**, **RB4** and **RB19**.

Occasionally the dyes have been used for labeling antibodies, for example, antiparathyroid markers which then permit identification of such cells in whole animals (King *et al.*, 1999: RB2). The dyes are also used microscopically as tracers, for example, to evaluate apical leakage in dental fillings (Tamse *et al.*, 1998: RB4).

More routine are applications using the dyes to stain proteins on various electrophoretic gels (Chu and Whitesides, 1993: RB4), and as self-indicating substrates, both directly to detect ligninolytic enzyme activity in fungi (Minussi *et al.*, 2001: RB19), or to label fatty acid monoesters for detection of lipolytic organisms (Wirth, 1992: RB19). However, the most widespread application of such dyes in biology is as immobilized ligands in systems for protein separations, see for instance reviews of partitioning of blood proteins (Birkenmeier, 1994) and of enzyme purification by dye-affinity chromatography (Clonis *et al.*, 2000).

NAPHTHALIMIDES

Lucifer yellow CH

There are two chemically distinct Lucifer yellow dyes, with suffixes CH and VS. The latter is little used, and the many applications papers referring merely to 'Lucifer yellow' probably used the former dye. This entry describes **Lucifer yellow CH**, with **Lucifer yellow VS** and Lucifer yellow derivatives briefly noted. Derivatives are not necessarily based on Lucifer yellow CH (or VS), but merely possess the same substituted naphthalimide chromophore. Suppliers' documentation and the research literature do not always provide unequivocal identifications.

Lithium salt: $C_{13}H_9Li_2N_5O_9S_2$, FW 457

Absorption maximum: 428 nm in water, in which solvent the emission maximum is 536 nm

Solubilities: extremely soluble (>10%) in water

Lucifer yellow CH has amino, sulfo and phenyl-*N*-hydrazinocarbonylamino substituents on the naphthalimide chromophore, the anion consequently being very hydrophilic. The hydrazine group allows the dye to be covalently connected to aldehyde groups, permitting both labeling of biomolecules and immobilization of the dye within cells and tissues using aldehyde fixation. The overall size is

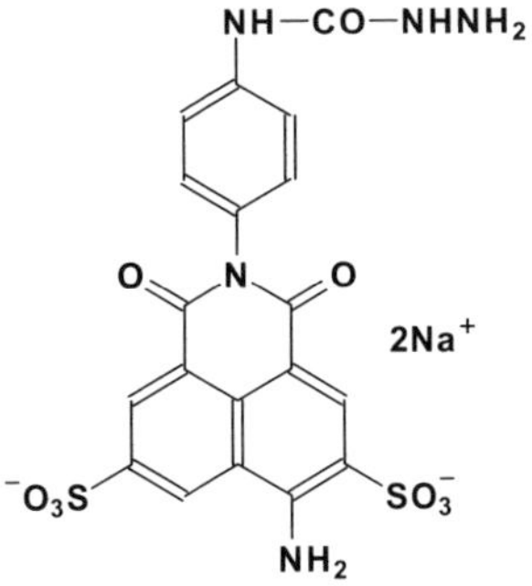

NH—CO—NHNH₂

$$2Na^+$$

⁻O₃S   SO₃⁻

NH₂

moderate, and the conjugated system is small. The dye is available commercially as the ammonium, lithium (illustrated) and potassium salts.

Applications: The very hydrophilic character of Lucifer yellow CH underlies its numerous applications as a stain of living cells and tissues. As has been repeatedly described in reviews, manuals and the research literature the dye is used microscopically to demonstrate fluid-phase endocytosis (Celis, 1998; Pringle *et al.*, 1989; Wiederkehr *et al.*, 2001), and to detect gap junctions (Gabrion *et al.*, 1998; Trosko *et al.*, 2000). Other recent applications using the dye as a hydrophilic tracer include testing for the presense of a blood–retina barrier in insects (Shaw and Varney, 1999), assessment of the degree of cellular permeabilization (Celis, 1998; Schapiro and Grinstein, 2000), and measurement of penetration of water into skin (Serrato-Valenti *et al.*, 2000).

In fixed brain slices Lucifer yellow CH is used for intracellular filling of individual neurons with dye, for example, using iontophoretic injection, to allow investigation of neuronal morphology and intraneuronal connectivities (Elston *et al.*, 1999; Mikkonen *et al.*, 2000).

Lucifer yellow CH is also used as a marker, for example, of individual neurons following their experimental manipulation (Molnar and Nadler, 1999); and as a fluorescent label, such as of the plant growth substance abscisic acid (Chen and Zhang, 1999); and occasionally as a light microscopic stain (Rogers *et al.*, 1992).

Purity of commercial batches: The dye content is up to 90%

Stability: The dry powder is hygroscopic; aqueous solutions are stable if kept in the dark

Lucifer yellow dextrin has been used to probe cell–cell connectivity (Bohrmann and Schill, 1997), and as a cell-marker to follow embryological development (Render, 1997).

26

Phthalocyanines, porphyrins and related aza[18]annulenes

R.W. Horobin

BACKGROUND INFORMATION

Structural features and issues of nomenclature

This chapter describes [18]annulenes, in which chains of methine groups form 18-member ring systems in which all double bonds are conjugated. The dyes to be discussed have varying numbers of methine groups replaced by nitrogen (–N=), and so are termed aza[18]annulenes.

The porphyrins include porphine and its derivatives. The porphine molecule is composed of four pyrrole rings cyclized by four methine groups. Porphyrins are thus tetraaza[18]annulenes. Some dyes have phenyl rings attached to these bridging methines, and are termed meso-tetraphenylporphines. Note that several important natural products, e.g. heme, the iron-complexed oxygen carrying component of vertebrate hemoglobin are porphyrins. Phthalocyanine dyes (e.g. alcian blue 8G) are octaaza[18]annulenes because the methine bridges of porphine are replaced with aza units; a benzo group fused to each pyrrole ring enlarges the structure. Variants (e.g. cuprolinic blue) have four fused pyridyl rings. Examples of porphyrin, phthalocyanine and fused-pyridyl annulene and aza annulene structures are shown in Figure 26.1.

Homologs and isomers

Many of these dyes contain mixtures of species, varying in the numbers or positions of heterocyclic nitrogen atoms (e.g. cupromeronic blue), side-chains (e.g. alcian blue 8G) or sulfonate substituents. Consequently the symmetry of these molecules varies markedly, influencing properties such as dye aggregation and surfactancy.

Metal complexes

Most dyes described are copper complexes in which the metal ion is bound by the nitrogen atoms of the pyrrole units. Note: in structural formulae all

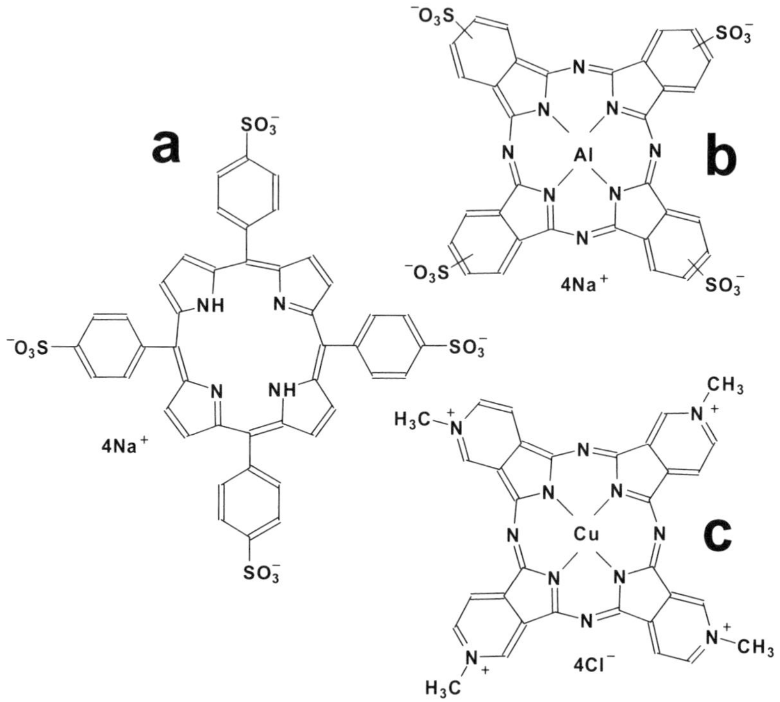

Figure 26.1 Aza annulene structures. **a** is the tetraaza[18]annulene meso-tetraphenyl-porphine tetrasulfonate; **b** is the octaaza[18]annulene aluminum phthalocyanine tetrasulfonate; **c** is the octaaza fused pyridyl aza[18]annulene cupromeronic blue.

metal–ligand bonds are shown using continuous lines with formal charges on separate atoms omitted. This is the appropriate formalism because it is not usually known whether such bonds are primarily electrostatic or covalent. Related compounds are complexes of other metal ions (e.g. aluminum, palladium and technetium) or are metal-free.

Properties influencing staining

Electric charge – The aza[18]annulene chromophore carries no innate electric charge, which property is thus controlled by any ionic groupings present. Monastral blue B is nonionic; dyes with sulfonate substituents (e.g. copper phthalocyanine tetrasulfonate) are anionic, and dyes with *N*-methyl quaternary nitrogens (e.g. cuprolinic blue) or cationic side-chains (e.g. alcian blue 8G) are cationic.

Acid–base properties – Nearly all dyes are quaternary ammonium salts or salts of strong acids, which groups retain their charges at any usage pH. Dyes with pyridyl heterocycles, such as cuprolinic and cupromeronic blues, could however form pseudobases; and indeed such dyes have been described as 'unstable under alkaline conditions'.

Hydrophilic/lipophilic properties – Vary widely, from very hydrophilic (cuprolinic blue) to very lipophilic (monastral blue B).

Extent of conjugation – This ranges from moderate to large; all the dyes have a large number of aromatic rings but in some (e.g. meso-substituted phenyl porphines) these do not form a single conjugated system.

Dye size – Even the smallest dye (monastral blue B) has a formula weight of 576, and many are much larger.

How to find the dye that interests you

The usual sequence of nonionic dyes, anions, and cations is followed. Within each subgroup dyes are listed in order of increasing formula weight, although certain anionic and cationic dyes which are only occasionally used in biology have generic not individual entries.

NONIONIC DYE

Monastral blue B

CI 74160, CI Pigment blue 15
Synonym: copper phthalocyanine
$C_{32}H_{16}CuN_8$, FW 576

Absorption maximum: 694 nm in concentrated sulfuric acid
Solubilities: insoluble in water, ethanol or hydrocarbons; applied as an aqueous suspension

This dye is unsubstituted copper phthalocyanine. The overall size is moderate, that of the conjugated system large. This nonionized dye is strongly lipophilic.

Applications: Monastral blue B is widely applied as a tracer to facilitate the light microscopic observation of phenomena such as lymphatic transport (Fritz and Waag, 1999), and of leakage from the microvasculature (Kwan *et al.*, 2001). Such leakage is also observed macroscopically using this dye (Albassam *et al.*, 2001). Another application of monstral blue B is as a marker; such as of pericytes in developmental studies on osteoblast origins (Diaz-Flores *et al.*, 1992), and of macrophages (Niehaus and Mehendale, 1998). The copper-containing dye can be used in transmission electron microscopy as a tracer of, for example, the interconnections of insect eye compartments (Schraermeyer *et al.*, 1999), and also as a marker for cells such as macrophages (Sone *et al.*, 1997).

Industrial uses – as a pigment in inks, paints, and polypropylene sutures
Purity of commercial batches: Dye content is up to 97%
Stability: Light fastness is excellent

ANIONIC DYES

Luxol fast blue MBS

CI 74180, CI Solvent blue 38
Synonym: often described in the biomedical literature merely as luxol fast blue, with no identifying suffix. Unfortunately this term has also been used to describe the azo dyes luxol fast blue G and luxol fast blue ARN (see Chapter 11). Currently the phthalocyanine dye is the compound readily available commercially; imprecision therefore usually implies use of the phthalocyanine.
$C_{32}H_{14}CuN_8O_6S_2$.diarylguanidine; if di-*o*-tolylguanidine, FW 1214

Absorption maximum: 666 nm in methanol
Solubilities: 3% in ethanol, very slightly soluble (0.2%) in water, insoluble in xylene

Luxol fast blue MBS is a sulfonated copper phthalocyanine in which the counterions are lipophilic arylguanidines, probably di-*o*-tolylguanidine as illustrated. The resulting salt is not soluble in water. The most common species is disulfonated, and is likely to be present as a mixture of geometric isomers. Overall size, like that of the conjugated system, is large.
Applications: Routine usage as a stain for myelin is described in manuals (Bancroft and Gamble, 2002; Kiernan, 1999; Presnell and Schreibman, 1997). Such procedures can be used for counterstaining after immunostaining (Yao *et al.*, 1994), may be quantified (Rodman *et al.*, 1998), are applicable to semi-thin epoxy sections (Deprez *et al.*, 1999), and can be microwave-accelerated (Churukian, 2000). The mechanism of myelin staining was studied by Clasen *et al.* (1973).

Luxol fast blue MBS is an anionic dye, albeit with unusual solubility properties, and it has also been used as an acid dye to demonstrate mitochondria in paraffin sections (Takaya, 1967) and cross-striations in skeletal muscle (Muller and Firsching, 1991).

Industrial uses – to color lacquer, leather, and ballpoint pen inks
Purity of commercial batches: There is a chromatographic method for the sodium salt of this dye (Sekiguchi *et al.*, 1969).
Stability: Light fastness is good

Other sulfonated phthalocyanines and porphines

Several anionic aza[18]annulenes that have occasional biomedical applications are noted here. A brief account of their chemistry is followed by more systematic information concerning a representative dye, and then by notes of applications.
Chemistry of dyes – Sulfonated copper phthalocyanines include disulfonated **sirius light blue** (CI 74180, Direct blue 86, durazol blue 8G) and various **tetra-sulfonated copper phthalocyanines**. Mono-, di-, tri- and tetra-**sulfonated aluminum phthalocyanines** (AlPcS$_n$) and metal-free **sulfonated tetraphenyl porphines** (TPPS$_n$) are also used in biomedicine; along with **palladium** and **technetium tetraphenylporphine tetrasulfonates**. All these anionic dyes have large conjugated systems and are of large overall size. All tetrasulfonates are extremely hydrophilic, trisulfonates are less so, and the monosulfonates are lipophilic. Asymmetrically substituted disulfonates are amphiphilic.

A representative dye – Copper phthalocyanine tetrasulfonic acid

CI 74220, CI Acid blue 249
Sodium salt: $C_{32}H_{12}CuN_8Na_4O_{12}S_4$, FW 984

Absorption maxima: 610 nm and 694 nm in water
Solubilities: 0.7% in water; trivially soluble (0.02%) in ethanol

The overall size, and that of the conjugated system, is large. The highly hydrophilic tetraanion is sold as the sodium salt, illustrated. Both 3, 4′, 4″, 4‴- and 4, 4′, 4″, 4‴-tetrasulfonates are prepared by direct synthesis. Products manufactured by sulfonation of copper phthalocyanine contain mixtures of geometric isomers.

Purity of commercial batches: Homogeneity varies with mode of synthesis. Overall dye contents are up to 85%. There is a thin-layer chromatographic method (Sekiguchi *et al.*, 1969).

Stability: Light fastness is good

Staining applications – Several dyes are used as stains of fixed tissues. *Applications:* examples include use of **copper phthalocyanine tetrasulfonate** to stain proteins in tissue prints (Bickar and Reid, 1992), and to stain gray matter on the surfaces of large slices of brain (Wu and Kiernan, 2001). **Sirius light blue** has been used to stain eosinophil granules in epoxy sections prior to imaging in the transmission electron microscope (Tato *et al.*, 1990).

Vital staining using anionic aza[18]annulenes has also been carried out. Such applications arise from use of metal-free tetraphenylporphines and aluminum phthalocyanines as experimental photodynamic therapy drugs. Investigation of mechanisms of such treatments have involved microscopic imaging of dye-live cell interactions (for representative observations see Berg and Moan, 1994; Berg *et al.*, 1994; Malik, Amit and Rothman, 1997 and Peng *et al.*, 1991).

Other biomedical applications – **Copper phthalocyanine tetrasulfonate** has been used to visualize atheroscleromatous plaques in artery walls (Eldar *et al.*, 1990), and as a marker of phagocytosis (Thanawongnuwech *et al.*, 2000). **Aluminum phthalocyanine disulfonate** is used to produce localized photodamage of membranes (Selbo *et al.*, 2000). **Technetium tetraphenylporphine tetrasulfonate** has been applied as a skeletal scintigraphy agent to identify osteomyelitis (Ali *et al.*, 1997), and **palladium tetraphenylporphine tetrasulfonate** is used to image oxygen in epicardial blood vessels (Barlow *et al.*, 1998).

CATIONIC DYES

Cuprolinic blue

Synonym: quinolinic phthalocyanin

$C_{36}H_{36}CuN_{12}O_{16}S_4$, FW 1085

Absorption maxima: 625 nm and 640 nm in pH 5.9 aqueous solution

Solubilities: at least 0.5% in water

This dye has four pyridyl rings fused to the aza[18]annulene chromophore. As the pyridyl nitrogens are symmetrically oriented, only a single geometric isomer is present, as illustrated. The tetracationic dye is hydrophilic, and both overall size and the conjugated system are large. The dye is sold as the methyl sulfate.

Applications: Applied from a solution with high inorganic salt concentration, cuprolinic blue stains RNA but not DNA (Tas *et al.*, 1983) and it is recommended in manuals for selective staining of RNA-rich structures such as nucleoli (Boon and Drijver, 1986). Less selectively, as a nucleic acid stain, the dye is used as a counterstain after enzyme histochemistry and immunocytochemistry (Van Ginneken *et al.*, 1999), and to assess neuronal numbers after wholemount staining (Heinicke *et al.*, 1987; Schafer *et al.*, 1999). The latter application can be microwave accelerated (Van Ginneken *et al.*, 1998).

Cuprolinic blue is also used to stain glycosaminoglycans, for instance, to permit quantitation (Kiraly *et al.*, 1996b). Such staining can be carried out both on paraffin and water-miscible resin sections. Indeed demonstration of glycosaminoglycans can be achieved with epoxy resin sections (Juarranz *et al.*, 1987), though in this case staining is slow. The dye is also used in this way as an electron stain (Ohma *et al.*, 2000).

Stability: Staining solutions are stable stored in refrigerated opaque containers, but are unstable under alkaline conditions, in strong light, or if exposed to metal such as stainless steel staining racks

Related dye: **Cupromeronic blue** (synonym: **cinchomeronic phthalocyanin**) is a similar dye, except that as the N-methylated ring nitrogens may be substituted either at the 2- or 3-position, mixtures of geometric isomers are possible; one of which is shown in Figure 26.1. The less symmetrical structure results in staining characteristics markedly different from those of cuprolinic blue (Scott, 1996). The dye is used to demonstrate proteoglycans in sections of specimens embedded in paraffin and water-miscible resins (Kiraly *et al.*, 1996b), in light and electron microscopy (Miyagawa *et al.*, 2001).

Alcian blue 8G

CI 74240, CI Ingrain blue 1
Synonyms: suffix often omitted, suffixes such as 8GN, 8GS and 8GX have also been used.
$C_{56}H_{68}Cl_4CuN_{16}S_4$, FW 1299

Absorption maximum: 615 nm in water
Solubilities: Considerable batch variation occurs, with published values from 0.8–9.5% in water and 0.1–6.0% in ethanol. This reflects both presence of additives and formation of insoluble pigment

Alcian blue 8G is a copper phthalocyanine substituted by four cationic guanidinium groups. These are attached to the ring system via thioether links, which are broken by hydroxide ions or nucleophiles such as tissue amines to give an insoluble pigment. Though not often commented on, such decomposition occurs spontaneously in routine staining, leading to irreversible coloration (Goldstein and Horobin, 1974); if induced by alkali treatment and followed by restaining, it can form the basis of a stain amplification procedure (Scott, 1996).

The overall size of alcian blue 8G is large, as is the conjugated system. Commercial products may contain variable amounts of the less cationic species; the tetracationic compound, illustrated, is strongly hydrophilic and is sold as the chloride.

Alcian blue 7G is similar but the average number of cationic substitutents per molecule has been said to be less, while **alcian blues 5G** and **2G** contain the same dye mixture plus a purple dye. While alcian blue was manufactured by ICI Ltd., suffixes attached to the name indicated the additives present. For instance, alcian blue 8GS contained inorganic salts but no stabilizer, whereas alcian blue 8GX contained boric acid and pentaborate. However the significance of suffixes on current samples is unknown. **Alcian green 2G** and **3B** are mixtures of alcian blue 8G with the azo dye alcian yellow (see Chapter 10).

Applications: Alcian blue 8G is recommended in manuals and reviews for a variety of staining methods. For demonstration of glycosaminoglycans, for example, in mucins, using pH and critical electrolyte-controlled procedures see Bancroft and Cook (1994) and Pearse (1985). Such methods may be microwave accelerated (Churukian, 2000), and have been applied to epoxy resin sections (Litwin, 1985) and in cytology (Boon and Drijver, 1986). Manuals also note use of alcian blue 8G for amyloid staining, for cystine following oxidation, and in polychromatic alcian blue–safranine methods for mast cells (Bancroft and Cook, 1994; Bancroft and Gamble, 2002). Another widely applied polychrome method is the double staining of fetal skeletons in wholemounts with alizarin red S and alcian blue 8G (Klymkowsky and Hanken, 1991). Alcian blue 8G is also used to stain anionic polysaccharides and the xylem vessels in plants (Ruzin, 1999).

Other applications of alcian blue 8G as a microscopic stain include identification of *Cryptococcus neoformans* (Cook, 1974), detection of bacterial capsules (Karlyshev and Wren, 2001), mucin staining of cytological fine-needle aspirates (Gupta *et al.*, 2000), and demonstration of sulfatide storage (Schott *et al.*, 2001). Alcian blue 8G is also used as a stain for glycosaminoglycans on electropherograms (De Muro *et al.*, 2001).

In other biological applications, alcian blue 8G is a marker for the long-term identification of individual trout (Bridcut, 1993), a tracer to assess leakage around dental fillings (Hosoya *et al.*, 2000), and a stain for gelatin that is to be injected into blood vessels (Vellar, 2001).

Industrial uses – manufactured in Asia for use as a textile dye

Purity of commercial batches: various homologs and geometrical isomers are routinely present, with variable amounts of an insoluble decomposition product. Overall dye content varies from 25–95%. Minimum dye content for Commission certification is 50%. A high dye content is at the expense of less stabilizer. Consequently, purer samples may have reduced shelf-lives (Churukian *et al.*, 2000). There are thin-layer chromatographic methods (Horobin and Goldstein, 1972). Regarding Commission testing methods see Chapter 28.

Stability: the dry powder decomposes only slowly in batches containing stabilizing borates; acidic staining solutions are stable for months, although salty solutions used in critical electrolyte concentration methods often precipitate after some hours; light fastness is excellent

Other cationic phthalocyanines

At least two phthalocyanines, namely **alcec blue** and **alcian blue pyridine variant** (probable synonym: **alcian blue–tetrakis(methylpyridinium) chloride**), are biological stains similar to alcian blue 8G. Notes on their chemistry and published applications follow.

Chemistry of the dyes – Hydrophilic, tetracationic copper phthalocyanine derivatives, with large conjugated systems and of large overall size. Alcian blue pyridine variant carries four pyridinium rings linked to the phthalocyanine chromophore by methylene bridges. The dye is available commercially as the chloride, illustrated. Alcec blue carries four aliphatic side-chains with terminal trimethylammonium groups, and is sold as the methyl sulfate as illustrated. These dyes, unlike alcian blue 8G, are stable under alkaline conditions and do not form pigments in contact with tissue sections. The solubilities and adsorption spectra of these dyes are similar to those of alcian blue 8G.

alcian blue pyridine variant

$R = -CH_2-\overset{+}{N}$ ... Cl^-

alcec blue

$R = -SO_2NHCH_2CH_2CH_2\overset{+}{N}-CH_3$ ... $CH_3OSO_3^-$ ... CH_3, CH_3

Staining applications – Both dyes are used as stable alternatives to alcian blue 8G for staining glycosaminoglycans and proteoglycans in such material as cartilage matrix, mucins and mast cell granules. The dyes are suitable for Mowry-type acidic staining solutions (Williams *et al.*, 1999 and Henwood, 2002 for alcec blue and alcian blue pyridine variant respectively) and for the critical electrolyte concentration procedure (Churukian *et al.*, 2000; using alcian blue pyridine variant). The latter methodology has also been used with alcec blue in staining for electron microscopy (Scott and Thomlinson, 1998).

27

Miscellaneous inorganic and organic substances used as biological stains

J.A. Kiernan and R.W. Horobin

BACKGROUND INFORMATION

This chapter describes colored, fluorescent and chromogenic reagents not fitting into the chemical classes previously discussed. Inorganic colorants are described first, each being recognized by the element (usually a metal) whose presence is necessary for absorption or emission of light. These colorants are listed in order of the groups (columns) of the periodic table, moving from left to right. Compounds used only as stains in electron microscopy are outside the scope of this book; for reviews see Glauert and Lewis (1998) and Hayat (2000). Diverse organic colorants are then described, sorted into several categories, namely a dye, reactive fluorochromes, including metal complexing agents, and chromogenic compounds used histochemically. Within each group compounds are placed in order of increasing formula weight.

INORGANIC COLORANTS

Lanthanide salts and complexes

The 15 lanthanide metals are commonly used as salts of trivalent cations (collectively designated Ln^{3+}) and as complexes with organic ligands. Of these metals europium (Eu) and terbium (Tb) occur in fluorescent analytical and histochemical reagents, while cerium salts are valuable in enzyme histochemistry.

Optical properties: Multiple sharp absorption peaks occur from near-ultraviolet to near-infrared. These are strongly influenced by pH, solvents and co-solutes. Emission maxima are less variable, typically 610–620 nm (red) for Eu^{3+} and 545 nm (green) for Tb^{3+}. Emission continues for many microseconds (often milliseconds) after the end of a pulse of exciting light. This delayed emission (phosphorescence) is exploited in time-resolved fluorescence microscopy and

389

spectroscopy, techniques that remove background fluorescence, which lasts less than 200 nanoseconds after stopping the exciting light.

Ln^{3+} ions form stable organic chelates, some absorbing in the near UV. Both synthetic chelates and compounds formed by combination of Ln^{3+} with components of cells or dissolved analytes occur. Eu^{3+} or Tb^{3+} complexes can absorb UV or visible light due to the ligand (the 'antenna') while emission is at or near the emission wavelength of the lanthanide cation. This process is termed **fluorescence** (or **luminescence**) **resonance energy transfer (FRET)** or **intramolecular energy transfer**. A lanthanide chelate generates visible light more efficiently than simple fluorescence, especially if the molecule has more than one antenna for each complexed lanthanide atom.

Solubilities: Lanthanide halides, nitrates and sulfates are freely soluble in water. Lanthanide chelates used in biomedical work are generally water-soluble.

Lanthanide salts

Eu^{3+} and Tb^{3+} ions bind to muscle fiber surfaces (Leung, 1997), bacterial endospores (Rosen, 1999) and nucleic acids. A fluorescent stain for nuclei or bacteria is obtained by applying EuCl$_3$, followed by thenoyltrifluoroacetonate, which serves as an antenna (Dyer and Mori, 1969). As colloidal labels for immunoreagents, inorganic phosphors such as Y$_2$O$_2$S:Eu^{3+} are notable for their brightness and resistance to fading (Beverloo *et al.*, 1993).

Lanthanide chelates

Labeling immunoreagents or nucleic acid probes with such complexes requires multidentate ligands that do not release hydrated Ln^{3+} when diluted. The ligand must be attached to a reactive group that can combine with proteins or nucleic acids, and to an aromatic ring system (the antenna) that absorbs in the near UV. In some instances the antenna is also the chelating part of the molecule.

Although a number of fluorescent and phosphorescent Eu and Tb complexes have been used in cell biology and biochemistry, publications often identify them only by trade names. Consequently identities of ligands, which largely determine properties of reagents, are often not stated. The following examples describe chelates of disclosed composition.

DTPAA-pAS-Tb: Tb chelate of the *bis*-cyclic anhydride of diethylenetriaminepentaacetic acid and p-aminosalicylic acid. Absorption maximum 312 nm; emission peaks at 488 nm and 545 nm. Phimphivong and Saavedra (1998) covalently linked DTPAA-pAS-Tb to a phospholipid, making a selective fluorochrome for cell membranes.

Eu-BCPDA-SO$_2$Cl: Eu chelate of 4,7-*bis*(chlorosulfophenyl)-1,10-phenanthroline-2,9-dicarboxylic acid. Absorption maximum 325 nm; emission 615 nm. The ligand can also be conjugated to streptavidin and complexed with Tb, for detection of bound biotinylated immunoreagents in immunohistochemistry and immunoassays (Scorilas *et al.*, 2000).

Tb-PCTMB: The Tb complex of 3,6,9-tris(methylene phosphonic acid *n*-butyl ester)-3,6,9,1,5-tetraazabicyclo[9.3.1]pentadeca-1(15),11,13-triene. This has been

tested as a fluorescent marker of tumor tissue, to assist in endoscopic examination (Bornhop *et al.*, 1999).

Industrial uses – rare earths are largely applied in light-emitting phosphors,
for example, in television screens.

Purity of commercial batches: Lanthanide chlorides are sold with stated purities of
99.9% and higher.

Stability: Inorganic salts are indefinitely stable, and some Eu and Tb chelates are
stable for long periods.

Prussian blue

Synonyms: Berlin blue, ferric ferrocyanide, ferric hexacyanoferrate(II), Turnbull's
blue.

CI 77510, CI Pigment blue 27

$Fe_4[Fe(CN)_6]_3$, FW 859 – As water and alkali metal ions are also present, these
values are approximate;

Solubilities: Dissolves in aqueous oxalic acid, re-precipitates if exposed to bright
light. Insoluble in water, dilute mineral acids and organic solvents.

A Prussian blue crystal has a cubic lattice with iron atoms at the corners, in alternating Fe(II) and Fe(III) oxidation states, and cyanide ions in their edges. K^+ (or
Na^+) ions or water molecules occupy the centers of most cubes. Prussian blue
precipitates from mixtures of aqueous ferric and ferrocyanide ions; ferrous and
ferricyanide ions make a chemically identical pigment termed Turnbull's blue or
'ferrous ferricyanide'.

Applications: Prussian blue, with gelatin or glycerol to increase viscosity, is a
traditional injection mass for filling blood vessels, lymphatics (Lubach *et al.*,
1991), cerebrospinal fluid drainage pathways (Caversacchio *et al.*, 1996), and
insect tracheoles (Gray, 1954). It is also the visible product of histochemical
methods for iron (Fritz *et al.*, 1996; Lillie and Fullmer, 1976), acid mucosubstances (Churukian, 2000; Hayat, 1993; McManus and Mowry, 1960), and
reducing groups such as thiols, but also catechols and others, in tissues (Lillie
and Fullmer, 1976; Gisbert and Sarasquete, 2000).

Purity of commercial batches: The method of preparation affects particle size and
shade of blue.

Stability: Light fastness is excellent. Prussian blue is changed by alkalis to a
brown iron oxide.

A related pigment, **copper ferrocyanide** (Hatchett's brown, $CuFe(CN)_6$), is the
insoluble product in histochemical methods for choline esterases (Karnovsky
and Roots, 1964) and some dehydrogenases (Ishibashi *et al.*, 1999).

Ruthenium red

Synonym: Ammoniated ruthenium oxychloride.

$Ru_3H_{42}N_{14}O_2Cl_6$, FW 786; exists as tetrahydrate, FW 858.

Absorption maximum: 540 nm in water.
Solubilities: Soluble in water; sparingly soluble in alcohol.

$$\left[\begin{array}{c} H_3N \quad NH_3 \; H_3N \quad NH_3 \; H_3N \quad NH_3 \\ H_3N - Ru - O - Ru - O - Ru - NH_3 \\ H_3N \quad NH_3 \; H_3N \quad NH_3 \; H_3N \quad NH_3 \end{array} \right]^{6+} \quad 6Cl^-$$

Ruthenium red is a large, cationic coordination complex containing three metal atoms linked by bridging oxygens, plus multiple amine ligands.

Applications: This wholly inorganic cationic dye has been used in light microscopy to stain pectins (Jensen, 1962) and in a two-color stain with alcian blue for glycosaminoglycans (Yamada, 1970). At higher pH ruthenium red acts as a typical basic dye, coloring nuclei, Nissl substance, cartilage matrix and mast cell granules. Other light microscopic applications include staining resin-embedded material, as summarized by Litwin (1985). Ruthenium red is an essential ingredient of fixatives of the bacterial glycocalyx (Fassel and Edmiston, 1999).

Purity of commercial batches: Much batch variation, the nominal compound often accounting for less than 30% of the sample. Colored contaminants include an oxidation product **ruthenium brown** (Blanquet, 1976), and a polymer **ruthenium violet** (Luft, 1971b). Luft (1971a) described methods for purifying and assaying the colored components.

Stability: The powder is stable in tightly capped containers. Neutral aqueous solutions are stable for several hours.

Osmium tetroxide

Synonyms: osmium(VIII) oxide, osmium tetraoxide, osmic acid
OsO_4, FW 254

The old name 'osmic acid', still occasionally seen in scientific and medical literature, is inappropriate. Aqueous OsO_4 solutions are neutral, and in osmates the oxidation number of Os is +6, not +8.

Solubilities and appearance: 7.24% at 25°C in water. In carbon tetrachloride: 250% at 20°C. OsO_4 is reduced by oxidizable solvents, such as ethanol, to a black lower oxide present largely as a colloidal suspension in water. Vapor is evolved from the solid and from solutions, even at room temperature. Solutions are colorless.

Osmium tetroxide molecules are small, lipophilic and uncharged. They react rapidly with double bonds of lipids, forming Os(VI) diesters and, eventually, insoluble black OsO_2 (Korn, 1967). OsO_4 also reacts with proteins, some of which become cross-linked (Hopwood, 2002; Nielson and Griffith, 1979).

Applications: The principal use of OsO_4 is as a fixative or post-fixative of specimens to be examined by electron microscopy (Glauert and Lewis, 1998; Hunter, 1993). In pathology OsO_4 is used to make black-stained teased preparations of peripheral nerves (Frankl and Denaro, 1998), and to detect and quantify fat emboli in lungs (Mudd *et al.*, 2000). All lipids are stained black. In the presence of a polar oxidizing agent such as potassium dichromate (Marchi, 1892) or chlorate (Adams and Bayliss 1968; Swank and Davenport, 1935), only the most hydrophobic lipids are stained, including fats and cholesterol esters. The latter lipids occur in atheromatous lesions and in degenerating myelin. Other modern applications include staining lipids (Bal, 1990) and phenolic compounds (Schoenwaelder and Clayton, 1999) in plants; and intensifying the brown oxidation product of diaminobenzidine (DAB), the end-product of many histochemical techniques (Krueger *et al.*, 1999).

In the presence of iodide ions, preferably from ZnI_2 (Maillet, 1963), the variety of structures blackened by OsO_4 is extended. Iodide-osmium solutions are valuable for staining such structures as unmyelinated autonomic axons (Maillet, 1968; Watanabe *et al.*, 1995), sprouting axon terminals in muscle (Fagg *et al.*, 1981; Tian *et al.*, 1995), and dendritic cells in lymphoid tissue (Crivellato and Mallardi, 1997).

Osmium tetroxide is used in analytical chemistry of lipids (Holm *et al.*, 1996), and a medical application is chemical synovectomy by intra-articular injection of OsO_4 (Molho *et al.*, 1999).

Purity of commercial batches: OsO_4 is reported by suppliers to be 98% to 99.8%.

Stability: The solid is indefinitely stable in sealed glass vials. Solutions lose OsO_4 vapor, which is best contained by polyethylene or black rubber stoppers (Gabe, 1976). Protection from light is not necessary.

Hazards: The acrid vapor of osmium tetroxide can be smelled at concentrations lower than those harming the respiratory system, conjunctiva or cornea. Exposure to high concentrations causes transient visual symptoms. Osmium tetroxide is not an environmental hazard because it is rapidly reduced to the insoluble dioxide.

Osmeth is a commercially available osmium(VIII) complex with hexamethylenetetramine (Hanker *et al.*, 1976). It does not smell of OsO_4, but if dissolved in dimethylformamide and then diluted with water it provides an OsO_4 solution.

Silver nitrate

$AgNO_3$, FW 169.9

Appearance: White translucent solid; solutions are colorless.
Solubilities: 122% at 0°C, 952% at 100°C in water. Solubility in alcohol 3.3%; in acetone 0.4%; also soluble in ether and glycerol.
Applications of silver nitrate as a biological stain exploit the strong colors of colloidal metallic silver. This is formed when Ag^+ ions or complexes such as

silver diammine $[Ag(NH_3)_2]^+$, are reduced in the presence of organic macro-molecules. Colors vary with the sizes and local concentrations of particles. Mechanisms of histological silver staining have been reviewed by Grizzle (1996) and Kiernan (1999).

A colloidal particle of silver (or of gold, or of certain sulfides or selenides) can serve as a catalytic nucleus for deposition of silver from solutions containing Ag^+ and a reducing (developing) agent. Spontaneous reduction is retarded by complexing ions such as sulfite or citrate, together with macromolecules such as proteins or polysaccharides. Such solutions, called **physical developers**, enlarge the catalytic nuclei, amplifying their visibility. Alternative names for physical development include **argyrophil III reaction** (Gallyas, 1982), **auto-metallography** (Danscher and Norgaard, 1985) and **silver enhancement**. The last term is commonly applied to enlargement of colloidal gold particles used as labels in immunohistochemistry; or to darkening the oxidation product of diaminobenzidine (Smiley and Goldman-Racic, 1993), a reagent used to localize peroxidase activity. Physical developers can be made in the laboratory (Gallyas, 1979a,b; Newman and Jasani, 1998) or purchased as proprietary kits of undisclosed composition.

Industrial uses – form manufacturing photographic emulsions and silver compounds, such as those used in electroplating.

Purity of commercial batches: Even reagent grade $AgNO_3$ is over 99%.

Stability: The solid, and aqueous solutions, are indefinitely stable if kept clean. $AgNO_3$ is photosensitive only when contaminated.

Hazards: The solid is caustic. Solutions cause delayed black staining of skin or clothing. Silver diammine (ammoniacal silver nitrate) solutions must not be allowed to evaporate because silver fulminate, a detonating explosive, can form.

Related compounds: **Silver acetate**, CH_3COOAg, FW 166.9 or **Silver lactate**, $CH_3CH(OH)COOAg.H_2O$, FW 214.6, is often used instead of $AgNO_3$ in physical developers for immunohistochemistry (Lah *et al.*, 1990; Polak and Van Noorden, 1997) and in a modified von Kossa stain for calcified deposits (Rungby *et al.*, 1993). A review of such developers (Danscher *et al.*, 1993) includes practical instructions.

Protargol-S

Synonyms: silver proteinate strong, strong silver protein.

This yellow to brown powder contains about 8% silver, some of which is precipitated as the chloride by adding sodium chloride to an aqueous solution. Protargol-S must not be confused with mild silver protein. This latter is usually darker and contains about 20% silver, which is more tightly bound to protein and cannot be precipitated as the chloride. The name protargol-S is used by the Biological Stain Commission for preparations intended for staining. In publications, the name protargol is usually used without qualification.

Solubilities. Freely soluble in water; almost insoluble in alcohol or chloroform.

Applications: Bodian's (1936) original procedure for staining axons in paraffin sections has several modifications (see manuals of Churukian, 2000; Clark, 1981 and Gray, 1954), and new variants continue to be published (Deprez *et al.*, 1999). The method is not selective for axons. Duyckaerts *et al.* (1987), for example, used the protargol method in a quantitative study of the neurofibrillary tangles of Alzheimer's disease. Nuclei also stain, as do some endocrine secretory granules (Scopsi and Larsson, 1986) and bone canaliculi (Kusuzaki *et al.*, 2000). Protargol-S produces colloidal silver in histochemical demonstrations of DNA and carbohydrates (Hayat, 1993; Tsukise *et al.*, 1990), and amplifies deposits of Prussian blue (Hong *et al.*, 2000). Protargol is much used for staining protozoa (Zagon *et al.*, 1970), especially in quantitative (Kepner *et al.*, 1999) and taxonomic (Song *et al.*, 2001) studies of ciliates.

Strong and mild silver protein solutions have been used in human and veterinary medicine as antiseptics for application to mucous membranes (Blacow, 1972; Budavari, 1996).

Purity of commercial batches: Preparations vary in composition and suitability for staining. In catalogs the names protargol and protargol-S are sometimes wrongly applied to mild silver protein. Regarding the Commission's testing of protargol-S see Chapter 28.

Stability. Protargol-S should be protected from light.

Colloidal gold

Au, Atomic weight 196.97

Synonym: Gold sol

Appearance: Clear liquid, red in transmitted light; purple if the particle size exceeds about 80 nm.

Gold sols are made by reduction of dilute aqueous solutions of gold(III) ions, derived from $HAuCl_4$, in the presence of stabilizing organic anions. Typically colloidal gold consists of spherical, hydrophobic particles that are negatively charged because of anions adhering to their surfaces. Particle size is determined by the preparative technique (Bendayan, 2000; Hayat, 1993). Macromolecules such as proteins can adhere to particle surfaces, forming a monomolecular shell (de Roe *et al.*, 1987) and increasing colloid stability. Biological properties of macromolecules are largely unaffected by binding to colloidal gold. Gold sols with particles 1–20 nm in diameter are commonly used in biological work, with the smaller sizes being preferred for microscopy and the larger for staining proteins or nucleic acids on membranes.

Applications: Colloidal gold is used to prepare labeled antibodies (Roth, 1982), lectins (Benhamou *et al.*, 1988; Horisberger and Rosset, 1977), the antibody-binding reagent protein A (Bendayan and Duhr, 1986; Roth *et al.*, 1978) and enzymes (Bendayan, 1981a,b). A colloid with positively charged particles, **cationic colloidal gold**, is available for demonstrating anionic sites such as cell surfaces (Horisberger, 1992) and glycoprotein secretions (Yang *et al.*, 1998). Bound colloidal gold is hardly visible in light microscopy, but particles can be enlarged by physical development, see under **silver nitrate**).

Stability: Gold-labeled reagents such as antibodies and lectins are commonly sold with shelf lives of one year. A color change from red to blue indicates aggregation (flocculation) of the particles.

Gold chloride

This name is applied to various compounds that have similar properties and uses:

Chloroauric acid

Synonyms: brown gold chloride; gold trichloride acid
$HAuCl_4.H_2O$, FW 358 (55.0% Au)
$HAuCl_4.3H_2O$, FW 394 (50.0% Au)
$HAuCl_4.4H_2O$, FW 412 (47.8% Au)

Sodium chloroaurate

Synonyms: sodium tetrachloroaurate(III); yellow gold chloride
$NaAuCl_4.2H_2O$, FW 398 (49.5% Au)

Appearance: Yellow deliquescent crystals when obtained from laboratory supply houses. Chloroauric acid can also exist as a brown crystalline mass. Solutions are golden-yellow.
Solubilities: Very soluble in water and alcohol. $HAuCl_4$ is even more soluble in nonpolar solvents such as ether and ethyl acetate.
Applications: True gold chlorides ($AuCl_3$ and $AuCl$) are not used as biological stains. Older manuals of microtechnique made much of using brown rather than yellow gold chloride in some staining methods (Gray, 1954; McClung, 1929); more recent texts (Cook, 1974; Kiernan, 1999) do not support the assertion that there is a difference.

Some of the oldest gold staining methods, such as Ranvier's, are still used in research to display peripheral nerve endings, especially in joints (Gomez-Barrena *et al.*, 1999) and the cornea (Jacot *et al.*, 1995), and to demonstrate elusive objects such as perisinusoidal cells of the liver (Noyan *et al.*, 2000). Another old technique, Cajal's gold-sublimate for astrocytes, is employed in neuropathology (Baloyannis *et al.*, 2001) despite the availability of antisera for immunostaining of different neuroglial cell-types. A frequent application of gold chloride in microtechnique is **gold toning**. Deposits of silver from a staining or histochemical method are replaced by gold derived from an aqueous solution of $HAuCl_4$ or $NaAuCl_4$, providing increased contrast, and resistance to oxidizing agents including osmium tetroxide (Leitinger *et al.*, 2000).

Industrial and commercial uses – Gold chloride is used to prepare gold plating solutions and photographic toners.

Purity of commercial batches: Water of crystallization is variable, especially for HAuCl₄. Perhaps for this reason gold contents declared by vendors vary somewhat.

Stability: Solid gold chloride keeps indefinitely in sealed glass ampoules. Aqueous solutions can be stored for several years in clean containers. Exposure to organic materials or bright light leads to reduction, with a change in color followed by precipitation of flakes of gold. Old solutions and precipitates can be easily recycled in the laboratory (Kiernan, 1977).

Hazards: Solid gold chloride can cause skin blistering followed by violet or brown staining.

Carbon

Synonyms: carbon black, lampblack.
CI 77266, CI Pigment black 6 and 7 (carbon black)
C, atomic weight 12.01

Absorption: Black forms of carbon absorb all visible wavelengths; appearance varies with particle size.

Solubilities: Ordinary carbon is insoluble in water and organic liquids.

Carbon blacks are made by burning hydrocarbon fuels with a limited air supply; carbon particles (20–120 nm diameter) condense where the flame meets a cool surface. **Indian ink (China ink)** is an aqueous dispersion containing carbon black and glue.

Applications: Carbon black, as Indian ink, provides a black background in Duguid's negative staining method for showing capsules of bacteria (Murray *et al.*, 1994). It is injected into living animals, locally or intravenously, to demonstrate cells that phagocytose the particles (Foot, 1929). It is widely used as a physiological tracer, for example, to show branching blood vessels (Ansari, 2001), to map movements of cerebrospinal fluid around blood vessels and nerve roots and into blood and lymph vessels (Miura *et al.*, 1998), and to observe movements of mucus on ciliated epithelial surfaces (Hassab and Kennedy, 2001). Applied to electrophoretic gels, Indian ink provides a permanent general protein stain that can be used in conjunction with immunofluorescent detection (Eynard and Lauriere, 1998).

As a marker, carbon black is fed to animals to determine transit times in the alimentary tract (Jiang and Claussen, 1993), and used by surgeons to detect sentinel nodes (Lucci *et al.*, 1999).

Charcoal and coarsely grained carbon black are extensively used to absorb impurities from liquids. An example is removal of brown solutes from Schiff's reagent and from 3,3'-diaminobenzidine tetrahydrochloride (Ros Barcelo *et al.*, 1989).

Industrial uses – Added to rubber to increase hardness, and for manufacture of printing inks. Grades of high purity are used as food colorants.

Stability: Carbon particles are stable unless combusted or exposed to strong oxidants such as chromic acid (CrO_3 with H_2SO_4).

Sulfides

Insoluble brown or black metal sulfides are the end-products of several micro-scopical techniques, most notably histochemical methods for phosphatases and choline esterases. Soluble salts of the corresponding metals are therefore chro-mogenic reagents in such methods. Examples are lead sulfide, formed from lead phosphate in methods for acid phosphatase; cobalt sulfide formed from cobalt phosphate in calcium–cobalt methods for alkaline phosphatase or ATPase activity; and copper sulfide formed from a copper–thiocholine complex in methods for choline esterases. Minute amounts of metals that form insoluble sulfides can be detected histochemically in tissues by Timm's method and its variants (Danscher, 1996); amplification is brought about by a physical developer (see under **silver nitrate**).

Iodine

I, atomic weight 126.90 or I_2, FW 253.81

Appearance: A dark gray solid, that evolves purple gas (I_2) when heated. Solutions are violet or yellow–brown, see below.
Solubilities: 0.03% in water, 21% in ethanol (brown), 20% in xylene (violet). Solutions in hydrocarbons or chlorohydrocarbons are violet, whereas those in oxygen- or nitrogen-containing organic solvents are yellow to brown, depending on the concentration. Iodine is very soluble (over 60%) in aqueous solutions of iodides (usually KI), giving brown solutions containing I_2 in equi-librium with triiodide (I_3^-) and other polyiodide (e.g. I_5^-, I_7^-) ions.

According to Presnell and Schreibman (1997) the aqueous iodine (KI_3) solu-tions used in bacteriology and histotechnology are **Gram's** (0.33% I_2 in 0.67% KI) and **Weigert's** (1% I_2 in 2% KI). **Lugol's** iodine is 6% I_2 in 4% KI (McClung, 1929; Gray, 1954). Different formulations for these solutions are given in other manuals of microtechnique.

Dissolved in aqueous KI or a solvent such as alcohol, iodine is present as an equilibrium mixture of an uncharged and somewhat hydrophobic diatomic molecule and either polyiodide anions (I_3^- etc.) or iodine–solvent complexes. The interactions of iodine solutions with biological macromolecules are often due to the oxidizing action of the free I_2 (Chilean Iodine Education Bureau, 1951; Sykes, 1958).
Applications: Yellow staining is due to formation of easily dissociated complexes with proteins, cellulose and other materials. Differently colored complexes are formed with starch (blue) and glycogen (brown–red). Newly formed glycogen, the product of histochemical methods for phosphorylase activity, is stained blue. In the starch–iodine complex, linear arrays of I_2 mole-cules or I_3^- ions occupy the central cavity of the helical macromolecule (Rawlings and Schneider, 1970).

The I_3^- ion forms insoluble precipitates with several cationic dyes, notably crystal violet and methyl violet. In Gram staining this immobilizes them within bacteria (Popescu and Doyle, 1996). There are also methods in which crystal violet is applied to iodine-stained plant tissues (Berlyn and Miksche, 1976; Gray, 1954).

Other uses of iodine in microtechnique include fixation of cultured cells (Denning and Fulton, 1986) and protozoa and diatoms, which are also stained (Baker, 1958; Belcher and Swale, 1979; Bowers *et al.*, 2000). Solutions of iodine in KI or alcohol are used to remove insoluble deposits formed in tissues by mercuric chloride-containing fixatives. Colloidal silver deposits can be similarly removed. The color of iodine is removed from tissues by treatment with aqueous sodium thiosulfate.

The most notable medical application of iodine solutions is disinfection, especially of intact skin. Iodine inactivates most viruses (Payan *et al.*, 2001).

Purity of commercial batches: 'Iodine, resublimed' from laboratory suppliers is typically 99.5% pure.

Stability: Containers should be tightly closed to avoid evaporative loss.

Hazards: Solid iodine strongly irritates skin and mucous membranes. Ingestion of 2–3 g may be fatal. Solid and gaseous iodine vigorously attack many metals and organic materials.

ANIONIC ORGANIC COLORANT

Pyranine

CI 59040, CI Solvent green 7

Synonyms: HPTS, 8-hydroxypyrene-1,3,6-trisulfonic acid; do not confuse with pyronine basic dyes

Sodium salt: $C_{16}H_7Na_3O_{10}S_3$, FW 524

Absorption maxima: 403 nm in acidic aqueous solutions; 454 nm in aqueous alkali, emission maximum in this being 511 nm.

Fluorescent pH indicator; changes from blue to green at pH 6.5–7.5; *pK value* 7.2

Solubilities: Highly soluble in water, slightly soluble in glacial acetic acid

Pyranine carries three sulfonate substituents on a pyrene ring system, the anion being extremely hydrophilic. Overall size is moderate, that of the conjugated system small. The dye is sold as the trisodium salt, illustrated.

Applications: Pyranine has occasionally been used as a stain of fixed tissues, for instance eosinophil granules in blood smears (Trigoso *et al.*, 1995). More commonly it is applied as a vital stain, for instance to measure intracellular pH (Gan *et al.*, 1998), and widely as a tracer. Such applications include assessment of the integrity of macrophage vacuoles (Beauregard *et al.*, 1997), and identification of solute pathways in plants (Gisel *et al.*, 1999). Although pyranine is usually applied as the salt, it can be generated *in situ* by photolysis of a caged precursor (Xia *et al.*, 1998).

Industrial uses – a solvent dye, and for coloration of cosmetics and pharmaceuticals

Purity of commercial batches: Dye contents are up to 97%. There is a thin-layer chromatographic method (Marmion, 1991).

Stability: Photofades in solution

COMPOUNDS FORMING COVALENT BONDS WITH TISSUE OR CO-SOLUTES

Dithiooxamide

Synonyms: ethanedithioamide; rubeanic acid
$C_2H_4N_2S_2$ FW 120.2

Solubilities: Slightly soluble in water; soluble in alcohols and alcohol–water mixtures.

This compound is an orange–red solid, whose solutions form insoluble polymeric complexes with many metals.

Applications: The one with copper(II) is dark green, and dithiooxamide is used to detect pathological deposits of this metal histochemically (Cook, 1974; Kiernan, 1999 – note that many manuals use the name rubeanic acid). There are recent studies of the livers of patients with Wilson's disease (Jonas *et al.*, 2001). A related application is to darken deposits of copper ferrocyanide formed as the product of a histochemical method for acetylcholinesterase activity (Nakao *et al.*, 2001). Used together, silver nitrate and dithiooxamide provide a reagent for staining calcium oxalate deposits in the eye (Pecorella *et al.*, 1995). In root canals of teeth, Pecora *et al.* (2000) used dithiooxamide to detect leakage of a $CuSO_4$ tracer from dentinal tubules.

Purity of commercial batches: 98% to over 99% purity is usually claimed by suppliers.
Stability: The solid compound is stable at room temperature. Solutions for histochemical use can be kept for several weeks.

NBD chloride

Synonyms: 4-chloro-7-nitrobenz-2-oxa-1,3-diazole, and 4-chloro-7-nitrobenzo-furazan
$C_6H_2ClN_3O_3$, FW 200

Absorption maxima: 336 nm in methanol; aliphatic primary amine derivatives absorb at around 465 nm in methanol and emit at 535 nm
Solubilities: soluble in acetonitrile, dimethylformamide, and methanol

A benzofurazan heterocyclic system carries a reactive chloro group. Overall size, and that of the conjugated system, is small. Links to nucleophilic groups such as amino, hydroxy and thiol on biomolecules or xenobiotics are formed via the choro group; which also reacts with water.
Applications: NBD chloride is used to generate fluorescent biomolecules and xenobiotics, which are then applied as stains, tracers or substrates in living cells. Representative examples follow. **NBD-phospholipids** – Labeled phosphatidic acid has been used to trace lipid metabolic pathways in yeasts (Trotter, 2000). **NBD-steroids** – A labeled bile acid has been used in investigations of secretion mechanisms of hepatocytes (Miller *et al.*, 1996a). **NBD-peptides** – Interaction of a labeled basic peptide with cultured cells has been used to probe the mechanism of adsorptive endocytosis (Sai *et al.*, 1998). **NBD-xenobiotics** – Labeled cyclosporin A can be used to demonstrate the ATP-driven P-glycoprotein ion transporter (Masereeuw *et al.*, 2000). Other reactive species based on the NBD system include **iodoacetyl NBD**, used to label a calmodulin-binding peptide for demonstrating protein in living cells (Hulvershorn *et al.*, 2001).
Stability: The dry powder should be stored under desiccation, aqueous solutions are unstable; solutions photofade

Glyoxal-*bis*(2-hydroxyanil)

Synonyms: GBHA; 2,2'-(ethanediylidenedinitrilo)diphenol
$C_{14}H_{12}N_2O_2$ FW 240.3

Solubilities: The light brown powder dissolves in alkaline water or dioxane to give a pink to red–brown solution.

Glyoxal-*bis*(2-hydroxyanil) (GBHA)   Calcium chelate of GBHA

Applications: This compound is a histochemical reagent for detecting calcium. Techniques using GBHA are much more sensitive than those using alizarin red S (Wolters *et al.*, 1979), however, the red color of the calcium complex fades after a few days. Both extra- and intracellular calcium can be localized, for example, in hepatocytes following toxic injury (Itoh *et al.*, 1990), in resins bonded to dentine (Hanaizumi *et al.*, 1998), and in cerebral white matter of hydrocephalic rats (Del Bigio, 2000).

Purity of commercial batches: 98% purity is claimed by suppliers.

Stability: The solid compound is stable; storage below 0°C is recommended. Solutions should be prepared immediately before using.

p-Dimethylaminobenzylidenerhodanine

Synonyms: *p*-dimethylaminobenzalrhodanine; 5-(4-dimethylaminobenzyl-idene)rhodanine. Shortened in some staining manuals to rhodanine, unfortunately this is also the name of another compound. Do not confuse with the unrelated rhodamine dyes.

$C_{12}H_{12}N_2OS_2$, FW 264.4

Absorption maximum: 451 nm in methanol.

Solubilities: 0.05% in water; 0.03% in ethanol; more soluble in acetone; soluble in strong acids (yellow solution).

Applications: This compound, which forms colored complexes with several metals, is used histochemically to localize copper in tissues. It is more sensitive than dithiooxamide for this purpose (Irons *et al.*, 1977), but the red–brown product is less stable (Churukian, 2000; Pearse, 1985). Histochemical applications include detection of copper in the livers of poisoned animals (Fuentealba *et al.*, 2000) and of patients with Wilson's disease (Pilloni *et al.*, 1998), and in the toxic milk mutant mouse (Deng *et al.*, 1998).

Purity of commercial batches: 97% purity is claimed by suppliers
Stability: The solid can be stored at room temperature; solutions are kept for only one day.

Dansyl chloride

Synonym: 5-dimethylaminonaphthalene-1-sulfonyl chloride
$C_{12}H_{12}ClNO_2S$, FW 270

Absorption maximum: 372 nm in chloroform, in which solvent fluorescent aliphatic amine derivatives emit at ca 429 nm
Solubilities: Soluble in acetonitrile, chloroform and dimethylformamide

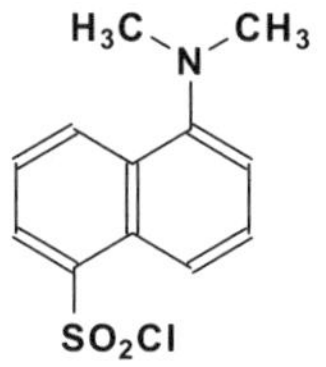

A naphthalene ring system is substituted with dimethylamino and sulfonyl chloride groups. The overall size, like that of the conjugated system, is small.
Applications: Dansyl chloride has been used directly to generate fluorescent products with tissue proteins and peptides, including adhesion peptides on glass surfaces (Kouvroukoglou *et al.*, 2000). Biomolecules labeled for subsequent use as markers and vital stains include G-protein in vesicular membranes (Luan *et al.*, 1995). The reagent is used to label antibodies, for example, for detection of proteins on Western blots (Abuharfeil *et al.*, 1991), and as a general fluorescent stain for proteins on electrophoresis gels (Beeley *et al.*, 1996). A widely used diagnostic application is the assessment of stratum corneum turnover time (Hood *et al.*, 1999; Pierard, 1992). Dansyl chloride is a biochemical reagent in peptide sequencing (see Walker, 1994).
Purity of commercial batches: Lower methylated homologs can be present. There is a thin-layer chromatographic method (Mildenstein, 1971).
Stability: Reactive with dimethylsulfoxide and water; reaction products with biomolecules photofade
Hazards: Harmful if ingested, destructive to mucous membranes

BODIPY and BODIPY derivatives

Nomenclature – The term BODIPY describes the 4,4-difluoro-4-bora-3a,4a-diaza-s-indacene fluorophore, shown below. A wide variety of commercially available reactive derivatives are used to label biomolecules and xenobiotics for application as stains and tracers. Unfortunately such labels are usually described merely as BODIPY in the literature.

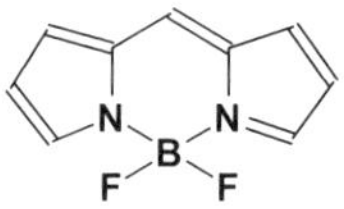

Chemical features – The absorption/emission properties of reactive BODIPY labels are controlled by the substituents on the fluorophore. Ranges of BODIPY derivatives are sold whose absorption maxima cover the visible spectrum. The illustrative examples, below, include an alkylated BODIPY-FL (i.e. spectrally fluorescein-like) and a phenylated BODIPY TR (Texas red-like) derivative. BODIPY-R6G (rhodamine 6G-like) and BODIPY-TMR (tetramethylrhodamine-like) derivatives are also available (Haugland, 1996).

Most reactive BODIPY fluorochromes are rather lipophilic, suitable for preparing labeled peptides or pharmaceuticals. For labeling polynucleotides and proteins hydrophilic compounds are desirable. An example described below, BODIPY-FL CASE, is water soluble due to a sulfonic acid substituent on the side-chain.

Binding of BODIPY dyes to target molecules is often achieved via an *N*-succinimidyl ester substituent, which reacts with nucleophilic groups such as amino, hydroxy or thiol in biomolecules or xenobiotics. Reactive esters are attached to the fluorophore by side-chains of varying length. Extended spacers are present in both examples described below. In other compounds there are shorter spacers, or the reactive group is directly attached to the fluorophore.

Properties of two representative BODIPY dyes are sketched below. **BODIPY-FL CASE** is hydrophilic and used primarily for labeling proteins. **BODIPY-TR-X SE** is lipophilic with a long spacer between the fluorophore and the reactive ester group. Following the physicochemical thumbnails, applications of a variety of BODIPY-labeled biomolecules and xenobiotics are summarized.

BODIPY-FL CASE

$C_{27}H_{38}BF_2N_5O_8S$, FW 641

BODIPY-TR-X SE

$C_{31}H_{29}BF_2N_4O_6S$, FW 634

Absorption/emission maxima in methanol

504/511 nm

588/616 nm

Solubilities

Soluble in water, acetonitrile and dimethylsulfoxide (DMSO)

Insoluble in water, soluble in acetonitrile and DMSO

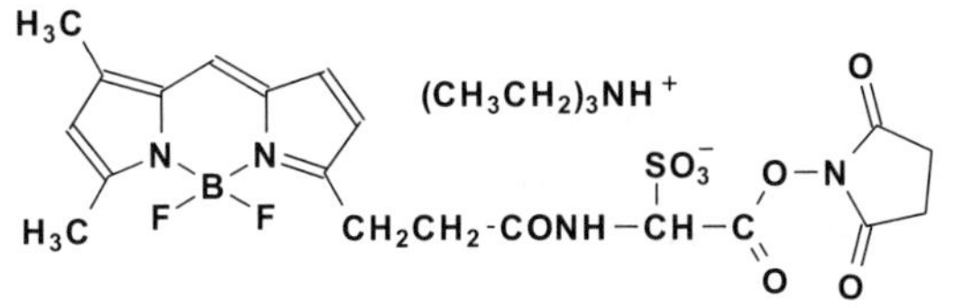

BODIPY- FL, CASE

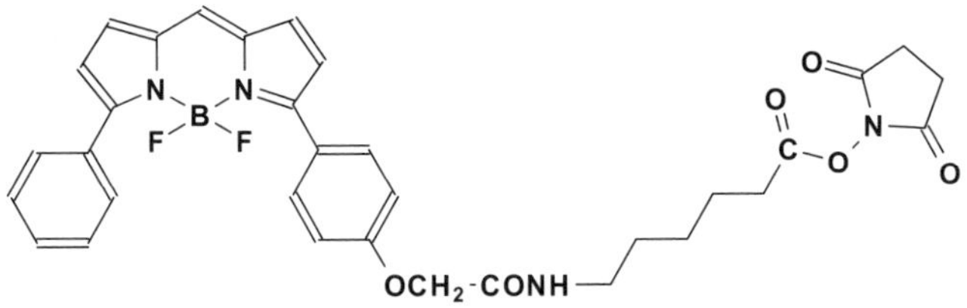

BODIPY-TR-X, SE

Stability: Dry powders should be stored tightly sealed, aqueous staining solutions slowly decompose; photofading occurs, especially in solution

Hazards: These compounds are protein-reactive so they may give rise to hypersensitivity responses

Applications: BODIPY reactive dyes are used to generate fluorescent biomolecules or xenobiotics, which are applied as stains and tracers in living cells. Representative examples follow. BODIPY-proteins – Labeled high density lipoprotein has been used to investigate cholesteryl ester transport within intact cells (Reaven *et al.*, 2001). Labeled opiate peptides have been used to follow the uptake into and trafficking within neurons (Arttamangkul *et al.*, 2000). BODIPY-polynucleotides – These have been used to determine the intracellular localizations of antiherpetic nucleotides (Kulka and Aurelian, 1995), and by making use of FRET the occurrence of hybridization has been monitored intracellularly (Tsuji *et al.*, 2000). BODIPY compounds have also been used to label dUTP for assessing DNA strand breaks using flow cytometry (Celis, 1998). BODIPY-lipids – Labeled phospholipids have been used to study lipid metabolism in wholemount zebrafish (Farber *et al.*, 2001). Labeled ceramide is used as a marker of the Golgi apparatus (Bracho *et al.*, 2001). BODIPY-pharmaceuticals – These have been widely used, for example, to trace the release of the antihypertensive prazosin from intracellular stores (Al-Damluji and Shen, 2001).

Cascade blue

Synonyms: cascade blue hydrazide; a related compound, cascade blue acetyl azide, is also routinely referred to as cascade blue
Trisodium salt: $C_{18}H_{11}N_2Na_3O_{11}S_3$, FW 596

Absorption maxima: ca. 376 and 399 nm in water; in which solvent the emission maximum is 421 nm
Solubilities: The potassium and sodium salts are soluble to 1% in water, and the lithium salt to 8%.

Cascade blue is a trisulfonated pyrene, also carrying a hydrazide group. This latter can form covalent links to aldehydes. Overall size is moderate, that of the conjugated system small. The dye is sold as the trisodium salt, illustrated, whose anion is extremely hydrophilic. The trilithium and tripotassium salts are also commercially available.

Applications: The moderate size and hydrophilic character of cascade blue underlies its use as a tracer for detection of gap junctions between live cells (Di *et al.*, 2001). More often cascade blue is applied to label biomolecules which are then used as tracers or stains, for instance to label nucleotides for *in situ* hybridization (Tanke *et al.*, 1998). Proteins may also be labeled, for example, cascade blue-bungerotoxin is used to stain neuromuscular junctions (Huard *et al.*, 1992). Labeled dextrins of various sizes are widely used, for instance, to assess permeability of the stratum corneum (Grewal *et al.*, 2000). Another application is as a marker, for example, in the form of cascade blue labeled latex beads for identification of transplanted neurons (Pyapali *et al.*, 1992). The dye has been used to label xenobiotics such as selective dopamine-receptor ligands, which are then used to determine anatomical and cellular distributions of dopaminergic receptor subtypes (Ariano *et al.*, 1991).

Cascade blue is applied as a stain in flow cytometry. It has been used directly as a marker of membrane integrity (Boutonnat *et al.*, 1999); and indirectly as a label of dUTP, to assess DNA strand breaks (Celis, 1998), and of antibodies to allow purification of hemopoetic stem cells (Donahue *et al.*, 1999).
Stability: Photofades, especially in solution

Tetracyclines

The tetracyclines are primarily used in treatment of bacterial infections (see textbooks such as Trevor *et al.*, 2002). Their fluorescence and metal complexing characteristics have also resulted in their use as stains. Such applications are noted following the summaries of physicochemical properties of four tetracyclines used as fluorochromes, whose structures are illustrated.

Tetracycline	$R_1=H, R_2=OH, R_3=H$
Oxytetracycline	$R_1=H, R_2=OH, R_3=OH$
Chlortetracycline	$R_1=Cl, R_2=OH, R_3=H$
Doxycycline	$R_1=H, R_2=H, R_3=OH$

Tetracycline
Synonym: achromycin; an antibiotic from *Streptomyces viridifaciens*; also made by dechlorination of chlortetracycline.
Base trihydrate: $C_{22}H_{24}N_2O_8 \cdot 3H_2O$, FW 498.4.
Hydrochloride: $C_{22}H_{25}N_2ClO_8$, FW 480.9.

Absorption maxima: 220, 268 and 355 nm (in 0.1M HCl)
Solubilities: of the trihydrate are 1.7 mg/ml in water, over 20 mg/ml in methanol. The hydrochloride is more than 2% soluble in water, less so in methanol and ethanol, insoluble in ether and hydrocarbons.

Oxytetracycline
Synonyms: 5-hydroxytetracycline; terramycin; an antibiotic from *Streptomyces rimosus.*
Base dihydrate: $C_{22}H_{28}N_2O_{11}$, FW 496.5.
Disodium salt dihydrate, $C_{22}H_{22}N_2Na_2O_9 \cdot 2H_2O$, FW 553.5.
Hydrochloride: $C_{22}H_{25}N_2O_9Cl$, FW 496.9.

Absorption maxima: 249 276 and 353 nm (in 0.1 M phosphate buffer, pH 4.5)
Solubilities: Base dihydrate water soluble to 3.1% at pH 1.2; 0.1% at pH 7.0; 3.9% at pH 9.0; 1.2% soluble in ethanol. The hydrochloride is very soluble in water (1 g/ml); 1.2% in 100% ethanol.

Chlortetracycline
Synonyms: 7-chlorotetracycline; aureomycin; biomycin; an antibiotic from *Streptomyces aureofaciens.*
Base: $C_{22}H_{23}ClN_2O_8$, FW 478.9
Hydrochloride: $C_{22}H_{24}Cl_2N_2O_8$, FW 496.9

Absorption maxima: in 0.1 M HCl: 230, 264 and 388 nm; in 0.1 M NaOH: 255, 285 and 345 nm.
Solubilities: base slightly soluble (0.05%) in water but becomes very soluble above pH 8.5. Freely soluble in ethylene glycol ethers (cellosolves), slightly soluble in ethanol and acetone. The hydrochloride is soluble in water (0.86%; solution has pH 2.8–2.9) and ethanol (0.17%).

Doxycycline
Synonym: vibramycin; a synthesized antibacterial drug notable for being more slowly metabolized and excreted than other tetracyclines.
Base monohydrate: $C_{22}H_{24}N_2O_8.H_2O$, FW 462.5.
Hydrochloride, crystallizes with ethanol and water as a hemiethanolate hemi-hydrate, also called doxycycline hyclate: $C_{22}H_{25}N_2O_8Cl.\frac{1}{2}C_2H_6O.\frac{1}{2}H_2O$, FW 512.9. This is the form present in pharmaceutical products.

Absorption maxima: of doxycycline hyclate, 267 and 351 nm (in methanol with HCl). Yellow–green fluorescent emission.
Solubilities: Doxycycline hyclate is soluble in water.

Applications of tetracyclines: Tetracycline, oxytetracycline and chlortetracycline are taken up by mitochondria following addition to cell culture media or injection into mice (du Buy and Showacre, 1961). In more recent studies chlor-tetracycline has been the preferred compound for this purpose. Tetracyclines have long been administered to living animals to study the growth of bones and teeth. The fluorescent calcium chelate formed in mineralizing tissue persists indefinitely as fluorescent lines. Recent examples include using oxytetracycline to examine alveolar bone remodeling associated with orthodontic treatment (Verna *et al.*, 1999), and an 8-year study of growth of the humerus in a sea turtle (Coles *et al.*, 2001). For technical instructions concerning tetracycline labeling of bone, see Nakamura *et al.* (2000). Chlortetracycline is taken up into the acro-somes when spermatozoa are capacitated in artificial media. This simple test has been used in studies of preservation of semen (Quan *et al.*, 2001).

Tetracycline is used as a histochemical reagent with tissue sections for detecting calcium (Timmers *et al.*, 1996), although it is not specific for this metal. Oxytetracycline has been used, with other dyes, to demonstrate initiation and growth of microcracks in mechanically stressed bones (Lee *et al.*, 2000a,b). Applied to sections of fixed tissue with an excess of aluminum sulfate, tetra-cycline provides a fluorescent nuclear stain (Kiernan, 1999).

Nonmicroscopic fluorescent marking of fishes and other marine animals is achieved by immersion in dilute solutions, usually of oxytetracycline (Hastein *et al.*, 2001). The label persists for at least 8 months and does not enter the edible tissues (Unkenholz *et al.*, 1997).

Purity of commercial batches: Vendors claim purity of 99% for the hydrated base and 95% for the hydrochloride of tetracycline; 80% for the hydrochloride of chlor-tetracycline, plus 20% tetracycline. Methods for assaying tetracyclines, especially in foodstuffs, typically involve initial separation by HPLC (Blanchflower *et al.*, 1997) or TLC (Choma, 2000).

Stability: Solutions are generally stable in acidic conditions, less so in neutral or alkaline media. Neutral solutions of chlortetracycline are less stable than those of tetracycline or oxytetracycline.

Fura-2

Both Fura-2 and its acetoxymethyl (AM) ester Fura-2 AM are described in this entry. Following physicochemical data for the two compounds is an applications section. Fura-2 AM is often misleadingly described merely as Fura-2.

FURA-2	**FURA-2 AM**
Potassium salt: $C_{29}H_{22}K_5N_3O_{14}$	$C_{44}H_{47}N_3O_{24}$
FW 832	FW 1002

Absorption maxima

363 nm in water, in which solvent the emission maximum is 512 nm. In aqueous calcium salts the corresponding values are 335 and 505 nm

370 nm in ethyl acetate, in which solvent the emission maximum is 476 nm

Metal complexation

chelates
calcium ions

Solubility

Very soluble in aqueous solutions above pH 6

Insoluble in water
soluble in dimethylsulfoxide

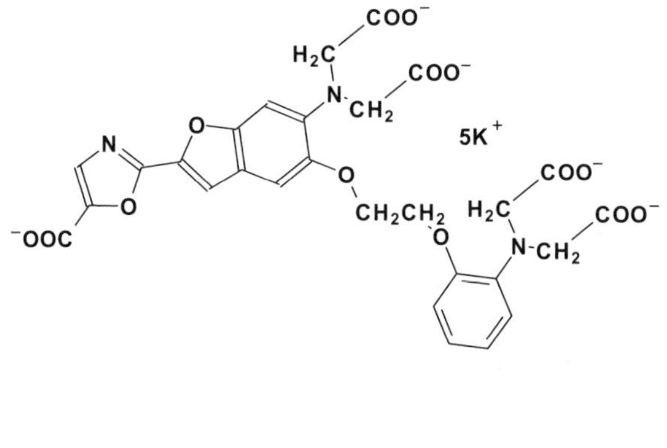

Fura-2 carries two calcium-chelating methyliminodiacetic acid moieties, attached to a complicated organic framework. The hydrophilic pentacarboxylate, illustrated, has a small conjugated system but a large overall size. The superlipophilic penta-acetoxymethyl ester derivative fura-2 AM is even larger. The free acid, the penta-potassium salt, and the acetoxymethyl ester are available commercially.

Purity of commercial batches. There is a HPLC method for both fura-2 and fura-2 AM (Tran *et al.*, 1995).

Stability. Both salt and ester photofade, especially in solution; de-esterification of fura-2 AM can occur in methanolic solution, especially in glass containers

Applications: Fura-2 is widely used to detect and assay intracellular calcium ions. The calcium chelating species is hydrophilic and membrane impermeant, the Fura-2 AM ester often being used to facilitate entry into live cells. Following uptake this is converted by intracellular esterases to the active species, Fura-2. Image-based variants of such applications have been repeatedly described in manuals and reviews (Celis, 1998; Dupont *et al.*, 2000; Silver, 1998). Use of fura-2 as a stain for cytosolic calcium is also described in flow cytometry manuals (Darzynkiewicz and Crissman, 1990; Macey, 1994).

   Fura-2 derivatives include **fura-2-FF**, used as a low affinity calcium stain (Raza *et al.*, 2001). Its modified chelating properties result from fluoro substitution of the phenyl ring system. **Fura-2-FF-C18** carries a lipophilic alkyl chain, and is an amphiphile used to assay near-membrane calcium ions (Wang *et al.*, 2001).

Mag-fura-2

Both mag-fura-2 and its acetoxymethyl (AM) ester, mag-fura-2 AM, are described in this entry. Following physicochemical data for the two compounds, is an applications section. Mag-fura-2 is sometimes termed furaptra

MAG-FURA-2	**MAG-FURA-2 AM**
Potassium salt: $C_{18}H_{10}K_4N_2O_{11}$ FW 587	$C_{30}H_{30}N_2O_{19}$ FW 723

Absorption maxima:

369 nm in water, in which solvent emission maximum is 511 nm. In aqueous calcium salts the corresponding values are 329 and 508 nm	366 nm in ethyl acetate

Metal complexation:

Chelates calcium, magnesium and related ions	

Solubility:

Soluble in aqueous solutions above pH 6	Insoluble in water, soluble in dimethylsulfoxide

410

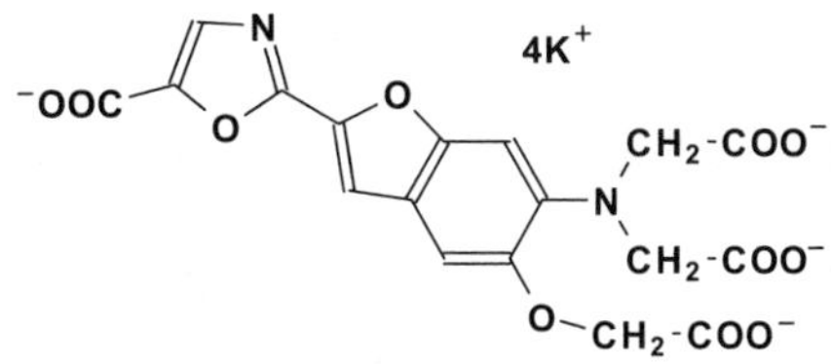

Mag-fura-2 carries a metal ion-chelating methyliminodiacetic acid group, plus two other carboxy groups, attached to a complex organic framework. The hydrophilic tetracarboxylate, illustrated, has a small conjugated system and is of moderate overall size. The tetra acetoxymethyl ester derivative mag-fura-2 AM is lipophilic, and its overall size is large. The tetrapotassium salt of mag-fura-2 and the acetoxymethyl ester are available commercially.

Purity of commercial batches: An HPLC method is available applicable to both the salt and the ester (Castle and Neuteboom, 1995).

Stability – both salt and ester photofade in solution; decomposition can occur in methanolic solution, especially in glass containers

Applications: Mag-fura-2 is used to detect and assay intracellular magnesium and, despite the name, calcium and other metal ions. The hydrophilic chelating species is membrane impermeant, so to assist entry into live cells the ester mag-fura-2 AM is often used. Following uptake this is converted by intracellular esterases to the active species mag-fura-2. For accounts of such applications see laboratory manuals such as Celis (1998) and Mason (1999). Interestingly, mag-fura-2 has been used to report on calcium content in a spatially restricted region of cells, namely endoplasmic reticulum (Wang *et al.*, 2001).

Fura red

Both fura red and its acetoxymethyl (AM) ester, fura red AM, are described in this entry. Initially physicochemical data for the two compounds are given, following which is an applications section.

Fura Red

Potassium salt: $C_{29}H_{24}K_4N_4O_{12}S$
FW 809

Fura Red AM

$C_{41}H_{44}N_4O_{20}S$
FW 1089

Absorption maxima:

473 nm in water, in which solvent the emission maximum is 670 nm. In aqueous calcium salts the corresponding values are 436 and 655 nm

458 nm in methanol, in which solvent the emission maximum is 597 nm

Metal complexation:

Chelates calcium ions

411

Solubility:

Soluble in aqueous
solutions > pH 6, and in methanol

Insoluble in water, soluble in
dimethylsulfoxide and methanol

Fura red carries two metal ion-chelating methyliminodiacetic acid groups, attached to an organic framework containing phenyl, benzofuran and imidazolidine rings. The hydrophilic tetracarboxylate, illustrated, has a small conjugated system and is of substantial overall size. The tetra acetoxymethyl ester derivative fura red AM is strongly lipophilic, and its overall size is large. The tetrapotassium salt and the acetoxymethyl ester of fura red are available commercially.

Fura red is used to detect and assay intracellular calcium. As the calcium-chelating species is membrane impermeant, the AM ester is often used to facilitate probe entry into live cells. Following uptake the active species is generated by intracellular esterases. Such applications are described in laboratory manuals (Celis, 1998; Mason, 1999). Fura red is also used in this way with plant cells (Walczysko *et al.*, 2000). In flow cytometry fura red is applied both to measure intracellular calcium, as reviewed by Burchiel *et al.* (1999), and to assay cell adhesion (Edwards *et al.*, 2001).

Stability: Both salt and ester photofade, especially in solution

PEROXIDASE CHROMOGENS

Peroxidase from horseradish, *Armoracia rusticana*, (HRP) is a frequently used label for macromolecular reagents such as antibodies, lectins and avidin. The enzyme catalyzes oxidation of many organic compounds by its substrate, which is hydrogen peroxide. Organic compounds yielding colored or fluorescent products when oxidized by H_2O_2 at sites of peroxidase activity are termed chromogens.

Diaminobenzidine

Synonym: DAB
Base: $C_{12}H_{14}N_4$, FW 214
Tetrahydrochloride dihydrate: $C_{12}H_{22}Cl_4N_4O_2$, FW 396

Appearance: White solid that forms colorless solutions. The oxidation product is brown.
Solubilities: The base is insoluble in water, the tetrahydrochloride is soluble. Aqueous solutions may be prepared by dissolving the salt in water before adding a neutral or slightly basic buffer.

diaminobenzidine

The brown oxidation product, assumed to be a polymer crosslinked by multiple aza groups (Stoward and Pearse, 1991, p. 98), is insoluble in water and other solvents used in microtechnique.
Applications: Techniques for histochemical localization of endogenous enzymes (peroxidase, catalase and several oxidases) are given in texts such as Kiernan (1999), Stoward and Pearse (1991) and Van Noorden and Frederiks (1992). The same methods are used to detect horseradish peroxidase, whether introduced into animals as a tracer or used as a label for antibodies (Bancroft and Gamble, 2002; Polak and Van Noorden, 1997; van der Loos, 1999) or lectins (Brooks *et al.*, 1997). Nonenzymatic oxidation of DAB occurs in techniques that detect production of oxygen radicals by living cells (Karnovsky, 1994; Kerver *et al.*, 1997).

DAB is also used in procedures that increase the sensitivity of Perls' Prussian blue reaction for iron (Connor *et al.*, 1995), and in enzyme histochemistry to enhance the visibility of such end products as cerium perhydroxide and cerium phosphate (Nakos and Gossrau, 1993; Van den Munckhof, 1996) and copper ferrocyanide (Tago *et al.*, 1986). Intense illumination of fluorescently stained preparations in the presence of DAB and hydrogen peroxide deposits a permanent brown product at sites of fluorescence, a procedure known as photoconversion (Kacza *et al.*, 1997).

The ordinarily brown DAB end-product can be made dark blue by inclusion of cobalt, copper or nickel salts in the incubation medium (Hsu and Soban, 1982), or it can be intensified by silver deposition and gold toning (Gallyas *et al.*, 1982; Newman, 1998).

DAB reacts with aldehydes forming colorless but fluorescent products. This occurs simultaneously with enzymatic oxidation and provides, for example, a built-in counterstain for myelinated nerve fibers in immunostained frozen sections of brain (von Bohlen und Halbach and Kiernan, 1999).

Purity of commercial batches: Suppliers specify up to 98% purity. If solid DAB is brown or its solution is not colorless, there will be nonspecific brown staining of tissue. Brown impurities are removed by shaking solutions with activated charcoal (Ros Barcelo *et al.*, 1989).

Stability: Light sensitive. The solid oxidizes with aging but can be kept for several years if stored in small airtight containers in the freezer.

Hazards: Skin irritant. DAB is a possible carcinogen; related compounds such as benzidine are known carcinogens. DAB is destroyed by oxidation. Sodium hypochlorite (household bleach) is frequently used but acidified $KMnO_4$ may be preferable (Lunn and Sansone, 1990).

benzidine

tetramethylbenzidine

Related compounds: **Benzidine**, either as the free base ($C_{12}H_{12}N_2$, FW 184.2) or dihydrochloride (FW 257.2) is a sensitive chromogen for detection of peroxidase activity, generating a blue dimeric oxidation product. It has been used to detect axonally transported HRP in neuroanatomical studies (Lynch *et al.*, 1973) and, with DAB, in two-color immunostaining for different antigens (Levey *et al.*, 1986). **Tetramethylbenzidine** (TMB) is preferred as a chromogen that provides a blue reaction product because it is probably not carcinogenic (Chung, 2000). TMB is supplied as the water-insoluble base ($C_{16}H_{20}N_2$, FW 313.2), which must be dissolved in a small volume of alcohol or dimethylsulfoxide before being diluted with an aqueous buffer solution (Gomez-Segade, 1980). TMB is used mainly for localizing the activity of axonally transported HRP and HRP-labeled lectins (Olsson *et al.*, 1983). The blue reaction products of benzidine and TMB are prone to over-oxidation, giving light brown polymers. The blue color can be stabilized, for example, by adding tungstate ions to the incubation medium (Llewellyn-Smith *et al.*, 1993).

Tyramide reagents

Tyramine (2-*p*-hydroxyphenylethylamine, $C_8H_{11}NO$, FW 137.2) and its hydrochloride (FW 173.7) are soluble in water and alcohol. The amino group can be linked to labeling reagents such as a reactive fluorochrome or sulfo-*N*-succinimidobiotin, providing tyramide reagents that can be used to detect peroxidase labels with great sensitivity in immunohistochemistry and nucleic acid hybridization (Adams, 1992; Bobrow *et al.*, 1989). Such reagents may be made in the laboratory (Hopman *et al.*, 1998; Speel *et al.*, 1999) or purchased as components of kits, with the exact identity and nature of components usually not stated.

For detection of peroxidase-labeled antibodies and nucleotides, tyramides provide greater sensitivity than any other chromogenic system (Wiedorn *et al.*, 1999). This is not always advantageous however, especially for immunohisto-chemistry (Mengel *et al.*, 1999). Moreover, intensity of staining by this and other amplifying techniques is not proportional to quantity of antigen or of bound primary antibody (Watanabe *et al.*, 1999).

28

Methods for testing biological stains

D.P. Penney and J.M. Powers, with the assistance of
C. Willis, M. Frank and C. Churukian

Procedures were worked out for evaluating the most common stains by Peterson *et al.* (1933a,b,c; 1934a,b). These methods, improved and extended over the years, are used in the examination of dyes submitted to the Biological Stain Commission for certification. In addition to determining the dye content, stains are tested in the applications for which they are supplied, and some are tested for purity by examination of the complete absorption spectrum and by chromatographic methods.

The following is a summary of the methods used in the Commission's laboratory in Rochester, NY shortly before the publication of this 10th Edition of *Conn's Biological Stains*. A more complete and regularly updated account will be printed as a special issue of the Biological Stains Commision's Journal, *Biotechnic & Histochemistry*. An electronic version will be available at www.connsstains.com. This provides for each certified stain detailed accounts of the assay and associated calculations, and technical instructions for all the staining methods and other biological tests used in the Commission's laboratory.

GENERAL METHODOLOGY

Spectral characteristics

A UV/visible absorption spectrum is obtained from a solution of the dye in water or other appropriate solvent. Elaborate methods have been used to express the shape of a spectral curve independently of the peak position (Lillie and Roe, 1942), but we have chosen the simple ratio of the color densities at minus and plus 15 nm of the peak wavelength **(P–15/P+15)** found with the given sample. Several dyes are assayed by spectrophotometry.

Titanous chloride assay

Many dyes are reduced by Ti^{3+} ions to colorless or differently colored substances. Ti^{3+} salts are readily oxidized to Ti^{4+} by atmospheric oxygen. Consequently titanous chloride solutions must be standardized before using. This is accomplished by titration against a commercially available standard (0.1N) ceric sulfate solution, which is completely stable. The indicator used in the titration of $TiCl_3$ against $CeCl_4$ is ferrous-*o*-phenanthroline (ferroin), which gives a sharp end point. The titrations are carried out in an atmosphere of nitrogen, using the apparatus shown in Figure 28.1.

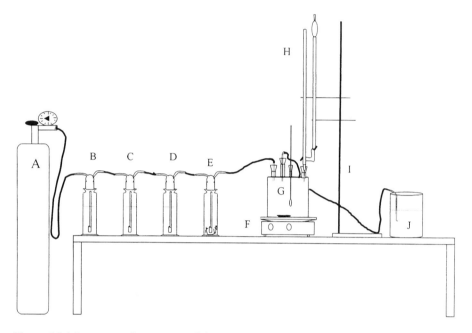

Figure 28.1 Apparatus for titanous chloride titration. **A**: Nitrogen source (Open valve to allow a steady flow of bubbles in the gas washing bottles B, C and D. A steady flow of bubbles in the collection beaker J verifies that the system is not leaking). **B, C, D**: Gas washing bottles, each two-thirds filled with $TiCl_3$-HCl-safranine solution. **E**: Dry bottle with cotton in bottom, to capture any fluids caught in tubing. **F**: Hotplate-magnetic stirrer. **G**: Custom-made 500 ml titration chamber with four ports (for buret, thermometer, gas inlet and gas outlet). **H**: Buret, filled with $TiCl_3$ solution. **I**: Ring stand. **J**: Washing beaker (500 ml, filled with water to rinse exhaust gas from titration chamber).

The percentage of dye is calculated by the general formula:

$$\text{Percentage of dye} = \frac{\text{ml TiCl}_3 \text{ used} \times \text{normality of TiCl}_3 \times \text{mol. wt. of dye} \times 100}{\text{weight of sample} \times \text{no. of hydrogen equivalents} \times 1000}$$

The molecular weight of the dye is that of its organic ion with the usual counter-ion (usually Cl⁻ or Na⁺) and does not include water of crystallization or other inorganic components such as half-$ZnCl_2$.

Chromatography

Thin-layer chromatography (TLC) and high-performance liquid chromatography (HPLC) are used for samples expected to contain two or more colored components. Dyes routinely examined by TLC and HPLC are azure B, azure C, Giemsa stain, Jenner's stain, MacNeil's tetrachrome stain, methylene blue, methylene violet Bernthsen, oil red O, thionine, toluidine blue O and Wright's stain.

Other techniques

Percentages of dye have been calculated by comparing observed with ideal *elemental analyses*, to provide factors applicable to spectrophotometric assay of these dyes. Eosin and some related xanthene dyes are assayed *gravimetrically*, after precipitation of the color acid. *Biological tests* for dyes are staining methods applied to sections of animal or plant material or bacteria. Instructions for most of the staining methods used can be found in the manuals of Clark (1981) and Churukian (2000).

TESTS FOR INDIVIDUAL STAINS

The following tabulation does not include all the tests carried out in the Commission's laboratory.

Table 28.1 Testing of stains by the Biological Stain Commission. The type of assay, minimum dye content (percent by weight) for certification and principal biological tests are shown.

Dye Biological tests	Assay	Minimum dye content

Acid fuchsine (C.I. 42685)
  Van Gieson's stain[1]
  Mallory's stain[1]
  Altmann's mitochondrial stain[2]
	TiCl$_3$ titration	60%

Alcian blue 8G (C.I. 74240)
  Scott's alcian blue method[3]
  Mowry's alcian blue (pH 2.5)[4]
  Kreyberg's method for keratin
  and mucus[5]
	Spectrophotometric	50%

Alizarin red S (C.I. 58005) Dahl's method for calcium[1,6]	Not assayed	

Aniline blue (C.I. 42780)
  Mallory's stain[1]
  Gomori's trichrome stain[1]
	TiCl$_3$ titration	12.0 ml of
		0.1 N TiCl$_3$
		per g of dye[7]

Auramine O (C.I. 41000) Modified Truant's stain[8]	Spectrophotometric	80%
Azocarmine B (C.I. 50090) Heidenhain's AZAN[1]	TiCl$_3$ titration	80%
Azocarmine G (C.I. 50085) Heidenhain's AZAN[1]	TiCl$_3$ titration	75%

Azure A (C.I. 52005)
  Modified Nocht method[10]
  Wright, Jenner and NCCLS
  blood stains[1]
	TiCl$_3$ titration	75%[9]

Azure B (C.I. 52010)
  Modified Nocht method[10]
  Wright, Jenner and NCCLS
  blood stains[1]
	TiCl$_3$ titration	89%

Azure C (C.I. 52002)
  Modified Nocht method[10]
  Wright, Jenner and NCCLS
  blood stains[1]
	TiCl$_3$ titration	70%

Basic fuchsine
  Acid-fast staining[1]
  Periodic acid-Schiff[1]
  Feulgen stain for DNA[1]
  ENDO culture medium
  Leifson's method for
  bacterial flagella[11]
	TiCl$_3$ titration	88%

Bismarck brown Y (C. I. 21000) Stain for nuclei and mucus[12]	TiCl₃ titration	45%
Brilliant cresyl blue (C.I. 51010) Supravital stain for reticulocytes[2]	TiCl₃ titration	60%
Brilliant green (C.I. 42040) Bacteriostatic potency	TiCl₃ titration	90%
Carmine (C.I. 75470) Orth's method for nuclei[2] Southgate's method for mucus[1] Best's method for glycogen[1,2]	Not assayed	
Chlorazol black E (C. I. 30235) Stain for plant tissues	Not assayed	
Congo red (C. I. 22120) Counterstain to alum-hematoxylin Highman's amyloid stain[13] Puchtler's amyloid stain[1]	TiCl₃ titration	85%
Cresyl violet Stain for Nissl substance and nuclei in nervous tissue[1]	TiCl₃ titration	65%
Crystal violet (C.I. 42555) Gram stain for bacteria Flemming's triple stain[14] Newton's technique[2,15] Lieb's amyloid stain[1,16]	Spectrophotometric	90%
Darrow red Nissl staining of neurons[17]	TiCl₃ titration	65%
Eosin B (C.I. 45400) Counterstain for alum-hematoxylin Modified Nocht method[10]	Gravimetric	85%
Eosin Y (C.I. 45380) Counterstain for alum-hematoxylin In Wright's blood stain Modified Nocht method[10] In eosin-methylene blue culture medium	Gravimetric	88%
Erythrosin B (C.I. 45430) Jackson's plant anatomy stain[18] Counterstain for alum-hematoxylin Method for keratin and mucus[1,5]	Gravimetric	80%
Ethyl eosin (C.I. 45386) Stovall and Black method for Negri bodies[2]	Spectrophotometric	80%
Ethyl green (C.I. 42590) ('methyl green')[19] Altmann's mitochondrial stain[2] Ethyl green-pyronine method for DNA and RNA[1,20]	TiCl₃ titration	80%

Fast green FCF (C.I. 42053)	$TiCl_3$ titration	85%
With safranine O, for plant tissues[21]		
Gomori's trichrome stain[1]		
Fluorescein isothiocyanate	Not assayed	
Tested for protein conjugate (FITC)		
formation[22]		
Giemsa stain	TLC and HPLC	
Stain for blood films		
Wolbach's Giemsa method		
for sections[23]		
Hematoxylin (C. I. 75290)	Not assayed	
Weigert's iron-hematoxylin[1]		
Mayer's hemalum[1]		
Harris's hemalum[1]		
Heidenhain's technique[2]		
Indigocarmine (C.I. 73015)	$TiCl_3$ titration	80%
Cajal's picro-indigocarmine method[2]		
Janus green B (C.I. 11050)	$TiCl_3$ titration	50%
Doan & Ralph supravital method		
for mitochondria in blood cells[2]		
Jenner's stain	TLC and HPLC	
Stain for blood films		
Modified Nocht method[10]		
Light green SF (C.I. 42095)	$TiCl_3$ titration	65%
With safranine O, for plant tissues[21]		
Gomori's trichrome stain[1]		
Malachite green (C.I. 42000) (oxalate or	$TiCl_3$ titration	90% (oxalate)
chloride)		
Method with safranine O for		75% (chloride)
fungi parasitic in plants		
Spore stain for bacteria[24]		
Martius yellow (C.I. 10315)	$TiCl_3$ titration	Uncertain[25]
Counterstain in lacmoid method		
for pollen tubes[26]		
Yellow counterstain for		
alum-hematoxylin		
Methyl orange (C.I. 13025)	$TiCl_3$ titration	85%
As substitute for orange G in		
Wilson & Ezrin's pituitary		
stain[27], Kreyberg's method[5],		
Flemming's triple stain[14],		
and as a counterstain in		
iron-hematoxylin techniques		
Methyl violet (C.I. 42535)	Spectrophotometric	75%
Flemming's triple stain[14]		
Newton's technique[2,15]		
Lieb's amyloid stain[1,16]		

Methylene blue (C.I. 52015) TiCl$_3$ titration 82%
    Mallory's phloxin-methylene blue[4]
    Levine & Black stain for bacteria
    in milk[2]
    As a component of Wright's
    blood stain

Methylene violet Bernthsen (C.I. 52041) Spectrophotometric 65%
    As an ingredient of MacNeal's
    tetrachrome stain for blood

Neutral red (C.I. 50040) Spectrophotometric 50%
    Doan & Ralph supravital method
    for leukocyte granules[2]

Nigrosine, water soluble (C.I. 50420) Not assayed
    Dorner's bacterial spore
    stain[28]

Nile blue A (C.I. 51180) TiCl$_3$ titration 75%
    Method for melanin and lipofuscin[29]
    Method for fatty acids[29]

Oil red O (C.I. 26125) Spectrophotometric 70%
    Stain for fat[30]

Orange II (C.I. 15510) TiCl$_3$ titration 85%
    As substitute for orange G in
    Wilson & Ezrin's pituitary stain[27],
    Flemming's triple stain[14], and as a
    counterstain in iron-hematoxylin
    techniques

Orange G (C.I. 16230) TiCl$_3$ titration 80%
    Mallory's stain[1]
    Wilson & Ezrin's pituitary stain[27]
    Kreyberg's method for keratin
    and mucus[5]
    Flemming's triple stain[14],
    Counterstain in Heidenhain's
    iron-hematoxylin technique[2]

Orcein Not assayed
    Darrow's elastic fiber stain[2]
    Stain for hepatitis B surface antigen[31]

Pararosaniline (C.I. 42500) HPLC and 88%
    Acid-fast staining[1] TiCl$_3$ titration
    Periodic acid-Schiff[1]
    Feulgen stain for DNA[1]
    ENDO culture medium
    Leifson's method for
    bacterial flagella[11]
    Gomori's aldehyde fuchsine
    method[32] for pancreatic
    islet beta cells

Phloxin B (C. I. 45410)   Gravimetric   80%
    Mallory's phloxin-methylene
    blue stain[4]
    Kreyberg's method for keratin
    and mucus[1,5]

Protargol-S (and similar silver proteinates)   Not assayed
    Modified Bodian's method
    for nerve fibers[1]

Pyronine B (C.I.45010)   Spectrophotometric   40%
    Ethyl green-pyronine method
    for DNA and RNA[1,20]

Pyronine Y (C.I. 45005)   Spectrophotometric   50%
    Ethyl green-pyronine method
    for DNA and RNA[1,20]

Resazurin   Spectrophotometric
    No test currently conducted
    by the Biological Stain Commission

Rose Bengal (C.I. 45440)   Gravimetric   80%
    Conn's technique for bacteria in soil[1]
    Kreyberg's stain for keratin
    and mucus[5]

Safranine O (C.I. 50240)   Spectrophotometric   80%
    Flemming's triple stain[14]
    Gram stain for bacteria
    With fast green FCF or light
    green SF, for plant tissue[21]

Sudan black B (C.I. 26150)   Not assayed
    Burdon's stain for bacterial lipid[33]
    In place of oil red O in
    Lillie-Ashburn technique[30]

Sudan III (C. I. 26100) $TiCl_3$ titration   80%
    In place of oil red O in   after sulfonation
    Lillie-Ashburn technique[30]

Sudan IV (C. I. 26105) $TiCl_3$ titration   80%
    In place of oil red O in   after sulfonation
    Lillie-Ashburn technique[30]

Tetrachrome stain (MacNeal)   TLC and HPLC
    Stain for blood films

Thionine (C.I. 52000)   Spectrophotometric   85%
    Churukian-Schenk method for
    cartilage and mast cells[34]
    Stoughton's method for fungi
    in plant tissue[35]

Toluidine blue (C.I. 52040) $TiCl_3$ titration   50%
    Metachromatic stain for mast
    cells and cartilage[34]
    Modified Nocht method[10]

Wright's stain   TLC and HPLC
    Stain for blood films
    Modified Nocht method[10]

Footnotes to Table 28-1.

[1] see Churukian (2000)

[2] see Clark (1981)

[3] Scott & Dorling (1965)

[4] see McManus & Mowry (1960)

[5] Kreyberg (1961), as modified by Churukian & Schenk (1984)

[6] Dahl (1952)

[7] The variable nature of aniline blue precludes the correlation of this standard with a percentage by weight of dye

[8] Churukian (1991)

[9] At least 33% must be azure A as determined by HPLC

[10] A Nocht-type (azure-eosin) solution for application to paraffin sections of formaldehyde-fixed tissue contains a blue cationic thiazine and a red anionic xanthene dye, dissolved in an acetone-water mixture buffered to pH 4.1 (see Lillie & Fullmer, 1976). Pre-made blood stains are used in this way by 8-fold dilution of a stock solution in methanol with 10% acetone in an aqueous phosphate-citrate buffer.

[11] Leifson (1951) This test is carried out only for batches of basic fuchsine designated by the vendor as 'special, for flagella'.

[12] With 1% aqueous solution of dye

[13] Highman (1946)

[14] Margolena (1935)

[15] Newton (1925)

[16] Lieb (1947)

[17] Powers *et al.* (1960)

[18] Jackson (1926)

[19] See Chapter 14 for explanation of this dye's identity.

[20] Potvin (1979)

[21] Haynes (1928)

[22] Cherry *et al.* (1969); NCCLS (1975)[Q2]

[23] see Luna (1968)

[24] Schaeffer and Fulton (1933)

[25] Acceptable minimum dye content not yet established for martius yellow, which is sold as Na or Ca salt.

[26] Nebel (1931); Ruzin (1999)

[27] Wilson and Ezrin (1954)

[28] see Gray (1954)

[29] see Lillie and Fullmer (1976)

[30] Lillie and Ashburn (1943), as modified by Churukian (1999)

[31] Deodhar (1975[Q1]); Churukian (2000)

[32] Shikata *et al.* (1974)

[33] Burdon *et al.* (1947)

[34] Churukian and Schenk (1981)

[35] Stoughton (1930)

BIBLIOGRAPHY

Abdel-Hamid KM, Tymianski M (1997) Mechanisms and effects of intracellular calcium buffering on neuronal survival in organotypic hippocampal cultures exposed to anoxia/aglycemia or to excitotoxins. *Journal of Neuroscience* 17:3538–3553.

Abe S, Satoh T, Tokuda Y, Tansho S, Yamaguchi H (1994) A rapid colorimetric assay for determination of leukocyte-mediated inhibition of mycelial growth of *Candida albicans*. *Microbiology and Immunology* 38:385–388.

Abolhoda A, Yu SM, Oyarzun JR, Allen KR, McCormick JR, Han SG, Kemp FW, Bogden JD, Lu Q, Gabbay S (1996) No-react detoxification process: a superior anticalcification method for bioprostheses. *Annals of Thoracic Surgery* 62:1724–1730.

Aboussaouira T, Gerard A, Gerard H (1998) Effect of in utero infusion route on lymphocyte distribution in fetal rat tissues. *Fetal Diagnosis and Therapy* 13:216–222.

Abrahart EN (1977) *Dyes and Their Intermediates*, 2nd Edn. Edward Arnold, London.

Abramsson-Zetterberg L, Zetterberg G, Bergqvist M, Grawe J (2000) Human cytogenetic biomonitoring using flow-cytometric analysis of micronuclei in transferrin-positive immature peripheral blood reticulocytes. *Environmental and Molecular Mutagenesis* 36:22–31.

Abreu MA, Baroza LGV, Rossi MA (1993) Toluidine blue-basic fuchsin stain for glycol-methacrylate embedded tissue. *Journal of Histotechnology* 16:139–140.

Abuharfeil NM, Atmeh RF, Abo-Shehada MN, el-Sukhon SN (1991) Detection of proteins after immunoblotting on nitrocellulose using fluorescent antibodies. *Electrophoresis* 12:683–684.

Accini L, Hsu KC, Speile H, de Martino C (1974) Picric acid-formaldehyde fixation for immunoferritin studies. *Histochemistry* 42:257–264.

Adamchik Y, Frantseva MV, Weisspapir M, Carlen PL, Velazquez JLP (2000) Methods to induce primary and secondary traumatic changes in organotypic hippocampal slice cultures. *Brain Research Protocols* 5:153–158.

Adamo HL, Buruiana R, Schertzer L, Boylan RJ (1999) A comparison of MTA, Super-EBA, composite and amalgam as root-end filling materials using a bacterial microleakage model. *International Endodontic Journal* 32:197–203.

Adams CWM, Bayliss OB (1968) Reappraisal of osmium tetroxide and OTAN histo-chemical reactions. *Histochemie* 16:162–166.

Adams JC (1992) Biotin amplification of biotin and horseradish peroxidase signals in histochemical stains. *Journal of Histochemistry and Cytochemistry* 40:1457–1463.

Adaskaveg JE (1995) Conidial morphology, host colonization, and development of shot hole of almond caused by *Wilsonomyces carpophilus*. *Canadian Journal of Botany* 73:432–444.

Adler J, Baldwin PM, Melia CD (1995) Starch damage. 2. Types of damage in ball-milled potato starch upon hydration observed by confocal microscopy. *Starch* 47:252–256.

Agmon A, Yang LT, Jones EG, O'Dowd DK (1995) Topological precision in the thalamic projection to neonatal mouse barrel cortex. *Journal of Neuroscience* 15:549–561.

Akashi K, Miyata H, Itoh H, Kinosita K (1996) Preparation of giant liposomes in physiological conditions and their characterisation under an optical microscope. *Biophysical Journal* 71:3242–3250.

Akura K, Takenaka M (1991) Modified EA staining solution using fast green in the Papanicolaou method. *Diagnostic Cytopathology* 7:317.

Al Mehdi AB, Shuman H, Fisher AB (1997) Intracellular generation of reactive oxygen species during nonhypoxic lung ischemia. *American Journal of Physiology – Lung Cellular and Molecular Physiology* 16:L294–L300.

Al-Damluji S, Shen WB (2001) Release of amines from acidified stores following accumulation by Transport-P. *British Journal of Pharmacology* 132:851–860.

Albassam MA, Metz AL, Potoczak RE, Gallagher KP, Haleen S, Hallak H, McGuire EJ (2001) Studies on coronary arteriopathy in dogs following administration of CI-1020, and endothelin A receptor antagonist. *Toxicologic Pathology* 29:277–284.

Albert A (1966) *The Acridines. Their preparation, physical, chemical and biological properties and uses.* Arnold, London.

Alcantara Licudine JP, Kawate MK, Li QX (1997) Method for the analysis of phloxine B, uranine, and related xanthene dyes in soil using supercritical fluid extraction and high performance liquid chromatography. *Journal of Agricultural and Food Chemistry* 45:766–773.

Ali SA, Cesani F, Nusynowitz ML, Briscoe EG, Shirtliff ME, Mader JT (1997) Skeletal scintigraphy with technetium-99m-tetraphenyl porphyrin sulfonate for the detection and determination of osteomyelitis in an animal model. *Journal of Nuclear Medicine* 38:1999–2002.

Ali TT (1992) Altered collagen (fibrinoid change) at the site of postmortem injuries. *Medicine Science and the Law* 32:218–224.

Aliotta G, Roso E (1971) Cromatografia sobre papel y en capa delgada de indicadores derivados del la fenolsulfontaleina. *Anales de la Acociasion Quimica Argentina* 59:437–439.

Allchin JP, Evans GO (1986) A simple rapid method for the detection of rat urinary proteins by agarose electrophoresis and nigrosine staining. *Laboratory Animals* 20:202–205.

Allen DT, Kiernan JA (1994) Penetration of proteins from the blood into peripheral nerves and ganglia. *Neuroscience* 99:755–764.

Allen EM (1993) Acute iodine ingestion increases intrathyroidal glutathione. *Journal of Endocrinological Investigation* 16:265–270.

Allen RLM (1971) *Colour Chemistry.* Nelson, London.

Alles AJ, Waldron MA, Sierra LS, Mattia AR (1995) Prospective comparison of direct immunofluorescent and conventional staining methods for detection of *Giardia* and *Cryptosporidium* in human fecal specimens. *Journal of Clinical Microbiology* 33:1632–1634.

Allison RT (1995) Picro-thionin (Schmorl) staining of bone and other hard tissues. *British Journal of Biomedical Science* 52:162–164.

Altman FP (1974) Studies on the reduction of tetrazolium salts. 1. The products of chemical and enzymic reduction. *Histochemistry* 38:155–171.

Altman FP (1976) The quantification of formazan in tissue sections by microdensitometry. II. The use of BPST, a new tetrazolium salt. *Histochemical Journal* 8:501–506.

Altman FP (1980) *How Thick is Your Section?* Abstracts of the VIth International Histochemistry and Cytochemistry Congress. Oxford: Royal Microscopical Society. p. 6.

Altman FP, Butcher RG (1973) Studies on the reduction of tetrazolium salts. 1. The isolation and characterisation of a half-formazan intermediate produced during the reduction of neotetrazolium chloride. *Histochemie* 37:333–350.

Altman SA, Randers L, Rao G (1993) Comparison of trypan blue-dye exclusion and fluorometric assays for mammalian-cell viability determinations. *Biotechnology Progress* 9:671–674.

428

Altmann R (1890) *Die Elementarorganismen und ihre Beziehungen zu den Zellen.* Veit, Leipzig.

Alvarez Hernandez X, Smith M, Glass J (1998) The effect of apotransferrin on iron release from Caco-2 cells, an intestinal epithelial cell line. *Blood* 91:3974–3979.

Alvarez-Buylla A, Ling CY, Kirn JR (1990) Cresyl violet: a red fluorescent Nissl stain. *Journal of Neuroscience Methods* 33:129–133.

Andersen AP, Jakobsen P, Lyon H, Hoyer PE (1986) Purification of methyl green using polyamide. *Histochemical Journal* 18:461–462.

Anderson GM, Nukada H, McMorran PD (1997) Carbonyl histochemistry in rat reperfusion nerve injury. *Brain Research* 772:156–160.

Anderson TJ, Lapp CA, Billman MA, Schuster BS (1998) Effects of transforming growth factor-beta and platelet-derived growth factor on human gingival fibroblasts grown in serum-containing and serum-free medium. *Journal of Clinical Periodontology* 25:48–55.

Anderson WM, Delinck DL (1987) Inhibition of bovine heart mitochondrial NADH-ubiquinone reductase by tinopal AN, a cationic benzoxazole. *Federation Proceedings* 46:1965.

Anderson WM, Wood JM, Anderson AC (1993) Inhibition of mitochondrial and *Paraccus denitrificans* NADH-ubiquinone reductase by oxacarbocyanine dyes. A structure-activity study. *Biochemical Pharmacology* 45:2115–2122.

Andrade WN, Johnston MG, Hay JB (1998) The relationship of blood lymphocytes to the recirculating lymphocyte pool. *Blood* 91:1653–1661.

Andreasen K, Nielsen PH (1997) Application of microautoradiography to the study of substrate uptake by filamentous microorganisms in activated sludge. *Applied and Environmental Microbiology* 63:3662–3668.

Anklam E, Muller A, Schmalfuss J (1995) High performance liquid chromatographic analysis of patent blue V in cheese. *Chromatographia* 41:431–434.

Ansari A (2001) Anatomy and clinical significance of ventricular Thebesian veins. *Clinical Anatomy* 14:102–110.

Antonio C, Gonzalez-Garcia JM, Page J, Suja JA, Stockert JC, Rufas JS (1996) The osmium tetroxide-p-phenylenediamine procedure reveals the chromatid cores and kinetochores of meiotic chromosomes by light and electron microscopy. *Journal of Histochemistry and Cytochemistry* 44:1279–1288.

Antopol W, Glaubach S, Goldman L (1948) Effects of a new tetrazolium derivative on tissue, bacteria and onion root tips. *Public Health Reports* 61:1231–1238.

Aparicio SR, Marsden PA (1969) A rapid Methylene Blue–Basic Fuchsin stain for semi-thin sections of peripheral nerve and other tissues. *Journal of Microscopy* 89:139–141.

Apgar DA, Patel JA (1969) Determination of toluidine blue O in aqueous solution. *American Journal of Hospital Pharmacy* 26:541–542.

Apgar JM, Juarranz A, Espada J, Villanueva A, Canete M, Stockert JC (1998) Fluorescence microscopy of rat embryo sections stained with haematoxylin-eosin and Masson's trichrome method. *Journal of Microscopy* 191:20–27.

Arakawa T, Kamimura M, Furuta Y, Miyazawa M, Kato M (2000) Peroral infection of nuclear polyhedrosis virus budded particles in the host, *Bombyx mori* L., enabled by an optical brightener, Tinopal UNPA-GX. *Journal of Virological Methods* 88:145–152.

Arellano J, Dinis MT, Sarasquete C (1999) Histomorphological and histochemical characteristics of the intestine of the Senegal Sole, *Solea senegalensis. European Journal of Histochemistry* 43:121–133.

Ariano MA, Kang HC, Haugland RP, Sibley DR (1991) Multiple fluorescent ligands for dopamine receptors. II. Visualization in neural tissues. *Brain Research* 547:208–222.

Arienti G, Carlini E, Palmerini CA (1997) Fusion of human sperm to prostasomes at acidic pH. *Journal of Membrane Biology* 155:89–94.

Armed Forces Institute of Pathology (1992) *AFIP Laboratory Methods in Histotechnology*. American Registry of Pathology, Washington.

Arndt-Jovin DJ, Jovin TM (1989) Fluorescence labeling and microscopy of DNA. *Methods in Cell Biology* 30:417–448.

Arnoldus EPJ, Wiegant J, Noordermeer IA, Wessels JW, Beverstock GC, Grosveld GC, van der Ploeg M, Raap AK (1990) Detection of the Philadelphia chromosome in interphase nuclei. *Cytogenetics and Cell Genetics* 54:108–111.

Arregui CO, Balsamo J, Lilien J (1998) Impaired integrin-mediated adhesion and signaling in fibroblasts expressing a dominant-negative mutant PTP1B. *Journal of Cell Biology* 143:861–873.

Arshid FM, Connelly RF, Desai JN, Fulton RG, Giles CH, Kefalas JC (1954) A study of certain natural dyes. II. The structure of the metallic lakes of the brazilwood and logwood colouring matters. *Journal of the Society of Dyers and Colourists* 70:402–412.

Arshid FM, Desai NF, Giles CH, McLintock GK (1953) Quantitative analysis of azo and other dyes and intermediates. *Journal of the Society of Dyers and Colourists* 69:11–18.

Arttamangkul S, Alvarez-Maubecin V, Thomas G, Williams JT, Grandy DK (2000) Binding and internalization of fluorescent opioid peptide conjugates in living cells. *Molecular Pharmacology* 58:1570–1580.

Asan E, Kugler P (1995) Qualitative and quantitative detection of alkaline-phosphatase coupled to an oligonucleotide probe for somatostatin messenger-RNA after in-situ hybridization using unfixed rat-brain tissue. *Histochemistry and Cell Biology* 103:463–471.

Aschoff A, Hollander H (1982) Fluorescent compounds as retrograde tracers compared with horseradish peroxidase (HRP). I. A parametric study in the central nervous system of the albino rat. *Journal of Neuroscience Methods* 6:179–197.

Asikoglu M, Ertan G, Cosar G (1995) The release of isoconazole nitrate from different suppository bases – *in vitro* dissolution, physicochemical and microbiological studies. *Journal of Pharmacy and Pharmacology* 47:713–716.

Atkins PW (1987) *Physical Chemistry*, 6th Edn. Oxford University Press, Oxford.

Augsburger HR (1997) Elastic fibre system of the female canine urethra. Histochemical identification of elastic, elaunin and oxytalan fibres. *Anatomia Histologia Embryologica – Journal of Veterinary Medicine Series C* 26:297–302.

Augulis V, Sepinwall J (1971) Use of gallocyanin as a myelin stain for brain and spinal cord. *Stain Technology* 46:137–143.

Autio K, Mattila-Sandholm T (1992) Detection of active yeast cells (*Saccharomyces cerevisiae*) in frozen dough sections. *Applied and Environmental Microbiology* 58:2153–2157.

Awonusonu F, Srinivasan S, Strange J, Al-Jumaily W, Bruce MC (1999) Developmental shift in the relative percentages of lung fibroblast subsets: role of apoptosis post-septation. *American Journal of Physiology – Lung Cellular and Molecular Physiology* 277:L848–859.

Ayllon L, Silva M, Perez Bendito D (1994) Improved automatic kinetic method for the determination of various corticosteroids. *Journal of Pharmaceutical Sciences* 83:1135–1141.

Ayres WW, Duda J (1993) Differential diagnosis of *Staphylococcus aureus* from *Staphylococcus epidermis* and *Staphylococcus saprophyticus* by alphazurine A dye. *Military Medicine* 158:571–572.

Azhar M, Manogaran PS, Kennady PK, Pande G, Nanjundiah V (1996) A Ca^{2+}-dependent early functional heterogeneity in amoebae of *Dictyostelium discoideum*, revealed by flow cytometry. *Experimental Cell Research* 227:344–351.

Baba K, Adachi K, Take T, Yokoyama T, Itoh T, Nakamura T (1995) Induction of tension wood in GA(3)-treated branches of the weeping type of Japanese cherry, *Prunus spachiana*. *Plant and Cell Physiology* 36:983–988.

Baba Y, Miyazono N, Ueno K, Kanetsuki I, Nishi H, Inoue H, Nakajo M (2000) Hepatic falciform artery – Angiographic findings in 25 patients. *Acta Radiologica* 41:329–333.

Babb RR (1995) Giardiasis. Taming this pervasive parasitic infection. *Postgraduate Medicine* 98:155–158.

Bach PH, Reynolds CH, Clark JM, Mottley J, Poole PL (1993) *Biotechnology Applications of Microinjection, Microscopic Imaging, and Fluorescence*. Plenum Press, New York.

Bachmaier N, Graf R (1999) The anchoring zone in the human placental amnion: bunches of oxytalan and collagen connect mesoderm and epithelium. *Anatomy and Embryology* 200:81–90.

Baden HP, Goldaber ML, Kvedar JC (1992) Keratinocytes stimulate prostaglandin-12 synthesis by 3T3-cells and exhibit enhanced cornification when exposed to prostaglandin-12 analogs. *Journal of Cellular Physiology* 150:269–275.

Bago B, Azcon Alguilar C (1997) Changes in the rhizospheric pH induced by arbuscular mycorrhiza formation in onion (*Alium cepa* L). *Zeitschrift fur Pflanzenernahrung und Bodenkunde* 160:333–339.

Bahar S, Payuk D, Somjen GG, Aitken PG, Turner DA (2000) Mitochondrial and intrinsic optical signals imaged during hypoxia and spreading depression in rat hippocampal slices. *Journal of Neurophysiology* 84:311–324.

Bainton DF, Ullyott IL, Farquhar MG (1971) The development of neutrophilic polymorphonuclear leukocytes in human bone marrow. Origin and content of azurophil and specific granules. *Journal of Experimental Medicine* 134:907–913.

Bajno L, Peng XR, Schreiber AD, Moore HP, Trimble WS, Grinstein S (2000) Focal exocytosis of VAMP3-containing vesicles at sites of phagosome formation. *Journal of Cell Biology* 149:697–705.

Baker CJ, Mock NM (1994) An improved method for monitoring cell death in cell suspension and leaf disk assays using Evans blue. *Plant Cell Tissue and Organ Culture* 39:7–12.

Baker JR (1941) Chlorazol black E as a vital dye. *Nature* 147:744.

Baker JR (1958) *Principles of Biological Microtechnique* (Reprinted 1970, with corrections). Methuen, London.

Baker JR (1966) *Cytological Technique*, 5th Edn. Methuen, London.

Bal AK (1990) Localization of plant lipids for light microscopy using p-phenylenediamine in tissues of *Arachis hypogaea* L. *Stain Technology* 65:91–94.

Balazs I, Beekman J, Neuweiler J, Liu H, Watson E, Ray B (2001) Molecular typing of HLA-A, -B, and DRB using a high throughput micro array format. *Human Immunology* 62:850–857.

Balfour-Paul J (1998) *Indigo*. British Museum Press, London.

Balice-Gordon RJ, Bone LJ, Scherer SS (1998) Functional gap junctions in the Schwann cell myelin sheath. *Journal of Cell Biology* 142:1095–1104.

Ballanti P, Wedard BM, Mazzaferro S, Coen G, Costantini S, Giordano R, Bonucci E (1995) Comparison between aluminon and solochrome azurine techniques for the histochemical detection of aluminium in bone of patients with chronic renal failure. *Italian Journal of Mineral and Electrolyte Metabolism* 9:145–151.

Ballardini G, Groff P, DeGiorgi LB, Schuppan D, Bianchi FB (1994) Ito cell heterogeneity – Desmin-negative Ito cells in normal rat liver. *Hepatology* 19:440–446.

Baloyannis SJ, Manolidis SL, Manolidis LS (2001) The acoustic cortex in frontal dementia. *Acta Oto-Laryngologica* 121:289–292.

Bameul F (1992) Revision of the genus *Psalitrus dorchymont* from southern India and Sri-Lanka (Coleoptera, Hydrophilidae, Omocrini). *Systematic Entomology* 17:1–20.

Bancroft JD, Cook HC (1994) *Manual of Histological Techniques*. Churchill-Livingstone, Edinburgh.

Bancroft JD, Gamble M, eds (2002) *Theory and Practice of Histological Techniques,* 5th Edn. Churchill-Livingstone, London.

Bangle R (1954) Gomori's paraldehyde-fuchsin stain. I. Physico-chemical and staining properties of the dye. *Journal of Histochemistry and Cytochemistry* 2:291–299.

Bannister NJ, Publicover SJ (1995) Interacting effects of Ca^{2+} and hypoxia in the induction of sarcolemmal damage in mouse diaphragm *in vitro. Acta Neuropathologica* 90:411–414.

Barber DL, Lott JNA, Harris DA (1991) Practical methods for identification of rice endosperm protein bodies and fecal protein particles in light microscopy. *Food Structure* 10:137–144.

Barber L, Prince HM, Rossi R, Bertoncello I (1999) Fluoro-gold: an alternative viability stain for multicolor flow cytometric analysis. *Cytometry* 36:349–354.

Bardi L, Delloro V, Delfini C, Marzona M (1993) A rapid spectrophotometric method to determine esterase activity of yeast cells in an aqueous medium. *Journal of the Institute of Brewing* 99:385–388.

Barer MR, Burdess D, Freeman R (1992) A study into the mechanism of the crystal violet reaction in *Staphylococcus aureus. Epidemiology and Infection* 109:87–96.

Barlow CH, Rorvik DA, Kelly JJ (1998) Imaging epicardial oxygen. *Annals of Biomedical Engineering* 26:76–85.

Barnabas AD (1994) Apoplastic and symplastic pathways in leaves and roots of the seagrass *Halodule uninervis* (Forssk) Aschers. *Aquatic Botany* 47:155–174.

Barnes BI, Cassar CA, Halablab MA, Parkinson MH, Miles RJ (1996) An *in situ* method for determining bacterial survival on food preparation surfaces using a redox dye. *Letters in Applied Microbiology* 23:325–328.

Baron-Epel O, Hernandez D, Jiang L-W, Meiners S, Schindler M (1988) Dynamic continuity of cytoplasmic and membrane compartments between plant cells. *Journal of Cell Biology* 106:715–721.

Barr DB, Duncan JA, Kiernan JA, Soper BD, Tepperman BL (1989) Binding and biological actions of prostaglandin E2 and a stable prostacyclin analogue (Iloprost) in cells isolated from rabbit gastric mucosa. *Journal of Physiology* 405:39–55.

Barrow GI, Feltham RKA, eds (1993) *Cowan and Steel's Manual for the Identification of Medical Bacteria,* 3rd Edn. Cambridge University Press, Cambridge, UK.

Barsony J, Renyi I, McKoy W (1997) Subcellular distribution of normal and mutant vitamin D receptors in living cells studies with a novel fluorescent ligand. *Journal of Biological Chemistry* 272:5774–5782.

Basanez G, Ruiz-Arguello MB, Alonso A, Goni FM, Karlsson G, Edwards K (1997) Morphological changes induced by phospholipase C and by sphingomyelinase on large unilamellar vesicles: A cryo-transmission electron microscopy study of liposome fusion. *Biophysical Journal* 72:2630–2637.

Baskaya MK, Rao AM, Dogan A, Donaldson D, Dempsey RJ (1997) The biphasic opening of the blood–brain barrier in the cortex and hippocampus after traumatic brain injury in rats. *Neuroscience Letters* 226:33–36.

Basle MF, Bertrand G, Guyetant S, Chappard D, Lesourd M (1996) Migration of metal and polyethylene particles from articular prostheses may generate lymphadenopathy with histiocytosis. *Journal of Biomedical Materials Research* 30:157–164.

Bassnett S (1992) Mitochondrial dynamics in differentiating fiber cells of the mammalian lens. *Current Eye Research* 11:1227–1232.

Bassnett S (1997) Fiber cell denucleation in the primate lens. *Investigative Ophthalmology and Visual Science* 38:1678–1687.

Battaglia M, Pozzi D, Grimaldi S, Parasassi T (1994) Hoechst 33258 staining for detecting mycoplasma contamination in cell cultures. *Biotechnic and Histochemistry* 69:152–156.

Baumann EW (1995) Colorimetric determination of low pH with malachite green. *Talanta (Oxford)* 42:457–462.

Bayliss OB, Adams CWM (1972) Bromine Sudan black, a general stain for lipids including free cholesterol. *Histochemical Journal* 4:505–515.

Beach JM, McGahren ED, Xia J, Duling BR (1996) Ratiometric measurement of endothelial depolarization in arterioles with a potential-sensitive dye. *American Journal of Physiology* 270:H2216–H2227.

Beale LS (1860) On the distribution of nerves to the elementary fibres of striped muscle. *Philosophical Transactions of the Royal Society* 1860:611–618.

Beauregard KE, Lee KD, Collier RJ, Swanson JA (1997) pH-dependent perforation of macrophage phagosomes by listeriolysin O from *Listeria monocytogenes*. *Journal of Experimental Medicine* 186:1159–1163.

Becher S (1921) *Untersuchungen uber Echtfarbung der Zellkerne mit kunstlichen Beizenfarbstoffen.* Gebruder Borntrager, Berlin.

Becker B, Clapper J, Harkins KR, Olson JA (1994) *In situ* screening assay for cell viability using a dimeric cyanine nucleic acid stain. *Analytical Biochemistry* 221:78–84.

Becker DE, Ancin H, Szarowski DG, Turner JN, Roysam B (1996) Automated 3-D montage synthesis from laser-scanning confocal images: application to quantitative tissue-level cytological analysis. *Cytometry* 25:235–245.

Becker GF, Busso CA, Montani T (1997) Effects of defoliating *Stipa tenuis* and *Piptochaetium napostaense* at different phenological stages: axillary bud viability and growth. *Journal of Arid Environments* 35:233–250.

Beckmann HJ, Dierichs R (1982) Lipid extracting properties of 2,2′-dimethoxypropane as revealed by electron microscopy and thin layer chromatography. *Histochemistry* 76:407–412.

Bedi KS, Horobin RW (1976) An alcohol-soluble Schiff's reagent: a histochemical application of the complex between Schiff's reagent and phosphotungstic acid. *Histochemistry* 48:153–159.

Bedi KS, Horobin RW (1980) The chemical nature of de Tomasi Schiff Reagent. Changes of certain physical-chemical and histochemical properties during ageing. *Histochemistry* 68:197–209.

Bedlack RS, Wei MD, Fox SH, Gross E, Loew LM (1994) Distinct electric potentials in soma and neurite membranes. *Neuron* 13:1187–1193.

Bedner E, Li X, Gorczyca W, Melamed MR, Darzynkiewicz Z (1999) Analysis of apoptosis by laser scanning cytometry. *Cytometry* 35:181–195.

Bedrick AE (1970) Differential staining of malignant and benign tumors with pontacyl dark green B. *Stain Technology* 45:273–276.

Beecken H, Gottschalk E-M, Gizycki U, Kramer H, Maassen D, Matthies H-G, Musso H, Rathjen C, Zahorszky UI (1961) *Orcein und Lackmus. Angewandte Chemie* 73:665–675.

Beeley JA, Newman F, Wilson PH, Shimmin JC (1996) Sodium dodecyl sulphate-polyacrylamide gel electrophoresis of human parotid salivary proteins: comparison of dansylation, coomassie blue R-250 and silver detection methods. *Electrophoresis* 17:505–506.

Beers MF (1996) Inhibition of cellular processing of surfactant protein C by drugs affecting intracellular pH gradients. *Journal of Biological Chemistry* 271:14361–14370.

Beesley JE, ed. (1993) *Immunocytochemistry: A Practical Approach.* IRL Press, Oxford.

Belai A, Burnstock G (1994) Evidence for coexistence of ATP and nitric oxide in noradrenergic, non-cholinergic (NANC) inhibitory neurones in the rat ileum, colon and anococcygeus muscle. *Cell and Tissue Research* 278:197–200.

Belant JL, Seamans TW (1993) Evaluation of dyes and techniques to color-mark incubating herring gulls. *Journal of Field Ornithology* 64:440–451.

Belcher H, Swale E (1979) *An Illustrated Guide to River Phytoplankton.* Her Majesty's Stationery Office, London.

Bellido M, Rubiol E, Ubeda J, Lopez O, Estivill C, Carnicer MJ, Munoz L, Bordes R, Sierra J, Nomdedeu J (2001) Flow cytometry using the monoclonal antibody CD10-Pe/Cy5 is a useful tool to identify follicular lymphoma cells. *European Journal of Haematology* 66:100–106.

Belloc F, Belaud-Rotureau MA, Lavignolle V, Bascans E, Braz-Pereira E, Durrieu F, Lacombe F (2000) Flow cytometry detection of caspase 3 activation in preapoptotic leukemic cells. *Cytometry* 40:151–160.

Belluscio L, Koentges G, Axel R, Dulac C (1999) A map of pheromone receptor activation in the mammalian brain. *Cell* 97:209–220.

Belz GT, Heath TJ (1995) Lymphatic drainage from the tonsil of the soft palate in pigs. *Journal of Anatomy* 187:491–495.

Benda C (1901) Die Mitochondriafarbung und andere Methoden zur Untersuchung der Zellsubstanzen. *Anatomischer Anzeiger* 19:155–174.

Bendayan M (1981a) Electron microscopical localization of nucleic acids by means of nuclease-gold complexes. *Histochemical Journal* 13:699–710.

Bendayan M (1981b) Ultrastructural localization of nucleic acids by the use of enzyme–gold complexes. *Journal of Histochemistry and Cytochemistry* 29:531–541.

Bendayan M (2000) A review of the potential and versatility of colloidal gold cyto-chemical labeling for molecular morphology. *Biotechnic and Histochemistry* 75:203–242.

Bendayan M, Duhr MA (1986) Modification of the protein A-gold immunocytochemical technique for the enhancement of its efficiency. *Journal of Histochemistry and Cytochemistry* 34:569–575.

Bender CM, Pretorius ZA, Kloppers FJ, Spies JJ (2000) Histopathology of leaf rust infection and development in wheat genotypes containing Lr12 and Lr13. *Journal of Phytopathology* 148:65–76.

Benderitter M, Vincent-Genod L, Berroud A, Voisin P (2000) Simultaneous analysis of radio-induced membrane alteration and cell viability by flow cytometry. *Cytometry* 39:151–157.

Beneke FW (1862) (Note without title). *Korrespondenzblatt der Vereins fur gemeinschaftlichen Arbeiten* 59:980.

Benhamou N, Gilboa-Garber N, Trudel J, Asselin A (1988) A new lectin–gold complex for ultrastructural localization of galacturonic acids. *Journal of Histochemistry and Cytochemistry* 36:1403–1411.

Benians THC (1916) Relief staining for bacteria and spirochaetes. *British Medical Journal* 1916–2:722.

Benjaminson MA, Katz IJ (1970) Properties of SITS (4-acetamido-4'-isothiocyanostilbene-2,2'-disulfonic acid): fluorescence and biological staining. *Stain Technology* 45:57–62.

Benne CA, Benaissa Trouw B, Van Strijp JAG, Kraaijeveld CA, Van Iwaarden JFF (1997) Surfactant protein A, but not surfactant protein B, is an opsonin for influenza A phago-cytosis by rat alveolar macrophages. *European Journal of Immunology* 27:886–890.

Bennhold H (1922) Eine spezifische Amyloidfarbung mit Kongorot. *Munchen medizinischer Wochenschrift* 2:1537–1538.

Bennion PJ, Horobin RW (1974) Some effects of salts on staining: use of the Donnan equilibrium to describe staining of tissue sections with acid and basic dyes. *Histochemistry* 39:71–82.

Bennion PJ, Horobin RW, Murgatroyd LB (1975) The use of a basic dye (azure A or toluidine blue) plus a cationic surfactant for selective staining of RNA: a technical and mechanistic study. *Stain Technology* 50:307–313.

Bensley RR (1911) Studies on the pancreas of the guinea pig. *American Journal of Anatomy* 12:297–388.

Bensley SE (1952) Pinacyanol erythrosinate as a stain for mast cells. *Stain Technology* 27:269–273.

Bentolila V, Boyce TM, Fyhrie DP, Drumb R, Skerry TM, Schaffler MB (1998) Intracortical remodeling in adult rat long bones after fatigue loading. *Bone* 23:275–281.

Bereiter-Hahn J, Voth M (1994) Dynamics of mitochondria in living cells: shape changes, dislocations, fusion, and fision of mitochondria. *Microscopy Research and Techniques* 27:198–219.

Berg K, Moan J (1994) Lysosomes as photochemical targets. *International Journal of Cancer* 59:814–822.

Berg K, Peng Q, Nesland JM, Moan J (1994) Cellular responses to photodynamic therapy. *SPIE Proceedings of Photodynamic Therapy of Cancer* 2078:278–285.

Berger M, Schawalder P, Stich H, Lussi A (1996) Differential diagnosis of resorptive dental lesions (FORL) and caries [German]. *Schweizer Archiv fur Tierheilkunde* 138:546–551.

Bergonzini C (1891) Uber das Vorkommen von granulierten basophilen und acidophilen Zellen im Bindegewebe und ber die Art sie sichtbar zu machen. *Anatomischer Anzeiger* 6:595–600.

Bergvinson DJ, Arnason JT, Pietr-Zak LN (1994) Localization and quantification of cell wall phenolics in European corn-borer resistant and susceptible maize inbreds. *Canadian Journal of Botany* 72:1243–1249.

Beris P, Solenthaler M, Deutsch S, Darbellay R, Tobler A, Bochaton-Pialat ML, Gabbiani G (1999) Severe inclusion body beta-thalassaemia with haemolysis in a patient double heterozygous for beta degrees-thalassaemia and quadruplicated alpha-globulin gene arrangement of the anti-4.2 type. *British Journal of Haematology* 105:1074–1080.

Berks BC, McEwan AG, Ferguson SJ (1993) Membrane-associated NADH dehydro-genase-activities in *Rhodobacter capsulatus* – purification of a dihydrolipoyl dehydro-genase. *Journal of General Microbiology* 139:184–1851.

Berlin RB, Farnsworth P, Grothvasselli B, Kuo JA (1992) Procion yellow fluorescent microscopy – a new method of studying arterial and venous pathology. *Angiology* 43:893–898.

Berlyn GP, Miksche JP (1976) *Botanical Microtechnique and Cytochemistry.* Iowa State University Press, Ames.

Berman RS, Martin W (1993) Arterial endothelial barrier dysfunction – actions of homo-cysteine and the hypoxanthene-xanthine oxidase free radical generating system. *British Journal of Pharmacology* 108:920–926.

Bernthsen A (1906) Uber die chemische Natur des Methylenazurs. *Berichte der Deutschen Chemischen Gesellschaft* 39:1804–1809.

Bertolesi GE, Trigoso CI, Espada J, Stockert JC (1995) Cytochemical application of tris(2,2′-bipyridine)ruthenium(II): fluorescence reaction with sulfated polyanions of mast cell granules. *Journal of Histochemistry and Cytochemistry* 43:537–543.

Berube GR, Powers MM, Clark G (1964) The cationic chelate of chromium and aluminon as a selective nuclear stain. *Stain Technology* 39:337–338.

Berzas-Nevado JJ, Guiberteau-Cabanillas C, Contento-Salcedo AM (1998) A reverse phase HPLC method to determine six food dyes using buffered mobile phase. *Analytical Letters* 31:2513–2535.

Bettarel Y, Sime-Ngando T, Amblard C, Laveran H (2000) A comparison of methods for counting viruses in aquatic systems. *Applied and Environmental Microbiology* 66:2283–2289.

Bettinger C, Zimmermann HW (1991) New investigations on hematoxylin, hematein and hematein-aluminium complexes. 2. Hematein-aluminium complexes and hemalum staining. *Histochemistry* 96:215–228.

Beveridge TJ (2001) Use of the Gram stain in microbiology. *Biotechnic and Histochemistry* 76:111–118.

Beverloo HB, Tanke HJ (1991) Inorganic phosphors: new luminescent immunolabels. *Applied Fluorescence Technology* 3:27–29.

Beverloo HB, van Schadewijk A, Zijlmans HJ, Verwoerd NP, Bonnett J, Vrolijk H, Tanke HJ (1993) A comparison of the detection sensitivity of lymphocyte membrane antigens using fluorescein and phosphor immunoconjugates. *Journal of Histochemistry and Cytochemistry* 41:719–725.

Bickar D, Reid PD (1992) A high-affinity protein stain for western blots, tissue prints, and electrophoretic gels. *Analytical Biochemistry* 203:109–115.

Bickmore W, Craig J (1997) *Chromosome Bands: Pattern in the Genome.* Springer-Verlag, Berlin.

Bird CL, Boston WS, eds (1975) *The Theory of Coloration of Textiles.* Dyers' Company Publication Trust, Bradford, UK.

Birkenmeier G (1994) Partitioning of blood proteins using immobilized dyes. *Methods in Enzymology* 228:154–167.

Bishop E, ed. (1972) *Indicators.* Pergamon Press, Oxford.

Blacow NW, ed. (1972) *Martindale's Extra Pharmacopeia*, 26th Edn. Pharmaceutical Press, London.

Blais R (1995) Colorant and exciter filter for better measurement of annual ring widths in the incremental core of certain hardwoods [French]. *Forestry Chronicle* 71:211–212.

Blanchflower WJ, McCracken RJ, Haggan AS, Kennedy DG (1997) Confirmatory assay for the determination of tetracyclines. *Journal of Chromatography B* 692:351–360.

Blanquet PR (1976) Ultrahistochemical study on the ruthenium red surface staining. II. Nature and affinity of the electron dense marker. *Histochemistry* 47:175–189.

Blivet D, Salvat G, Humbert F, Colin P (1998) Development of a new culture medium for the rapid detection of *Salmonella* by indirect conductance measurements. *Journal of Applied Microbiology* 84:399–403.

Bobrow MN, Harris TD, Shaugnessy KJ, Litt GJ (1989) Catalyzed reporter deposition, a novel method of signal amplification. Application to immunoassays. *Journal of Immunological Methods* 125:279–285.

Bobtelsky B, Ben-Bassat A (1956) The aurintricarboxylates of aluminium, iron and chromium. Composition, structure and analytical use: A heterometric study. *Analytica Chimica Acta* 14:344–355.

Bodian D (1936) A new method for staining nerve fibers and nerve endings in mounted paraffin sections. *Anatomical Record* 65:89–97.

Boerboom LE, Olinger GN, Almassi GH, Skrinska VA (1997) Both dietary fish-oil supplementation and aspirin fail to inhibit atherosclerosis in long-term vein bypass grafts in moderately hypercholesterolemic nonhuman primates. *Circulation* 96:968–974.

Bohmer F (1865) Zur pathologischen Anatomie der Meningitis cerebro-medularis epidemica. *Aerzlich Intelligenzblatt (Munchen)* 12:539–550.

Bohrmann J, Schill S (1997) Cytoplasmic transport in *Drosophila* ovarian follicles: the migration of microinjected fluorescent probes through intercellular bridges depends neither on electrical charge nor on external osmolarity. *International Journal of Developmental Biology* 41:499–507.

Bonartseva GA (1985) Testing for activity of nodule bacteria in terms of poly-beta-hydroxybutyrate accumulation following vital staining of the colonies with phosphine 3R. *Microbiology* 54:371–374.

Bonhivers M, Plancon L, Ghazi A, Boulanger P, le Maire M, Lambert O, Rigaud JL, Letellier L (1998) FhuA, an *Escherichia coli* outer membrane protein with a dual function of transporter and channel which mediates the transport of phage DNA. *Biochimie* 80:363–369.

Bonner BA, Gifford EM, Reed NMR (1991) Differential RNA-synthesis in the shoot apex of *Pharbitis nil* during floral evocation. *American Journal of Botany* 78:401–407.

Bonnett R (1995) Photosensitizers of the porphyrin and phthalocyanine series for photodynamic therapy. *Chemical Society Reviews* 24:19–33.

Boon ME, Drijver JS (1986) *Routine Cytological Staining Techniques. Theoretical Background and Practice.* Elsevier, New York.

Boon ME, Wittekind D (1986) Theoretical and practical aspects of standardization of staining in diagnostic cytology. Ch. 14 In: ME Boon, LP Kok (eds) *Standardization and Quantitation of Diagnostic Staining in Cytology.* Coulomb Press, Leiden. pp. 100–110.

Borg K, Colenbrander B, Fazeli A, Parevliet J, Malmgren L (1997) Influence of thawing method on motility, plasma membrane integrity and morphology of frozen-thawed stallion spermatozoa. *Theriogenology* 48:531–536.

Borgstein PJ, Meijer S, Pipers R (1997) Intradermal blue dye to identify sentinel lymphnode in breast cancer. *Lancet* 349:1668–1669.

Bornhop DJ, Hubbard DS, Houlne MP, Adair C, Kiefer GE, Pence BC, Morgan DL (1999) Fluorescent tissue site-selective lanthanide chelate, Tb-PCTMB for enhanced imaging of cancer. *Analytical Chemistry* 71:2607–2615.

Boston DW, Cotmore JM, Sperrazza L (1995) Caries diagnosis with dye-staining at amalgam restoration margins. *American Journal of Dentistry* 8:280–282.

Bouffard G (1906) Injection des couleurs de benzidine aux animaux normaux. *Annales de l'Institute Pasteur* 20:539–546.

Boullier A, Gillotte KL, Horkko S, Green SR, Friedman P, Dennis EA, Witztum JL, Steinberg D, Quehenberger O (2000) The binding of oxidized low density lipoprotein to mouse CD36 is mediated in part by oxidized phospholipids that are associated with both the lipid and protein moieties of the lipoprotein. *Journal of Biological Chemistry* 275:9163–9169.

Boutonnat J, Barbier M, Muirhead K, Mousseau M, Ronot X, Seigneurin D (1999) Optimized fluorescent probe combinations for evaluation of proliferation and necrosis in anthracycline-treated leukaemic cell lines. *Cell Proliferation* 32:203–214.

Bowers HA, Tengs T, Glasgow HB, Burkholder JM, Rublee PA, Oldach DW (2000) Development of real-time PCR assays for rapid detection of *Pfiesteria piscicida* and related dinoflagellates. *Applied and Environmental Microbiology* 66:4641–4648.

Boyce BF, Byars J, McWilliams S, Mocan MZ, Elder HY, Boyle IT, Junor BJR (1992) Histological and electron-microprobe studies of mineralization in aluminum-related osteomalacia. *Journal of Clinical Pathology* 45:502–508.

Boyce TM, Fyhrie DP, Glotkowski MC, Radin EL, Schaffler MB (1998) Damage type and strain-mode associations in human compact bone bending fatigue. *Journal of Orthopaedic Research* 16:322–329.

Boyle D (1998) Nutritional factors limiting the growth of *Lentinula edodes* and other white-rot fungi in wood. *Soil Biology and Biochemistry* 30:817–823.

Braakman I, Pijning T, Verest O, Weert B, Meijer DK, Groothuis GM (1989) Vesicular uptake system for the cation lucigenin in the rat hepatocyte. *Molecular Pharmacology* 36:537–542.

Bracegirdle B (1986) *A History of Microtechnique*, 2nd Edn. Science Heritage, Chicago.

Bracho C, Dunia I, Romano M, Benedetti EL, Perez HA (2001) *Plasmodium vivax* and *Plasmodium chabaudi:* intraerythrocytic traffic of antigenically homologous proteins involves a brefeldin A-sensitive secretory pathway. *European Journal of Cell Biology* 80:164–170.

Bradbeer JN, Riminucci M, Bianco P (1994) Giemsa as a fluorescent stain for mineralized bone. *Journal of Histochemistry and Cytochemistry* 42:677–680.

Bradbury JM, Abdul Wahab OMS, Yavari CA, Dupiellet JP, Bove JM (1993) *Mycoplasma imitans* sp-nov is related to *Mycoplasma gallisepticum* and found in birds. *International Journal of Systematic Bacteriology* 43:721–728.

Braselton JP, Wilkinson MJ, Clulow SA (1996) Feulgen staining of intact plant tissues for confocal microscopy. *Biotechnic and Histochemistry* 71:84–87.

Brattig NW, Medina-De la Garza CE, Tischendorf FW (1993) Improved Randolph stain for direct leukocyte differentiation and determination of total eosinophil count in a hemocytometer. *Biotechnic and Histochemistry* 68:255–259.

Breccia JD, Baigori MD, Castro GR, Sineriz F (1995) Detection of endo-xylanase activities in electrophoretic gels with Congo red staining. *Biotechnology Techniques* 9:145–148.

Breeuwer P, de Reu JC, Drocourt JL, Rombouts FM, Abee T (1997) Nonanoic acid, a fungal self-inhibitor, prevents germination of *Rhizopus oligosporus* sporangiospores by dissipation of the pH gradient. *Applied and Environmental Microbiology* 63:178–185.

Breininger JF, Baskin DG (2000) Fluorescence *in situ* hybridization of scarce leptin receptor mRNA using the enzyme-labeled fluorescent substrate method and tyramide signal amplification. *Journal of Histochemistry and Cytochemistry* 48:1593–1600.

Brelje TC, Wessendorf MW, Sorenson RL (1993) Multicolor laser-scanning confocal immunofluorescence microscopy: practical application and limitations. *Methods in Cell Biology* 38:97–181.

Bretagne S, Foulet F, Alkassoum W, Fleuryfeith J, Develoux M (1993) Prevalence of *Enterocytozoon bieneusi* spores in stools of AIDs patients and of HIV-negative African children [French]. *Bulletin de la Société de Pathologie Exotique* 86:351–357.

Bridcut EE (1993) A coded alcian blue marking technique for the identification of individual brown trout, *Salmo trutta* L – An evaluation of its use in fish biology. *Biology and Environment – Proceedings of the Royal Irish Academy* 93B:107–110.

Brinkman MA, Duffy WG (1996) Evaluation of four wetland aquatic invertebrate samplers and four sample sorting methods. *Journal of Freshwater Ecology* 11:193–200.

Brinkman UA, De Vries G (1972) Small-scale thin-layer chromatography. *Journal of Chemical Education* 49:545–546.

Broadus J, Doe CQ (1997) Extrinsic cues, intrinsic cues and microfilaments regulate asymmetric protein localization in *Drosophila* neuroblasts. *Current Biology* 7:827–835.

Brodin B, Nielson R (2000) Evidence for P2Y-type ATP receptors on the serosal membrane of frog skin epithelium. *Pflugers Archiv – European Journal of Physiology* 439:234–239.

Brooker BE (1991) The study of food systems using confocal laser scanning microscopy. *Microscopy and Analysis* (November issue):13–115.

Brooker LGS (1966) Sensitizing and desensitizing dyes. Ch. 11 In: TH James, *The Theory of the Photographic Process.* Macmillan, New York. pp. 198–232.

Brooks SA, Leathem AJC, Schumacher U (1997) *Lectin Histochemistry. A Concise Practical Handbook.* BIOS Scientific Publishers, Oxford.

Brown H, Cauchi DM, Holden JL, Wrobel H, Cordner S (1999) Image analysis of gunshot residue on entry wounds. I. – The technique and preliminary study. *Forensic Science International* 100:163–177.

Brown JC (1969) The chromatography and identification of dyes. *Journal of the Society of Dyers and Colourists* 85:137–146.

Brownlee C (2000) Cellular calcium imaging: so, what's new? *Trends in Cell Biology* 10:451–457.

Bruck I, Bezard J, Baltsen M, Synnestvedt B, Couty I, Greve T, Duchamp G (2000) Effect of administering a crude equine gonadotrophin preparation to mares on follicular development, oocyte recovery rate and oocyte maturation *in vivo*. *Journal of Reproduction and Fertility* 118:351–360.

Bruning JW, Kardol MJ, Arentzen R (1980) Carboxyfluorescein fluorochromasia assays. 1. Non-radioactively labeled cell mediated lympholysis. *Journal of Immunological Methods* 33:33–44.

Bruno JG, Sincock SA, Stopa PJ (1996) Highly selective acridine and ethidium staining of bacterial DNA and RNA. *Biotechnic and Histochemistry* 71:130–136.

Bryan JHD (1970) An eosin-fast green-naphthol yellow mixture for differential staining of cytologic components in mammalian spermatozoa. *Stain Technology* 45:231–236.

Bryant V, Watson JHL (1967) Comparison of light microscopy staining methods applied to a polyester and three epoxy resins. *Henry Ford Hospital Medical Bulletin* 15:65–67.

Bucaretchi F, Vinagre AM, Chavez-Olortegui C, Collares EF (1999) Effect of toxin-Y from *Tityus serrulatus* scorpion venom on gastric emptying in rats. *Brazilian Journal of Medical and Biological Research* 32:431–434.

Bucherer HT (1914) *Lehrbuch der Farbenchemie*. Otto Spamer, Leipzig.

Buck JW, Andrews JH (1999) Attachment of the yeast *Rhodosporidium toruloides* is mediated by adhesives localized at sites of bud cell development. *Applied and Environmental Microbiology* 65:465–471.

Budavari S, ed. (1996) *The Merck Index: An Encyclopedia of Chemicals, Drugs and Biologicals*. 12th Edn. Merck and Co., Rahway, NJ.

Buel E, Schwartz M (1993) DAPI, a simple sensitive alternative to ethidium bromide staining of DNA in agarose gels. *Applied and Theoretical Electrophoresis* 3:253–255.

Bugnon MP (1919) Sur une nouvelle methode de coloration elective des membranes vegetales lignifiees. *C. R. Acad. Comptes Rendus de l'Academie des Sciences (Paris)* 168:62–64.

Bullock WL (1980) The use of the Kohn Chlorazol Black fixative-stain in an intestinal parasite survey in rural Costa Rica. *Journal of Parasitology* 66:811–813.

Burchiel SW, Lauer FT, Gurule D, Mounho BJ, Salas VM (1999) Uses and future applications of flow cytometry in immunotoxicity testing. *Methods* 19:28–35.

Burdon KL, Stokes JC, Kimbrough CE (1947) Studies of the common aerobic spore-forming bacilli. I. Staining for fat with Sudan black B-safranin. *Journal of Bacteriology* 42:717–724.

Burgauer SA, Stockert JC (1975) Observations on the selective demonstration of nucleolar material by protein staining techniques in epon thick sections. *Histochemistry* 41:241–247.

Burkhardt JE, Hill MA, Carlton WW (1992) Morphologic and biochemical changes in articular cartilages of immature beagle does dosed with difloaxacin. *Toxicologic Pathology* 20:246–252.

Burkholder GD (1982) Dansyl chloride-stained nucleolar organizers and core-like structures in Chinese hamster metaphase chromosomes. *Experimental Cell Research* 142:485–489.

Burrin DH (1976) Spectroscopic techniques. In: BL Williams, K Wilson (eds) *A Biologist's Guide to Principles and Techniques of Practical Biochemistry*. Edward Arnold, London. pp. 128–169.

Burstone MS (1957a) Studies on calcification. I. The effect of inhibition of enzyme activity on developing bone and dentine. *Journal of Pathology* 33:1229–1236.

Burstone MS (1957b) The cytochemical localization of esterase. *Journal of the National Cancer Institute* 18:167–173.

Burstone MS (1958a) Histochemical comparison of naphthol AS-phosphates for the demonstration of phosphatases. *Journal of the National Cancer Institute* 20:601–615.

Burstone MS (1958b) Histochemical demonstration of acid phosphatases with naphthol AS-phosphates. *Journal of the National Cancer Institute* 21:523–529.

Burstone MS (1960) Postcoupling, noncoupling and fluorescence techniques for the demonstration of alkaline phosphatase. *Journal of the National Cancer Institute* 24:1199–1207.

Burstone MS (1962) *Enzyme Histochemistry*. Academic Press, New York.

Burton MP, Schneider BG, Brown R, Escamilla-Ponce N, Gulley ML (1998) Comparison of histologic stains for use in PCR analysis of microdissected, paraffin-embedded tissues. *BioTechniques* 24:86–92.

Burton SJ, Stead CV, Ansell RJ, Lowe CR (1996) An artificial redox coenzyme based on a triazine dye template. *Enzyme and Microbial Technology* 18:570–580.

Bush CE, Di Michele LJ, Peterson WR, Sherman DG, Godsey JH (1992) Solid-phase time-resolved fluorescence detection of human immunodeficiency virus polymerase chain reaction amplification products. *Analytical Biochemistry* 202:146–151.

Cain JC, Thorpe JF (1923) *The Synthetic Dyestuffs and the Intermediate Products from which they are Derived*, 6th Edn. Charles Griffin, London.

Cain JC, Thorpe JF (1933) *The Synthetic Dyestuffs*, 7th Edn. Griffin, London.

Cake MA, Read RA, Guillou B, Ghosh P (2000) Modification of articular cartilage and subchondral bone pathology in an ovine meniscectomy model of osteoarthritis by avocado and soya unsaponifiables (ASU). *Osteoarthritis and Cartilage* 8:404–411.

Calba H, Jaillard B (1997) Effect of aluminium on ion uptake and H^+ release by maize. *New Phytologist* 137:607–616.

Callebaut M, Vakaet L (1981) Fluorescent yolk marking of the primary gonocytes in quail blastoderms by administration of trypan blue, during late oogenesis. *IRCS Medical Science – Biochemistry* 9:458.

Campagnola PJ, Wei MD, Lewis A, Loew LM (1999) High-resolution nonlinear optical imaging of live cells by second harmonic generation. *Biophysical Journal* 77:3341–3349.

Cannon HG (1937) A new biological stain for general purposes. *Nature* 139:549.

Cannon HG (1950) The technique of biological staining. *Endeavour* 9:188–195.

Cannon MS, Stuth NR (1993) Combination hematoxylin-eosin-rose bengal and resorcinol-fuchsin connective tissue stain for glycol methacrylate embedded sections. *Journal of Histotechnology* 16:141–142.

Cannon MS, Stuth NR, McGuinn D (1992) A connective tissue stain for glycol methacrylate-embedded tissue sections. *Laboratory Medicine* 23:185–186.

Cansell M, Parisel C, Jozefonvicz J, Letourneur D (1999) Liposomes coated with chemically modified dextran interact with human endothelial cells. *Journal of Biomedical Materials Research* 44:140–148.

Cappelier JM, Lazaro B, Rossero A, Fernandez Astorga A, Federighi M (1997) Double staining (CTC-DAPI) for detection and enumeration of viable but non-culturable *Campylobacter jejuni* cells. *Veterinary Research* 28:547–555.

Cappell DF (1929) Intravitam and supravital staining. *Journal of Pathology and Bacteriology* 32:595–708.

Caputo V, Marchegiani F, Olmo E (1996) Karyotype differentiation between two species of carangid fishes, genus *Trachurus* (Perciformes: Carangidae). *Marine Biology* 127:193–199.

Carls FR, Schupbach P, Sailer HF, Jackson IT (1997) Distraction osteogenesis for lengthening of the hard palate. 2. Histological study of the hard and soft palate after distraction. *Plastic and Reconstructive Surgery* 100:1648–1654.

Carlsson C, Jonsson M, Akerman B (1995) Double bands in DNA gel-electrophoresis caused by bis-intercalating dyes. *Nucleic Acids Research* 23:2413–2420.

Carri NG, Ebendal T (1989) Staining of developing neurites with coomassie blue. *Stain Technology* 64:50–52.

Carruba G, Webber MM, Bello-Deocampo D, Amodio R, Notarbartolo M, Decampo ND, Trosko JE, Castagnetta LAM (1999) Laser scanning analysis of cell–cell communication in cultured human prostate tumor cells. *Analytical and Quantitative Cytology and Histology* 21:54–58.

Carson FL (1997) *Histotechnology. A Self-Instructional Text.* 2nd Edn. American Society of Clinical Pathologists, Chicago.

Carter JS (1933) Reactions of *Stenostumum* to vital staining. *Journal of Experimental Zoology* 65:159–179.

Caspersson T, Farber S, Foley GE, Kudynowski J, Modest EJ, Simonsson E, Wagh U, Zech L (1968) Chemical differentiation along metaphase chromosomes. *Experimental Cell Research* 49:219–222.

Castaneda F, Kinne RKH (2000) Short exposure to millimolar concentrations of ethanol induces apoptotic cell death in multicellular HepG2 spheroids. *Journal of Cancer Research and Clinical Oncology* 126:305–310.

Castejon O, Sims P (1999) Cytoarchitectonic arrangement and intra cortical circuits of hamster cerebellum. A study by means of confocal scanning laser microscopy. *Biocell* 23:187–196.

Castilho RF, Ward MW, Nicholls DG (1999) Oxidative stress, mitochondrial function, and acute glutamate excitotoxicity in cultured cerebellar granule cells. *Journal of Neurochemistry* 72:1394–1401.

Castle M, Neuteboom E (1995) High-performance liquid chromatography of the fluorescent dyes Fura-2 and Mag-Fura. Stability in organic solvents. *Journal of Chromatography* A 696:93–99.

Cavalcante LA, Barradas PC, Vieira AM (1996) The regional distribution of neuronal glycogen in the opossum brain, with special reference to hypothalamic systems. *Journal of Neurocytology* 25:455–463.

Caversaccio M, Peschel O, Arnold W (1996) The drainage of cerebrospinal fluid into the lymphatic system of the neck in humans. *Journal of Otorhinolaryngology and Related Specialties* 58:164–166.

Cawood AH, Potter U, Dickinson HG (1978) An evaluation of coomassie brilliant blue as a stain for quantitative microdensitometry of protein in sections. *Journal of Histochemistry and Cytochemistry* 26:645–650.

Celis JE (1998) *Cell Biology:A Laboratory Handbook.* 2nd Edn. v. 1. Academic Press, San Diego.

Celluzzi CM, Falo LD (1998) Physical interaction between dendritic cells and tumor cells results in an immunogen that induces protective and therapeutic tumor rejection. *Journal of Immunology* 160:3081–3085.

Certes MA (1881) Sur un procede de coloration des Infusoires et des elements anatomiques, pendant la vie. *Comptes Rendus des Seances de l'Academie des Sciences* 92:424–426.

Challener WA, Ollmann RR, Kam KK (1999) A surface plasmon resonance gas sensor in a 'compact disc' format. *Sensors and Actuators B – Chemical* 56:254–258.

Chaloupka R, Plasek J, Slavik J, Siglerova V, Sigler K (1997) Measurement of membrane potential in *Saccharomyces cerevisiae* by the electrochromic probe di-4-ANEPPS: effect of intracellular probe distribution. *Folia Microbiologica (Praha)* 42:451–456.

Chamberlain CJ (1901) *Methods in Plant Histology.* Chicago University Press, Chicago.

Chambers R (1935) Disposal of dyes by proximal tubule cells of chick mesonephros in tissue culture. *Proceedings of the Society for Experimental Biology and Medicine* 32:1199–1200.

Chanda B, Mathew MK (1999) Functional reconstitution of bacterially expressed human potassium channels in proteoliposomes: membrane potential measurements with JC-1 to assay ion channel activity. *Biochemica et Biophysica Acta* 1416:92–100.

Chang CH, Beer M, Marzilli LG (1977) Osmium-labeled polynucleotides. The reaction of osmium tetroxide with deoxyribonucleic acid and synthetic polynucleotides in the presence of tertiary nitrogen donor ligands. *Biochemistry* 16:33–38.

Chang EY, Mao SY, Metzger H, Holowka D, Baird B (1995) Effects of subunit mutation on the rotational-dynamics of FC-epsilon-ri, the high-affinity receptor for IgE, in trans-fected cells. *Biochemistry* 34:6093–6099.

Chang WC, Lin YL, Lee MJ, Shiow SJ, Wang CJ (1996) Inhibitory effect of crocetin on benzo[a]pyrene genotoxicity and neoplastic transformation in C3H10T1/2 cells. *Anticancer Research* 16:3603–3608.

Chantawiboonchai P, Warita H, Ohya K, Soma K (1998) Confocal laser scanning microscopic observations on the three dimensional distribution of oxytalan fibres in mouse peridontal ligament. *Archives of Oral Biology* 43:811–817.

Chaplin AJ, Grace SR (1976) An evaluation of some complexing methods for the histochemistry of calcium. *Histochemistry* 47:263–269.

Chapman DM (1975) Dichromatism of bromphenol blue, with an improvement in the mercuric bromphenol blue technic for protein. *Stain Technology* 50:25–30.

Chapman J, Zhou MJ (1999) Microplate-based fluorometric methods for the enzymatic determination of L-glutamate: application in measuring L-glutamate in food samples. *Analytica Chimica Acta* 402:47–52.

Chapman M, Mulcahy DL (1997) Confocal optical sectioning for meiotic analysis in *Oenothera* species and hybrids. *Biotechnic and Histochemistry* 72:105–110.

Chappard D, Retailleau N, Filmon R, Basle MF, Rebel A (1996) Nucleolar organizer regions (AgNORs) staining on undecalcified bone embedded in resin – light and TEM methodologies. *Journal of Histotechnology* 19:27–32.

Chartier S, Faucher L, Tousignant J, Rochette L (1997) Acquired cutis laxa associated with cutaneous angiocentric T-cell lymphoma. *International Journal of Dermatology* 36:772–775.

Chatton J-Y, Spring KR (1994) Acidic pH of the lateral intercellular spaces of MDCK cells cultured on permeable supports. *Journal of Membrane Biology* 140:89–99.

Chavan RB (1976) Use of mixtures of solvents and non-solvents to purify anionic dyes. *Textile Research Journal* 46:435–437.

Chaves AH, da Silva JF, Pinheiro AJR, de Campos OF, Valadares SD (1999) Isolation of *Lactobacillus acidophilus* from calves feces [Portugese]. *Revista Brasileira de Zootecnia* 28:1086–1092.

Chayen J, Bitensky L (1991) *Practical Histochemistry*. 2nd Edn. Wiley, Chichester.

Chen CW, Gomez LE, Chen CL, Jacobsen DB (1991) Investigation of beach contamination using tracer. *Journal of Environmental Engineering* 117:101–115.

Chen JY, Cheung NH, Fung MC, Wen JM, Leung WN, Mak NK (2000) Subcellular localization of merocyanine 540 (MC540) and induction of apoptosis in murine myeloid leukemia cells. *Photochemistry and Photobiology* 72:114–120.

Chen QG, Li DH, Zhao Y, Yang HH, Zhu QZ, Xu JG (1999) Interaction of a novel red-region fluorescent probe, Nile blue, with DNA and its application to nucleic acids assay. *Analyst* 124:901–906.

Chen RF (1967) Fluorescence of dansyl amino acids in organic solvents and protein solutions. *Archives of Biochemistry and Biophysics* 120:609–620.

Chen SW, Zhang DP (1999) Subcellular localization of ABA binding sites in tissues of apple. [Chinese]. *Acta Botanica Sinica* 41:353.

Chen T, Segall EM, Langer R, Weaver JC (1998) Skin electroporation: rapid measurements of the transdermal voltage and flux of four fluorescent molecules show a transition to large fluxes near 50 V. *Journal of Pharmaceutical Sciences* 87:1368–1374.

Chen WL, Fujishige S (1996) Bilateral structure of wool fiber studied with differential staining method. *Seni-i Gakkaishi* 52:A48–53.

Chenciner R (2000) *Madder Red. A history of luxury and trade. Plant dyes and pigments in world commerce and art.* Curzon Caucasus World, Richmond, UK.

Cheng DK, Tung L, Sobie EA (1999) Nonuniform responses of trans-membrane potential during electric field stimulation of single cardiac cells. *American Journal of Physiology* 277:H351–H362.

Cherry WB, McKinney RM, Emmel VM, Spillane JT, Herbert GA, Pittman B (1969) Evaluation of commercial fluorescein isothiocyanates used in fluorescent antibody studies. *Stain Technology* 44:179–186.

Chesher BK, Stone JM, Rowe WF (1992) Use of the Omniprint 1000 alternate light-source to produce fluorescence in cyanoacrylate-developed latent fingerprints stained with biological stains and commercial fabric dyes. *Forensic Science International* 57:163–168.

Chiang H-C, Lin S-L (1969) Polyamide-kieselguhr thin-layer chromatography of yellow food dyes. *Journal of Chromatography* 44:203–204.

Chieco P, Boor PJ (1983) Use of low temperatures for glutathione histochemical stain. *Journal of Histochemistry and Cytochemistry* 31:975–976.

Chilean Iodine Educational Bureau (1951) *Iodine. Its Properties and Technical Applications.* Chilean Iodine Educational Bureau, Inc., New York.

Choi JK, Yoon SH, Hong HY, Choi DK, Yoo GS (1996) A modified coomassie blue staining of proteins in polyacrylamide gels with bismark brown R. *Analytical Biochemistry* 236:82–84.

Choma IM (2000) TLC determination of tetracyclines in milk. *Journal of Planar Chromatography – Modern TLC* 13:261–265.

Christensen P, Whitfield CH, Parkinson TJ (1996) *In vitro* induction of acrosome reactions in stallion spermatozoa by heparin and A23187. *Theriogenology* 45:1201–1210.

Chronszczewsky N (1864) Zur Anatomie der Niere. *Virchows Archiv fur pathologische Anatomie und Physiologie und fur klinische Medizin* 31:153–199.

Chu HW, Wang JM, Boutet M, Boulet LP, Laviolette M (1995) Increased expression of intercellular-adhesion molecule-1 (ICAM-1) in a murine model of pulmonary eosinophilia and high IgE level. *Clinical and Experimental Immunology* 100:319–324.

Chu YH, Whitesides GM (1993) A convenient procedure for transfer blotting of coomassie blue stained proteins from PAGE gels to transparencies. *BioTechniques* 14:925–930.

Chung KT (2000) Mutagenicity and carcinogenicity of aromatic amines metabolically produced from azo dyes. *Environmental Carcinogenesis and Ecotoxicology Reviews* C18:51–74.

Chung KT, Cerniglia CE (1992) Mutagenicity of azo dyes: structure–activity relationships. *Mutation Research* 277:201–220.

Chung WC, Fan PC, Lin CY, Wu CC (1991) Poor efficacy of albendazole for the treatment of human taeniasis. *International Journal for Parasitology* 21:269–270.

Chung WH, Miller A (1994) Film dosimeters based on methylene blue and methyl orange in polyvinyl alcohol. *Nuclear Technology* 106:261–264.

Churukian CC, Schenk EA (1981) A toluidine blue O method for demonstrating mast cells. *Journal of Histotechnology* 4:85–86.

Churukian CC, Schenk EA (1984) A modification of Kreyburg's method for demonstrating keratin and mucin. *Journal of Histotechnology* 7:146–148.

Churukian CC (1999) Lillie's oil red O method for neutral lipids. *Journal of Histotechnology* 22:309–311.

Churukian CJ, Frank M, Horobin RW (2000) Alcian blue pyridine variant – a superior alternative to alcian blue 8GX: staining performance and stability. *Biotechnic and Histochemistry* 75:147–150.

Churukian CJ (1991) Demonstration of mycobacteria – a brief review with special emphasis on fluorochrome staining. *Journal of Histotechnology* 14:117–121.

Churukian CJ (2000) *Manual of the Special Stains Laboratory* (2000 Edn.). University of Rochester Medical Center, Rochester, NY.

Churukian CJ, Schenk EA (1983) Staining with basic fuchsin. *Laboratory Medicine* 14:431–434.

Churukian CJ, Rubio A, Lapham LW (2000) A simple colloidal silver method (autometallographic technique) for demonstrating inorganic mercury in brain sections. *Journal of Histotechnology* 23:337–339.

Ciapetti G, Granchi D, Verri E, Savarino L, Cavedagna D, Pizzoferrato A (1996) Application of a combination of neutral red and amido black staining for rapid, reliable cytotoxicity testing of biomaterials. *Biomaterials* 17:1259–1264.

Cihalikova J, Dolezel J, Novak F (1985) Cytofluorometric determination of nuclear DNA in plant cells using auramine O. *Acta Histochemica* 76:151–156.

Cinelli AR, Kauer JS (1992) Voltage-sensitive dyes and functional activity in the olfactory pathway. *Annual Review of Neuroscience* 15:321–351.

Clamme JP, Bernacchi S, Vuilleumier C, Duportail G, Mely Y (2000) Gene transfer by cationic surfactants is essentially limited by the trapping of the surfactant/DNA complexes onto the cell membrane: a fluorescence investigation. *Biochemica et Biophysica Acta* 1467:347–361.

Clark DA, Wang SL, Rogers P, Vince G, Affandi B (1996) Endometrial lymphomyeloid cells in abnormal uterine bleeding due to levo-norgestrel (norplant). *Human Reproduction* 11:1438–1444.

Clark G (1969) Neuron staining by basic dyes versus basic metal–dye complexes; differences shown by histochemical blocking reactions. *Stain Technology* 44:15–20.

Clark G (1979) Displacement. *Stain Technology* 54:111–119.

Clark G, ed. (1981) *Staining Procedures used by the Biological Stain Commission.* 4th Edn. Williams and Wilkins, Baltimore.

Clark G, Kasten FH (1983) *History of Staining.* 3rd Edn. Williams and Wilkins, Baltimore.

Clark G, Meischen S (1978) A comparison of methods for blocking staining of nucleic acids. *Acta Histochemica* 61:S192–S196.

Clasen RA, Simon G, Scott R, Pandolfi S, Lesak A (1973) The staining of myelin sheath by luxol dye techniques. *Journal of Neuropathology and Experimental Neurology* 32:271–283.

Clemens HJ, Toepfer K (1968) Physikalisch-chemische Eigenschaften von kommerziellen Thiazinfarbstoffen. 2. Qualitative und quantitative Untersuchungen ber die Verunreinigungen der Farbstoffe und das 'Umlosen' zum Erhalt eines hoheren Reinheitsgrades. *Acta Histochemica* 31:126–134.

Clement Lacroix P, Ormandy C, Lepescheux L, Ammann P, Damotte D, Goffin V, Bouchard B, Amling M, Gaillard Kelly M, Binart N, Baron R, Kelly PA (1999) Osteoclasts are a new target for prolactin: analysis of bone formation in prolactin receptor knockout mice. *Endocrinology* 140:96–105.

Clonis YD, Labrou NE, Kotsira VP, Mazitsos C, Melissis S, Gogolas G (2000) Biomimetic dyes as affinity chromatography tools in enyzme purification. *Journal of Chromatography A* 891:33–44.

Cochilla AJ, Angleson JK, Betz WJ (1999) Monitoring secretory membrane with FM1–43 fluorescence. *Annual Review of Neuroscience* 22:1–10.

Cohnheim J (1866) Ueber die Endigung der sensiblen Nerven in der Hornhaut. *Virchows Archiv fur pathologische Anatomie und Physiologie und fur klinische Medizin* 38:343–386.

Cole MB, Ellinger J (1981) Glycol methacrylate in light microscopy: nucleic acid cytochemistry. *Journal of Microscopy* 123:75–88.

Coles WC, Musick JA, Williamson LA (2001) Skeletochronology validation from an adult loggerhead (*Caretta caretta*). *Copeia* 2001–1:240–242.

Collard JG, De Wildt A (1978) Localization of the lipid probe 1,6-diphenyl-1,3,5-hexatriene (DPH) in intact cells by fluorescence microscopy. *Experimental Cell Research* 116:447–450.

Colour Index. See under Society of Dyers and Colourists for information about printed and CD-ROM versions.

Cominacini L, Garbin U, Pasini AF, Davoli A, Campagnola M, Pastorino AM, Gaviraghi G, LoCascio V (1998) Oxidized low-density lipoprotein increases the production of intracellular reactive oxygen species in endothelial cells: inhibitory effect of lacidipine. *Journal of Hypertension* 16:1913–1919.

Conklin MW, Ahern CA, Vallejo P, Sorrentino V, Takeshima H, Coronado R (2000) Comparison of Ca^{2+} sparks produced independently by two ryanodine receptor isoforms (type 1 or type 3). *Biophysical Journal* 78:1777–1785.

Conn HJ (1922a) An investigation of American gentian violets. Report of Committee on Bacteriological Technic. *Journal of Bacteriology* 7:529–536.

Conn HJ (1922b) An investigation of American stains. Report of Committee on Bacteriological Technic. *Journal of Bacteriology* 7:127–148.

Conn HJ (1946) Development of histochemical staining. *Ciba Symposia* 7:270–300.

Conn HJ (1980a) The history of the Stain Commission. I. *Stain Technology* 55:269–279.

Conn HJ (1980b) The history of the Stain Commission. II. *Stain Technology* 55:327–352.

Conn HJ (1981a) The history of the Stain Commission. III. *Stain Technology* 56:1–17.

Conn HJ (1981b) The history of the Stain Commission. IV. *Stain Technology* 56:59–66.

Conn HJ (1981c) The history of the Stain Commission. V. *Stain Technology* 56:135–142.

Connell C, Rutter A, Hill B, Suller M, Lloyd D (2001) Encystation of *Acanthamoeba castellanii*: dye uptake for assessment by flow cytometry and confocal laser scanning microscopy. *Journal of Applied Microbiology* 90:706–712.

Connolly JH, Berlyn GP (1996) Cytochemical assay for differential respiratory activity in roots and root hairs. *Biotechnic and Histochemistry* 71:197–201.

Connor JR, Pavlick G, Karli D, Menzies SL, Palmer C (1995) A histochemical study of iron-positive cells in the developing rat brain. *Journal of Comparative Neurology* 355:111–123.

Constantino PJ, Budd DE, Gare NF (1995) Enumeration of viable *Candida albicans* blastospores using tetrabromofluorescein (eosin Y) and flow cytometry. *Cytometry* 19:370–375.

Cook HC (1974) *Manual of Histological Demonstration Methods*. Butterworths, London.

Cook HC, Lamb RA (1969) Alkali blue 5B – a rapid stain for elastin. *Journal of Medical Laboratory Technology* 26:361–362.

Cooper JD, Payne JN, Horobin RW (1988) Accurate counting of neurons in frozen sections: some necessary precautions. *Journal of Anatomy* 157:13–21.

Cope GH, Williams MA (1969) Quantitative studies on the preservation of choline and ethanolamine phosphatides during tissue preparation for electron microscopy. 2. Other preparative methods. *Journal of Microscopy* 90:47–60.

Corash L (1999) Inactivation of viruses, bacteria, protozoa, and leukocytes in platelet concentrates: current research perspectives. *Transfusion Medicine Reviews* 13:18–30.

Corbel MJ (1991) Identification of dye-sensitive strains of *Brucella melitensis*. *Journal of Clinical Microbiology* 29:1066–1968.

Cornelisse CJ, Ploem PS (1976) A new type of two-color fluorescence staining for cytology specimens. *Journal of Histochemistry and Cytochemistry* 24:72–81.

Cornfield DN, Stevens T, McMurtry IF, Abman SH, Rodman DM (1994) Acute hypoxia causes membrane depolarization and calcium efflux in fetal pulmonary artery smooth muscle cells. *American Journal of Physiology* 266:L469–L475.

Cornil MV (1875) Sur la dissociation du violet de methylaniline et sa separation en deux coleurs sous l'influence de certain tissus normaux et pathologiques, en particulier par les tissus en degenerescence amyloide. *Comptes Rendus des Seances de l'Academie des Sciences* 80:1288–1291.

Corti A (1851) Recherches sur l'organe de l'oue des mammiferes. *Zeitschrift fur wissenschaftiche Zoologie* [Q]:109–169.

Corti P, Mazzei E, Ferri S, Franchi GG, Dreassi E (1996) High performance thin layer chromatographic quantitative analysis of picrocrocin and crocetin active principles of saffron (*Crocus sativus* L – Iridaceae): a new method. *Phytochemical Analysis* 7:201–203.

Coscione AR, de Andrade JC, van Raij B (1998) Revisiting titration procedures for the determination of exchangeable acidity and exchangeable aluminum in soils. *Communications in Soil Science and Plant Analysis* 29:1973–1982.

Cosnier S, LeLous K (1996) A new strategy for the construction of amperometric dehydrogenase electrodes based on laponite gel-methylene blue polymer as the host matrix. *Journal of Electroanalytical Chemistry* 406:243–246.

Costanzo EM, Barry JA, Ribchester RR (1999) Co-regulation of synaptic efficacy at stable polyneuronally innervated neuromuscular junctions in reinnervated rat muscle. *Journal of Physiology* 521:365–374.

Cousin MA, Nicholls DG (1997) Synaptic vesicle recycling in cultured granule cells: role of vesicular acidification and refilling. *Journal of Neurochemistry* 69:1927–1935.

Cousin MA, Robinson PJ (1999) Mechanisms of synaptic vesicle recycling illuminated by fluorescent dyes. *Journal of Neurochemistry* 73:2227–2239.

Cowden RR, Curtis SK (1970) Demonstration of protein-bound sulfhydryl and disulfide groups with fluorescent mercurials. *Histochemie* 22:247–255.

Cowden RR, Curtis SK (1974) Interactions of fluorescent mercurial compounds and acidic fluorochromes with isolated living cells and nuclei. *Histochemistry* 40:253–262.

Creagh TA, Gleeson M, Travis D, Grainger R, McDermott TED, Butler MR (1995) Is there a role for *in vivo* methylene blue staining in the prediction of bladder tumour recurrence? *British Journal of Urology* 75:477–479.

Creasy GL, Price SF, Lombard PB (1993) Evidence for xylem discontinuity in pinot-noire and merlot grapes – dye uptake and mineral-composition during berry maturation. *American Journal of Enology and Viticulture* 44:187–192.

Crescenzi A, Muda AO, Origgi P, Bonfichi R, Faraggiana T (1991) Physicochemical analysis of resorcinol-fuchsin reagents and its relevance to staining quality of elastic fibers. *European Journal of Basic and Applied Histochemistry* 35:37–44.

Crew JP, Fellows GJ (1996) Epidural needle and catheter for bladder instillation of a marker dye. *British Journal of Urology* 77:914–915.

Crissman HA, Hirons GT (1994) Staining DNA in live and fixed cells. *Methods in Cell Biology* 41:211–217.

Crist KA, Kim K, Goldblatt PJ, Boone CW, Kelloff GJ, You M (1996) DNA quantification in cervical intraepithelial neoplasia thick tissue sections by confocal laser scanning microscopy. *Journal of Cellular Biochemistry* S25:49–56.

Crivellato E, Mallardi F (1997) Stromal cell organisation in the mouse lymph node. A light and electron microscopic investigation using the zinc iodide-osmium technique. *Journal of Anatomy* 190:85–92.

Crough HB, Becker ER (1931) A method of staining the oocysts of *Coccidia*. *Science* 73:212–213.

Crowe R, Burnstock G (1984) Quinacrine-positive neurones in some regions of the guinea-pig brain. *Brain Research Bulletin* 12:387–391.

Cserni G (1998) Elimination of phenol from the auramine fluorescence staining method for acid fast bacilli. *Journal of Histotechnology* 21:241–242.

Cuellar T, Gosalvez J, Del Castillo P, Stockert JC (1991) Fluram induces species-dependent C and G bands in mammalian chromosomes, revealing heterogeneous distribution of chromosomal proteins. *Genome* 34:772–776.

Cuellar T, Orellana J, Belhassen E, Bella JL (1999) Chromosomal characterization and physical mapping of the 5S and the 18S-5.8S-25S ribosomal DNA in *Helianthus argophyllus*, with new data from *Helianthus annuus*. *Genome* 42:110–115.

Cui JK, Hsu CY, Liu PK (1999) Suppression of postischemic hippocampal nerve growth factor expression by a c-fos antisense oligodeoxynucleotide. *Journal of Neuroscience* 19:1335–1344.

Culling CFA, Allison RT, Barr WT (1985) *Cellular Pathology Technique*. 4th Edn. Butterworths, London.

Cumley RW (1935) Negative stains in the demonstration of bacteria. *Stain Technology* 10:53–56.

Cunningham RS, Tompkins EH (1938) The supravital method of studying blood cells. In: H Downey (ed.) *Handbook of Hematology*. Hoeber, New York.

Curtis DJ, Horobin RW (1975) Staining banded human chromosomes with Romanowsky dyes: some practical consequences of the nature of the stain. *Humangenetik* 26:99–104.

Curtis F (1905) Methode de coloration elective du tissu conjonctif. *Comptes Rendus des Seances de la Société de Biologie* 58:1038–1040.

Curtis SK, Cowden RR, Benner DB (1987) Reaction of seven basic fluorochromes with unfixed cells obtained from the salivary glands of the dipteran fly *Megaseiia scalaris* Loew (Phoridae). *Histochemistry* 85:475–481.

Czyz J, Irmer U, Schulz G, Mindermann A, Hulser DF (2000) Gap-junctional coupling measured by flow cytometry. *Experimental Cell Research* 255:40–46.

Daddi L (1896) Nouvelle methode pour colorer la graisse dans les tissus. *Archives Italiennes de Biologie* 26:143–146.

Dahl JK (1952) A simple and sensitive histochemical method for calcium. *Proceedings of the Society for Experimental Biology and Medicine* 80:474–479.

Dahlan AN, Gordh G (1996) Development of *Trichogramma australicum* Girault (Hymenoptera: Trichogrammatidae) on *Helicoverpa armigera* (Hubneer) eggs (Lepidoptera: Noctuidae). *Australian Journal of Entomology* 35:337–344.

Dale GL (1988) Use of the janus green to facilitate quantitative analysis of phospho-inositides. *Journal of Chromatography - Biomedical Applications* 424:445–448.

Dall'Asta V, Gatti R, Orlandini G, Rossi PA, Rotoli BM, Sala R, Bussolati O, Gazzola GC (1997) Membrane potential changes visualized in complete growth media through confocal laser scanning microscopy of bis-oxonol-loaded cells. *Experimental Cell Research* 231:260–268.

Daniel FB, Behrman EJ (1976) Osmium(VI) complexes of the 3′,5′-dinucleoside monophosphates, ApU and UpA. *Biochemistry* 15:565–568.

Danielli JF (1953) *Cytochemistry. A Critical Approach*. Wiley, New York.

Danilatou V, Lydaki E, Dimitriou H, Papazoglou T, Kalmanti M (2000) Bone marrow purging by photodynamic treatment in children with acute leukemia – Cytoprotective action of Amifostine. *Leukemia Research* 24:427–435.

Danscher G, Norgaard JOR (1985) Ultrastructural autometallography: a method for silver amplification of catalytic metals. *Journal of Histochemistry and Cytochemistry* 33:706–710.

Danscher G, Hacker GW, Grimelius L, Norgaard JOR (1993) Autometallographic silver amplification of colloidal gold. *Journal of Histotechnology* 16:201–207.

Danscher G (1996) The autometallographic zinc-sulphide method. A new approach involving *in vivo* creation of nanometer-sized zinc sulphide crystal lattices in zinc-enriched synaptic and secretory vesicles. *Histochemical Journal* 28:361–373.

Darrow MA (1952) Synthetic orcein as an elastic tissue stain. *Stain Technology* 27:329–332.

Darzynkiewicz Z, Crissman HA, eds (1990) *Flow Cytometry* (Methods in Cell Biology, Vol. 33). Academic Press, San Diego.

Davey HM, Kell DB (1996) Flow cytometry and cell sorting of hetergeneous microbial populations: The importance of single-cell analyses. *Microbiological Reviews* 60:641–698.

Davey HM, Kell DB (1997) Fluorescent brighteners: novel stains for the flow cytometric analysis of microorganisms. *Cytometry* 28:311–315.

David BP, Purushothaman V, Venkatesan RA (1992) Resistogram typing as an epidemiological tool for *Escherichia coli* isolates of poultry origin. *Letters in Applied Microbiology* 15:96–99.

Davies JD, Young EW (1982) Congo blue: a rapid stain for elastic fibres. *Journal of Clinical Pathology* 35:789–791.

De Carvalho HF, Vidal BD (1995) The elastic system of a pressure-bearing tendon of the bullfrog *Rana catesbeiana. Annals of Anatomy – Anatomische Anzeiger* 177:397–404.

De Jong JP, Voerman JSA, Leenen PJM, Van Der Sluijs-Gelling AJ, Ploemacher RE (1991) Improved fixation of frozen lympho-haemopoietic tissue sections with hexazotized pararosaniline. *Histochemical Journal* 23:392–401.

De la Lande IS, Waterson JG (1968) Modification of autofluorescence in the formaldehyde treated rabbit ear artery by Evans blue. *Journal of Histochemistry and Cytochemistry* 16:281–282.

de la Torre J, Gosalvez J, Stockert JC (1990) Cytochemical observations on the centriolar adjunct of grasshopper spermatids in relation to its silver stainability. *Journal of Microscopy* 159:109–112.

de Maziere AM, Hage WJ, Ubbels GA (1996) A method for staining of cell nuclei in *Xenopus laevis* embryos with cyanine dyes for whole-mount confocal laser scanning microscopy. *Journal of Histochemistry and Cytochemistry* 44:399–402.

De Muro P, Faedda R, Formato M, Re F, Satta A, Cherchi GM, Carcassi A (2001) Urinary glycosaminoglycans in patients with systemic lupus erythematosus. *Clinical and Experimental Rheumatology* 19:125–130.

De Ondarza J, Hootman SR (1997) Confocal microscopic analysis of intracellular pH regulation in isolated guinea pig pancreatic ducts. *American Journal of Physiology – Gastrointestinal and Liver Physiology* 35:G124–G134.

de Paiva A, Meunier FA, Molgo J, Aoki KR, Dolly JO (1999) Functional repair of motor endplates after botulinum neurotoxin type A poisoning: biphasic switch of synaptic activity between nerve sprouts and their parent terminals. *Proceedings of the National Academy of Sciences of the United States of America* 96:3200–3205.

de Roe C, Courtoy PJ, Baudhuin P (1987) A model of protein–colloidal gold interactions. *Journal of Histochemistry and Cytochemistry* 35:1191–1198.

De Ronde JA, Van der Mescht A (1997) 2,3,5-triphenyltetrazolium chloride reduction as a measure of drought tolerance and heat tolerance in cotton. *South African Journal of Science* 93:431–433.

Dean WW, Lubrano GJ, Heinsohn HG, Stastny M (1976) The analysis of Romanowsky blood stains by high-performance liquid chromatography. *Journal of Chromatography* 124:287–301.

Decho AW, Kawaguchi T (1999) Confocal imaging of *in situ* natural microbial communities and their extracellular polymeric secretions using Nanoplast(R) resin. *BioTechniques* 27:1246–1252.

Declerck LS, Bridts CH, Mertens AM, Moens MM, Stevens WJ (1994) Use of fluorescent dyes in the determination of adherence of human-leukocytes to endothelial-cells and the effect of fluorochromes on cellular function. *Journal of Immunological Methods* 172:115–124.

Deerinck TJ, Martone ME, Lev-Ram V, Green DPL, Tsien RY, Spector DL, Huang S, Ellisman MH (1994) Fluorescence photooxidation with eosin: a method for high resolution immunolocalization and *in situ* hybridization detection for light and electron microscopy. *Journal of Cell Biology* 126:901–910.

Defendi V, Pearse AGE (1955) Significance of coupling rate in histochemical azo dye methods for enzymes. *Journal of Histochemistry and Cytochemistry* 3:203–211.

Del Bigio MR (2000) Calcium-mediated proteolytic damage in white matter of hydrocephalic rats? *Journal of Neuropathology and Experimental Neurology* 59:946–954.

Del Castillo P, Gomez A, Stockert JC (1988) Reaccion de fluorescencia de la cromatina por lacas preformadas de alumino. *Citologia* 10:53–61.

Del Castillo P, Llorente AR, Gomez A, Gosalvez J, Goyanes VJ, Stockert JC (1990) New fluorescence reactions in DNA cytochemistry. 2. Microscopic and spectroscopic studies on fluorescent aluminium complexes. *Analytical and Quantitative Cytology and Histology* 12:11–20.

Del Giorgio PA, Prairie YT, Bird DF (1997) Coupling between rates of bacterial production and the abundance of metabolically active bacteria in lakes, enumerated using CTC reduction and flow cytometry. *Microbial Ecology* 34:144–154.

Denffer H, Heidbrink V (1974) Dunnschichtchromatographische Untersuchung vewrshiedener Aldehydfuchsin. *Acta Histochemica* 48:62–68.

Deng C, Rogers LJ (1999) Differential sensitivities of the two visual pathways of the chick to labelling by fluorescent retrograde tracers. *Journal of Neuroscience Methods* 89:75–86.

Deng DX, Ono S, Koropatnick J, Cherian MG (1998) Metallothionein and apoptosis in the toxic milk mutant mouse. *Laboratory Investigation* 78:175–183.

Denizli A, Kocakulak M, Piskin E (1998a) Alkali blue 6B-derivitized poly (EGDMA/HEMA) microbeads for bilirubin removal from human plasma. *Journal of Macromolecular Science – Pure and Applied Chemistry* A35:137–149.

Denizli A, Salih B, Piskin E (1998b) New chelate-forming polymer microspheres carrying dyes as chelators for iron overload. *Journal of Biomaterials Science - Polymer Edition* 9:175–187.

Denning G, Fulton AB (1986) A simple trypsin resistance assay for muscle and other cell fusion. *Journal of Histochemistry and Cytochemistry* 34:959–962.

Denny WA, Cain BF, Atwell GJ, Hansch C, Panthananickal A, Leo A (1982) Potential antitumor agents. 36. Quantitative relationships between experimental antitumor activity, toxicity, and structure for the general class of 9-anilinoacridine antitumor agents. *Journal of Medicinal Chemistry* 25:276–281.

Denton J, Oughton DH (1993) The use of an acid solochrome-azurine stain to detect and assess the distribution of aluminum in *Sphagnum* moss. *Ambio* 22:19–21.

Deodhar KP, Tapp E, Scheuer PJ (1975) Orcein staining of hepatitis B antigen in paraffin sections of liver biopsies. *Journal of Clinical Pathology* 28:66–70.

Deprez M, Ceuterick-de Groote C, Fumal A, Reznik M, Martin JJ (1999) A new combined Bodian-luxol technique for staining unmyelinated axons in semithin, resin-embedded peripheral nerves: a comparison with electron microscopy. *Acta Neuropathologica* 98:323–329.

Derenzini M, Farabegoli F, Trere D (1992) Relationship between interphase AgNOR distribution and nucleolar size in cancer-cells. *Histochemical Journal* 24:951–956.

Derias NW, Adams CWM (1982) Macroscopic enzyme histochemistry in myocardial infarction – Use of coenzyme, cyanide, and phenazine methosulfate. *Journal of Clinical Pathology* 35:410–413.

Di WL, Lachelin GC, McGarrigle HH, Thomas NS, Becker DL (2001) Oestriol and oestradiol increase cell to cell communication and connexin 43 protein expression in human myometrium. *Molecular Human Reproduction* 7:671–679.

Diaz-Flores L, Gutierrez R, Lopez-Alonso A, Gonzalez R, Varela H (1992) Pericytes as a supplementary source of osteoblasts in periosteal osteogenesis. *Clinical Orthopaedics and Related Research* 275:280–286.

Diaz-Marta GLA, Fernandez MRS, Fernandez MAS, Alonso JG (1998) Mineral composition of saffron from Spain and other countries – Applications to differentiation. [Spanish]. *Agrochimica* 42:263–272.

DiCesare J, Grossman B, Katz E, Picozza E, Ragusa R, Woudenberg T (1993) A high-sensitivity electrochemiluminescence-based detection system for automated PCR product quantitation. *BioTechniques* 15:152–157.

Dieckmann-Schuppert A, Schnittler HJ (1997) A simple assay for quantification of protein in tissue sections, cell cultures, and cell homogenates, and of protein immobilized on solid surfaces. *Cell and Tissue Research* 288:119–126.

Dietz TH, Udoetok AS, Cherry JS, Silverman H, Byrne RA (2000) Kidney function and sulfate uptake and loss in the freshwater bivalve *Toxolasma texasensis*. *Biological Bulletin* 199:14–20.

Dinkelaker B, Marschner H (1992) *In vivo* demonstration of acid phosphatase activity in the rhizosphere of soil-grown plants. *Plant and Soil* 144:199–205.

Dinkelaker B, Hahn G, Romheld V, Wolf GA, Marschner. (1993). Non-destructive methods for demonstrating chemical changes in the rhizosphere (2001) Description of methods. *Plant and Soil* 156:67–70.

Disteche C, Bontemps J, Houssier C, Frederic J, Fredericq E (1980) Quantitative analysis of fluorescent profiles of chromosomes. Influence of DNA base composition on banding. *Experimental Cell Research* 125:251–264.

Diwu ZJ, Zimmerman J, Meyer T, Lown JW (1994) Design, synthesis and investigation of mechanisms of action of novel protein-kinase-C inhibitors: perylenequinonoid pigments. *Biochemical Pharmacology* 47:373–385.

Dixon EJ, Grisley LM, Sawyer R (1970) The purification of solochrome cyanine R. *Analyst* 95:945–949.

Dodge JD (1964) Cytochemical staining of sections from plastic-embedded flagellates. *Stain Technology* 39:381–386.

Doerner KC, White BA (1990) Detection of glycoproteins separated by nondenaturing polyacrylamide-gel electrophoresis using the periodic acid-Schiff stain. *Analytical Biochemistry* 187:147–150.

Dogra GS, Tandan BK (1964) Adaptation of certain histological technics for *in situ* demonstration of the neuro-endocrine systems of insects and other animals. *Quarterly Journal of Microscopical Science* 105:455–466.

Dolleman van der Weel MJ, Wouterlood FG, Witter MP (1994) Multiple anterograde tracing, combining *Phaseolus vulgaris*-leukoagglutinin with rhodamine-conjugated and biotin-conjugated dextran amine. *Journal of Neuroscience Methods* 51:9–21.

Donahue CJ, Fennie C, Villacorta R, La H, Lasky LA, Ohneda O (1999) Multicolor immunofluorescence and flow cytometry utilizing cascade blue to purify murine hematopoietic stem cells from fetal liver and bone marrow. *Cytometry* 37:60–67.

Dong Z, Patel Y, Saikumar P, Weinberg JM, Venkatachalam MA (1998) Development of porous defects in plasma membranes of adenosine triphosphate-depleted Madin-Darby canine kidney cells and its inhibition by glycine. *Laboratory Investigation* 78:657–668.

Doostzadeh-Cizeron J, Yin S, Goodrich DW (2000) Apoptosis induced by the nuclear death domain protein p84N5 is associated with caspase-6 and NF-kappa B activation. *Journal of Biological Chemistry* 275:25336–25341.

Dougherty MM, King JS (1984) A simple, rapid staining procedure for methacrylate embedded tissue sections using chromotrope 2R and methylene blue. *Stain Technology* 59:149–153.

Drury RAB, Wallington EA (1980) *Carleton's Histological Technique*. 5th Edn. Oxford University Press, Oxford.

du Buy HG, Showacre JL (1961) Selective localization of tetracycline in mitochondria of living cells. *Science* 133:196–197.

Duckett JG, Read DJ (1991) The use of the fluorescent dye 3,3'-dihexyloxacarbocyanine iodide, for selective staining of ascomycete fungi associated with liverwort rhizoids and ericoid mycorrhizal roots. *New Phytologist* 118:259–272.

Ducruet VJ, Rasse A, Feigenbaum AE (1996) Food and packaging interactions – use of methyl red as a probe for pvc swelling by fatty acid esters. *Journal of Applied Polymer Science* 62:1745–1752.

Duncan MJ, Jennes L, Jefferson JB, Brownfield MS (2000) Localization of serotonin(5A) receptors in discrete regions of the circadian timing system in the Syrian hamster. *Brain Research* 869:178–185.

Dunnigan MG (1968) Chromatographic separation and photometric analysis of the components of Nile blue sulphate. *Stain Technology* 43:243–248.

Dupont G, Swillens S, Clair C, Tordjmann T, Combettes L (2000) Hierarchical organization of calcium signals in hepatocytes: from experiments to models. *Biochimica et Biophysica Acta* 1498:134–152.

Dupuis M, Denis-Mize K, Woo C, Goldbeck C, Selby MJ, Chen M, Otten GR, Ulmer JB, Donnelly JJ, Ott G, McDonald DM (2000) Distribution of DNA vaccines determines their immunogenicity after intramuscular injection in mice. *Journal of Immunology* 165:2850–2858.

Dux L, Meszaros MG, Rohan J, Gajdos L, Jakab G, Guba F (1981) The value of simple lipid stains for typing skeletal muscle fibres. *Histochemical Journal* 13:63–71.

Duyckaerts C, Brion JP, Hauw J-J, Flament-Durand J (1987) Quantitative assessment of the density of neurofibrillary tangles and senile plaques in senile dementia of the Alzheimer type. Comparison of immunocytochemistry with a specific antibody and Bodian's protargol method. *Acta Neuropathologica* 73:167–170.

Dyer DL, Mori K (1969) Fluorescent nuclear staining with europium thenoyltrifluoro-acetonate. *Journal of Histochemistry and Cytochemistry* 17:755–756.

Ebbinghaus H (1902) Eine neue Method zur Farbung von Hornsubstanzen. *Zentralblatt fur allgemeine Pathologie* 13:422–425.

Ebener U, Wehner S (1993) Microwave stimulated cell marker analysis. Possibilities for more rapid immune diagnosis [German]. *Klinische Paediatrie* 205:34–40.

Edidin M (1989) Fluorescent labelling of cell surfaces. *Methods in Cell Biology* 29:87–102.

Editorial (1926) Constitution of Commission on Standardization of Biological Stains. *Stain Technology* 1:4–10.

Edwards BS, Kuckuck FW, Prossnitz ER, Okun A, Ransom JT, Sklar LA (2001) Plug flow cytometry extends analytical capabilities in cell adhesion and receptor pharmacology. *Cytometry* 43:211–216.

Egami C, Suzuki Y, Sugihara O, Fujimura H, Okamoto N (1997) Wide range pH fibre sensor with congo red and methyl red doped poly(methyl methacrylate) cladding. *Japanese Journal of Applied Physics* Part 1 36:2902–2905.

Egerton GS, Gleadle JM, Uffindell ND (1967) Purification of organic dyes, particularly of the anthraquinone series. *Journal of Chromatography* 26:62–71.

Ehrenberg B, Montana V, Wei M-D, Wuskell JP, Loew LM (1988) Membrane potential can be determined in individual cells from the Nernstian distribution of cationic dyes. *Biophysical Journal* 53:785–794.

Ehrlich P, Krause R, Mosse M, Rosin H, Weigert C, eds (1903) *Enzyklopadie der mikroskopischen Technik mit besonderer Berucksichtigung der Farblehre.* Urban and Schwarzenberg, Berlin.

Ehrlich P (1877) Beitrage zur Kenntniss der Anilinfarbungen und ihrer Verwendung in der mikroskopischen technik. *Archiv fur mikroskopische Anatomie* 13:263–277.

Ehrlich P (1879) Methodolische Beitrage zur Physiologie und Pathologie der verschiedenen Formen der Leukocyten. *Zeitschrift fur klinische Medizin* 1:553–560.

Ehrlich P (1881) Uber das Methylenblau und seine klinishbakterioskopische Verwendung. *Zeitschrift fur klinische Medizin* 2:710–713.

Ehrlich P (1883) Sulfodiazobenzol, ein Reagens auf Bilirubin. *Zentralblatt fuer klinische Medizin* 4:721–723.

Ehrlich P, Lazarus A (1909) *Die Anamie. I. Normale und pathologische Histologie des Blutes.* Alfred Holder, Wien and Leipzig.

Ekelund M, Ahren B, Hakanson R, Lundqvist I, Sundler F (1980) Quinacrine accumulates in certain peptide hormone-producing cells. *Histochemistry* 66:1–9.

Ekman R, Hakanson R, Sundler F, Thorell J (1980) High content of tyrosine and arginine in peptides and proteins from the growth hormone producing cells of the adenohypophysis. *Journal of Histochemistry and Cytochemistry* 28:401–407.

El Ahmer OR, Essery SD, Saadi AT, Raza MW, Ogilvie MM, Weir DM, Blackwell CC (1999) The effect of cigarette smoke on adherence of repiratory pathogens to buccal epithelial cells. *FEMS Immunology and Medical Microbiology* 23:27–36.

Eldar M, Yerushalmi Y, Kessler E, Scheinowitz M, Goldbourt U, Ben Hur E, Rosenthal I, Battler A (1990) Preferential uptake of a water-soluble phthalocyanine by atherosclerotic plaques in rabbits. *Atherosclerosis* 84:135–139.

Elder HY, Owen G (1967) Occurrence of 'elastic' fibres in the invertebrates. *Journal of Zoology (London)* 152:1–8.

Elfgang C, Eckert R, Lichtenberg-Frate H, Butterweck A, Traub O, Klein RA, Hulser DF, Willecke K (1995) Specific permeability and selective formation of gap junction channels in connexin-transfected HeLa cells. *Journal of Cell Biology* 129:805–817.

Ellinger P, Hirt A (1929) Mikroskopische Untersuchungen an lebenden Organen. 1. Mitteil. Methodik. Intravitalmikroskopie. *Zeitschrift fur Anatomie und Entwicklungs-Geschichte* 90:791–802.

Ellwart JW, Dormer P (1990) Vitality measurement using spectrum shift in Hoechst 33342 stained cells. *Cytometry* 11:239–243.

Elston GN, Tweedale R, Rosa MGP (1999) Cellular heterogeneity in cerebral cortex: a study of the morphology of pyramidal neurones in visual areas of the marmoset monkey. *Journal of Comparative Neurology* 415:33–51.

Emery AJ, Stotz E (1953) Spectrophotometric characteristics and assay of biological stains. IV. The phenyl methane dyes. *Stain Technology* 28:235–244.

Emery AJ, Hazen FH, Stotz E (1950) Spectrophotometric characteristics and assay of biological stains. III. The xanthenes. *Stain Technology* 25:201–208.

Emig WH (1941) *Stain Technique.* Science Press Printing Company, Pittsburgh.

Emmel VM, Cowdry EV (1964) *Laboratory Technique in Biology and Medicine.* 4th Edn. Williams and Wilkins, Baltimore.

Eneström S, Kniola B (1994) Cryofixation combined with physical dehydration for quantitative immunoelectron cytochemistry. *Biotechnic and Histochemistry* 69:89–98.

Eng LK, Cole AL (1976) Tinopal AN in fluorescent microscopic detection of bacteria within plant tissues. *Stain Technology* 51:277–278.

Engel CR, Sawicki E (1967) A superior thin-layer chromatographic procedure for the separation of aza arenes and its application to air pollution. *Journal of Chromatography* 31:109.

Erokhina IL, Selivanova GV, Vlasova TD, Komarova NI, Emeljanova OI, Soroka VV (1992) Ultrastructure and biosynthetic activity of polyploid atrial myocytes in patients with mitral valve disease. *Acta Histochemica Suppl.* 42:293–299.

Espada J, Stockert JC (1994) Fluorescence of bisazo dye reactions products from the coupled tetrazonium method for proteins. *Acta Histochemica* 96:315–324.

Espada J, Horobin RW, Stockert JC (1997) Fluorescent cytochemistry of acid phosphatase and demonstration of fluid-phase endocytosis using an azo dye method. *Histochemistry and Cell Biology* 108:481–487.

Estil S, Primo EJ, Wilson G (2000) Apoptosis in shed human corneal cells. *Investigative Ophthalmology and Visual Science* 41:3360–3364.

European Committee for Clinical Laboratory Standards [ECCLS] S on RM for TS[SRMTS (1992a) Dye standards, Part I. Terminology and general principles. *Histochemical Journal* 24:217–219.

European Committee for Clinical Laboratory Standards [ECCLS] S on RM for TS[SRMTS (1992b) Dye standards, Part II.1: Pyronin Y (CI 45005). *Histochemical Journal* 24:220–223.

European Committee for Clinical Laboratory Standards [ECCLS] S on RM for TS[SRMTS (1992c) Dye standards, Part II.2: Methyl green (CI 42585) and ethyl green (CI 42590). *Histochemical Journal* 24:224–227.

European Committee for Clinical Laboratory Standards [ECCLS] S on RM for TS[SRMTS (1992d) Dye standards, Part II.3: Thionin (CI 52000). *Histochemical Journal* 24:228–229.

European Committee for Clinical Laboratory Standards [ECCLS] S on RM for TS[SRMTS (1992e) Dye standards, Part II.4: Victoria blue B (CI 44045). *Histochemical Journal* 24:230–232.

European Committee for Clinical Laboratory Standards [ECCLS] S on RM for TS[SRMTS (1992f) Dye standards, Part II.5: Pararosaniline (CI 42500). *Histochemical Journal* 24:233–235.

European Committee for Clinical Laboratory Standards [ECCLS] S on RM for TS[SRMTS (1992g) Dye standards, Part II.6: Rosaniline (CI 42510). *Histochemical Journal* 24:236–237.

European Committee for Clinical Laboratory Standards [ECCLS] S on RM for TS[SRMTS (1992h) Dye standards, Part II.7: Magenta II (no CI number). *Histochemical Journal* 24:238–239.

European Committee for Clinical Laboratory Standards [ECCLS] S on RM for TS[SRMTS (1992i) Dye standards, Part II.8: New fuchsin (CI 42520). *Histochemical Journal* 24:240–242.

Evangelista RA, Pollak A, Templeton EFG (1991) Enzyme-amplified lanthanide luminescence for enzyme detection in bioanalytical assays. *Analytical Biochemistry* 197:213–224.

Evans HM, Schulemann W (1915) Uber Natur und Genese durch saure Farbstoffe entstehenden Vitalfarbungsgranula. *Folia Haematologica* 19:207–219.

Ewen AB (1962) An improved aldehyde fuchsin staining technique for neurosecretory products in insects. *Transactions of the American Microscopical Society* 81:94–96.

Eyden BP, Richmond I, Hale R, Buckley CH, Yamazaki K (1991) Lipid-rich residual bodies in human myometrium – qualitative observations. *Journal of Submicroscopic Cytology and Pathology* 23:585–594.

Eynard L, Lauriere M (1998) The combination of Indian ink staining with immunochemiluminescence detection allows precise identification of antigens on blots. Application to the study of glycosylated barley storage proteins. *Electrophoresis* 19:1394–1396.

Ezzughayyar A (1993) Neurosecretory cells in the central nervous system of the red slug *Arion rufus* L. *Israel Journal of Zoology* 39:1–9.

Fabian J, Hartmenn H (1980) *Light Absorption of Organic Colorants: Theoretical and Empirical Rules*. Springer-Verlag, New York.

Fagg GE, Scheff SW, Cotman CW (1981) Axonal sprouting at the neuromuscular junction of adult and aged rats. *Experimental Neurology* 74:847–854.

Fagrell D, Enestrom S, Berggren A, Kniola B (1996) Fat cylinder transplantation: an experimental comparative study of three different kinds of fat transplants. *Plastic and Reconstructive Surgery* 98:90–96.

Fallin CW, Christopher MM (1996) *In vitro* effects of ketones and hyperglycemia on feline hemoglobin oxidation and D- and L-lactate production. *American Journal of Veterinary Research* 57:463–467.

Farber E, Sternberg WH, Dunlap CE (1956) Histochemical localization of specific oxidative enzymes. 1. Tetrazolium stains for diphosphopyridine nucleotide diaphorase and triphosphopyridine nucleotide diaphorase. *Journal of Histochemistry and Cytochemistry* 4:254–265.

Farber SA, Pack M, Ho SY, Johnson ID, Wagner DS, Dosch R, Mullins MC, Hendrickson HS, Hendrickson EK, Halpern ME (2001) Genetic analysis of digestive physiology using fluorescent phospholipid reporters. *Science* 292:1385–1388.

Faris HA (1924) Neutral red and Janus green as histological stains. *Anatomical Record* 27:241–244.

Fassel TA, Edmiston CE (1999) Ruthenium red and the bacterial glycocalyx. *Biotechnic and Histochemistry* 74:194–212.

Fattorossi A, Nisini R, Lemoli S, Depetrillo G, Damelio R (1990) Flow cytometric evaluation of nitro blue tetrazolium (NBT) reduction in human polymorphonuclear leukocytes. *Cytometry* 11:907–912.

Favel A, Peyron F, De Meo M, Michel-Nguyen A, Carriere J, Chastin C, Regli P (1999) Amphotericin B susceptibility testing of *Candida lusitaniae* isolates by flow cytofluorometry: comparison with the Etest and the NCCLS broth macrodilution method. *Journal of Antimicrobial Chemotherapy* 43:227–232.

Fawcett JM, Harrison SM, Orchard CH (1998) A method for reversible permeabilization of isolated rat ventricular myocytes. *Experimental Physiology* 83:293–303.

Fay IW (1911) *Chemistry of the Coal-Tar Dyes.* Springer-Verlag, New York.

FDA (Food and Drug Administration) (1996a) Proposed rules to classify/reclassify analyte specific reagents. Federal Register 61, No. 051.

FDA (Food and Drug Administration) (1996b) Proposed rules to classify/reclassify immunohistochemical reagents and kits. Federal Register 61, No. 116.

Fedorko ME, Morse FL (1965) Isolation, characterization and distribution of acid mucopolysaccharides in rabbit leukocytes. *Journal of Experimental Medicine* 121:39–48.

Ferey L, Herlin P, Marnay J, Mandard AM, Catania R, Lubet P, Lande R, Bloyet D (1986) Pararosaniline or acriflavine-Schiff staining of epoxy embedded tissue after periodic acid oxidation in ethanol. A method suitable for morphometric and fluorometric analysis of glycogen. *Stain Technology* 61:107–110.

Ferlini C, De Angelis C, Biselli R, Distefano M, Scambia G, Fattorossi A (1999) Sequence of metabolic changes during X-ray-induced apoptosis. *Experimental Cell Research* 247:160–167.

Fernandez H, Eller G, Freymuller E, Vivanco T (1997) Detection of *Campylobacter jejuni* invasion of HEp-2 cells by acridine orange-crystal violet staining. *Memorias do Instituto Oswaldo Cruz* 92:509–511.

Ferreira SLC, Banderia MLSF, Lemos VA, dos Santos HC, Costa ACS, de Jesus DS (1997) Sensitive spectrophotometric determination of ascorbic acid in fruit juices and pharmaceutical formulations using 2-(5-bromo-2-pyridylazo)-5-diethylaminophenol (Br-PADAP). *Fresenius' Journal of Analytical Chemistry* 357:1174–1178.

Feulgen R, Rossenbeck H (1924) Mikroskopisch-chemischer Nachweis einer Nucleinsaure von Typus der Thymonucleinsaure und die darauf beruhende elektive Farbung von Zellkernen in mikroskopischer Praparater. *Hoppe-Seylers Zeitschrift fur physiologische Chemie* 135:2203–2248.

Ffrench M, Morel F, Souchier C, Benchaib M, Catallo R, Bryon PA (1994) Choice of fixation and denaturation for the triple labelling of intra-cytoplasmic antigen, bromo-deoxyuridine and DNA. Application to bone marrow plasma cells. *Histochemistry* 101:385–390.

Field MS, Wilhelm RG, Quinlan JF, Aley TJ (1995) An assessment of the potential adverse properties of fluorescent tracer dyes used for groundwater tracing. *Environmental Monitoring and Assessment* 38:75–96.

Fierz-David HE (1949) *Fundamental Processes of Dye Chemistry*. Interscience, New York.

Filipski GT, Wilson MVH (1985) Staining nerves in whole cleared amphibians and reptiles using Sudan black B. *Copeia* 2:500–502.

Fineran BA (1997) Cyto- and histochemical demonstration of lignins in plant cell walls: an evaluation of the chlorine water ethanolamine silver nitrate method of Coppick and Fowler. *Protoplasma* 198:186–201.

Fischer-Parton S, Parton RM, Hickey PC, Dijksterhuis J, Atkinson HA, Read ND (2000) Confocal microscopy of FM4-64 as a tool for analysing endocytosis and vesicle trafficking in living fungal hyphae. *Journal of Microscopy* 198:246–259.

Fisher DB (1968) Protein staining of ribboned epon sections for light microscopy. *Histochemie* 16:92–96.

Fishov I, Woldringh CL (1999) Visualization of membrane domains in *Escherichia coli*. *Molecular Microbiology* 32:1166–1172.

Fiszman ML, Novotny EA, Lange GD, Barker JL (1990) Embryonic and early postnatal hippocampal cells respond to nanomolar concentrations of muscimol. *Developmental Brain Research* 53:186–193.

Flanagan MT, Hesketh TR, Chung SH (1974) Procion yellow M-4RS binding to neuronal membranes. *Journal of Histochemistry and Cytochemistry* 22:952–961.

Flemming W (1891) Uber Theilung und Kernformen bei Leukocyten und uber deren Attractionsspharen. *Archiv fur mikroskopische Anatomie* 37:249–298.

Flinn CL, Ashworth EN (1994) Seasonal changes in ice distribution and xylem development in blueberry flower buds. *Journal of the Americal Society for Horticultural Science* 119:1176–1184.

Flint FO, Moss R (1970) Selective staining of protein and starch in wheat flour and its products. *Stain Technology* 45:75–79.

Flint FO, Pickering K (1984) Demonstration of collagen in meat products by an improved picro-sirius red polarisation method. *Analyst* 109:1505–1506.

Flint O (1994) *Food Microscopy*. Royal Microscopical Society Handbook No. 30. Bios, Oxford.

Flock A, Scarfone E, Ulfendahl M (1998) Vital staining of the hearing organ: visualization of cellular structure with confocal microscopy. *Neuroscience* 83:215–228.

Florenzano F, Bentivoglio M (2000) Degranulation, density, and distribution of mast cells in the rat thalamus: a light and electron microscopic study in basal conditions and after intracerebroventricular administration of nerve growth factor. *Journal of Comparative Neurology* 424:651–669.

Flores TR, Hoffmann EO (1997) Improved high resolution light microscopy and transmission electron microscopy techniques in diagnostic pathology. *Journal of Histotechnology* 20:45–52.

Foissner W (1991) Basic light and scanning electron microscopic methods for taxonomic studies of ciliated protozoa. *European Journal of Protistology* 27:313–330.

Fonagy A, Yokoyama N, Okano K, Tatsuki S, Maeda S, Matsumoto S (2000) Pheromone-producing cells in the silkmoth, *Bombyx mori*: identification and their morphological changes in response to pheromonotropic stimuli. *Journal of Insect Physiology* 46:735–744.

Foot NC (1929) Chemical agents: vital stains. In: CE McClung (ed.) *Handbook of Microscopical Technique*. Hoeber, New York. pp. 74–80.

Foot NC (1933) The Masson trichrome staining methods in routine laboratory use. *Stain Technology* 8:101–110.

Forney JR, DeWald DB, Yang S, Speer CA, Healey MC (1999) A role for host phospho-inositide 3-kinase and cytoskeletal remodeling during *Cryptosporidium parvum* infection. *Infection and Immunity* 67:844–852.

Foster AC, McInnes JL, Skingle DC, Symons RH (1985) Non-radioactive hybridization probes prepared by the chemical labellings of DNA and RNA with a novel reagent, photobiotin. *Nucleic Acids Research* 13:745–761.

Frank F, Ives DJG (1966) The structural properties of alcohol–water mixtures. *Quarterly Reviews* 20:1–44.

Franke C, Westerholm H, Niessner R (1997) Solid-phase extraction (SPE) of the fluorescence tracers uranine and sulphorhodamine B. *Water Research* 31:2633–2637.

Frankl SM, Denaro FJ (1998) Peripheral nerve teasing: two protocols for diagnosis and research. *Journal of Histotechnology* 21:39–43.

Franzen C, Muller A, Salzberger B, Fatkenheuer G, Eidt S, Mahrle G, Diehl V, Schrappe M (1995) Tissue diagnosis of intestinal microsporidiosis using a fluorescent stain with uvitex 2B. *Journal of Clinical Pathology* 48:1009–1010.

Franzen C, Muller A, Salzberger B, Hartmann P, Diehl V, Fatkenheuer G (1996) Uvitex 2B stain for the diagnosis of *Isospora belli* infections in patients with the acquired immunodeficiency syndrome. *Archives of Pathology and Laboratory Medicine* 120:1023–1025.

Fraser SE (1996) Iontophoretic dye labeling of embryonic cells. *Methods in Cell Biology* 51:147–160.

Frater R (1981) Neutralization of acid in glycol methacrylate and the use of cyclohexanol as a plasticizer. *Stain Technology* 56:99–101.

Frederiks WM, Bosch KS (1993) Quantitative aspects of enzyme histochemistry on sections of freeze substituted glycol methacrylate-embedded rat liver. *Histochemistry* 100:297–302.

Frederiks WM, Bosch KS, Kooij A (1995) Quantitative *in situ* analysis of xanthine oxido-reductase activity in rat liver. *Journal of Histochemistry and Cytochemistry* 43:723–726.

Freifelder D (1982) *Physical Biochemistry: Applications to Biochemistry and Molecular Biology*. 2nd Edn. Freeman, New York.

French RW (1926a) Azure C tissue stain. *Stain Technology* 1:79–80.

French RW (1926b) Standardization of biological stains as a problem of the medical department of the army. *Stain Technology* 1:11–16.

Frens G (1973) Controlled nucleation for the regulation of the particle size in mono-disperse gold suspensions. Nature Physical Science 241:20–22.

Frey H (1863) Die Lymphwege einer Peyer'schen Plaque beim Menschen. *Virchows Archiv fur pathologische Anatomie und Physiologie und fur klinische Medizin* 26:344–257.

Fricker M, Plieth C, Knight H, Blancaflor E, Knight MR, White NS, Gilroy S (1999) Fluorescence and luminescence techniques to probe ion activities in living plant cells. In: WT Mason (ed.) *Fluorescent and Luminescent Probes for Biological Activity*. Academic Press, San Diego. pp. 569–596.

Fricker MD, White NS, Obermeyer G (1997) pH gradients are not associated with tip growth in pollen tubes of *Lilium longiflorum*. *Journal of Cell Science* 110:1729–1740.

Friedberg SH, Goldstein DJ (1969) Thermodynamics of orcein staining of elastic fibres. *Histochemical Journal* 1:361–376.

Friedrich K, Seiffert W, Zimmermann HW (1990) Romanowsky dyes and Romanowsky-Giemsa effect. 5. Structural investigations of the purple DNA-AB-EY dye complexes of Romanowsky-Giemsa staining. *Histochemistry* 93:247–256.

456

Fritz DL, Waag DM (1999) Transdiaphragmatic lymphatic transport of intraperitoneally administered marker in hamsters. *Laboratory Animal Science* 49:522–529.

Fritz P, Saal JG, Wicherek C, Konig A, Laschner W, Rautenstrauch H (1996) Quantitative photometrical assessment of iron deposits in synovial membranes in different joint diseases. *Rheumatology International* 15:211–216.

Frodyma MM, Frei RW (1969) Reflectance spectroscopic analysis of dyes separated by thin layer chromatography. *Journal of Chemical Education* 46:522–524.

Fu YS, Tseng GF, Yin HS (1996) Extrinsic inhibitory innervation to rubral neurons in rat brain-stem slices. *Experimental Neurology* 137:142–150.

Fuchs BM, Syutsubo K, Ludwig W, Amann R (2001) *In situ* accessibility of *Escherichia coli* 23S rRNA to fluorescently labeled oligonucleotide probes. *Applied and Environmental Microbiology* 67:961–968.

Fuentealba IC, Mullins JE, Aburto EM, Lau JC, Cherian GM (2000) Effect of age and sex on liver damage due to excess dietary copper in Fischer 344 rats. *Journal of Toxicology – Clinical Toxicology* 38:709–717.

Fujii H, Cody SH, Seydel U, Papadimitriou JM, Wood DJ, Zheng MH (1997) Recording of mitochondrial transmembrane potential and volume in cultured rat osteoclasts by confocal laser scanning microscopy. *Histochemical Journal* 29:571–581.

Fujii T, Shimomura T, Fujimoto TT, Kimura A, Fujimura K (2000) A new approach to detect reticulated platelets stained with thiazole orange in thrombocytopenic patients. *Thrombosis Research* 97:431–440.

Fukuzawa K, Noguchi Y, Yoshikawa T, Saito A, Doi C, Makino T, Takanashi Y, Ito T, Tsuburaya A (1999) High incidence of synchronous cancer of the oral cavity and the upper gastrointestinal tract. *Cancer Letters* 144:145–151.

Furlong ST, Thibault KS, Rogers RA (1992) Fluorescent phospholipids preferentially accumulate in sub-tegumental cells of schistosomula of *Schistosoma mansoni*. *Journal of Cell Science* 103:823–830.

Fyson A, Oaks A (1992) Rapid methods for quantifying VAM fungal infections in maize roots. *Plant and Soil* 147:317–319.

Gabbett M (1887) Rapid staining of the tubercle bacillus. *Lancet* 1887–1:757.

Gabe M (1976) *Histological Techniques* (English Edn., transl. E. Blackith and A. Kavoor). Masson, Paris.

Gabrion JB, Herbute S, Bouille C, Maurel D, Kuchler-Bopp S, Laabich A, Delaunoy JP (1998) Ependymal and choroidal cells in culture: characterization and functional differentiation. *Microscopy Research and Techniques* 41:124–157.

Gadea J, Matas C, Lucas X (1998) Prediction of porcine semen fertility by homologous *in vitro* penetration (hIVP) assay. *Animal Reproduction Science* 54:95–108.

Gagliardi L, De Orsi D, Cavazzutti G, Multari G, Tonelli D (1996) HPLC determination of rhodamine B (CI 45170) in cosmetic products. *Chromatographia* 43:76–78.

Gagliardi L, De Orsi D, Multari G, Cavazzutti G, Tonelli D (1999) Quantification of phenolphthalein in cosmetic products. *Analusis* 27:163–165.

Gahan PB (1984) *Plant Histochemistry and Cytochemistry. An Introduction*. Academic Press, London.

Galassi L (1993) A simple procedure for crystallization of the Schiff Reagent. *Biotechnic and Histochemistry* 68:175–179.

Gallyas F (1979a) Factors affecting the formation of metallic silver and the binding of silver ions by the tissue components. *Histochemistry* 64:97–109.

Gallyas F (1979b) Light insensitive physical developers. *Stain Technology* 54:173–176.

Gallyas F (1982) Physicochemical mechanism of the argyrophil III reaction. *Histochemistry* 74:409–421.

Gallyas F, Gorcs T, Merchenthaler I (1982) High-grade intensification of the end-product of the diaminobenzidine reaction for peroxidase. *Journal of Histochemistry and Cytochemistry* 30:183–184.

Gan BS, Krump E, Shrode LD, Grinstein S (1998) Loading pyranine via purinergic receptors or hypotonic stress for measurement of cytosolic pH by imaging. *American Journal of Physiology* 275:C1158–C1166.

Gan WB, Grutzendler J, Wong WT, Wong ROL, Lichtman JW (2000) Multicolor "DiOdistic" labeling of the nervous system using lipophilic dye combinations *Neuron* 27: 219–225.

Gandor DW, Meyer J (1988) A simple two-dye basic stain facilitating recognition of mitosis in plastic mebedded tissue sections. *Stain Technology* 63:75–81.

Ganter P, Jolles G (1969,1970) *Histochimie Normale et Pathologique* (2 vols). Gaulthier-Villars, Paris.

Ganz J, Jork H (1994) Quantitative determination of erythrosine in hard gelatin capsules. *Journal of Planar Chromatography* 7:18–21.

Garcia-Ruiza C, Colell A, Mari M, Morales A, Fernandez-Checa JC (1997) Direct effect of ceramide on the mitochondrial electron transport chain leads to generation of reactive oxygen species. *Journal of Biological Chemistry* 272:11369–11377.

Gardner RL, Cockcroft DL (1998) Complete dissipation of coherent clonal growth occurs before gastrulation in mouse epiblast. *Development* 125:2397–2402.

Garfield S (2000) *Mauve*. Faber, London.

Garrett SD (1937) Bromothymol blue in aqueous sodium hydroxide as a clearing and staining agent for fungus-infected roots. *Annals of Botany* 1:563.

Garrote RL, Bertone RA, Silva ER, Coutaz VR, Avalle A (1998) Heat and NaOH penetration during chemical peeling of potatoes. *Food Science and Technology International* 4:23–32.

Garvey W, Fathi A, Bigelow F, Jimenez C, Carpenter B (1992) Combined modified periodic acid-Schiff and batch staining method. *Journal of Histotechnology* 15:117–120.

Garvey W, Fathi A, Bigelow F, Jimenez C, Carpenter B (1993) Use of heat to improve connective tissue stains: modifications of Masson, Movat and Fibrin stains. *Journal of Histotechnology* 16:349–353.

Garvey W, Bigelow F, Fathi A, Jimenez C, Carpenter B (1996) Modified Gomori trichrome stain for frozen skeletal-muscle and paraffin-embedded sections. *Journal of Histotechnology* 19:329–333.

Gasparic J (1966) Papier- und dunnschichtchromatographie der Echtfabebasen. LXVI. Mitteilung uber die Identifizierrung organischer Verbindugen. *Zeitschrift fur analytische Chemie* 218:113–118.

Gatenby JB, Beams HW, eds (1950) *The Microtomist's Vade-mecum* (Bolles Lee). 11th Edn. Churchill, London.

Gavin J, Button NF, Watson-Craik A, Logan NA (2000) Observation of soft contact lens disinfection with fluorescent metabolic stains. *Applied and Environmental Microbiology* 66:874–875.

Gavrilov VB, Kravchenko ON, Konev SV (2000) Dye sorption as an indicator of erythrocyte membrane damage and prehemolytic state of erythrocytes. *Bulletin of Experimental Biology and Medicine* 129:306–308.

Gelsleichter J, Cortes E, Manire CA, Hueter RE, Musick JA (1997) Use of calcein as a fluorescent marker for elasmobranch vertebral cartilage. *Transactions of the American Fisheries Society* 126:862–865.

Gerard P (1935) Sur l'emploi du noir soudane B pour reconnaitre les inclusions de vaseline liquide. *Bulletin d'Histologie Appliquée* 12:92–93.

Gerrits PO, Horobin RW (1996) Glycol methacrylate embedding for light microscopy: basic principles and trouble-shooting. *Journal of Histotechnology* 19:297–311.

Gerrits PO, Horobin RW, Wright DJ (1990) Staining sections of water-miscible resins. 1. The effects of the molecular size of the stain, and of resin cross-linking, on the staining of glycol methacrylate embedded tissues. *Journal of Microscopy* 160:279–290.

Gerrits PO, Brekelmans-Bartels M, Mast L, Gravenmade EJ, Horobin RW, Holstege G (1992) Staining myelin and myelin-like degradation products in the spinal cords of chronic experimental allergic encephalomyelitis (Cr-EAE) rats using Sudan black B staining of glycol methacrylate-embedded material. *Journal of Neuroscience Methods* 45:99–105.

Gersak K, Lavrencak J, Us-Krasovec M (2000) DNA ploidy of human granulosa cells from natural and stimulated *in vitro* fertilization cycles. *Fertility and Sterility* 74:158–161.

Geue L, Schluter H (1998) A Salmonella monitoring programme in egg production farms in Germany. *Journal of Veterinary Medicine, Series B – Infectious Diseases and Veterinary Public Health* 45:95–103.

Gibbes H (1880) On the double and treble staining of animal tissues for microscopical investigation. *Journal of the Royal Microscopical Society* 3:390–393.

Giebel J, Arends H, Fanghanel J, Cetin Y, Thiedemann KI, Schwenk M (1995) Suitability of different staining methods for the identification of isolated and cultured-cells from guinea-pig (*Cavia aperea porcellus*) stomach. *European Journal of Morphology* 33:359–372.

Giemsa G (1902a) Farbenmethoden fur Malariaparasiten. *Zentralblatt fur Bakteriologie* 31:429–430.

Giemsa G (1902b) Farbenmethoden fur Malariaparasiten. *Zentralblatt fur Bakteriologie* 32:307–313.

Giemsa G (1904) Eine Vereinfachung und Vervollkommnung meiner methylenazur-methylenblau-eosin Farbemethode zur Erzielung der Romanowsky-Nochtschen Chromatin-Farbung. *Zentralblatt fur Bakteriologie* 37:308–311.

Giemsa G (1910) Uber eine neue Schnellfarbung mit meiner Azureosinmethode. *Munchner medizinischer Wochenschrift* 47:1–3.

Giemsa G (1924) Zur Praxis der Giemsa-Farbung. *Zentralblatt fur Bakteriologie* 91:343–346.

Giemsa G (1934) Geschichte, Theorie und Weiterentwicklung der Romanowsky-Farbung. *Die medizinische Welt* 8:1432–1434.

Gilbert P, Kettenmann H, Orkand KK, Schachner M (1982) Immunocytochemical cell identification in nervous system culture combined with intracellular injection of a blue fluorescing dye (SITS). *Neuroscience Letters* 34:123–128.

Gilchrist JSC, Palahniuk C, Bose R (1997) Spectroscopic determination of sarcoplasmic reticulum Ca^{2+} uptake and Ca^{2+} release. *Molecular and Cellular Biochemistry* 172:159–170.

Giles CH, Greczek JJ (1962) A review of methods of purifying and analyzing water-soluble dyes. *Textile Research Journal* 32:506–515.

Gill JE, Jotz MM (1976) Further observations on the chemistry of pararosaniline-Feulgen staining. *Histochemistry* 46:146–160.

Gingrich JC, Davis DR, Nguyen Q (2000) Multiplex detection and quantitation of proteins on western blots using fluorescent probes. *BioTechniques* 29:636–642.

Giordano PA, Mazzini G, Riccardi A, Montecucco CM, Ucci G, Danova M (1985) Propidium iodide staining of cytoautoradiographic preparations for the simultaneous determination of DNA content and grain count. *Histochemical Journal* 17:1259–1270.

Gisbert E, Sarasquete C (2000) Histochemical identification of the black-brown pigment granules found in the alimentary canal of Siberian sturgeon (*Acipenser baeri*) during the lecithotrophic stage. *Fish Physiology and Biochemistry* 22:349–354.

Gisel A, Barella S, Hempel FD, Zambryski PC (1999) Temporal and spatial regulation of symplastic trafficking during development in *Arabidopsis thaliana* apices. *Development* 126:1879–1889.

Gitaitis RD, Chang CJ, Sijam K, Dowler CC (1991) A differential medium for semi-selective isolation of *Xanthomonas campestris* PV *vesticatoria* and other cellulolytic xanthomonads from various natural sources. *Plant Disease* 75:1274–1278.

Glattstein B, Shor Y, Levin N, Zeichner A (1996) pH indicators as chemical reagents for the enhancement of footwear marks. *Journal of Forensic Sciences* 41:23–26.

Glauert AM, Lewis PR (1998) *Biological Specimen Preparation for Transmission Electron Microscopy.* Portland Press, London.

Gluth WP, Kaliwoda G, Dann O (1986) Determination of fluorescent trypanocidal diamidines by quantitative thin-layer chromatography. *Journal of Chromatography* 378:183–193.

Goffin V, Letawe C, Pierard GE (1997) Effect of organic solvents on normal human stratum corneum: evaluation by the corneoxenometry bioassay. *Dermatology* 195:321–324.

Gohla A, Eckert K, Maurer HR (1996) A rapid and sensitive fluorometric screening assay using YO-PRO-1to quantify tumour cell invasion through Matrigel. *Clinical and Experimental Metastasis* 14:451–458.

Goland P, Engel M (1963) Fixation of cells and tissues by chloro-s-triazines. *Journal of Histochemistry and Cytochemistry* 11:751–762.

Gold D (1997) Assessment of the viability of *Schistosoma mansoni* schistosoma by comparative uptake of various vital dyes. *Parasitology Research* 83:163–169.

Goldmann EE (1909) Die dussere und innere Sekretion des gesunden Organismus im Lichte der vitalen Farbung. *Beitrage fur klinische Chirurgie (Tubingen)* 64:192–204.

Goldring JPD, Ravaioli L (1996) Solubilization of protein–dye complexes on nitrocellulose to quantify proteins spectrophotometrically. *Analytical Biochemistry* 242:197–201.

Goldstein DJ (1963a) An approach to the thermodynamics of histological dyeing, illustrated by experiments with azure A. *Quarterly Journal of Microscopical Science* 104:413–439.

Goldstein DJ (1963b) Selective staining of eosinophil granules in sections by alkaline orcein in a concentrated urea solution. *Stain Technology* 38:49–51.

Goldstein DJ, Horobin RW (1974) Rate factors in staining by alcian blue. *Histochemical Journal* 6:157–174.

Golgi C (1886) Sulla fina anatomia degli organi centrali del sistema nervoso. *Archives Italiennes de Biologie* 7:15–47.

Gomez-Barrena E, Martinez-Moreno E, Masso RB, Perez DM, Martinez LM (1999) Gold chloride technique to study articular innervation. A protocol validated through computer-assisted colorimetry. *Histology and Histopathology* 14:69–79.

Gomez-Segade LA (1980) A modification of the tetramethylbenzidine method for the histochemical demonstration of horseradish peroxidase (HRP) under the light microscope. *Trabajos del Instituto Cajal* 71:219–221.

Gomori G (1954) Histochemistry of the enterochromaffin substance. *Journal of Histochemistry and Cytochemistry* 2:50–53.

Gong Y, Wang Y, Chen F, Han J, Miao J, Shao N, Fang Z, Ou Yang R (2000) Identification of the subcellular localization of daunorubicin in multidrug-resistant K562 cell line. *Leukemia Research* 24:769–774.

Gordon PF, Gregory P (1983) *Organic Chemistry in Colour.* Springer-Verlag, Berlin.

Gori P (1977) Ponceau 2R staining on semi-thin sections of tissues fixed in glutaraldehyde-osmium tetroxide and embedded in epoxy resins. *Journal of Microscopy* 110:163–165.

Gorlenko MV, Kozheniv PA (1994) Differentiation of soil microbial communities by multisubstrate testing. *Microbiology* 63:158–161.

Gossner W (1958) Histochemischer Nachweis hydrolytischer Enzyme mit Hilfe der Azofarbstoffmethode. Untersuchungen zur Methodik und vergleichenden Histotopik der Esterasen und Phosphatasen bei Wirbeltieren. *Histochemie* 1:48–69.

Gossrau R (1990) Indoxyl alfa-D-galactoside as the temporarily last substrate for glycosidase histochemistry. The present state-of-the-art in histochemical glycosidase research using indoxyl glycosidas. *Folia Histochemica et Cytobiologica* 28:129–143.

Gossrau R (1991) Histochemical and biochemical studies of dipeptidyl peptidase I (DPP-1) in laboratory rodents. *Acta Histochemica* 91:85–100.

Gothot A, Pyatt R, McMahel J, Rice S, Srour EF (1997) Functional heterogeneity of human CD34(+) cells isolated in subcompartments of the G(0)/G(1) phases of the cell cycle. *Blood* 90:4384–4393.

Goyena M, Ortiz JM, Alonso FD (1997) Influence of different systems of feeding in the appearance of cryptosporidiosis in goat kids. *Journal of Parasitology* 83:1182–1185.

Graham ET, Joshi PA (1996) Plant cuticle staining with bismark brown Y and azure B or toluidine blue O before paraffin extraction. *Biotechnic and Histochemistry* 71:92–95.

Graham RJT (1968) The thin layer chromatographic separation of nitrophenols on polyamide surfaces. *Journal of Chromatography* 33:118–119.

Graham RK, Caiger P (1969) Fluorescence staining for the determination of cell viability. *Applied Microbiology* 17:489–490.

Grauman W, Arnold M (1969) Farberische Darstellung der Zymogengranula des Pankreas auf Grund ihres Gehalts and Chymotrypsinogen. *Histochemie* 20:355–362.

Gravance CG, Garner DL, Miller MG, Berger T (2001) Fluorescent probes and flow cytometry to assess rat sperm integrity and mitochrondrial function. *Reproductive Toxicology* 15:5–10.

Gray JD, Kolesik P, Hoj PB, Coombe BG (1999) Confocal measurement of the three-dimensional size and shape of plant parenchyma cells in a developing fruit tissue. *Plant Journal* 19:229–236.

Gray P (1954) *The Microtomist's Formulary and Guide.* Blakiston, New York.

Gray P (1973) *The Encyclopedia of Microscopy and Microtechnique.* Van Nostrand Reinhold, New York.

Gray-Keller M, Denk W, Shraiman B, Detwiler PB (1999) Longitudinal spread of second messenger signals in isolated rod outer segments of lizards. *Journal of Physiology* 519:679–692.

Green FJ (1966) A comparision of pre-World War II cresyl echt violet with various similarly named products. *Stain Technology* 41:115–120.

Green FJ (1990) *The Sigma-Aldrich Handbook of Stains, Dyes and Indicators.* Aldrich Chemical Company, Milwaukee.

Green FJ (2000) Coming full circle: a brief history of the domestic synthetic dye and biological stain industries. *Biotechnic and Histochemistry* 75:167–175.

Green HA, Bua D, Anderson RR, Nishioka NS (1992) Burn depth estimation using indo-cyanine green fluorescence. *Archives of Dermatology* 128:43–49.

Green LK, Ansari MQ, Schwartz MR, Ro JY, Alpert LC (1994) Non-specific fluorescent whitener stains in the rapid recognition of pulmonary dirofilariasis – A report of 20 cases. *Thorax* 49:590–593.

Greenhalgh CW (1976) Aspects of anthraquinone dyestuff chemistry. *Endeavour* 35:134–140.

Gregersen MI, Shiro H (1938) The behavior of the dye T1824 with respect to its absorption by red blood cells and its fate in blood undergoing coagulation. *American Journal of Physiology* 121:293–309.

Gregory GE (1973) Simple fluorescent staining of insect control nerve fibers with Procion yellow. *Stain Technology* 48:85–87.

Gregory P (1990) Classification of dyes by chemical structure. Ch. 2 In: DH Waring, G Hallas (eds) *The Chemistry and Application of Dyes.* New York, Plenum Press. pp. 17–47.

Gregory P (1991) *High Technology Applications of Organic Colorants.* Plenum Press, New York.

Grewal BS, Naik A, Irwin WJ, Gooris G, de Grauw CJ, Gerritsen HG, Bouwstra JA (2000) Transdermal macromolecular delivery: real-time visualization of iontophoretic and chemically enhanced transport using two-photon excitation microscopy. *Pharmaceutical Research* 17:788–795.

Gribbon LT, Barer MR (1995) Oxidative metabolism in nonculturable *Helicobacter pylori* and *Vibrio vulnificus* cells studied by substrate enhanced tetrazolium reduction and digital image processing. *Applied and Environmental Microbiology* 61:3379–3384.

Griesbach HA (1886) Weitere Untersuchungen uber Azofarbestoffe behufs Tinction menschlicher und thierischer Gewebe. *Zeitschrift fur wissenschaftliche Mikroskopie* 3:358–385.

Griesbach HA (1889) (Report of a demonstration by H. A. Griesbach in Wurzburg in 1888). *American Monthly Microscopical Journal* 10:30–33.

Griffiths J, ed. (1984) *Developments in the Chemistry and Technology of Organic Dyes* (Critical Reports on Applied Chemistry, Vol. 17). Blackwell, Oxford.

Grill V, Zweyer M, Bareggi R, Martelli AM, Basa M, Narducci P (1995) A simple and rapid staining technique for plastic-embedded cartilage and bone. *Biotechnic and Histochemistry* 70:75–80.

Grizzle W, Mowry RW (1994) FDA regulation of immunochemicals used in immuno-histochemistry: our view. *Biotechnic and Histochemistry* 69:348–350.

Grizzle WE (1996) Theory and practice of silver staining in histopathology. *Journal of Histotechnology* 19:183–195.

Groneberg DA, Doring F, Eynott PR, Fischer A, Daniel H (2001) Intestinal peptide transport: *ex vivo* uptake studies and localization of peptide carrier PEPT1. *American Journal of Physiology – Gastrointestinal and Liver Physiology* 281:G697–6704.

Grotz MRW, Deitch EA, Ding JY, Xu DZ, Huang QH, Regel G (1999) Intestinal cytokine response after gut ischemia – Role of gut barrier failure. *Annals of Surgery* 229:478–486.

Grouls V, Helpap B (1981) Selective staining of eosinophils and their immature precursor in tissue sections and autoradiographs with Congo red. *Stain Technology* 56:323–325.

Guffanti AA, Cheng J, Krulwich TA (1998) Electrogenic antiport activities of the Gram-positive Tet proteins include a Na+(K+)/K+ mode that mediates K+ uptake. *Journal of Biological Chemistry* 273:26447–26454.

Gulati GL, Hyun PH (1996) An unusual WBC scattergram and its possible causes. *Laboratory Medicine* 27:398–402.

Gulland JM (1938) Nucleic acids. *Journal of the Chemical Society* 1938:1722–1734.

Guntern R, Bouras C, Hof PR, Vallet PG (1992) An improved thioflavine S method for staining neurofibrillary tangles and senile plaques in Alzheimer's disease. *Experientia* 48:8–10.

Gupta BD, Sharma S (1998) A long-range fiber optic pH sensor prepared by dye doped sol-gel immobilization technique. *Optics Communications* 154:282–284.

Gupta RK, Kenwright D, Naran S, Lallu S, Fauck R (2000) Fine needle aspiration cyto-diagnosis of secretory carcinoma of the breast. *Cytopathology* 11:496–502.

Gupta VS, Kraft SC, Samuelson JS (1967) Purification and properties of acriflavine, proflavine and related compounds. *Journal of Chromatography* 26:158–163.

Gurr E (1960) *Encyclopaedia of Microscopical Stains*. Leonard Hill, London.

Gurr E (1965) *The Rational Use of Dyes in Biology and General Staining Methods*. Leonard Hill, London.

Gurr E (1971) *Synthetic Dyes in Biology Medicine and Chemistry*. Academic Press, London.

Gutmann M (1995) Improved staining procedures for photographic documentation of phenolic deposits in semithin sections of plant tissue. *Journal of Microscopy* 179:277–281.

Gutstein M (1937) New direct staining methods for elementary bodies. *Journal of Pathology and Bacteriology* 45:313.

Guttman A, Lengyel T, Szoke M, Sasvari-Szekely M (2000) Ultra-thin-layer agarose gel electrophoresis. II. Separation of DNA fragments on composite agarose-linear polymer matrices. *Journal of Chromatography* A 871:289–298.

Guyer MF (1906) *Animal Microbiology*. Chicago University Press, Chicago.

Haapala J, Arokoski J, Pirttimaki J, Lyyra T, Jurvelin J, Tammi M, Helminen HJ, Kiviranta I (2000) Incomplete restoration of immobilization induced softening of young beagle knee articular cartilage after 50-week remobilization. *International Journal of Sports Medicine* 21:76–81.

Haas F (1992) Serial sectioning of insects with hard exoskeleton by dissolution of the exocuticle. *Biotechnic and Histochemistry* 67:50–54.

Haghighi AZ, Wei R (1998) Measurement of superoxide dismutase in erythrocytes and whole blood using iodonitrotetrazolium violet. *Analytical Letters* 31:981–990.

Hagino S, Itagaki H, Kato S, Kobayashi T, Tanaka M (1991) Quantitative evaluation to predict the eye irritancy of chemicals – modification of chorioallantoic membrane test by using trypan blue. *Toxicology in Vitro* 5:301–304.

Hagler AN, Mendonca-Hagler LC (1991) A diazonium blue B test for yeasts grown 3 days on yeast carbon base-urea agar. *Revista de Microbiologia* 22:71–74.

Hahn D, Amann RI, Ludwig W, Akkermans ADL, Schleifer KH (1992) Detection of micro-organisms in soil after *in situ* hybridization with ribosomal RNA-targeted, fluores-cently labelled oligonucleotides. *Journal of General Microbiology* 138:879–887.

Hahnel C, Somodi S, Weiss DG, Guthoff RF (2000) The keratocyte network of human cornea: a three-dimensional study using confocal laser scanning fluorescence microscopy. *Cornea* 19:185–193.

Hakansson S, Morisaki H, Hueser J, Sibley LD (1999) Time-lapse video microscopy of gliding motility in *Toxoplasma gondii* reveals a novel, biphasic mechanism of cell loco-motion. *Molecular Biology of the Cell* 10:3539–3547.

Hale DC, Johnson CC, Kirkham MD (1985) *In vitro Giardia* cyst viability evaluation by fluorescent dyes. *Microecology and Therapy* 15:141–148.

Hall JO, Novakofski JE, Beasley VR (1998) Neutral red assay modification to prevent cytotoxicity and improve reproducibility using E-63 rat skeletal muscle cells. *Biotechnic and Histochemistry* 73:211–221.

Halliday BE, Silverman JF, Finley JL (1998) Fine-needle aspiration cytology of amyloid asso-ciated with nonneoplastic and malignant lesions. *Diagnostic Cytopathology* 18:270–275.

Hals E (1977) Selective fluorescence of rat elastic fibres stained with acid fuchsin or fast green FCF. *Scandinavian Journal of Dental Research* 85:542–548.

Hals E (1978) Fluorescence microscopic demonstration of rat mast cells stained with acid dyes. *Acta Odontologica Scandinavica* 36:57–66.

Hamberger A, Hamberger B (1966) Uptake of catecholamines and penetration of trypan blue after blood–brain barrier lesions. *Zeitschrift fur Zellforschung und mikroskopische Anatomie* 70:386–392.

Hamperl H (1943) Fluoreszenzmikroskopie. Wesen und Anwendung in der Medizin. *Medizinische Klinik* 39:781–784.

Hanaizumi Y, Maeda T, Takano Y (1998) Distribution of calcium ions at the interface between resin bonding materials and tooth dentin. Use of commercially available adhesive systems. *Journal of Electron Microscopy* 47:227–241.

Hanker JS, Romanovicz DK, Padykula HA (1976) Tissue fixation and osmium black formation with nonvolatile octavalent osmium compounds. *Histochemistry* 49:263–291.

Hansen CL, Chen G, Ebner TJ (2000) Role of climbing fibres in determining the spatial patterns of activation in the cerebellar cortex to peripheral stimulation: an optical imaging study. *Neuroscience* 96:317–331.

Hansen M, Thrane C, Olsson S, Sorensen J (2000) Confocal imaging of living fungal hyphae challenged with the fungal antagonist viscosinamide. *Mycologia* 92:216–221.

Harapanhalli RS, Howell RW, Rao DV (1994) Bis-benzimidazole dyes, Hoechst 33258 and Hoechst 33342: radioiodination, facile purification and subcellular distribution. *Nuclear Medicine Biology* 21:641–647.

Haresh K, Suresh K, Anuar AK, Saminathan S (1999) Isolate resistance of *Blastocystis hominis* to metronidazole. *Tropical Medicine and International Health* 4:274–277.

Harf C, Goffinet S, Meunier O, Monteil H, Colin DA (1997) Flow cytometric determination of endocytosis of viable labelled *Legionella pneumophila* by *Acanthamoeba palestinensis. Cytometry* 27:269–274.

Harms H (1965) *Handbuch der Farbstoffe fur die Mikroskopie.* Stauffen Verlag, Kamp-Lintfort.

Harris WH, Marshall JS, Yamashiro S, Shaikh N (1999) Mast cells of the bovine trachea: staining characteristics, dispersion techniques and response to secretagogues. *Canadian Journal of Veterinary Research* 63:5–12.

Hart RA, Mo O, Borius F, Fung DYC (1991) Comparative analysis of trypan blue agar and Congo red agar for the enumeration of yeast and mold using the HGMF system. *Journal of Food Safety* 11:227–230.

Hartemink R, Domenech VR, Rombouts FM (1997) LAMVAB – A new selective medium for the isolation of lactobacilli from faeces. *Journal of Microbiological Methods* 29:77–84.

Harvey RW, Kinner NE, Bunn A, MacDonald D, Metge D (1995) Transport behavior of groundwater protozoa and protozoan-sized microspheres in sandy aquifer sediments. *Applied and Environmental Microbiology* 61:209–217.

Hasegawa S, Kusuoka H, Fukuchi K, Hori M, Nishimura T (2000) Estimation of the area at risk in myocardial infarction of rats by means of I-123 beta-methyliodophenyl pentadecanoic acid imaging. *Annals of Nuclear Medicine* 14:347–352.

Hassab MH, Kennedy DW (2001) Effects of long-term induced ostial obstruction in the rabbit maxillary sinus. *American Journal of Rhinology* 15:55–59.

Hassner A, Birnbaum D, Loew LM (1984) Charge-shift probes of membrane potential. Synthesis. *Journal of Organic Chemistry* 49:2546–2551.

Hastein T, Hill BJ, Berthe F, Lightner DV (2001) Traceability of aquatic animals. *Revue Scientifique et Technique de l'Office International des Epizooties* 20:564–583.

Hasty RA, Lima FJ, Ottaway JM (1981) Bromate oxidation of methyl orange. 2. Effect of bromide and application to the determination of bromide. *Analyst* 106:76–84.

Haugen TS, Skjonsberg OH, Kahler H, Lyberg T (1999) Production of oxidants in alveolar macrophages and blood leukocytes. *European Respiratory Journal* 14:1100–1105.

Haugland RP (1995) Detecting enzymatic activity in cells using fluorogenic substrates. *Biotechnic and Histochemistry* 70:243–251.

Haugland RP (1996) *Handbook of Fluorescent Probes and Research Chemicals.* 6th Edn. *Molecular Probes.* Eugene, OR.

Haugland RP (2001) *Handbook of Fluorescent Probes and Research Products.* CD-ROM 8th Edn. *Molecular Probes,* Eugene, OR.

Hauptmann G, Gerster T (1996) Multicolor whole-mount *in situ* hybridization to *Drosophila* embryos. *Development Genetics and Evolution* 206:292–295.

Havemeister W, Rehbein H, Steinhart H, Gonzales Sotelo C, Krogsgaard-Nielsen M, Jorgensen B (1999) Visualization of the enzyme trimethylamine oxide demethylase in isoelectric focusing gels by an enzyme-specific staining method. *Electrophoresis* 20:1934–1938.

Hay J, Khan W, Sugden JK (1994) Photochemical decomposition of phenol red (phenol-sulfophthalein). *Dyes and Pigments* 24:305–312.

Hay J, Ranson C, Sugden JK (1995) Photochemical stability of sulphorhodamine B. *Dyes and Pigments* 27:55–61.

Hayat MA (1975) *Positive Staining for Electron Microscopy.* Van Nostrand Reinhold, New York.

Hayat MA (1981) *Fixation for Electron Microscopy.* Academic Press, New York.

Hayat MA (1993) *Stains and Cytochemical Methods.* Plenum Press, New York.

Hayat MA (2000) Positive staining. In: MA Hayat (ed.) *Principles and Techniques of Electron Microscopy: Biological Applications.* Cambridge University Press, Cambridge, UK, pp. 242–366.

Hayes WP, Nyaku NY, Burns DT (1973) Separation and identification of food dyes. 5. Examination of ponceau 6R dyes: extraction of dyes from confectionary products (cakes, cake mixtures and pastries). *Journal of Chromatography* 84:195–199.

Hayhoe FGJ, Quaglino D (1988) *Haematological Cytochemistry.* 2nd Edn. Churchill-Livingstone, Edinburgh.

Haynes R (1928) Fast green, a new substitute for light green SF yellowish. *Stain Technology* 3:40.

Haynes RJ, Williams PH (1999) Influence of stock camping behaviour on the soil micro-biological and biochemical properties of grazed pastoral soils. *Biology and Fertility of Soils* 28:253–258.

He CM, Roach MR (1994) The composition and mechanical properties of abdominal aortic aneurysms. *Journal of Vascular Surgery* 20:6–13.

He ZH, Bottinelli R, Pellegrino MA, Ferenczi MA, Reggiani C (2000) ATP consumption and efficiency of human single muscle fibers with different myosin isoform composition. *Biophysical Journal* 79:945–961.

Hebda PA, Dohar JE (1999) Transplanted fetal fibroblasts: survival and distribution over time in normal adult dermis compared with autogenic, allogenic, and xenogenic adult fibroblasts. *Otolaryngology, Head and Neck Surgery* 121:245–251.

Hed J, Dahlgren C, Rundqvist I (1983) A simple fluorescence technique to stain the plasma membrane of human neutrophils. *Histochemistry* 79:105–110.

Heidelberg JF, Shahamat M, Levin M, Rahman I, Stelma G, Grim C, Colwell RR (1997) Effect of aerosolization on culturability and viability of gram-negative bacteria. *Applied and Environmental Microbiology* 63:3585–3588.

Heidenhain M (1916) Uber die Mallorysche Bindegewebsfarbung mit Karmin und Azokarmin als Vorfarben. *Zeitschrift fur wissenschaftliche Mikroskopie* 32:361–362.

Heidenhain R (1888) Beitrage zur Histologie und Physiologie der Dunndarmschleimhaut. *Pflugers Archhiv fur gesamte Physiologie des Menschen und der Tiere* 42 (Suppl.):1–103.

Heimstadt O (1911) Das Fluoreszenzmikroskop. *Zeitschrift fur wissenschaftliche Mikroskopie* 28:330–337.

Heinicke EA, Kiernan JA, Wijsman J (1987) Specific, selective, and complete staining of neurons of the myenteric plexus, using cuprolinic blue. *Journal of Neuroscience Methods* 21:45–54.

Heinicke EA, Kiernan JA (1978) Vascular permeability and axonal regeneration in skin autotransplanted into the brain. *Journal of Anatomy* 125:409–420.

Heiny JA, Jong DS (1990) A nonlinear electrostatic potential change in the T-system of skeletal muscle detected under passive recording conditions using potentiometric dyes. *Journal of General Physiology* 95:147–175.

Heiss J, Zeller K-P (1969a) Massenspektrometrische Untersuchungen an Heterocyclen III: Thioxanthen, Thioxanthon und Phenothiazin sowie deren S-oxide. *Organic Mass Spectrometry* 2:819–828.

Heiss J, Zeller K-P (1969b) Massenspektrometrische Untersuchungen an Heterocyclen IV: Zur Wasserstoffabspaltung aus Xanthen und Thioxanthen. *Organic Mass Spectrometry* 2:829–833.

Hemar A, Olivo J-C, Williamson E, Saffrich R, Dotti CG (1997) Dendroaxonal transcytosis of transferrin in cultured hippocampal and sympathetic neurons. *Journal of Neuroscience* 17:9026–9034.

Hemmila IA (1989) Fluorescent labels for use in fluorescence immunoassays. *Applied Fluorescence Technology* 1:1–8.

Henderson B (1983) Sensitivity to light of solutions of phenazine methosulfate – reply. *Histochemical Journal* 15:275–276.

Hendin BN, Patel B, Levin HS, Thomas AJ, Agarwal A (1998) Identification of spermatozoa and round spermatids in the ejaculates of men with spermatogenic failure. *Urology* 51:816–819.

Hendrickson DA, Southwood LL, Lopez MJ, Johnson R, Kruse-Elliott KT (1998) Cranial migration of different volumes of new methylene blue after caudal epidural injection in the horse. *Equine Practice* 20:12–14.

Henley N, Baron C, Roberts KD (1994) Flow cytometric evaluation of the acrosome reaction of human spermatozoa – A new method using a photoactivated supravital stain. *International Journal of Andrology* 17:78–84.

Henning JA, Teuber LR (1992) Identification of pollenkitt variation among alfalfa germplasm sources. *Crop Science* 32:653–656.

Henry JB (1968) Vital staining by procion navy blue M3RS in laboratory mammals. *Stain Technology* 43:297–302.

Hentz RV, Traina GC, Cadossi R, Zucchini P, Muglia MA, Giordani M (2000) The protective efficacy of surgical latex gloves against the risk of skin contamination: how well are the operators protected? *Journal of Materials Science – Materials in Medicine* 11:825–832.

Henwood A (2002) Mast cell staining with alcian blue tetrakis(methylpyridinium) chloride. *Biotechnic and Histochemistry* 77:93–94.

Heppert KE, Davies MI (1999) Using a microdialysis shunt probe to monitor phenolphthalein glucuronide in rats with intact and diverted bile flow. *Analytica Chimica Acta* 379:359–366.

Hermann E (1875) Uber eine neue Tintionsmethode. *Tageblatt der 48 Naturfverein (Graz)* 1875:105.

Herrera AA, Qiang H, Ko CP (2000) The role of perisynaptic Schwann cells in development of neuromuscular junctions in the frog (*Xenopus laevis*). *Journal of Neurobiology* 45:237–254.

Hetherington DC (1936) Pinacyanol as a supra-vital mitochrondrial stain for blood. *Stain Technology* 11:153–154.

Hetland G, Lovik M, Wiker GH (1998) Protective effect of beta-glucan against *Mycobacterium bovis,* BCG infection in BALB/c mice. *Scandinavian Journal of Immunology* 47:548–553.

Higa F, Kusano N, Tateyama M, Shinzato T, Arakaki N, Kawakami K, Saito A (1998) Simplified quantitative assay system for measuring activities of drugs against intracellular *Legionella pneumophila*. *Journal of Clinical Microbiology* 36:1392–1398.

Higashi H, Sato K, Ohtake A, Omori A, Yoshida S, Kudo Y (1997) Imaging of cAMP-dependent protein kinase activity in living neural cells using a novel fluorescent substrate. *FEBS Letters* 414:55–60.

Higashiyama T, Kuroiwa H, Kawano S, Kuroiwa T (1997) Kinetics of double fertilization in *Torenia fournieri* based on direct observations of the naked embryo sac. *Planta* 203:101–110.

Highman B (1945) Staining of mucins with buffered solutions of toluidine blue and new methylene blue N. *Stain Technology* 20:85–87.

Highman B (1946) Improved methods for demonstrating amyloid in paraffin sections. *Archives of Pathology* 41:559–562.

Hill A (1896) The chrome-silver method. A study of the conditions under which the reaction occurs and a criticism of its results. *Brain* 19:1–42.

Hill BF, Borror AC (1992) Redefinition of the genera *Diophrys* and *Paradiophrys* and establishment of the genus *Diophryopsis* NG (Ciliophora, hypotrichida) – Implications for the species problem. *Journal of Protozoology* 39:144–153.

Hill JJ (1770) *The Construction of Timber*. Privately published for the author, London.

Hiltbrunner E, Fluckiger W (1996) Manganese deficiency of silver fir trees (*Abies alba*) at a reforested site in the Jura mountains, Switzerland: Aspects of cause and effect. *Tree Physiology* 16:963–975.

Himpens B, De Smedt H, Casteels R (1994) Relationship between [Ca^{2+}] changes in nucleus and cytosol. *Cell Calcium* 16:239–246.

Hiratsuka M, Miyashiro I, Ishikawa O, Furukawa H, Motomura K, Ohigashi H, Kameyama M, Sasaki Y, Kabuto T, Ishiguro S, Imaoka S, Koyama H (2001) Application of sentinel node biopsy to gastric cancer surgery. *Surgery* 129:335–340.

Hirnle P (1991) Histological findings in rabbit lymph nodes after endolymphatic injection of liposomes containing blue dye. *Journal of Pharmacy and Pharmacology* 43:217–218.

Hiroi J, Kaneko T, Tanaka M (1999) *In vivo* sequential changes in chloride cell morphology in the yolk-sac membrane of the Mozambique tilapia (*Oreochromis mossambicus*) embryos and larvae during seawater adaptation. *Journal of Experimental Biology* 202:3485–3495.

Hiroshiba N, Ogura Y, Sasai K, Nishiwaki H, Miyamoto K, Hamada M, Tsujikawa A, Honda Y (1999) Radiation-induced leukocyte entrapment in the rat retinal microcirculation. *Investigative Ophthalmology and Visual Science* 40:1217–1222.

Hirpara JL, Seyed MA, Loh KW, Dong H, Kini RM, Pervaiz S (2000) Induction of mitochondrial permeability transition and cytochrome C release in the absence of caspase activation is insufficient for effective apoptosis in human leukemia cells. *Blood* 95:1773–1780.

Hitzman MW (1999) Routine staining of drill core to determine carbonate mineralogy and distinguish carbonate alteration textures. *Mineralum Deposita* 34:794–798.

Hobson J, Wright J, Churg A (1991) Histochemical evidence for generation of active oxygen species on the apical surface of cigarette-smoke exposed tracheal explants. *American Journal of Pathology* 139:573–580.

Hodges E, Boddy SM, Thomas S, Smith JL (1997) Modification of IgH PCR clonal analysis by the addition of sucrose and cresol red directly to PCR reaction mixes. *Journal of Clinical Pathology – Molecular Pathology* 50:164–166.

Hoefakker S, Balk HP, Boersma WJA, Van Joost T, Notten WRF, Claassen E (1995) Migration of human antigen-presenting cells in a human skin graft onto nude mice model after contact sensitization. *Immunology* 86:296–303.

Hoeksma EA, Vanderlei B, Jonkman MF (1988) Sudan black B as a histological stain for polymeric biomaterials embedded in glycol methacrylate. *Biomaterials* 9:463–465.

Hoffmeister ER (1953) Trypan blue as a nuclear stain for plant material. *Stain Technology* 28:309–310.

Hohage H, Stachon A, Feidt C, Hirsch JR, Schlatter E (1998) Regulation of organic cation transport in IHKE-1 and LLC-PK1 cells. Fluorometric studies with 4-(4-dimethyl-aminostyryl)-N-methylpyridinium. *Journal of Pharmacology and Experimental Therapeutics* 286:305–310.

Holland MS, Mackenzie CD, Bull RW, Silva RF (1996) A comparative study of histological conditions suitable for both immunofluorescence and ISH in the detection of *Herpes* virus and its antigens in chicken tissues. *Journal of Histochemistry and Cytochemistry* 44:259–265.

Hollander H (1963) A staining method for cerebroside-sulphuric-esters in brain tissue. *Journal of Histochemistry and Cytochemistry* 11:118–119.

Holm BA, Wang ZD, Egan EA, Notter RH (1996) Content of dipalmitoyl phosphatidyl-choline in lung surfactant: ramifications for surface activity. *Pediatric Research* 39:5–11.

Holmquist JG (1998) Permeability of patch boundaries to benthic invertebrates: influences of boundary contrast, light level, and faunal density and mobility. *Oikos* 81:558–566.

Holt SJ, Sadler PW (1958) Studies in enzyme cytochemistry. 3. Relationships between solubility, molecular association and structure in indigoid dyes. *Proceedings of the Royal Society B* 148:495–505.

Holtzman E (1989) *Lysosomes.* Plenum Press, New York.

Hong LJ, Mubarak WAE, Sunami Y, Murakami S, Fuyama Y, Ohtsuka A, Murakami T (2000) Enhanced visualization of weak colloidal iron signals with Bodian's protein silver for demonstration of perineuronal nets of proteoglycans in the central nervous system. *Archives of Histology and Cytology* 63:459–465.

Honma Y, Shigehisa H, Chiba A, Oka S (1998) Immunohistochemical demonstration of oxytocin-like, neuropeptide-Y-like and gonadotraphin-releasing hormone-like substances in the hypothalamo-hypophysial system of Ayo, *Plecoglossus altivelis altivelis*, in osmotically different environments. *Ichthyological Research* 45:35–42.

Honorato RS, Araujo MCU, Veras G, Zagatto EAG, Lapa RAS, Lima JLFC (1999) A mono-segmented flow titration for the spectrophotometric determination of total acidity in vinegar. *Analytical Sciences* 15:665–668.

Hood HL, Kraeling MEK, Robl MG, Bronaugh RL (1999) The effects of an alpha hydroxy acid (glycolic acid) on hairless guinea pig skin permeability. *Food and Chemical Toxicology* 37:1105–1111.

Hoodless RA, Pitman CG, Stewart TE, Thompson J, Arnold JE (1971) Separation and identification of food colours. 1. Identification of synthetic water soluble food colours using thin layer chromatography. *Journal of Chromatography* 54:393–404.

Hopman AHN, Claessen S, Speel EJM (1997) Multi-color brightfield *in situ* hybridization on tissue sections. *Histochemistry and Cell Biology* 108:291–298.

Hopman AHN, Ramaekers FCS, Speel EJM (1998) Rapid synthesis of biotin-, digoxigenin-, trinitrophenyl-, and fluorochrome-labeled tyramides and their application for *in situ* hybridization using CARD amplification. *Journal of Histochemistry and Cytochemistry* 46:771–777.

Hopp L, Megee SO, Lloyd JB (1998) Biphenols that stimulate cells to release alkali metal cations: a structure-activity study. *Journal of Medicinal Chemistry* 41:4421–4423.

Horan PK, Melnicoff MJ, Jensen BD, Slezak SE (1990) Fluorescent cell labeling for *in vivo* and *in vitro* cell tracking. *Methods in Cell Biology* 33:469–490.

Horisberger M (1992) Colloidal gold and its application in cell biology. *International Review of Cytology* 136:227–287.

Horisberger M, Rosset J (1977) Colloidal gold, a useful marker for transmission and scanning electron microscopy. *Journal of Histochemistry and Cytochemistry* 25:295–305.

Horobin RW (1968) Simple preparative thin-layer chromatography. *Journal of Chromatography* 37:354–356.

Horobin RW (1969) The impurities of biological dyes: their detection, removal, occurrence and histological significance – a review. *Histochemical Journal* 1:231–265.

Horobin RW (1970) Dextrin and salt as impurities of histological dyes. Estimation and removal. *Histochemie* 22:39–44.

Horobin RW (1971) Analysis and purification of biological stains by gel filtration. *Stain Technology* 46:297–304.

Horobin RW (1973) Effects on staining of the presence of various kinds of dye impurities. *Acta Histochemica Suppl. XIII*:269–280.

Horobin RW (1974) A preliminary quantitative study of the 'rate of staining' model for rationalising various mixed acid dye stains ('trichromes'). *Proceedings of the Royal Microscopical Society* 9:110.

Horobin RW (1980) *The Use of Pure Dyes for Biological Staining.* Koch-Light Laboratories Ltd, Colnbrook, Slough, UK.

Horobin RW (1981) Structure-staining relationships in histochemistry and biological staining. 3. Some comments on the intentional and artifactual staining of lipids. *Acta Histochemica Supplement* 24:237–246.

Horobin RW (1982) *Histochemistry: An Explanatory Outline of Histochemistry and Biophysical Staining.* Gustav Fischer, Stuttgart.

Horobin RW (1983) Staining plastic sections: a review of problems, explanations and possible solutions. *Journal of Microscopy* 31:173–186.

Horobin RW (1988) *Understanding Histochemistry: Selection, Evaluation and Design of Biological Stains.* Ellis Horwood, Chichester, UK.

Horobin RW (2001) Uptake, distribution and accumulation of dyes and fluorescent probes within living cells. A structure–activity modelling approach. *Advances in Colour Science and Technology* 4:101–107.

Horobin RW, Bancroft JD (1998) *Troubleshooting Histology Stains.* Churchill Livingstone, New York, Edinburgh.

Horobin RW, Bennion PJ (1973) The interrelation of the size and substantivity of dyes: the role of van der Waals attractions and hydrophobic bonding in biological staining. *Histochemie* 33:191–204.

Horobin RW, Flemming L (1980) Structure-staining relationships in histochemistry and biological staining. 1. Mechanistic and practical aspects of the staining of elastic fibres. *Journal of Microscopy* 119:357–372.

Horobin RW, Flemming L (1982) Is mercury orange a selective stain for thiols? *Histochemical Journal* 14:1004–1006.

Horobin RW, Flemming L (1988) One bath trichrome staining: investigation of a general mechanism based on a structure-staining correlation analysis. *Histochemical Journal* 20:29–34.

Horobin RW, Goldstein DJ (1972) Impurities and staining characteristics of alcian blue samples. *Histochemical Journal* 4:391–399.

Horobin RW, Goldstein DJ (1974) The influence of salt on the staining of tissue sections with basic dyes: an investigation into the general applicability of the critical electrolyte concentration theory. *Histochemical Journal* 6:599–609.

Horobin RW, Murgatroyd LB (1967) A comparison of rapid electrophoretic and chromatographic methods for the investigation of common histological dyes. *Histochemie* 11:141–151.

Horobin RW, Murgatroyd LB (1968) The composition and properties of gallocyanine-chrome alum stains. *Histochemical Journal* 1:36–54.

Horobin RW, Murgatroyd LB (1969) The identification and purification of pyronin and rhodamine dyes. *Stain Technology* 44:297–302.

Horobin RW, Murgatroyd LB (1971) The staining of glycogen with Best's carmine and similar hydrogen bonding dyes. A mechanistic study. *Histochemical Journal* 3:1–9.

Horobin RW, Proctor GB (1989) Acetyl Sudan B – a non-existent reagent? *Stain Technology* 64:257–258.

Horobin RW, Proctor GB (1990) A numerical method for selecting stains for epoxy resin-embedded tissues. *Transactions of the Royal Microscopical Society* 1:199–202.

Horobin RW, Kevill-Davies IM (1971) A mechanistic study of the histochemical reactions between aldehydes and basic fuchsin in acid alcohol used as simplified substitute for Schiff reagent. *Histochemical Journal* 3:371–378.

Horobin RW, Payne JN, Jakobsen P (1987) Histochemical implications of the chemical and biological properties of SITS and some related compounds. *Journal of Microscopy* 146:87–96.

Horobin RW, Gerrits PO, Wright DJ (1992) Staining sections of water-miscible resins. 2. Effects of staining-reagent lipophilicity on the staining of glycol-methacrylate-embedded tissues. *Journal of Microscopy* 166:199–205.

Horvath TG, Lamberti GA (1999) Mortality of zebra mussel, *Dreissena polymorpha*, veligers during downstream transport. *Freshwater Biology* 42:69–76.

Hosoya N, Cox CF, Arai T, Nakamura J (2000) The walking bleach procedure: an *in vitro* study to measure microleakage of five temporary sealing agents. *Journal of Endodontics* 26:716–718.

Hospelhorn AC, Faris B, Mogayzel PJ, Tan OT, Franzblau C (1988) Congo red binding in aortic smooth muscle cell cultures. *Journal of Histochemistry and Cytochemistry* 36:1353–1358.

Hostager BS, Catlett IM, Bishop GA (2000) Recruitment of CD40 and tumor necrosis factor receptor-associated factors 2 and 3 to membrane microdomains during CD40 signalling. *Journal of Biological Chemistry* 275:15392–15398.

Howard S (1993) Basic fuchsin – A guide to a one-step processing technique for black electrical taspe. *Journal of Forensic Sciences* 38:1391–1403.

Hoyer PE, Kirkeby S (1996) The impact of fixatives on the binding of lectins to N-acetyl glucosamine residues of human syncytiotrophoblast: A quantitative histochemical study. *Journal of Histochemistry and Cytochemistry* 44:855–863.

Hoyer PE, Lyon H, Jakobsen P, Andersen AP (1986) Standardized methyl green-pyronin Y procedures using pure dyes. *Histochemical Journal* 18:90–94.

Hoyer PE, Kayser L, Barer MR, Lyon H (1991) Quantitation in histochemistry. Ch. 28 In: H Lyon (ed.) *Theory and Strategy in Histochemistry*. Springer-Verlag, Berlin. pp. 397–442.

Hoyt KR, Reynolds IJ (1996) Localization of D1 dopamine receptors on live cultured striatal neurons by quantitative fluorescence microscopy. *Brain Research* 731:21–30.

Hrabovszky Z, Farmer PJ, Hutson JM (2001) Undescended testis is accompanied by calcitonin gene related peptide accumulation within the sensory nucleus of the genitofemoral nerve in trans-scrotal rats. *Journal of Urology* 165:1015–1018.

Hsiao K, Chapman P, Nilsen S, Eckman C, Harigaya Y, Younkin S, Yang FS, Cole G (1996) Correlative memory deficits, A-beta elevation and amyloid plaques in transgenic mice. *Science* 274:99–102.

Hsiu SL, Tsao CW, Tsai YC, Ho HJ, Chao PDL (2001) Determinations of morin, quercetin and their conjugate metabolites in serum. *Biological and Pharmaceutical Bulletin* 24:967–969.

Hsu S-M, Soban E (1982) Color modification of diaminobenzidine (DAB) precipitation by metallic ions and its application for double immunocytochemistry. *Journal of Histochemistry and Cytochemistry* 30:1079–1082.

Hu EH, Dacheux RF, Bloomfield SA (2000) A flattened retina-eyecup preparation suitable for electrophysiological studies of neurons visualized with trans-scleral infrared illumination. *Journal of Neuroscience Methods* 103:209–216.

Hu QH, Xia Y, Corda S, Zweier JL, Ziegelstein RC (1998) Hydrogen peroxide decreases pH(i) in human aortic endothelial cells by inhibiting Na+/H+ exchange. *Circulation Research* 83:644–651.

Huan CT, McFeters GA, Stewart PS (1996) Evaluation of physiological staining, cryo-embedding and autofluorescence quenching techniques on fouling biofilms. *Biofouling* 9:269–280.

Huard J, Fortier LP, Dansereau G, Labrecque C, Tremblay JP (1992) A light and electron microscopic study of dystrophin localization at the mouse neuromuscular junction. *Synapse* 10:83–93.

Huber JD, Parker F, Odland GF (1968) A basic fuchsin and alkalinised methylene blue rapid stain for epoxy-embedded tissue. *Stain Technology* 43:83–87.

Hubl W, Iturraspe J, Matinez GA, Hutcheson CE, Roberts CG, Fisk DD, Sugrue MW, Wingard JR, Braylan RC (1998) Measurement of absolute concentration and viability of CD34+ cells in cord blood and cord blood prodcuts using fluorescent beads and cyanine nucleic acid dyes. *Cytometry* 34:121–127.

Huglin D, Seiffert W, Zimmermann HW (1986) Spektroskopische und thermodynamische Untersuchungen zur Bindung von Azur B an Chondroitinsulfat und zur Bindungsgeometrie des metachromatischen Farbstoffkomplexes. *Histochemistry* 86:71–82.

Hulman G, Taylor LA (1987) *Entamoeba histolytica* – Demonstration by PAS martius yellow technique. *Medical Laboratory Sciences* 44:396–397.

Hulvershorn J, Gallant C, Wang CA, Dessy C, Morgan KG (2001) Calmodulin levels are dynamically regulated in living vascular smooth muscle cells. *American Journal of Physiology* 280:H1422–H1426.

Humason GL (1979) *Animal Tissue Techniques*. 4th Edn. Freeman, San Francisco.

Humphries JE, Fried B (1996) Histological and histochemical studies on the paraeosophageal glands in cercariae and metacercariae of *Echinostoma revolutum* and *E. trivolvis*. *Journal of Helminthology* 70:299–301.

Hunter EE (1993) *Practical Electron Microscopy. A Beginner's Illustrated Guide*. 2nd Edn. Cambridge University Press, Cambridge.

Huser J, Lipp P, Niggli E (1996a) Confocal microscopic detection of potential-sensitive dyes used to reveal loss of voltage control during patch-clamp experiments. *Pflugers Archiv – European Journal of Physiology* 433:194–199.

Huser M, Stegemann E, Kammermeier H (1996b) Is enzyme release a sign of irreversible injury of cardiomyocytes? *Life Sciences* 58:545–550.

Hutchison LJ (1991) Description and identification of cultures of ectomycorrhizal fungi found in North America. *Mycotaxon* 42:387–504.

Hutter KJ, Eipel HE (1978) Advances in determination of cell viability. *Journal of General Microbiology* 107:165–167.

Huxley VH, Williams DA (1996) Basal and adenosine-mediated protein flux from isolated coronary arterioles. *American Journal of Physiology – Heart and Circulatory Physiology* 271:H1099–H1108.

Hwang SY, Cho SY, Yoo ID (1995) Phenylalanyl-2-sulfanilylglycine as a substrate for leucine aminopeptidase assay. *Biotechnology* 5:319–323.

Iacono G, Carroccio A, Cavataio F, Montalto G, Balsamo V (1991) Fecal fat measurement – The steatocrit test and other simple methods to be used routinely. [Italian]. *Italian Journal of Pediatrics* 17:173–177.

Iijima S, Shiba K, Inoue J, Yoshida T, Kimura M (1997) Simultaneous analysis of microheterogeneity of immunoglobulins and serum protein fraction using high-voltage isoelectric focussing on six cellulose acetate membranes. *Journal of Clinical Laboratory Analysis* 11:220–224.

Ilgaz C, Kocabiyik H, Erdogan D, Ozogul C, Peker T (1999) Double staining of skeleton using microwave irradiation. *Biotechnic and Histochemistry* 74:57–63.

Illinger D, Kuhry JG (1994) The kinetic aspects of intracellular fluorescence labeling with TMA-DPH support the maturation model for endocytosis in L929 cells. *Journal of Cell Biology* 125:783–794.

Imbert D, Cullander C (1999) Buccal mucosa *in vitro* experiments. 1. Confocal imaging of vital staining and MTT assays for the determination of tissue viability. *Journal of Controlled Release* 58:39–50.

Imoto S, Hasebe T (1999) Initial experience with sentinel node biopsy in breast cancer at the National Cancer Center Hospital East. *Japanese Journal of Clinical Oncology* 29:11–15.

Inamdar JA, Murugan V, Subramanian RB (1987) Staining methods for the detection of laticifers in plant tissue. *Journal of Microscopy* 147:347–351.

Inoue M, Fujishiro N, Imanaga I (2000) Retardation of cation channel deactivation by mitochondrial dysfunction in adrenal medullary cells. *American Journal of Physiology* 278:C26–C32.

Inoue Y, Trevanichi S, Fukuda K, Izawa S, Wakai Y, Kimura A (1997) Roles of esterase and alcohol acetyltransferase on production of isoamyl acetate in *Hansenula mrakii*. *Journal of Agricultural and Food Chemistry* 45:644–649.

IntVeld PH, Havelaar AH, van Strijp Lockefeer NGWM (1999) The certification of a reference material for the evaluation of methods for the enumeration of *Bacillus cereus*. *Journal of Applied Microbiology* 86:266–274.

Irion G, Ochsenfeld L, Naujok A, Zimmermann HW (1993) The concentration jump method. Kinetics of vital staining of mitochondria in HeLa cells with lipophilic cationic fluorescent dyes. *Histochemistry* 99:75–83.

Irons RD, Schenk EA, Lee CK (1977) Cytochemical methods for copper. *Archives of Pathology and Laboratory Medicine* 101:298–301.

Isada M (1938) Geisselfarbung aus alter Bakterienkultur. *Zentralblatt fur Bakteriologie* 142:480–483.

Iseki K, Tatsuta M, Iishi H, Hiyama T, Tsukuma H, Yokota Y, Ikeda F (1997) *Helicobacter pylori*, atrophic fundal gastritis and risk for gastric adenocarcinoma. *Oncology Reports* 4:809–813.

Isenberg JS, Klaunig JE (2000) Role of the mitochondrial membrane permeability transition (MPT) in rotenone-induced apoptosis in liver cells. *Toxicological Science* 53:340–351.

Ishaque A, Al Rubeai M (1998) Use of intracellular pH and annexin V flow cytometric assays to monitor apoptosis and its suppression by bcl2 over-expression in hybridoma cell culture. *Journal of Immunological* Methods 221:43–57.

Ishibashi T, Takizawa T, Iwasaki H, Saito T, Matsubara S, Nakazawa E, Kanazawa K (1999) Glucose-6-phosphate dehydrogenase cytochemistry using a copper ferrocyanide method and its application to rapidly frozen cells. *Histochemistry and Cell Biology* 112:221–232.

Ishida M, Nakagawara G, Imamura Y, Fukuda M (1995) Iron and copper deposition in chronic active hepatitis and liver cirrhosis – Pathogenetic role in progressive liver cell-damage. *European Journal of Histochemistry* 39:221–236.

Ishii S, Mizoi T, Kawano K, Cay O, Thomas P, Nachman A, Ford R, Shoji Y, Krustal JB, Steele G, Jessup JM (1996) Implantation of human colorectal carcinoma cells in the liver studied by *in vivo* fluorescence videomicroscopy. *Clinical and Experimental Metastasis* 14:153–164.

Ishikawa M, Sekizuka E, Sato S, Yamaguchi N, Shimizu K, Kobayashi K, Bertalanffy H, Kawase T (1999) *In vivo* rat closed spinal window for spinal microcirculation: observation of pial vessels, leukocyte adhesion, and red blood cell velocity. *Neurosurgery* 44:156–161.

Issidorides MR, Arvanitis D (1993) Histochemical marker of human catecholamine neurons in ganglion-cells and processes of a temporal-lobe ganglioglioma. *Surgical Neurology* 39:66–67.

Issidorides MR, Mytilineou C, Panayotacopoulou MT, Yahr MD (1991) Lewy bodies in Parkinsonism share components with intraneuronal protein bodies of normal brains. *Journal of Neural Transmission* 3:49–61.

Ito S, Ohta T, Nakazato Y (1999) Characteristics of 5-HT-containing chemoreceptor cells of the chicken aortic body. *Journal of Physiology* 515:49–59.

Ito Y, Otsuki Y (1998) Localization of apoptotic cells in the human epidermis by an *in situ* DNA nick end-labeling method using confocal reflectant laser microscopy. *Journal of Histochemistry and Cytochemistry* 46:783–786.

Itoh J, Sanno N, Matsuno A, Itoh Y, Watanabe K, Osamura RY (1997) Application of confocal laser scanning microscopy (CLSM) to visualize prolactin (PRL) and PRL mRNA in the normal and estrogen-treated rat pituitary glands using non-fluorescent probes. *Microscopy Research and Techniques* 39:157–167.

Itoh S, Gohara S, Matsuyama Y, Yamagishi S (1990) Calcium staining by the glyoxal-bis-(2-hydroxyanil) method in the livers of rats treated with CCl4, diltiazem, and with both agents together. *Liver* 10:365–371.

Iwamoto Y, Murakami K, Danno M, Tsuchiya M, Masuzawa T, Shimizu T, Morita T, Yanagihara Y (1990) Singlet oxygen production and photobiological effects of pina-cyanol chloride on yeast *Saccharomyces cerevisiae*. *Journal of Pharmacobio-dynamics* 13:316–320.

Jackson G (1926) Crystal violet and erythrosin in plant anatomy. *Stain Technology* 1:33.

Jacobs AS, Pretorius ZA, Kloppers FJ, Cox TS (1996) Mechanisms associated with wheat leaf rust resistance derived from *Triticum monococcum*. *Phytopathology* 86:588–595.

Jacobsen KR, Fisher DG, Maretzki A, Moore PH (1992) Developmental changes in the anatomy of the sugarcane stem in relation to phloem unloading and sucrose storage. *Botanica Acta* 105:70–80.

Jacot JL, Glover JP, Robison WG (1995) Improved gold chloride procedure for nerve staining in whole mounts of rat corneas. *Biotechnic and Histochemistry* 70:277–284.

Jagadeeswaran R, Thirunavukkarasu C, Gunasekaran P, Ramamurty N, Sakthisekaran D (2000) *In vitro* studies on the selective cytotoxic effects of cocetin and quercetin. *Fitoterapia* 71:395–399.

Jahnke KD (1984) A simple technique for staining chrysocystidia with patent blue V. *Mycologia* 76:940–943.

Jakobsen P, Horobin RW (1989) Preparation and characterization of 4-acetamido-4'-iso-thiocyanatostilbene-2,2'-disulfonic acid (SITS) and related stilbene disulfonates. *Stain Technology* 64:301–313.

Jakobsen P, Andersen AP, Lyon H, Treppendahl S (1983) Preparation and characterization of pyronin Y. *Microscopica Acta* 87:41–47.

Jakobsen P, Andersen AP, Lyon H (1984a) Preparation and characterization of methyl green tetrafluoroborate. *Histochemistry* 81:177–179.

Jakobsen P, Lyon H, Treppendahl S (1984b) Spectrophotometric characteristics and assay of pure pyronin Y. *Histochemistry* 81:99–101.

James MR, Ewer AK (1999) Acid oro-pharyngeal secretions can predict gastro-oesophageal reflux in preterm infants. *European Journal of Pediatrics* 158:371–374.

Jan KY (1981) Reverse sister chromatid differential staining by coomassie brilliant blue R250. *Experimental Cell Research* 132:503–505.

Janecki AJ, Janecki M, Akhter S, Donowitz M (2000) Quantitation of plasma membrane expression of a fusion protein of Na/H exchanger NHE3 and green fluorescence protein (GFP) in living PS120 fibroblasts. *Journal of Histochemistry and Cytochemistry* 48:1479–1492.

Jarolim KL, McCosh JK, Howard MJ, John DT (2000) A light microscopy study of the migration of *Naegleria fowleri* from the nasal submucosa to the central nervous system during the early stages of primary amebic meningoencephalitis in mice. *Journal of Parasitology* 86:50–55.

Jarvis GN, Thiele JH (1997) Qualitative rhodamine B assay which uses tallow as a substrate for lipolytic obligately anaerobic bacteria. *Journal of Microbiological Methods* 29:41–47.

Jayabalan M, Shah JJ (1986) Histochemical techniques to localize rubber in guayule (*Parthenium argentatum* Gray). *Stain Technology* 61:303–308.

Jenner L (1899) A new preparation of rapidly fixing and staining blood. *Lancet* 1899–1:370.

Jensen C, Lysek G (1995) Fluorescence microscopy of fungi in native soil – improvement by additional substances. *Microscopy and Analysis* (September):7–9.

Jensen WA (1962) *Botanical Histochemistry*. Freeman, San Francisco.

Jesuthasan S (1998) Furrow-associated microtubule arrays are required for the cohesion of zebrafish blastomeres following cytokinesis. *Journal of Cell Science* 111:3695–3703.

Jiang SP, Claussen DL (1993) The effects of temperature on food passage time through the digestive tract in *Notophthalmus viridescens*. *Journal of Herpetology* 27:414–419.

Jiao L, Gray DWR, Gohde W, Flynn GJ, Morris PJ (1991) *In vitro* staining of Islets of Langerhans for fluorescence-activated cell sorting. *Transplantation* 52:450–452.

Jing J, Lai Z, Aston C, Lin J, Carucci DJ, Gardner MJ, Mishra B, Anantharaman TS, Tettelin H, Cummings LM, Hoffman SL, Venter JC, Schwartz DC (1999) Optical mapping of *Plasmodium falciparum* chromosome 2. *Genome Research* 9:175–181.

Joersbo M, Brunstedt J, Floto F (1990) Quantitative relationship between parameters of electroporation. *Journal of Plant Physiology* 137:169–174.

Johansen DA (1940) *Plant Microtechnique*. McGraw-Hill, New York.

John KR, Richards RH (1999) Characteristics of a new birnavirus associated with a warm water fish cell line. *Journal of General Virology* 80:2061–2065.

Johnson CA, Richner GA, Vitvar T, Schittli N, Eberhard M (1998) Hydrological and geochemical factors affecting leachate composition in municipal solid waste incinerator bottom ash. Part I. The hydrology of Landfill Lostorf, Switzerland. *Journal of Contaminant Hydrology* 33:361–376.

Johnson I (1998) Fluorescent probes for living cells. *Histochemical Journal* 30:123–140.

Johnson ID, Kang HC, Haugland RO (1991) Fluorescent membrane probes incorporating dipyrrometheneboron difluoride fluorophores. *Analytical Biochemistry* 198:228–237.

Johnson W (1994a) Final report on the safety assessment of 2,4-diaminophenol and 2,4-diaminophenol hydrochloride. *Journal of the American College of Toxicology* 13:330–343.

Johnson W (1994b) Final report on the safety assessment of N-phenyl-p-phenylenediamine hydrochloride and N-phenyl-p-phenylenediamine sulfate. *Journal of the American College of Toxicology* 13:374–394.

Joist JH, Bauman JE, Sutera SP (1998) Platelet adhesion and aggregation in pulsatile shear flow: effects of red blood cells. *Thrombosis Research* 92:S47–S52.

Jonas L, Fulda G, Salameh T, Schmidt W, Kroning GE, Hopt UT, Nizze H (2001) Electron microscopic detection of copper in the liver of two patients with morbus Wilson by EELS and EDX. *Ultrastructural Pathology* 25:111–118.

Jones GRN (1968) The purification and some properties of neotetrazolium chloride and its chief monotetrazolium salt contaminant. *Histochemical Journal* 1:59–67.

Jones GRN (1969) The synthesis of yellow tetrazolium. *Histochemie* 18:164–167.

Jones GRN (1973) Detection of primary arylamines on thin-layer chromagrams by diazotisation and coupling. Comparison of a new reagent with existing methods. *Journal of Chromatography* 77:357–367.

Jones JE, Corwin JT (1996) Regeneration of sensory cells after laser ablation in the lateral line system: hair cell lineage and macrophage behavior revealed by time-lapse video microscopy. *Journal of Neuroscience* 16:649–662.

Jones JM, Lorton SP, Bavister BD (1995) Measurement of intracellular pH in mammalian sperm cells under physiological conditions. *Cytometry* 19:235–242.

Jones M, Tussey L, Athanasou N, Jackson DG (2000) Heparan sulfate proteoglycan isoforms of the CD44 hyaluronan receptor induced in human inflammatory macrophages can function as paracrine regulators of fibroblast growth factor action. *Journal of Biological Chemistry* 275:7964–7974.

Jones R, Ryan AJ, Sternhill S, Wright SE (1963) The structure of some 5-pyrazolones and derived 4-arylazo-5-pyrazolones. *Tetrahedron* 19:1497–1507.

Jones RM, ed. (1950) *McClung's Handbook of Microscopical Technique*. 3rd Edn. Hoebner, New York.

Juan G, Darzynkiewicz Z (1998) Cell cycle analysis by flow and laser scanning cytometry. In: JE Celis (ed.) *Cell Biology: A Laboratory Handbook*. Academic Press, San Diego. pp. 261–274.

Juarranz A, Ferrer JM, Tato A, Canete M, Stockert JC (1987) Metachromatic staining and electron dense reaction of glycosaminoglycans by means of cuprolinic blue. *Histochemical Journal* 19:1–6.

Juliano RL (1974) Isothiocyanostilbene disulfonates are useful labels for the surface proteins of cultured mammalian cells. *Experimental Cell Research* 86:181–184.

Jullien L (1872) Sur une nouvelle methode de coloration des elements histologiques. *Lyon Medecine* 10:526–530.

June CH, Rabinovitch PS (1994) Intracellular ionized calcium. *Methods in Cell Biology* 41:149–174.

Jung M, Kiesslich R (1999) Chemoendoscopy and intravital staining techniques. *Best Practice and Research in Clinical Gastroenterology* 13:11–19.

Junqueira LC, Bignolas G, Brentani RR (1979) Picrosirius staining plus polarization microscopy, a specific method for collagen detection in tissue sections. *Histochemical Journal* 11:447–455.

Jurand A, Goel SC (1976) The use of methyl green-pyronin staining after glutaraldehyde fixation and paraffin or araldite embedding. *Tissue and Cell* 8:389–394.

Kaatz KW, Bazzett TJ, Albin RL (1992) A new, simple myelin stain. *Brain Research Bulletin* 29:697–698.

Kacza J, Hartig W, Seeger J (1997) Oxygen-enriched photoconversion of fluorescent dyes by means of a closed conversion chamber. *Journal of Neuroscience Methods* 71:225–232.

Kagayama M, Sasano Y, Tsuchiya M, Watanabe M, Mizoguchi I, Kamakura S, Motegi K (2000) Confocal microscopy of Tomes' granular layer in dog premolar teeth. *Anatomy and Embryology* 201:131–137.

Kahn E, Lizard G, Pelegrini M, Frouin F, Roignot P, Chardonnet Y, Paola R (1999) Four-dimensional factor analysis of confocal image sequences (FAMIS) to detect and characterize low copy numbers of human papillomavirus DNA by FISH in HeLa and SiHa cells. *Journal of Microscopy* 193:227–243.

Kalina M, Plapinger RE, Hoshino Y, Seligman AM (1972) Nonosmiophilic tetrazolium salts that yield osmiophilic, lipophobic formazans for ultrastructural localization of dehydrogenase activity. *Journal of Histochemistry and Cytochemistry* 20:685–695.

Kalter SS (1943) A quadruple staining method for tissues. *Journal of Laboratory Medicine* 28:995–997.

Kamalia N, McCulloch CAG, Tenenbaum HC, Limeback H (1992) Direct flow cytometric quantification of alkaline phosphatase activity in rat bone marrow stromal cells. *Journal of Histochemistry and Cytochemistry* 40:1059–1065.

Kamisaka Y, Noda N, Sakai Y, Kawasaki K (1999) Lipid bodies and lipid body formation in an oleaginous fungus, *Mortierella ramanniana* var *angulispora*. *Biochemica et Biophysica Acta* 1438:185–198.

Kang DH, Fung DYC (1999) Development of a medium for differentiation between *Escherichia coli* and 0 *Escherichia coli* 157:H7. *Journal of Food Protection* 62:313–317.

Kang XP, Wan HZ, Wang P, Wang SX, Chow LP (1990) Effectiveness of phenol-atabrine paste (PAP) instillation for female sterilization. *International Journal of Gynecology and Obstetrics* 33:49–57.

Kantar A, Littarru GP, Falcioni G, Cherubini V, Coppa GV, Fiorini R (1999) Plasma membrane fluidity and polarity of polymorphonuclear leukocytes from children with type I diabetes mellitus. *Journal of Diabetes and its Complications* 13:243–250.

Kapraun DF, Hinson TK, Lemus AJ (1991) Karyology and cytophotometric estimation of inter- and intraspecific nuclear-DNA variation in four species of *Porphyra* (Rhodophyta). *Phycologia* 30:458–466.

Kapuscinski J (1995) DAPI: a DNA-specific fluorescent probe. *Biotechnic and Histochemistry* 70:220–233.

Karlyshev AV, Wren BW (2001) Detection and initial characterization of novel capsular polysaccharide among diverse *Campylobacter jejuni* strains using alcian blue dye. *Journal of Clinical Microbiology* 39:279–284.

Karnovsky MJ (1994) Cytochemistry and reactive oxygen species: A retrospective. *Histochemistry* 102:15–27.

Karnovsky MJ, Roots L (1964) A 'direct coloring' thiocholine method for cholinesterases. *Journal of Histochemistry and Cytochemistry* 12:219–221.

Karsilayan H, Hemmila I, Takalo H, Toivonen A, Pettersson K, Lovgren T, Mukkala VM (1997) Influence of coupling method on the luminescence properties, coupling efficiency, and binding affinity of antibodies labeled with europium(III) chelates. *Bioconjugate Chemistry* 8:71–75.

Kass L (1980) Kallichrome: a new stain for erythroblasts. *Stain Technology* 55:31–33.

Kass L (1981) Staining of granulocyte cells by chlorazol black E. *American Journal of Clinical Pathology* 76:810–812.

Kass L (1986) Identification of normal and leukemic granulocytic cells with merocyanine 540. *Stain Technology* 61:7–15.

Kasten FH (1958) Additional Schiff-type reagents for use in cytochemistry. *Stain Technology* 33:39–45.

Kasten FH (1959) Schiff-type reagents in cytochemistry. 1. Theoretical and practical considerations. *Histochemie* 1:466–509.

Kasten FH (1962) Comparisons of pyronin dyes obtained from various commercial sources. 1. History. *Stain Technology* 37:265–275.

Kasten FH (1963) Schiff-type reagents in cytochemistry. 3. General applications. *Acta Histochemica* (Suppl.) 3:240–247.

Kasten FH (1964) The Feulgen reaction. An enigma in cytochemistry. *Acta Histochemica* 17:88–99.

Kasten FH (1967) Cytochemical studies with acridine orange and the influence of dye contaminants in the staining of nucleic acids. *International Review of Cytology* 21:141–202.

Kasten FH (1983) The development of fluorescence microscopy up through World War 2. In: G Clark, FH Kasten (eds) *History of Staining*. Williams and Wilkins, Baltimore. pp. 147–185.

Kasten FH (1989) The origins of modern fluorescence microscopy and fluorescence probes. In: *Cell Structure and Function by Microspectrophotometry*. Academic Press, San Diego. pp. 4–50.

Kasten FH (1993) Introduction to fluorescent probes: properties, history and applications. In: WT Mason (ed.) *Fluorescent and Luminescent Probes for Biological Activity*. Academic Press, London. pp. 12–32.

Kasten FH (1999) Introduction to fluorescent probes: properties, history and applications. In: WT Mason (ed.) *Fluorescence and Luminescent Probes for Biological Activity*. Academic Press, San Diego. pp. 17–39.

Kasten FH, Burton V, Glover P (1959) Fluorescent Schiff-type reagents for cytochemical detection of polyaldehyde moieties in sections and smears. *Nature* 184:1797.

Kasten FH, Burton V, Lofland S (1962) Schiff-type reagents in cytochemistry. 2. Detection of primary amine dye impurities in pyronin B and pyronin Y (G). *Stain Technology* 37:277–291.

Kater SB, Nicholson C, eds (1973) *Intracellular Staining in Neurobiology*. Springer, New York.

Katoh YY, Arai R, Benedek G (2000) Bifurcating projections from the cerebellar fastigial neurons to the thalamic suprageniculate nucleus and to the superior colliculus. *Brain Research* 864:308–311.

Katsube N, Sunaga K, Aishita H, Chuang DM, Ishitani R (1999) ONO-1603, a potential antidementia drug, delays age-induced apoptosis and suppresses overexpression of glyceraldehyde-3-phosphate dehydrogenase in cultured central nervous system neurons. *Journal of Pharmacology and Experimental Therapeutics* 288:6–13.

Kawamoto A, Ohashi K, Kishikawa H, Zhu LQ, Azuma C, Marata Y (1999) Two-color fluorescence staining of lectin and anti-CD46 antibody to assess acrosomal status. A highly sensitive method for evaluating stimulus-induced acrosome reaction. *Fertility and Sterility* 71:497–501.

Kawamura K, Tanaka T, Ikeda R, Fujikawa-Yamamoto K, Suzuki K (2000) DNA ploidy analysis of urinary tract epithelial tumors by laser scanning cytometry. *Analytical and Quantitative Cytometry and Histology* 22:26–30.

Kawamoto A, Gwon HC, Iwaguro H, Yamaguchi JI, Uchida S, Masuda H, Silver M, Ma H, Kearney M, Isner JM, Asahara T (2001) Therapeutic potential of *ex vivo* expanded endothelial progenitor cells for myocardial ischemia. *Circulation* 103:634–637.

Kawasaki Y, Saito T, Shirota-Someya Y, Ikegami Y, Komano H, Lee MH, Froelich CJ, Shinohara N, Takayama H (2000) Cell death-associated translocation of plasma membrane components induced by CTL. *Journal of Immunology* 164:4641–4648.

Kay AR, Alfonso A, Alford S, Cline HT, Holgado AM, Sakmann B, Snitsarev VA, Stricker TP, Takahashi M, Wu LG (1999) Imaging synaptic activity in intact brain and slices with FM1-43 in *C. elegans*, lamprey, and rat. *Neuron* 24:809–817.

Kelenyi G (1967) On the histochemistry of azo group-free thiazole dyes. *Journal of Histochemistry and Cytochemistry* 15:172–180.

Kempton JB, Rowe WF (1992) Contrast enhancement of cyanoacrylate-developed latent fingerprints using biological stains and commercial fabric dyes. *Journal of Forensic Sciences* 37:99–105.

Kennedy GY, Soper R, Lunn D (1970) A specific histochemical method for phospholipids and sulpholipids. *Histochemical Journal* 2:131–136.

Kennett CN, Cox SW, Eley BM (1997) Investigations into the cellular contribution to host tissue proteases and inhibitors in gingival crevicular fluid. *Journal of Clinical Periodontology* 24:424–431.

Kepner RL, Pratt JR (1994) Use of fluorochromes for direct enumeration of total bacteria in environmental samples: past and present. *Microbiological Reviews* 58:603–615.

Kepner RL, Wharton RA, Coats DW (1999) Ciliated protozoa of two Antarctic lakes: analysis by quantitative protargol staining and examination of artificial substrates. *Polar Biology* 21:285–294.

Kerver ED, Vogels IMC, Bosch KS, Vreeling-Sindelarova H, Van den Munckhof RJM, Frederiks WM (1997) *In situ* detection of spontaneous superoxide anion and singlet oxygen production by mitochondria in rat liver and small intestine. *Histochemical Journal* 29:229–237.

Keto RO (1984) Characterization of alkali blue pigment in counterfeit currency by high performance liquid chromatography. *Journal of Forensic Sciences* 29:198–208.

Khan KNM, Senese PB, Blomquist EM, Smith PF (1997) Culture requirements and cytochemical characteristics of colony forming units granulocyte/monocyte in cynomolgus monkeys. *Comparative Haematology International* 7:74–80.

Khelef N, Buton X, Beatini N, Wang H, Meiner V, Chang T-Y, Farese RV, Maxfield FR, Tabas I (1998) Immunolocalization of Acyl-Coenzyme A: Cholesterol O-acyl transferase in macrophages. *Journal of Biological Chemistry* 273:11218–11224.

Khoobehi B, Peyman GA (1999) Fluorescent labeling of blood cells for evaluation of retinal and choroidal circulation. *Ophthalmic Surgery and Lasers* 30:140–145.

Khoshyomn S, Penar PL, McBride WJ, Taatjes DJ (1998) Four-dimensional analysis of human brain tumor spheroid invasion into fetal rat brain aggregates using confocal scanning laser microscopy. *Journal of Neuro-oncology* 38:1–10.

Kiel EG, Kuyper GHA (1964) The paper chromatography of dyes. XIII. Azoic fast salts. *Tex* 23:229–231.

Kienzle F, Isler O (1978) Synthetic carotenoids as colorants for food and feed. Ch. 9 In: K Venkataraman, *The Chemistry of Synthetic Dyes*. Academic Press, New York. pp. 389–414.

Kiernan JA, Macpherson CM, Price A, Sun T (1998) A histochemical examination of the staining of kainate induced neuronal degeneration by anionic dyes. *Biotechnic and Histochemistry* 73:244–254.

Kiernan JA (1974) Effects of metabolic inhibitors on vital staining with methylene blue. *Histochemistry* 40:51–57.

Kiernan JA (1977) Recycling procedure for gold chloride used in neurohistology. *Stain Technology* 52:245–248.

Kiernan JA (1984a) Chromoxane cyanine R. I. Physical and chemical properties of the dye and of some of its iron complexes. *Journal of Microscopy* 134:13–23.

Kiernan JA (1984b) Chromoxane cyanine R. II. Staining of animal tissues by the dye and its iron complexes. *Journal of Microscopy* 134:25–39.

Kiernan JA (1985) The action of chromium (III) in fixation of animal tissues. *Histochemical Journal* 17:1131–1146.

Kiernan JA (1996) Vascular permeability in the peripheral autonomic and somatic nervous systems: Controversial aspects and comparisons with the blood–brain barrier. *Microscopy Research and Techniques* 35:122–136.

Kiernan JA (1999) *Histological and Histochemical Methods: Theory and Practice*. 3rd Edn. Butterworth-Heinemann (Reprinted 2000; London: Arnold Publishers), Oxford.

Kikui Y, Miki A (1995) A differential staining method for adenohypophyseal cells. *Archives of Histology and Cytology* 58:375–378.

Kim AM, Vergara JL (1998) Supercharging accelerates T-tubule membrane potential changes in voltage clamped frog skeletal muscle fibers. *Biophysical Journal* 75:2098–2116.

Kim P, Yoshimoto Y, Nakaguchi H, Mori T, Asai A, Sasaki T, Kirino T, Nonomura Y (1999) Increased sarcolemmal permeability in the cerebral artery during chronic spasm: an assessment using DNA-binding dyes and detection of apoptosis. *Journal of Cerebral Blood Flow and Metabolism* 19:889–897.

Kimura A, Ishikawa K, Ogawa I (1998) Myocardial salvage by reperfusion 12 hours after coronary ligation in dogs. *Japanese Circulation Journal – English Edition* 62:294–298.

Kimura S, Shiota K (1996) Sequential changes of programmed cell death in developing fetal mouse limbs and its possible roles in limb morphogenesis. *Journal of Morphology* 229:337–346.

Kimura Y, Takahashi-Sakai K, Wilder-Smith P, Krasieva TB, Liew LHL, Matsumoto K (2000) Morphological study of the effects of CO^2 laser emitted at 9.3 um on human dentin. *Journal of Clinical Laser Medicine and Surgery* 18:197–202.

King G, Chambers G, Murray GI (1999) Detection of immunoglobulin light chain mRNA by *in situ* hydridisation using biotinylated tyramine signal amplification. *Journal of Clinical Pathology – Molecular Pathology* 52:47–50.

King HG, Pruden G (1968) The purification of commercial alizarin red S for the determination of aluminium in silicate minerals. *Analyst* 93:601–605.

King MA (2000) Detection of dead cells and measurement of cell killing by flow cytometry. *Immunological Methods* 243:155–166.

King MS, Bradley RM (2000) Biophysical properties and responses to glutamate receptor agonists of identified subpopulations of rat geniculate ganglion neurons. *Brain Research* 866:237–246.

King RC, Mills SL, Medina JE (1999) Enhanced visualization of parathyroid tissue by infusion of a visible dye conjugated to an antiparathyroid antibody. *Head and Neck* 21:111–115.

King RL, Beams HW (1934) Somatic synapsis in *Chironomus* with special reference to the individuality of the chromosomes. *Journal of Morphology* 56:577–591.

Kinnunen J, Merikanto B (1955) *Chemist Analyst* 44:50–53.

Kippert F, Lloyd D (1995) The aniline blue fluorochrome specifically stains the septum of both live and fixed *Schizosaccharomyces pombe* cells. *FEMS Microbiology Letters* 132:215–219.

Kiraly K, Lammi M, Arokoski J, Lapvetelainen T, Tammi M, Helminen H, Kiviranta I (1996a) Safranin O reduces loss of glycosaminoglycans from bovine articular cartilage during histological specimen preparation. *Histochemical Journal* 28:99–107.

Kiraly K, Lapvetelainen T, Arokoski J, Torronen K, Modis L, Kiviranta I, Helminen HJ (1996b) Application of selected cationic dyes for the semiquantitative estimation of glycosaminoglycans in histological sections of articular cartilage by microspectrophotometry. *Histochemical Journal* 28:577–590.

Kirchner JG (1971) A new preparative thin-layer chromatographic technique for using thicker layers. *Journal of Chromatography* 63:45–57.

Kirkman H, Severinghaus AE (1938) A review of the Golgi apparatus. (Published in three parts). *Anatomical Record* 70:413–431; 70:557–573; 71:79–103:.

Kirsch AK, Subramaniam V, Striker G, Schnetter C, Arndt-Jovin DJ, Jovin TM (1998) Continuous wave two-photon scanning near-field optical microscopy. *Biophysical Journal* 75:1513–1521.

Kjellstrand P (1980) Mechanisms of the Feulgen acid hydrolysis. *Journal of Microscopy* 119:391–396.

Klebs G (1886) Ueber die Organisation der Gallerte bei einige Algen und Flagellaten. *Untersuchungen der botanische Institut Tubingen* 2:333–418.

Kleeman WP, Bailey LC (1985) Simplex optimization of the blue tetrazolium assay procedure for alpha-ketol steroids. *Journal of Pharmaceutical Sciences* 74:655–659.

Klinger RC, Van Den Avyle MJ (1993) Preservation of striped bass eggs: effects of formalin concentration, buffering, stain, and initial stage of development. *Copeia* 4:1114–1119.

Klut ME, Stockner J, Bisalputra T (1989) Further use of fluorochromes in the cytochemical characterization of phytoplankton. *Histochemical Journal* 21:645–650.

Kluve-Beckerman B, Manaloor J, Liepnieks JJ (2001) Binding, trafficking and accumulation of serum amyloid A in peritoneal macrophages. *Scandinavian Journal of Immunology* 53:393–400.

Klymkowsky MW, Hanken J (1991) Whole-mount staining of *Xenopus* and other verte-brates. *Methods in Cell Biology* 36:419–441.

Knaut H, Pelegri F, Bohmann K, Schwart H, Nusslein-Volhard C (2000) Zebrafish vasa RNA but not its protein is a component of the germ plasm and segregates asymmetri-cally before germline specification. *Journal of Cell Biology* 149:875–888.

Knebel W, Schnept E (1993) Confocal laser scanning microscopy of fluorescently stained wood cells – a new method for 3-dimensional imaging of xylem elements. *Trees – Structure and Function* 5:1–4.

Kobayashi K, Miyazu K, Fukutani Y, Nakamura I, Yamaguchi N (1992) Gallyas-Schiff stain for senile plaques. *Biotechnic and Histochemistry* 67:256–260.

Kobbert C, Apps R, Bechmann I, Lanciego JL, Mey J, Thanos S (2000) Current concepts in neuroanatomical tracing. *Progress in Neurobiology* 62:327–351.

Kobr M, Linhart I (1994) Geophysical survey as a basis for regeneration of waste dump Halde-10, Zwickau, Saxony. *Journal of Applied Geophysics* 31:107–116.

Koch H, Weisser P (1997) Exposure of honey bees during pesticide application under field conditions. *Apidologie* 28:439–447.

Koedel U, Pfister HW (1999) Superoxide production by primary rat cerebral endothelial cells in response to pneumococci. *Journal of Neuroimmunology* 96:190–200.

Kohen E, Hirschberg JG, eds (1989) *Cell Structure and Function by Microspectrophotometry.* Academic Press, San Diego.

Kohn J (1987) Nigrosine, the forgotten protein stain. *Clinica Chimica Acta* 166: 335–336.

Kominami T, Masui M (1996) A cyto-embryological study of gastrulation in the sand dollar, *Scaphechinus mirabilis*. *Development Growth and Differentiation* 38:129–139.

Kommareddi S, Abramowsky CR, Swinehart GL, Hrabak L (1984) Non-tuberculous mycobacterial infections: comparison of the fluorescent Auramine O and Ziehl-Neelsen techniques in tissue diagnosis. *Human Pathology* 15:1085–1089.

Konig K, Riemann I, Fischer P, Halbhuber KJ (2000) Multiplex FISH and three-dimen-sional DNA imaging with near infrared femtosecond laser pulses. *Histochemistry and Cell Biology* 114:337–345.

Koning AJ, Lum PY, Williams JM, Wright R (1993) DiOC6 staining reveals organelle structure and dynamics in living yeast cells. *Cell Motility and Cytoskeleton* 25:111–128.

Kooistra MJ (1991) A micromorphological approach to the interactions between soil structure and soil biota. *Agricultural Ecosystems and Environment* 34:315–328.

Kopke C, Cristovao A, Prata AM, Pereira CS, Marques JJF, San Romao MV (2000) Microbiological control of wine. The application of epifluorescence microscopy method as a rapid technique. *Food Microbiology* 17:257–260.

Korn ED (1967) A chromatographic and spectrophotometric study of the products of the reaction of osmium tetroxide with unsaturated lipids. *Journal of Cell Biology* 34:627–638.

Kornhauser SI (1954) Early use of paper chromatography in testing a histological staining mixture. *Stain Technology* 29:320.

Kosaki K, Suzuki H, Schmid-Schonbein GW, Nelson TR, Jones KL (1997) Parametric imaging of the chick embryonic cardiovascular system: a novel functional measure. *Pediatric Research* 41:451–456.

Kouvroukoglou S, Dee KC, Bizios R, McIntire LV, Zygourakis K (2000) Endothelial cell migration on surfaces modified with immobilized adhesive peptides. *Biomaterials* 21:1725–1733.

Kovacs A, Foote RH (1992) Viability and acrosome staining of bull, boar and rabbit spermatozoa. *Biotechnic and Histochemistry* 67:119–124.

Kozlowska H, Drewa G, Grzanka A (1995) Effect of trypan blue on the activity of lyso-somal enzymes, tumor-growth and cell ultrastructure in B16 melanotic melanoma in mice. *Neoplasma* 42:173–178.

Krasnow MA, Cumberledge S, Manning G, Herzenberg LA, Nolan GP (1991) Whole animal cell sorting of Drososphila embryos. *Science* 251:81–85.

Kraus JE, de Sousa HC, Rezende MH, Castro NM, Vecchi C, Luque R (1998) Astra blue and basic fuchsin double staining of plant materials. *Biotechnic and Histochemistry* 73:235–243.

Krause C. Haut. In Wagner, R (ed.) *Handworterbuch der Physiologie*. Brunswick: Vieweg.

Krause C, Werner T, Huber C, Wolfbels OS, Leiner MJP (1999) pH-insensitive ion selective optode: a coextraction-based sensor for potassium ions. *Analytical Chemistry* 71:1544–1548.

Krause R, ed. (1923) *Enzyklopadie der mikroskopischen Technik*, 3rd Edn. Urban and Schwarzenberg, Berlin and Wien.

Krause R (1926–1927) *Enzyklopadie der mikroskopischen Technik*. Urban and Schwarzenberg, Berlin.

Kremer JJ, Pallitto MM, Sklansky DJ, Murphy RM (2000) Correlation of beta-amyloid aggregate size and hydrophobicity with decreased bilayer fluidity of model membranes. *Biochemistry* 39:10309–10318.

Kreyberg L (1961) Main histological types of primary epithelial lung tumours. *British Journal of Cancer* 25:206–210.

Krishnankutty S, Kumari CKM, Mathew AG (1990) Microstructure of coconut haustorium. *Journal of Food Science and Technology (Mysore)* 27:302–303.

Kristensson K, Olsson Y, Sjostrand J (1971) Axonal uptake and retrograde transport of exogenous proteins in the hypoglossal nerve. *Brain Research* 32:399–406.

Krolenko SA, Amos WB, Lucy JA (1995) Reversible vacuolation of the transverse tubules of frog skeletal muscle: a confocal fluorescence microscopy study. *Journal of Muscle Research and Cell Motility* 16:401–411.

Krueger SK, Phillips DE, Frederick MM, Johnson RK (1999) Diaminobenzidine as a myelin stain in semithin plastic sections. *Biotechnic and Histochemistry* 74:105–109.

Kucheryavykh LE, Skopichev VG, Nozdrachev AD (1999) The structure of the initial inputs into the metasympathetic nervous system of the rat uterus. [Russian]. *Morfologiia* 116:22–25.

Kugler P (1982) Quantitative dehydrogenase histochemistry with exogenous electron carriers (PMS, MPMS, MB). *Histochemistry* 75:99–112.

Kuhle AV, Jespersen L (1998) Detection and identification of wild yeasts in lager breweries. *International Journal of Food Microbiology* 43:205–213.

Kulka M, Aurelian L (1995) Antiviral activity of an oligo(nucleoside methylphosphonate) that targets HSV-1 immediate-early pre-mRNA 4,5 is augmented by cotreatment with replication-defective adenovirus. *Antisense Research and Development* 5:243–249.

Kumar A, Rawlings RD, Beaman DC (1993) The mystery ingredients – sweeteners, flavorings, dyes and preservatives in analgesic antipyretic, antihistamine decongestant, cough and cold, antidiarrheal and liquid theophylline preparations. *Pediatrics* 91:927–933.

Kumar JR, Haberman HF, Ranadive NS (1997) Comparative studies on the tolerance to photoinduced cutaneous inflammatory reactions by psoralen and rose Bengal. *Journal of Photochemistry and Photobiology B – Biology* 37:245–253.

Kundu SK, Robey WG, Nabors P, Lopez MR, Buko A (1996) Purification of commercial coomassie brilliant blue R250 and characterization of the chromogenic fractions. *Analytical Biochemistry* 235:134–140.

Kurien BT, Jackson K, Scofield RH (1998) Immunoblotting of multiple antigenic peptides. *Electrophoresis* 19:1659–1661.

Kurnick NB (1955a) Histochemistry of nucleic acids. *International Review of Cytology* 4:221–268.

Kurnick NB (1955b) Pyronin Y in the methyl green-pyronin histological stain. *Stain Technology* 30:213–230.

Kuroiwa T, Kajimoto Y, Ohta T (1999) Comparison between operative findings on malignant glioma by fluorescein surgical microscopy and histological findings. *Neurological Research* 21:130–134.

Kurzweilova H, Sigler K (1995) Comparison of three different methods for determining yeast killer toxin K1 activity and standardisation of units. *Experientia* 51:26–28.

Kusuzaki K, Kageyama N, Shinjo H, Takeshita H, Murata H, Hashiguchu S, Asihara T, Hirasawa Y (2000) Development of bone canaliculi during bone repair. *Bone* 27:655–659.

Kuwae T, Hosokawa Y (1999) Determination of abundance and biovolume of bacteria in sediments by dual staining with 4′,6′-diamidino-2-phenylindole and acridine orange: relationship to dispersion treatment and sediment characteristics. *Applied and Environmental Microbiology* 65:3407–3412.

Kuypers HGJM, Catsman-Berrevoets CE, Padt RE (1977) Retrograde axonal transport of fluorescent substances in the rat's forebrain. *Neuroscience Letters* 6:127–135.

Kwan ML, Gomez AD, Baluk P, Hashizume H, McDonald DM (2001) Airway vasculature after mycoplasma infection: chronic leakiness and selective hypersensitivity to substance P. *American Journal of Physiology – Lung Cellular and Molecular Physiology* 280:L286–L297.

La Barre S, Singer S, Erard Le Denn E, Jozefowicz M (1999) Controlled cultivation of *Alexandrium minutum* and [P-33] orthophosphate cell labeling towards surface adhesion tests. *Journal of Biotechnology* 70:207–212.

Ladha S, Mackie AR, Clark DC (1994) Cheek cell membrane fluidity measured by fluorescence recovery after photobleaching and steady-state fluorescence anisotropy. *Journal of Membrane Biology* 142:223–228.

Lagardere F, Thibaudeau K, Begout Anras ML (2000) Feasibility of otolith markings in large juvenile turbot, *Scophthalmus maximus*, using immersion in alizarin-red S solutions. *ICES Journal of Marine Science* 57:1175–1181.

Lah JJ, Hayes DM, Burry RW (1990) A neutral pH silver development method for the visualization of 1-nanometer gold particles in pre-embedding electron microscopic immunocytochemistry. *Journal of Histochemistry and Cytochemistry* 38:503–508.

Lam KM (1998) A simple and fast way to count heterophils in chemotaxis. *Avian Diseases* 42:812–814.

Lam YW, Cohen LB, Wachowiak M, Zochowski MR (2000) Odors elicit three different oscillations in the turtle olfactory bulb. *Journal of Neuroscience* 20:749–762.

Landing BH, Hall HE (1956) Histochemical differentiation of anterior pituitary cell types by contrasting azo coupling and mucoprotein stains. *Stain Technology* 31:193–196.

Landra M, Acchiardi F, Pugno F, Forte G, Granetto C, Camuzzini GF (2000) Sentinel node mapping for malignant melanoma. *Tumori* 86:354–355.

Landriscina M, Prudovsky I, Carreira CM, Soldi R, Tarantini F, Maciag T (2000) Amlexanox reversibly inhibits cell migration and proliferation and induces the src-dependent disassembly of actin stress fibers *in vitro*. *Journal of Biological Chemistry* 275:32753–32762.

Langeron M (1925) *Precis de Microscopie*. 4th Edn. Masson, Paris.

Langmyhr FJ, Stumpe T (1965) Complex formation of iron(III) with Eriochrome cyanine R. *Analytica Chimica Acta* 32:535–543.

Lansink AGW (1968) Thin-layer chromatography and histochemistry of Sudan black B. *Histochemie* 16:68–84.

Lapertosa G, Baracchini P, Chiaramondia M, Fulcheri E, Picciotta A, Tanzi R (1984) Use of Victoria blue in the detection of intrahepatocyte HBS Ag – comparison with other methods. *Basic and Applied Histochemistry* 28:59–65.

Larouche I, Schiffrin EL (1999) Cardiac microvasculature in DOCA-salt hypertensive rats – Effect of endothelin ETA receptor antagonism. *Hypertension* 34:795–801.

Larramendy ML, Huhta T, Heinonen K, Vettenranta K, Mahlamaki E, Riikonen P, Saarinen Pihkala UM, Knuutila S (1998) DNA copy number changes in childhood acute lymphoblastic leukemia. *Haematologica* 83:890–895.

Larson DW, Doubt J, Matthes-Sears U (1994) Radially sectored hydraulic pathways in the xylem of *Thuja occidentalis* as revealed by the use of dyes. *International Journal of Plant Sciences* 155:569–582.

Larson JL, Miller DJ (1999) Simple histochemical stain for acrosomes on sperm from several species. *Molecular Reproduction and Development* 52:445–449.

Larsson LI (1988) *Immunocytochemistry: Theory and Practice.* CRC Press, Boca Raton, Florida.

Lauder RM, Beynon AD (1989) Evidence against the existence of acetylated sudan black B. *Histochemistry* 93:213–216.

Lavrova EA, Natochin IV (1985) Secretion of organic acids and bases by the kidneys of marine teleosts. [Russian]. *Archiv Anatomii Gistologii i Embriologii* 88:89–94.

Lawoko G, Tagerud S (1995) High endocytotic activity occurs periodically in the end-plate region of denervated mouse striated muscle fibers. *Experimental Cell Research* 219:598–603.

Le Leu RK, Young GP, McIntosh GH (2000) Folate deficiency diminishes the occurrence of aberrant crypt foci in the rat colon but does not alter global DNA methylation status. *Journal of Gastroenterology and Hepatology* 15:1158–1164.

Leach EH (1946) Curtis' substitute for Van Gieson stain. *Stain Technology* 2:107–109.

Lechevalier HA, Roisen FJ (1973) Stains for light microscopy. In: AI Lastin, HAJ Lechevalier, *Handbook of Microbiology.* CRC Press, Cleveland. pp. 680–686.

Lecuit T, Wieschuis E (2000) Polarized insertion of new membrane from a cytoplasmic reservoir during cleavage of the *Drosophila* embryo. *Journal of Cell Biology* 150:849–860.

Lecuona E, Saldias F, Comellas A, Ridge K, Guerrero C, Sznajder JI (1999) Ventilator-associated lung injury decreases lung ability to clear edema in rats. *American Journal of Respiratory and Critical Care Medicine* 159:603–609.

Lee AB (1885) *The Microtomist's Vade-mecum. A Handbook of the Methods of Microscopic Anatomy.* London: Churchill.

Lee C, Wu SS, Chen LB (1995) Photosensitization by 3,3′-dihexyloxacarbocyanine iodide: specific disruption of microtubules and inactivation of organelle motility. *Cancer Research* 55:2063–2069.

Lee TC, Arthur TL, Gibson LJ, Hayes WC (2000a) Sequential labelling of microdamage in bone using chelating agents. *Journal of Orthopaedic Research* 18:322–325.

Lee TC, O'Brien FJ, Taylor D (2000b) The nature of fatigue damage in bone. *International Journal of Fatigue* 22:847–853.

Leifson E (1951) Staining, shape, and arrangement of bacterial flagella. *Journal of Bacteriology* 62:377–389.

Leishman WB (1901) A simple and rapid method of producing Romanowsky staining in malarial and other blood films. *British Medical Journal* 1901–2:757–758.

Leith JT, Michelson S (1995) Levels of selected growth factors in viable and necrotic regions of xenografted HCT-8 human colon tumors. *Cell Proliferation* 28:279–286.

Leitinger G, Pabst MA, Kral K (2000) Gold toning preserves integrity of silver enhanced immunogold particles during osmium tetroxide treatment for demonstration of a biogenic amine. *Brain Research Protocols* 5:30–38.

Lelis AT (1992) The loss of intestinal flagellates in termites exposed to the juvenile hormone analog (JHA) – methoprene. *Material und Organismen* 27:170–178.

Lemasters JJ (1999) Mechanisms of hepatic toxicity – V. Necrapoptosis and the mitochondrial permeability transition: shared pathways to necrosis and apoptosis. *American Journal of Physiology – Gastrointestinal and Liver Physiology* 39:G1–G6.

Lenczewski ME, McGavin ST, Van Dyke K (1996) Comparison of automated and traditional minimum inhibitory concentration procedures for microbiological cosmetic preservatives. *Journal of AOAC International* 79:1294–1299.

Lendrum AC (1935) Celestin blue as a nuclear stain. *Journal of Pathology and Bacteriology* 40:415–416.

Lendrum AD, Fraser DS, Slidders W, Henderson R (1962) Studies on the character and staining of fibrin. *Journal of Clinical Pathology* 15:401–413.

Leung AF (1997) Fluorescence properties of frog skeletal muscles treated with lanthanide ions. *Spectroscopy Letters* 30:591–600.

Leung JK, Gibbon KJ, Vartanian RK (1996) Rapid staining method for *Helicobacter pylori* in gastric biopsies. *Journal of Histotechnology* 19:131–131.

Levey AI, Bolam JP, Rye DB, Hallanger AE, Demuth RM, Mesulam M-M, Wainer BH (1986) A light and electron microscopic procedure for sequential double antigen localization using diaminobenzidine and benzidine dihydrochloride. *Journal of Histochemistry and Cytochemistry* 34:1449–1457.

Levine BD, Zuckerman JH, Pawelczyk JA (1997) Cardiac atrophy after bed-rest deconditioning – a nonneural mechanism for orthostatic intolerance. *Circulation* 96:517–525.

Levine ND (1939) The dehydration of methylene blue stained material without loss of dye. *Stain Technology* 14:29–30.

Levine ND, Morrill CC (1941) Chlorazol black E a simple connective tissue stain. *Stain Technology* 16:121–122.

Levinson JW, Maher VM, McCormick JJ (1977) Purification of commercial acriflavine by Sephadex LH-20 column chromatography. *Journal of Histochemistry and Cytochemistry* 25:1275–1277.

Lew RR, Dearnaley JDW (2000) Extracellular nucleotide effects on the electrical properties of growing *Arabidopsis thaliana* root hairs. *Plant Science* 153:1–6.

Lewis GN (1945) Rules for the absorption spectra of dyes. *Journal of the American Chemical Society* 67:770–775.

Lewis PR, Knight DP (1977) *Staining Methods for Sectioned Material*. North-Holland, Amsterdam.

Lewis SM (1984) ICSH reference method for staining of blood and bone-marrow films by azure B and eosin Y (Romanowsky stain). *British Journal of Haematology* 57:707–710.

Li N, Lin G, Kwan YW, Min ZD (1999) Simultaneous quantification of five major biologically active ingredients of saffron by high-performance liquid chromatography. *Journal of Chromatography* A **849**:349–355.

Li S, Hartman GL, Widholm JM (1999) Viability staining of soybean suspension-cultured cells and a seedling stem cutting assay to evaluate phytotoxicity of *Fusarium solani* f. sp *glycines* culture filtrates. *Plant Cell Reports* 18:375–380.

Li YB, Stansbury KH, Zhu H, Trush MA (1999) Biochemical characterization of lucigenin (bis-N-methylacridinium) as a chemiluminescent probe for detecting intramitochondrial superoxide anion radical production. *Biochemical and Biophysical Research Communications* 262:80–87.

Liang XJ, Huang YG (2001) Alteration of membrane lipid biophysical properties and resistance of human lung adenocarcinoma A(549) cells to cisplatin. *Science in China Series C – Life Sciences* 44:25–32.

Lichtenstein SJ, Nettleton GS (1980) Effects of fuchsin variants in aldehyde fuchsin staining. *Journal of Histochemistry and Cytochemistry* 28:683–688.

Lieb E (1947) Permanent stain for amyloid. *American Journal of Clinical Pathology* 17:413–414.

Lieberkuhn N (1874) Ueber die Entwirkung von Alizarin auf die Gewebe des lebenden Korpers. *Marburger Sitzungsberichte* 1874:33.

Lii LJ, Wang CY, Lur HS (1999) A novel means of analyzing the soluble acidity of rice grains. *Crop Science* 39:1160–1164.

Lillie RD (1940) Further experiments with the Masson trichrome modification of Mallory's connective tissue stain. *Stain Technology* 15:17–22.

Lillie RD (1942) Studies on polychrome methylene blue. I. Eosinates, their spectra and staining capacity. *Stain Technology* 17:57–63.

Lillie RD (1943a) A Giemsa stain of quite constant composition and performance, made in the laboratory from eosin and methylene blue. *Public Health Reports* **58**:449–452.

Lillie RD (1943b) Studies on polychrome methylene blue. II. Acid oxidation methods of polychroming. *Stain Technology* **17**:97–110.

Lillie RD (1944a) Acetic methylene blue counterstain in staining tissues for acid-fast bacilli. *Stain Technology* **19**:45.

Lillie RD (1944b) Factors influencing the Romanowsky staining of blood films and the role of methylene violet. *Journal of Laboratory and Clinical Medicine* 29:1181–1197.

Lillie RD (1945) Studies on selective staining of collagen with acid anilin dyes. *Journal of Technical Methods, Bulletin of the International Association of Medical Museums* 25:1–47.

Lillie RD (1954) *Histopathologic Technic and Practical Histochemistry.* 2nd Edn. Blakiston, Philadelphia.

Lillie RD (1959) Preferred common names, formulae, *Colour Index* references and synonyms of stable diazonium salts used in histochemistry. *Journal of Histochemistry and Cytochemistry* 7:281–284.

Lillie RD (1962) The histochemical reaction of aryl amines with tissue aldehydes produced by periodic and chromic acids. *Journal of Histochemistry and Cytochemistry* 10:303–3314.

Lillie RD (1965) *Histopathologic Technic and Practical Histochemistry.* 3rd Edn. McGraw-Hill, New York.

Lillie RD (1977) *H. J. Conn's Biological Stains.* 9th Edn. Williams and Wilkins, Baltimore.

Lillie RD, Ashburn LL (1943) Supersaturated solutions of fat stains in dilute isopropanol for demonstration of acute fatty degeneration not shown by the Herxheimer technic. *Archives of Pathology* 36:432–435.

Lillie RD, Burtner HJ (1953) Stable sudanophilia of human neutrophil leucocytes in relation to peroxidase and oxidase. *Journal of Histochemistry and Cytochemistry* 1:8–26.

Lillie RD, Fullmer HM (1976) *Histopathologic Technic and Practical Histochemistry.* 4th Edn. McGraw-Hill, New York.

Lillie RD, Pizzolato P (1970) Histochemical azo coupling reactions of the pigments of obstructive icterus and of hematoidin. II. Effect of blockade, bleaching and extractive procedures on the pigments and their azo derivatives. *Journal of Histochemistry and Cytochemistry* 18:75–79.

Lillie RD, Roe MA (1942) Studies on polychrome methylene blue. I. Eosinates, their spectra and staining capacity. *Stain Technology* 17:57–63.

Lillie RD, Gilmer PR, Welsh RA (1961a) Black periodic and black Bauer methods for tissue polysaccharides. *Stain Technology* **36**:361–363.

Lillie RD, Henson JPG, Cason JC (1961b) Azocoupling rate of enterchromaffin with various diazonium salts. *Journal of Histochemistry and Cytochemistry* 99:11–21.

Lillie RD, Gutierrez A, Palmer RW (1967) Flavianic acid in place of picric acid in Van Gieson collagen fiber stains. Inclusion of iron and other metals in the Van Gieson stain. Note on a premixed ferrous sulfate hematoxylin. *Anatomical Record* 159:165–370.

Lin F, Fan W, Wise GE (1991) Eosin Y staining of proteins in polyacrylamide gels. *Analytical Biochemistry* 196:279–283.

Linnertz H, Urbanova P, Obsil T, Herman P, Amler E, Schoner W (1998) Molecular distance measurements reveal an (alpha beta) (2) dimeric structure of Na+/K+-ATPase – High affinity ATP binding site and K+-activated phosphatase reside on different alpha-subunits. *Journal of Biological Chemistry* 273:28813–28821.

Lintner F, Bosch P, Brand G (1982) The efficiency of the Sudan III staining to identify wear particles of PMMA-bone cement after total endoprostheses. *Archives of Orthopaedic and Traumatic Surgery* 100:79–81.

Liochev SI, Fridovich I (1995) Superoxide from glucose oxidase or from nitroblue tetrazolium. *Archives of Biochemistry and Biophysics* 318:408–410.

Lipp H (1940) Ersparnisse bei der Gonokokken- und Spirochatenfarbung. *Munchen medizinischer Wochenschrift* 87:888.

Lison L, Dagnelie J (1935) Methodes nouvelles de coloration de la myeline. *Bulletin d'Histologie Appliquée* 12:85–91.

Lison L (1955) Staining differences in cell nuclei. *Quarterly Journal of Microscopical Science* 96:227–237.

Lison L (1960) *Histochimie et Cytochimie Animales.* 3rd Edn. Gauthier-Villars, Paris.

Little AG, DeHoyos A, Kirgan DM, Arcomano TR, Murray KD (1999) Intraoperative lymphatic mapping for non-small cell lung cancer: the sentinel node technique. *Journal of Thoracic and Cardiovascular Surgery* 117:220–223.

Litwin JA (1985) Light microscopic histochemistry on plastic sections. *Progress in Histochemistry and Cytochemistry* 16:1–79.

Liu RM, Liu DJ, Sun AL, Liu GH (1995) Simultaneous determination of copper and zinc in the hair of children by pH gradient construction in a flow-injection system. *Analyst* 120:569–572.

Lizard G, Chignol MC, Roignot P, Souchier C, Chardonnet Y, Schmitt D (1997) Detection of human papillomavirus DNA in genital lesions by enzymatic *in situ* hybridization with fast red and laser scanning confocal microscopy. *Histochemical Journal* 29:545–554.

Llewellyn BD (1970) An improved sirius red method for amyloid. *Journal of Medical Laboratory Technology* 27:308–309.

Llewellyn-Smith IJ, Pilowsky P, Minson JB (1993) The tungstate-stabilized tetramethylbenzidine reaction for light and electron microscopic immunocytochemistry and for revealing biocytin-filled neurons. *Journal of Neuroscience Methods* 46:27–40.

Llorente-Cortes V, Martinez-Gonzalez J, Badimon L (2000) LDL receptor-related protein mediates uptake of aggregated LDL in human vascular smooth muscle cells. *Arteriosclerosis, Thrombosis and Vascular Biology* **20**:1572–1579.

Lloyd D, Thomas KL, Hayes A, Hill B, Hales BA, Edwards C, Saunders JR, Ritchie DA, Upton M (1998) Micro-ecology of peat: minimally invasive analysis using confocal laser scanning microscopy, membrane inlet mass spectrometry and PCR amplification of methanogen-specific gene sequences. *FEMS Microbiology and Ecology* 25:179–188.

Lloyd JB, Beck F (1963) An evaluation of acid disazo dyes by chloride determination and paper chromatography. *Stain Technology* 38:165–171.

Loew LM (1993) Confocal microscopy of potentiometric fluorescent dyes. *Methods in Cell Biology* 38:195–209.

Lofgren S, Soderberg PG (1998) Histochemical determination of lactate dehydrogenase activity in rat lens; influence of different parameters. *Acta Ophthalmologica Scandinavica* 76:555–560.

Lohr W, Grubhofer N, Sohmer I, Wittekind D (1975) The azure dyes: their purification and physicochemical properties. II. Purification of Azure B. *Stain Technology* 50:149–156.

Lohr W, Sohmer I, Wittekind D (1974) The azure dyes: their purification and physicochemical properties. I. Purification of Azure A. *Stain Technology* 49:359–366.

Lojda Z (1965) Remarks on the histochemical demonstration of dehydrogenases. II. Intracellular localization. *Folia Morphologica (Praha)* 13:84–97.

Lopez (1946) *Technical Bulletin* 7:53. Cited in Gray (1973), p. 479.

Lopez MK, Kornegay RW (1991) A multichromatic stain for Lowicryl K4M embedded tissues. *Biotechnic and Histochemistry* 65:35–36.

Lopez-Amoros R, Castel S, Comas-Riu J, Vives-Rego J (1997) Assessment of *E. coli* and *Salmonella* viability and starvation by confocal laser microscopy and flow cytometry using rhodamine 123, DiBAC4(3), propidium iodide and CTC. *Cytometry* 29:298–305.

Lorenz JN, Gruenstein E (1999) A simple, nonradioactive method for evaluating single-nephron filtration rate using FITC-inulin. *American Journal of Physiology – Renal, Fluid and Electrolyte Physiology* 276:F172–F177.

Lorincz AE, Kelly DR, Dobbins GC, Cardone VS, Fuchs SA, Schilleci JL (1999) Urinalysis: current status and prospects for the future. *Annals of Clinical and Laboratory Science* 29:169–175.

Lu HT, Mou SF, Yan Y, Tong SY, Riviello JM (1998) On-line pretreatment and determination of Pb, Cu and Cd at the microgram l-1 level in drinking water by chelation ion chromatography. *Journal of Chromatography* A **800**:247–255.

Luan P, Yang L, Glaser M (1995) Formation of membrane domains created during the budding of vesicular stomatitis virus. A model for selective lipid and protein sorting in biological membranes. *Biochemistry* 34:9874–9883.

Lubach D, Nissen S, Neukam D (1991) Demonstration of initial lymphatics in excised human skin using an extension technique and dye injection. *Journal of Investigative Dermatology* 96:754–757.

Lubke J (1993) Photoconversion of diaminobenzidine with different fluorescent neuronal markers into a light and electron microscopic dense reaction product. *Microscopy Research and Techniques* 24:2–14.

Lubrano GJ, Dean WW, Heinsohn HG, Stastny M (1977) The analysis of some commercial dyes and Romanowsky stains by high-performance liquid chromatography. *Stain Technology* 52:13–23.

Lubs HA, ed. (1955) *The Chemistry of Synthetic Dyes and Pigments.* Hafner, New York.

Lucci A, Turner RR, Morton DL (1999) Carbon dye as an adjunct to isosulfan blue dye for sentinel lymph node dissection. *Surgery* 126:48–53.

Luchtel DL, Embree L, Guest R, Albert RK (1991) Extra-alveolar veins are contiguous with, and leak fluid into, periarterial cuffs in rabbit lungs. *Journal of Applied Physiology* 71:1606–1613.

Luft JH (1971a) Ruthenium red and violet. I. Chemistry, purification, methods of use for electron microscopy and mechanism of action. *Anatomical Record* 171:347–368.

Luft JH (1971b) Ruthenium red and violet. III. Fine structure of the plasma membrane and extraneous coats in amoebae (*A. proteus* and *Chaos chaos*). *Anatomical Record* 171:417–442.

Lulai EC, Morgan WC (1992) Histochemical probing of potato periderm with neutral red: a sensitive cytofluorochrome for the hydrophobic domain of suberin. *Biotechnic and Histochemistry* 67:185–195.

Luna LG (1968) *Manual of Histologic Staining Methods of the Armed Forces Institute of Pathology.* 3rd Edn. McGraw-Hill, New York.

Lunn G, Sansone EB (1990) *Destruction of Hazardous Chemicals in the Laboratory.* Wiley Interscience, New York.

Lussenhop J, Fogel R (1993) Observing soil biota *in situ. Geoderma* 56:25–36.

Lussignoli S, Fraccaroli M, Andrioli G, Brocco G, Bellavite P (1999) A microplate-based colorimetric assay of the total peroxyl radical trapping capability of human plasma. *Analytical Biochemistry* 269:38–44.

Lustig DG, Herrick JL, Keifer J (1998) Comparison of cortically and subcortically controlled motor systems. I. Morphology of intracellularly filled rubrospinal neurons in rat and turtle. *Journal of Comparative Neurology* 396:521–530.

Lycette RM, Danforth WF, Koppel JL, Olwin JH (1970) The binding of luxol fast blue ARN by various biological lipids. *Stain Technology* 45:155–160.

Lyle SJ, Tehrani MS (1979) Thin-layer chromatographic separation and subsequent determination of some water-soluble dyestuffs. *Journal of Chromatography* 175:163–168.

Lynch G, Smith RL, Mensah P, Cotman C (1973) Tracing the dentate gyrus mossy fiber system with horseradish peroxidase histochemistry. *Experimental Neurology* 68:167–173.

Lynn A, Cochrane MP (1997) An evaluation of confocal microscopy for the study of starch granule enzymic digestion. *Starch* 49:106–110.

Lyon H (1991) *Theory and Strategy in Histochemistry*. Springer-Verlag, Berlin.

Lyon H, Andersen AP, Andersen I, Clausen PP, Herold B (1982) Purity of commercial non-certified European samples of pyronin Y. *Histochemical Journal* 14:621–630.

Lyon H, Jakobsen P, Hoyer P, Andersen AP (1987) An investigation of new commercial samples of methyl green and pyronin Y. *Histochemical Journal* 19:381–384.

Lyon H, Wittekind D, Schulte E (1991) Standardization of staining methods. In: H Lyon, *Theory and Strategy in Histochemistry*, Appendix A. Springer-Verlag, Berlin. pp. 509–517.

Lyon H, Schulte E, de Leenheer A, Lewis S, Friemert V, Struck C, Gadsdon D, Allison R, Brunk U, Van Liederkerke B, Hasselager E, Horobin RW, Husain O, Wittekind D, Zschoch H (1992a) Dye standards, Part 11.8: New fuchsin (CI 42520). *Histochemical Journal* 24:240–242.

Lyon H, Schulte E, De Leenheer A, Lewis S, Friemert V, Struck C, Gadsdon D, Allison R, Brunk U, Van Liederkerke B, Hasselager E, Horobin RW, Husain O, Wittekind D, Zschoch H (1992b) Dye standards. 2.5. Pararosaniline (CI 42500). *Histochemical Journal* 24:233–235.

Lyon HO, De Leenheer AP, Horobin RW, Lambert WE, Schulte EKW, Van Liedekerke B, Wittekind DH (1994) Standardization of reagents and methods used in cytological and histological practice with emphasis on dyes, stains and chromogenic reagents. *Histochemical Journal* 26:533–544.

Ma JF, Zheng SJ, Li XF, Takeda K, Matsumoto H (1997) A rapid hydroponic screening for aluminium tolerance in barley. *Plant and Soil* 191:133–137.

Macey M, ed. (1994) *Flow Cytometry. Clinical Applications*. Blackwell, Oxford.

MacNeal WJ (1925) Methylene violet and methylene azure A and B. *Journal of Infectious Diseases* 36:538–546.

MacPhee PJ, Michel CC (1995) Fluid uptake from the renal medulla into the ascending vasa recta in anaesthetized rats. *Journal of Physiology* 487:169–183.

Maddy AH (1964) A fluorescent label for the outer components of the plasma membrane. *Biochimica et Biophysica Acta* 88:390–399.

Maechler P, Jornot L, Wollheim CB (1999) Hydrogen peroxide alters mitochondrial activation and insulin secretion in pancreatic beta cells. *Journal of Biological Chemistry* **274**:27905–27913.

Maeda M, Kasornchandra J, Itami T, Suzuki N, Hennig O, Kondo M, Albaladejo JD, Takahashi Y (1998) Effect of various treatments on white spot syndrome virus (WSSV) from *Penaeus japonicus* (Japan) and *P. monodon* (Thailand). *Fish Pathology* 33:381–387.

Maezawa H, Manaka K, Yamakawa K, Ogawa K, Iizuki M (1997) Decreased sulfhydryl groups in the reperfused myocardial tissue of a rat model of myocardial infarction. *Japanese Circulation Journal* – English Edition **61**:151–160.

Mahieu I, Becq F, Wolfensberger T, Gola M, Carter N, Hollande E (1994) The expression of carbonic anhydrase-II and anhydrase-IV in the human pancreatic cancer cell line (Capan-1) is associated with bicarbonate ion channels. *Biology of the Cell* 81:131–141.

Maillet M (1963) Le reactif au tetraoxyde d'osmium-iodure du zinc. *Zeitschrift fur mikroskopische-anatomische Forschung* 76:397–425.

Maillet M (1968) Etude critique des fixatives au tetraoxyde d'osmium-iodure. *Comptes Rendus de l'Association des Anatomistes* 53:231–394.

Majumdar G, Samanta S, Ghosal SK (1996) Some histochemical observations on the cuticle of *Ascaridia galli*, a gastrointestinal nematode of poultry. *Helminthologia* 33:115–120.

Makarov AA, Dorofeev AG, Panikov NS (1998) Cell shape and size of starving micro-organisms as determined by computer image analysis. *Microbiology* 67:264–270.

Makowski AL, Ouellette EA, Puckett W, Marcillo A (1998) Large-scale cryosectioning technique allows *in vitro* correlation of magnetic resonance imaging and gross histo-logic findings in the wrist. *Journal of Histotechnology* 21:225–229.

Malachowski E (1891) Zur Morphologie der *Plasmodium malariae*. *Centralblatt fur klinische Medizin* 12:601–603.

Malik Z, Amit I, Rothmann C (1997) Subcellular localization of sulfonated tetraphenyl porphines in colon carcinoma cells by spectrally resolved imaging. *Photochemistry and Photobiology* 65:389–396.

Malkusch W, Rehn B, Bruch J (1995) Advantages of Sirius red staining for quantitative morphometric collagen measurements in lungs. *Experimental Lung Research* 21:67–77.

Mallolas J, Esteve M, Rius E, Cabre E, Gassull MA (2000) Antineutrophil antibodies associated with ulcerative colitis interact with the antigen(s) during the process of apoptosis. *Gut* 47:74–78.

Mallory FB (1900) A contribution to staining methods. 1. A differential stain for connective tissue fibrillae and reticulum. 2. Chloride of iron haematoxylin for nuclei and fibrin. 3. Phosphotungstic acid haematoxylin for neuroglia fibres. *Journal of Experimental Medicine* 5:15–20.

Mallory FB (1938) *Pathological Technique*. Hafner, New York.

Maltha JC, Bex JHM, Nottet SJAM (1977) Procion brilliant orange, a counterstain for decalcified sections vitally labelled with Procion brilliant red H-8BS. *Stain Technology* 52:211–215.

Maneval WE (1934) Rapid staining methods. *Science* 80:292–294.

Mann G (1894) Ueber die Behandlung der Nervenzellen fur experimentell-histologische Untersuchungen. *Zeitschrift fur wissenschaftliche Mikroskopie* 11:479–494.

Mann G (1902) *Physiological Histology. Methods and Theory*. Clarendon Press, Oxford.

Manyonda IT, Choy MY (1999) Collagen phagocytosis by human extravillous trophoblast: potential role in trophoblastic invasion. *Journal of the Society for Gynecologic Investigation* 6:158–166.

Marcaggi P, Thwaites DT, Deitmer JW, Coles JA (1999) Chloride-dependent transport of NH4+ into bee retinal glial cells. *European Journal of Neuroscience* 11:167–177.

Marchi V (1892) Sur l'origine et le cours des pedoncles cerebellaux et sur leurs rapports avec les autres centres nerveux. *Archives Italiennes de Biologie* 17:190–201.

Marczenko Z (1986) *Separation and Spectrophotometric Determination of Elements*. Ellis Horwood, Chichester, UK.

Margolena LA (1935) Lugol's solution for the Flemming triple stain. *Stain Technology* 10:35–36.

Maric D, Maric I, Barker JL (2000) Dual video microscopic imaging of membrane potential and cytosolic calcium of immunoidentified embryonic rat cortical cells. *Methods* 21:335–347.

Marin P, Maus M, Desagher S, Glowinski J, Premont J (1994) Nicotine protects cultured striatal neurons against N-methyl-D-aspartate receptor-mediated neurotoxicity. *NeuroReport* 5:1977–1980.

Marini M, Ferrari R (1998) A population survey of the Italian subterranean termite *Reticulitermes lucifugus lucifugus* Rossi in Bagnacavallo (Ravenna, Italy), using the triple mark recapture technique (TMR). *Zoological Science* 15:963–969.

Marko O, Cascieri MA, Ayad N, Strader CD, Candelore MR (1995) Isolation of a preadipocyte cell line from rat bone marrow and differentiation to adipocytes. *Endocrinology* 136:4582–4588.

Marmion DM (1991) *Handbook of US Colorants. Foods, drugs, cosmetics and medical devices.* 3rd Edn. Wiley, New York.

Mars MH, Van Deningh TSGAM, Hajer R, Wentink GH (1994) *In vitro* transport of carbon in the trachea of veal calves. *Veterinary Quarterly* 16:62–64.

Marshall PN (1975) Rules for the visible absorption spectra of halogenated fluorescein dyes. *Histochemical Journal* 7:299–303.

Marshall PN (1976a) The composition of erythrosins, fluorescein, phloxine and rose bengal: a study using thin-layer chromatography and solvent extraction. *Histochemical Journal* 8:487–499.

Marshall PN (1976b) The composition of stains produced by the oxidation of methylene blue. *Histochemical Journal* 8:431–442.

Marshall PN (1976c) Thin-layer chromatography of some cationic dyes commonly used in histology. *Journal of Chromatography* 129:277–285.

Marshall PN (1977) Thin-layer chromatography of Sudan dyes. *Journal of Chromatography* 136:353–357.

Marshall PN (1978a) Reticulation, polychromasia and stippling of erythrocytes. *Microscopica Acta* 81:89–106.

Marshall PN (1978b) Romanowsky-type stains in haematology. *Histochemical Journal* 10:1–29.

Marshall PN (1979a) Commercially available 'pure' azure dyes – caveat emptor. *Histochemical Journal* 11:489–493.

Marshall PN (1979b) Romanowsky staining: state of the art and 'ideal' techniques. Ch. 11 In: *Differential Leukocyte Counting.* College of American Pathologists, Stokie, IL. pp. 205–216.

Marshall PN, Horobin RW (1972a) The chemical nature of the gallocyanin-chrome alum staining complex. *Stain Technology* 47:155–161.

Marshall PN, Horobin RW (1972b) The oxidation products of haematoxylin and their role in biological staining. *Histochemical Journal* 4:493–503.

Marshall PN, Horobin RW (1973a) The mechanism of action of 'mordant' dyes – a study using preformed metal complexes. *Histochemie* 35:361–371.

Marshall PN, Horobin RW (1974a) A simple assay procedure for carmine and carminic acid samples. *Stain Technology* 49:19–28.

Marshall PN, Horobin RW (1974b) A simple assay procedure for mixtures of hematoxylin and hematein. *Stain Technology* 49:137–142.

Marshall PN, Horobin RW (1975) Thin layer chromatography of certain preformed metal complex dyes used in biological staining. *Stain Technology* 50:271–277.

Marshall PN, Lewis SM (1974a) Batch variations in commercial dyes employed for Romanowsky-type staining. *Stain Technology* 49:351–358.

Marshall PN, Lewis SM (1974b) The purification of methylene blue and azure B by solvent extraction and crystallization. *Stain Technology* 50:375–381.

Marshall PN, Lewis SM (1975) Metal contaminants in commercial thiazine dyes. *Stain Technology* 50:143–147.

Marshall PN, Bentley SA, Lewis SM (1975) A standardized Romanowsky stain prepared from purified dyes. *Journal of Clinical Pathology* 28:920–923.

Marshall PN, Galbraith W, Bacus JW (1979) Studies on Papanicolaou staining. 2. Quantitation of dye components bound to cervical cells. *Analytical and Quantitative Cytology* 1:169–178.

Martin A, Clynes M (1993) Comparison of five microplate colorimetric assays for *in vitro* cytotoxicity testing and cell proliferation assays. *Cytotechnology* 11:49–58.

Martin CA, Homaidan FR, Palaia T, Burakoff R, El Sabban ME (1998) Gap junctional communication between murine macrophages and intestinal epithelial cell lines. *Cell Adhesion and Communication* 5:437–451.

Martin J, Dinsdale D, White IN (1993) Characterization of Clara and type II cells isolated from rat lung by fluorescence-activated flow cytometry. *Biochemical Journal* 295:73–80.

Martin LM, Crensham CC, Dean JA, Dart MG, Purdy PH, Ericsson SA (1999) Determination of the number of motile sperm within an ovine semen sample using resazurin. *Small Ruminant Research* 32:161–165.

Marttin E, Verhoef JC, Cullander C, Romeijn SG, Nagelkerke JF, Merkus FWHM (1997) Confocal laser scanning microscopic visualization of the transport of dextrans after nasal administration to rats. Effects of absorption enhancers. *Pharmaceutical Research* 14:631–637.

Mascotti K, McCullough J, Burger SR (2000) HPC viability measurement: trypan blue versus acridine orange and propidium iodide. *Transfusion* 40:693–696.

Masereeuw R, Russel FGM, Miller DS (1996) Multiple pathways of organic anion secretion in renal proximal tubule revealed by confocal microscopy. *American Journal of Physiology – Renal Physiology* 40:F1173–F1182.

Masereeuw R, Terlouw SA, van Aubel RAMH, Russel FGM, Miller DS (2000) Endothelin B receptor-mediated regulation of ATP-driven drug secretion in renal proximal tubule. *Molecular Pharmacology* 57:59–67.

Mashberg A (1983) Final evaluation of tolonium chloride rinse for screening of high-risk patients with asymptomatic squamous carcinoma. *Journal of the American Dental Association* 106:319–323.

Maskiewicz R, Sogah D, Bruice TC (1979) Chemiluminescent reactions of lucigenin. 1. Reactions of lucigenin with hydrogen peroxide. *Journal of the American Chemical Association* 101:5347–5354.

Mason WT, ed. (1999) *Fluorescent and Luminescent Probes for Biological Activity*. 2nd Edn. Academic Press, San Diego and London.

Masson P (1911) Le safran en technique histologique. *Comptes Rendus des Seances de la Société de Biologie* 70:573–574.

Masson P (1929) Some histological methods. Trichrome stainings and their preliminary technique. *Journal of Technical Methods, Bulletin of the International Association of Medical Museums* 12:75–90.

Mathe G, Triana K, Pontiggia P, Blanquet D, Hallard M, Morette C (1998) Data of pre-clinical and early clinical trials of acriflavine and hydroxy-methyl-ellipticine reviewed, enriched by the experience of their use for 18 months to 6 years in combinations with other HIV1 virostatics. *Biomedicine and Pharmacotherapy* 52:391–396.

Mathews CK, van Holde KE (1996) *Biochemistry*. Benjamin-Cummings, Menlo Park, CA.

Mathur A, Hong Y, Kemp BK, Barrientos AA, Erusalimsky JD (2000) Evaluation of fluorescent dyes for the detection of mitochondrial membrane potential changes in cultured cardiomyocytes. *Cardiovascular Research* 46:126–138.

Matsuda R, Nishikawa A, Tanaka H (1995) Visualization of dystrophic muscle fibres in MDX mouse by vital staining with Evans blue – Evidence of apoptosis in dystrophin-deficient muscle. *Journal of Biochemistry* 118:959–964.

Matsui Y, Ohno K, Michi K, Hata H, Yamagata K, Ohtsuka S (1996) The evaluation of masticatory function with low adhesive colour-developing chewing gum. *Journal of Oral Rehabilitation* 23:251–256.

Matsumoto T, Komori K, Yonemitsu Y, Morishita R, Sueishi K, Kaneda Y, Sugimachi K (1998) Hemagglutinating virus of Japan-liposome-mediated gene transfer of endothelial cell nitric oxide synthase inhibits intimal hyperplasia of canine vein grafts under conditions of poor runoff. *Journal of Vascular Surgery* 27:135–144.

Matsumura J, Booker RE, Donaldson LA, Ridoutt BG (1998) Impregnation of radiata pine wood by vacuum treatment: identification of flow paths using fluorescent dye and confocal microscopy. *International Association of Wood Anatomists' Journal* 19:25–33.

Matsuoka H, Yang HC, Homma T, Nemoto Y, Yamada S, Sumita O, Takatori K, Kurata H (1995) Use of Congo red as a microscopic fluorescence indicator of hyphal growth. *Applied Microbiology and Biotechnology* 43:102–108.

Matsuoka M (1990) *Infrared Absorbing Dyes.* Plenum Press, New York.

Matsuura S (1925) Ueber die Farbung mit Kongorot. *Folia Anatomica Japonica* 3:107–110.

Matter HC, Schumacher CL, Kharmachi H, Hammami S, Tlatli A, Jemli J, Mrabet L, Meslin FX, Aubert MFA, Neuenschwander BE, El Hicheri K (1998) Field evaluation of two bait delivery sytems for the oral immunization of dogs against rabies in Tunisia. *Vaccine* 16:657–665.

Mattis AE, Bernhardt G, Lipp M, Forster R (1997) Analyzing cytotoxic T lymphocyte activity: a simple and reliable flow cytometry-based assay. *Journal of Immunological Methods* 204:135–142.

Matyas JR, Huang DQ, Adams ME (1999) A comparison of various 'housekeeping' probes for northern analysis of normal and osteoarthritic articular cartilage RNA. *Connective Tissue Research* 40:163–172.

Matysik G (1998) Densitometric method of determination of fluorescein in aqueous humor of the eye. *Chemia Analityczna* 43:719–723.

Matzke KH, Thiessen G (1976) The acridine dyes: their purification, physicochemical, and cytochemical properties. 1. A purity test of some commercial acriflavine samples and the identification of their components. *Histochemistry* 49:73–79.

Maurice DM (1987) Flow of water between aqueous and vitreous compartments in the rabbit eye. *American Journal of Physiology* 252:F104–F108.

Maurina FA, Deahl N (1943) A study of assay methods for methylene blue. *Journal of the American Pharmaceutical Association, Scientific Edition* 32:301–306.

Maus TL, Brubaker RF (1999) Measurement of aqueous humor flow by fluoro-photometry in the presence of a dilated pupil. *Investigative Ophthalmology and Visual Science* 40:542–546.

Maxwell A (1963) The alcian dyes applied to gastric mucosa. *Stain Technology* 38:286–287.

May R, Grunwald L (1902) Uber Blutfarbung. *Zentralblatt fur innere Medizin* 11:265–270.

Mayer P (1891) Ueber das Farben mit Hamatoxylin. *Mitteilungen aus dem Zoologische Station zu Neapel* 10:170–186.

Maymind M, Thomas JG, Abrons HL, Riley RS (1996) Laboratory implementation of a rapid three-stain technique for detection of microorganisms from lower respiratory specimens. *Journal of Clinical Laboratory Analysis* 10:104–109.

McAllister JC, Steelman CD, Skeeles JK, Gbur EE (1996) Reservoir competence of *Alphitobius diaperinus* (Coleoptera: Tenebrionidae) for *Escherichia coli* (Eubacteriales: Enterobacteriaceae). *Journal of Medical Entomology* **33**:983–987.

McBride JD, Stubberfield CR, Hayes DJ (1993) Electrophoretic detection of chitinase isoenzymes using the phastsystem. *Electrophoresis* 14:165–167.

McClung CE, ed. (1929) *Handbook of Microscopical Technique.* Hoeber, New York.

McConalogue K, Dery O, Lovett M, Wong H, Walsh JH, Grady EF, Bunnett NW (1999) Substance P-induced trafficking of beta-arrestins. The role of beta-arrestins in endocytosis of the neurokinin-1 receptor. *Journal of Biological Chemistry* 274:16257–16268.

McDougall JJ, Yeung G, Leonard CA, Sutherland C, Bray RC (2000) Adaption of post-traumatic angiogenesis in the rabbit knee by apposition of torn ligament ends. *Journal of Orthopaedic Research* 18:663–670.

McGahren ED, Beach JM, Duling BR (1998) Capillaries demonstrate changes in membrane potential in response to pharmacological stimuli. *American Journal of Physiology* 274:H60–H65.

McLaren K (1986) *The Colour Science of Dyes and Pigments.* 2nd Edn. Taylor and Francis, New York.

McLean JW, Fox EA, Baluk P, Bolton PB, Haskell A, Pearlman R, Thurston G, Umemoto EY, McDonald DM (1997) Organ-specific endothelial cell uptake of cationic liposome-DNA complexes in mice. *American Journal of Physiology* 273:H387–H404.

McLean PJ, Kawamata H, Ribich S, Hyman BT (2000) Membrane association and protein conformation of alpha-synuclein in intact neurons. Effect of Parkinson's disease-linked mutations. Journal of Biological Chemistry **275**:8812–8816.

McManus JFA, Mowry RW (1960) *Staining Methods. Histologic and Histochemical.* P. B. Hoeber, New York.

McManus JFA (1946) Histological demonstration of mucin after periodic acid. *Nature* 158:212.

McMaster PD, Parsons RJ (1938) Path of escape of vital dyes from the lymphatics into the tissues. *Proceedings of the Society for Experimental Biology and Medicine* 37:707–709.

McNary WF (1960) The histochemical demonstration of trace metals in leucocytes. *Journal of Histochemistry and Cytochemistry* 8:124–130.

Medina MB, Nagdy N (1993) Improved thin-layer chromagraphic detection of diethyl-stilbestrol and zeronal in plasma and tissues – isolated with alumina and ion-exchange membrane columns in tandem. *Journal of Chromatography – Biomedical Applications* 614:315–323.

Mehes G, Kalman E, Pajor L (1993) *In situ* fluorescent visualization of nucleolar organizer region-associated proteins with a thiol reagent. *Journal of Histochemistry and Cytochemistry* 41:1413–1417.

Meistrich H, Green LK (1989) The application of a combined Papanicolaou-tinopal CBS stain in cytopathology. *Laboratory Investigation* 60:A62.

Mellefowicz EJ, Riding RT, Little CHA (1993) Nucleolar activity in the fusiform cambial cells of *Abies balsamea* (Pinaceae) – effect of season and age. *American Journal of Botany* **80**:1168–1174.

Meloan SN, Puchtler H (1978) Demonstration of amyloid with Mesitol WLS – Congo red: application of a textile auxiliary to histochemistry. *Histochemistry* 58:163–166.

Meloan SN, Puchtler H (1986) On the structure and chemistry of leucofuchsin and Schiff's reagent. *Histotechnology* 9:119–122.

Meloan SN, Valentine LS, Puchtler H (1971) On the structure of carminic acid and carmine. *Histochemie* 27:87–95.

Meloan SN, Puchtler H, Valentine LS (1973) Staining of calcium deposits with acid dyes for lakes. Light and fluorescence microscopic studies. *Beitrage zur pathologischen Anatomie und zur allgemeinen Pathologie* 149:386–395.

Mene P, Fais S, Cinotti GA, Pugliese F, Luttmann W, Thierauch KH (1995) Regulation of U-937 monocyte adhesion to cultured human mesangial cells by cytokines and vasoactive agents. *Nephrology Dialysis Transplantation* 10:481–489.

Menegola E, Broccia ML, Prati M, Giavini E (1999) Morphological alterations induced by sodium valproate on somites and spinal nerves in rat embryos. *Teratology* 59:110–119.

Mengel M, Werner M, Von Wasielewski R (1999) Concentration dependent and adverse effects in immunohistochemistry using the tyramine amplification technique. *Histochemical Journal* 31:195–200.

Mengelers HJJ, Maikoe T, Raaijmakers JAM, Lammers JWJ, Koenderman L (1995) Cognate interaction between human lymphocytes and eosinophils is mediated by beta-2-integrins and very late antigen-4. *Journal of Laboratory and Clinical Medicine* 126:261–268.

Menghi G, Bondi AM, Marchetti L, Ballarini P, Materazzi G (1998) Confocal and electron microscopy to characterize sialoglycoconjugates in mouse sublingual acinar cells. *European Journal of Morphology* 36:222–229.

Menzies DW, Roberts JT (1963) Effect of age on the acidophilia of aortic elastic. *Nature* 198:1006–1007.

Merker MP, Bongard RD, Linehan JH, Okamoto Y, Vyprachticky D, Brantmeier BM, Roerig DL, Dawson CA (1997) Pulmonary endothelial thiazine uptake: Separation of cell surface reduction from intracellular reoxidation. *American Journal of Physiology – Lung Cellular and Molecular Physiology* 16:L673–L680.

Merritt K, Gaind A, Anderson JM (1998) Detection of bacterial adherence on biomedical polymers. *Journal of Biomedical Materials Research* 39:415–422.

Merton H (1932) Die Verwendung von Kupfersalzen zur Herstellung von Paramaecium. Preparate. *Archiv fur Protistenkunde* 76:171–187.

Mesulam M-M, Rosene DL (1979) Sensitivity in horseradish peroxidase neurohistochemistry: a comparative and quantitative study of nine methods. *Journal of Histochemistry and Cytochemistry* 27:763–773.

Metcalf RL, Patton RL (1944) Fluorescence microscopy applied to entomology and allied fields. *Stain Technology* 19:11–27.

Metzner F (1931) Einfache Einrichtungen zur Fluoreszenzmikroskopie und Fluoreszenzphotographie. *Mikrochemie* 9:72–91.

Meves F, Duesberg J (1908) Die Spermatozytenteilungen bei der Hornisse (*Vespa crabris* L.). *Archiv fur mikroskopische Anatomie und Entwicklungsgeschichte* 71:571–587.

Michael W, Cholodova VP, Ehwald R (1999) Gas and liquids in intercellular spaces of maize roots. *Annals of Botany* 84:665–673.

Michaelis L (1900) Die vitale Farbung, eine Darstellungsmethode der Zellgranula. *Archiv fur mikroskopische Anatomie und Entwicklungsgeschichte* 55:558–575.

Michaelis L (1901) Ueber Fett-Farbstoffe. *Virchows Archiv fur pathologische Anatomie und Physiologie und fur klinische Medizin* 164:263–270.

Miescher F (1871) Uber die chemische Untersuchung von Eiterzeller. *Hoppe-Seylers medizinsche und chemische Untersuchungen* 4:441.

Mihai R, Lai T, Schofield GJ, Farndon JR (2000) Changes in cytoplasmic calcium determine the secretory response to extracellular cations in human parathyroid cells: a confocal microscopy study using FM1–43 dye. *Biochemical Journal* 352:353–361.

Mikkonen M, Pitkanen A, Soininen H, Alafuzoff I, Miettinen R (2000) Morphology of spiny neurons in the human entorhinal cortex: intracellular filling with Lucifer yellow. *Neuroscience* 96:515–522.

Miklossy J, Van der Loos H (1991) The long-distance effects of brain lesions: visualization of myelinated pathways in the human brain using polarizing and fluorescence microscopy. *Journal of Neuropathology and Experimental Neurology* 50:1–15.

Mikolon AB, Gardner IA, De Anda JH, Hietala SK (1998) Risk factors for brucellosis seropositivity of goat herds in the Mexicali Valley of Baja California, Mexico. *Preventive Veterinary Medicine* 37:185–195.

Milacek P, Vitovec J (1985) Differential staining of cryptosporidia by aniline-carbol-methyl violet and tartrazine in smears from feces and scrapings of intestinal-mucosa. *Folia Parasitologica* 32:50.

Milanova E, Sithole BB (1997) A simple method for estimation of newsprint dyes in effluents and their migration from paper samples. *Tappi Journal* 80:121–128.

Mildenstein K (1971) Synthesis and determination of purity of 1-dimethylamino-naph-thalene-5-sulphonic acid chloride (DIS, DANSYL chloride). *Acta Histochemica* 40:29–50.

Mileykovskaya E, Dowham W (2000) Visualization of phospholipid domains in *Escherichia coli* by using the cardiolipin-specific fluorescent dye 10-N-nonyl acridine orange. *Journal of Bacteriology* 182:1172–1175.

Millen JW, Hess A (1958) The blood-brain barrier: an experimental study with vital dyes. *Brain* 81:248–257.

Miller CR, Bondurant B, McLean SD, McGovern KA, O'Brien DF (1998) Liposome-cell interactions *in vitro*: effect of liposome surface charge on the binding and endocytosis of conventional and sterically stabilized liposomes. *Biochemistry* 37:12875–12883.

Miller CR, Clapp PJ, O'Brien DF (2000) Visible light-induced destabilization of endo-cytosed liposomes. *FEBS Letters* 467:52–56.

Miller DS, Fricker G, Schramm U, Henson JH, Hager DN, Nundy S, Ballatori N, Boyer JL (1996a) Active microtubule-dependent secretion of a fluorescent bile salt derivative in skate hepatocyte clusters. *Journal of Physiology* 270:G887–G896.

Miller DS, Letcher S, Barnes DM (1996b) Fluorescence imaging study of organic anion transport from renal proximal tubule cell to lumen. *American Journal of Physiology – Renal, Fluid and Electrolyte Physiology* 271:F508–520.

Miller N, Hutt-Fletcher LM (1992) Epstein-Barr virus enters B cells and epithelial cells by different routes. *Journal of Virology* 66:3409–3414.

Millson CE, Wilson M, Macrobert AJ, Bedwell J, Bown SG (1996) The killing of *Helicobacter pylori* by low-power laser light in the presence of a photosensitiser. *Journal of Medical Microbiology* 44:245–252.

Minamitani M, Tanaka J, Maekawa K (1994) Peculiar eosinophilic inclusions within astrocytes in a patient with malformed brain. *Brain and Development* 16:309–314.

Minamiya Y, Tozawa K, Kitamura M, Saito S, Ogawa J (1998) Platelet-activating factor mediates intercellular adhesion molecule-1-dependent radical production in the nonhypoxic ischemia rat lung. *American Journal of Respiratory, Cell and Molecular Biology* 19:150–157.

Mindrup EA, Dubbel PA, Doughman DJ (1999) Evaluation and transplantation of corneas from pseudophakic donor eyes. *Cornea* 18:652–657.

Ming D, Ye H, Schaad NW, Roth DA (1991) Selective recovery of *Xanthomonas* spp from rice seed. *Phytopathology* 81:1358–1363.

Minier C, Moore MN (1996) Rhodamine B accumulation and MXR protein expression in mussel blood cells: effects of exposure to vincristine. *Marine Ecology – Progress Series* 142:165–173.

Minshall RD, Tiruppathi C, Vogel SM, Niles WD, Gilchrist A, Hamm HE, Malik AB (2000) Endothelial cell-surface gp60 activates vesicle formation and trafficking via G(i)-coupled Src kinase signalling pathway. *Journal of Cell Biology* 150:1057–1070.

Minussi RC, de Moraes SG, Pastore GM, Duran N (2001) Biodecolorization screening of synthetic dyes by four white-rot fungi in a solid medium: possible role of siderophores. *Letters in Applied Microbiology* 33:21–25.

Miskin I, Rhodes G, Lawlor K, Saunders JR, Pickup RW (1998) Bacteria in post-glacial freshwater sediments. *Microbiology UK* 144:2427–2439.

Mitchelmore CL, Birmelin C, Livingstone DR, Chipman JK (1998) Detection of DNA strand breaks in isolated mussel (*Mytilus edulis* L.) digestive gland cells using the 'comet' assay. *Ecotoxicology and Environmental Safety* 41:51–58.

Mitchinson TJ, Sawin KE, Theriot JA (1994) Caged fluorescent probes for monitoring cytoskeleton dynamics. In: JE Celis (ed.) *Cell Biology: A Laboratory Handbook*. Academic Press, New York. pp. 65–74.

Miura M, Kato S, von Ludinghausen M (1998) Lymphatic drainage of the cerebrospinal fluid from monkey spinal meninges with special reference to the distribution of the epidural lymphatics. *Archives of Histology and Cytology* 61:277–286.

Miura Y, Ichikawa Y, Ishikawa T, Ogura M, de Fries R, Shimada H, Mitsuhashi M (1996) Fluorometric determination of total mRNA with oligo(dT) immobilized on microtiter plates. *Clinical Chemistry* 42:1758–1764.

Miyagawa A, Kobayashi M, Fujita Y, Hamdy O, Hirano K, Nakamura M, Miyake Y (2001) Surface ultrastructure of collagen fibrils and their association with proteoglycans in human cornea and sclera by atomic force microscopy and energy-filtering transmission electron microscopy. *Cornea* 20:651–656.

Miyakawa H, Levram V, Lasserross N, Ross WN (1992) Calcium transients evoked by climbing fiber and parallel fiber synaptic inputs in guinea-pig cerebellar Purkinje neurons. *Journal of Neurophysiology* 68:1178–1189.

Modest EJ, Sengupta SK (1973) Chemical aspects of the fluorescence analysis of chromosomes. In: A Thaer, M Sernetz, *Fluorescence Techniques in Cell Biology.* Springer-Verlag, Berlin. pp. 125–134.

Modha J, Kusel JR, Kennedy MW (1995) A role for second messengers in the control of activation-associated modification of the surface of *Trichinella spiralis* infective larvae. *Molecular and Biochemical Parasitology* 72:141–148.

Moffat FL, Han T, Li ZM, Peck MD, Jy WC, Ahn YS, Chu AJ, Bourguignon LYW (1996) Supplemental L-arginine HCl augments bacterial phagocytosis in human polymorphonuclear leukocytes. *Journal of Cellular Physiology* 168:26–33.

Mohammad T, Morrison H (1997) Simultaneous determination of methylene violet, halogenated methylene violet and their photoproducts in the presence of DNA by high-performance liquid chromatography using an internal surface reversed-phase column. *Journal of Chromatography* B 704:265–275.

Mohr L, Schauer JI, Boutin RH, Moradpour D, Wands JR (1999) Targeted gene transfer to hepatocellular carcinoma cells *in vitro* using a novel monoclonal antibody-based gene delivery system. *Hepatology* 29:82–89.

Moldovan NI, Moldovan L, Simionescu N (1994) Binding of vascular anticoagulant alpha-(annexin V) to the aortic intima of the hypercholesterolemic rabbit – an autoradiographic study. *Blood Coagulation and Fibrinolysis* 5:921–928.

Molero ML, Hazen MJ, Stockert JC (1985) Observations on the orcein fluorescence. *Acta Histochemica* 76:77–79.

Molho P, Verrier P, Stieltjes N, Schacher JM, Ounnoughene N, Vassilieff D, Menkes CJ, Sultan Y (1999) A retrospective study on chemical and radioactive synovectomy in severe haemophilia patients with recurrent haemarthrosis. *Haemophilia* 5:115–123.

Mollier G (1938) Eine Vierfachfarbung zur Darstellung glatter und quergestreifter Musskulature und ihrer Beziehung zum Bindegewebe. *Zeitschrift fur wissenschaftliche Mikroskopie* 55:472–473.

Molnar P, Nadler JV (1999) Mossy fiber-granule cell synapses in the normal and epileptic rat dentate gyrus studied with minimal laser photostimulation. *Journal of Neurophysiology* 82:1883–1894.

Mommers JM, Goossen JW, van de Kerkhof PCM, van Erp PEJ (2000) Novel functional multiparameter flow cytometric assay to characterize proliferation in skin. *Cytometry* 42:43–49.

Monroe CW, Frommer J (1967) Neutral red-fast green FCF, a single stain for mammalian tissues. *Stain Technology* 42:262–264.

Montosi G, Garuti C, Iannone A, Pietrangelo A (1998) Spatial and temporal dynamics of hepatic stellate cell activation during oxidant-stress-induced fibrogenesis. *American Journal of Pathology* 152:1319–1326.

Moore AV, Kirk SM, Callister SM, Mazurek GH, Schell RF (1999) Safe determination of susceptibility of *Mycobacterium tuberculosis* to antimycobacterial agents by flow cytometry. *Journal of Clinical Microbiology* 37:479–483.

Moorthy JN, Shevchenko T, Magon A, Bohne C (1998) Paper acidity estimation: application of pH-dependent fluorescence probes. *Journal of Photochemistry and Photobiology A: Chemistry* 113:189–195.

Morel F, Mercier S, Roux C, Elmrini T, Clavequin MC, Bresson J (1998) Interindividual variations in the disomy frequencies of human spermatozoa and their correlation with nuclear maturity as evaluated by aniline blue staining. *Fertility and Sterility* 69:1122–1127.

Mori M, Masuda N (1993) Proteins of the vitelline membrane of quail (*Coturnix coturnix japonica*) eggs. *Poultry Science* 72:1566–1572.

Morimoto T, Suehiro M, Kodama H, Amemori T, Ohba S, Nemoto H, Hasegawa A, Yoshida H (1999) A new method for the identification and enumeration of carp granulocytes. *Fish Pathology* 34:85–86.

Morrison H, Mohammad T, Kurukulasuriya R (1997) Photobiological properties of methylene violet. *Photochemistry and Photobiology* 66:245–252.

Mosiman VL, Patterson BK, Canterero L, Goolsby CL (1997) Reducing cellular autofluorescence in flow cytometry: an *in situ* method. *Cytometry* 30:151–156.

Mosmann T (1983) Rapid colorimetric assay for cellular growth and survival: application to proliferation and cytotoxicity assays. *Journal of Immunological Methods* 65:55–63.

Moukova N, Kuban V, Sommer L (1979) Purity control of chromeazurol S and eriochrome azurol B and their spectrophotometric determination in commercial samples after TLC separation on Silufol. *Chemicke Listy* 73:1106–1111.

Mould MJR, Robb J (1992) The *Colletotricum trifolii–Medicago sativa* interface, in culture – a cytological analysis. *Canadian Journal of Botany* 70:114–124.

Moura H, Schwartz DA, Bournay Llinares F, Sodre FC, Wallace S, Visvesvara GS (1997) A new and improved 'quick-hot gram-chromotrope' technique that differentially stains microsporidian spores in clinical samples, including paraffin-embedded tissue sections. *Archives of Pathology and Laboratory Medicine* 121:888–893.

Movat HZ (1955) Demonstration of all connective tissue elements in a single section. *Archives of Pathology* 60:289–295.

Mowry RW (1980) The Biological Stain Commission: its goals, its past, its present status. *Stain Technology* 55:1–7.

Mowry RW, Emmel VM (1977) The production of aldehyde fuchsin depends on the pararosaniline (CI No 42500) content of basic fuchsins which is sometimes negligible and is sometimes mislabelled. *Journal of Histochemistry and Cytochemistry* 25:239.

Mowry RW, Kasten F (1975) The importance of dye purification and standardization in biomedicine. *Stain Technology* 50:65–81.

Mowry RW, Longley JB, Emmel VM (1980) Only aldehyde fuchsin made from pararosanilin stains pancreatic B cell granules and elastic fibers in unoxidized microsections: problems caused by mislabeling of certain basic fuchsins. *Stain Technology* 55:91–103.

Moxey PC, Yeomans ND (1976) Identification of cell types in semithin epoxy sections of gastric fundic mucosa. *Journal of Histochemistry and Cytochemistry* 24:755–756.

Mozes S, Lenhardt L, Martinkova A (1998) A quantitative histochemical study of alkaline phosphatase activity in isolated rat duodenal epithelial cells. *Histochemical Journal* 30:583–589.

Mshana RN, Tadesse G, Abate G, Miorner H (1998) Use of 3-(4,5-dimethylthiazol-2-yl)-2,5-diphenyltetrazolium bromide for rapid detection of rifampin-resistant *Mycobacterium tuberculosis*. *Journal of Clinical Microbiology* 36:1214–1219.

Mudd KL, Hunt A, Matherly RC, Goldsmith LJ, Campbell FR, Nichols GR, Rink RD (2000) Analysis of pulmonary fat embolism in blunt force fatalities. *Journal of Trauma – Injury, Infection and Critical Care* 48:711–715.

Muller G, Hanschke M (1996) Quantitative and qualitative analyses of proteoglycans in cartilage extracts by precipitation with 1,9-dimethylmethylene blue. *Connective Tissue Research* 33:243–248.

Muller HAC (1912) Kernstudien an Pflanzen. *Archiv fur Zellforschung* 8:1–51.

Muller J, Volksch B, Fritsche W (1997) Influence of pathogenic and non-pathogenic bacteria on soybean suspension cells. *Journal of Phytopathology* 145:117–122.

Muller T (1998) Methylene blue supravital staining: an evaluation of its applicability to the mammalian brain and pineal gland. *Histology and Histopathology* 13:1019–1026.

Muller W, Firsching R (1991) Magnesium in cross bands of injured myocytes. *Zentralblatt fur allegemeine Pathologie* 137:130–132.

Muller-Walz R, Zimmermann HW (1987) Uber Romanowsky-Farbstoffe und den Romanowsky-Giemsa-Effekt. 4. Mitteilung: Bindung von Azur B an DNA. *Histochemistry* 87:157–172.

Mundee Y, Bigelow NC, Davis BH, Porter JB (2001) Flow cytometric method for simultaneous assay of foetal haemoglobin containing red cells, reticulocytes and foetal haemoglobin containing reticulocytes. *Clinical and Laboratory Haematology* 23:149–154.

Mundkur B (1964) Electron microscopic studies of frozen-dried yeast. 5. Localisation of protein-bound sulphydryl. *Experimental Cell Research* 34:155–181.

Mundkur B, Greenwood H (1968) Amido black 10B as a nucleolar stain for lymph nodes in Hodgkin's disease. *Acta Cytologica* 12:218–226.

Muniesa Perez M, Jofre J, Blanch AR (1996) Identification of *Vibrio proteolyticus* with a differential medium and a specific probe. *Applied and Environmental Microbiology* 62:2673–2675.

Munoz-Barroso I, Durell S, Sakaguchi K, Appella E, Blumenthal R (1998) Dilation of the human immunodeficiency virus-1 envelope glycoprotein fusion pore revealed by the inhibitory action of a synthetic peptide from gp41. *Journal of Cell Biology* 140:315–323.

Munro CH, White PC (1995) Evaluation of diazonium salts as visualization reagents for the thin-layer chromatographic characterization of amphetamines. *Science and Justice* 35:37–44.

Murakami S, Muneta T, Ezura Y, Furuya K, Yamamoto H (1997) Quantitative analysis of synovial fibrosis in the infrapatellar fat pad before and after anterior cruciate ligament reconstruction. *American Journal of Sports Medicine* 25:29–34.

Murakami T, Marakami T, Mahmut N, Hitomi S, Ohtsuka A (1997) Dark and light neurons in the human brain, with special reference to their reactions to Golgi's silver nitrate, luxol fast blue MBS and azocarmine G. *Archives of Histology and Cytology* 60:265–274.

Murgatroyd LB (1982) 2-Hydroxystilbamidine isethionate – a new fluorochrome for use in general pathology. 1. The selective staining of DNA, mucosubstances and elastic fibers. *Histochemistry* 74:107–114.

Murgatroyd LB, Horobin RW (1969) Specific staining of glycogen with hematoxylin and certain anthraquinone dyes. *Stain Technology* 44:59–62.

Murphy CJ, Campbell S, Araki Sasaki K, Marfurt C (1998) Effect of norepinephrine on proliferation, migration and adhesion of SV-40 transformed human corneal epithelial cells. *Cornea* 17:529–536.

Murray GI, Ewen SW (1992) A new fluorescence method for alkaline phosphatase histochemistry. *Journal of Histochemistry and Cytochemistry* 40:1971–1974.

Murray RGE, Doetsch RN, Robinow CF (1994) Determinative and cytological light microscopy. Ch. 2 In: P Gerhardt, RGE Murray, WA Wood, NR Krieg (eds) *Methods for*

General and Molecular Bacteriology. American Society for Microbiology, Washington. pp. 21–41.

Musso H, Seeger M, Zahorszky UI (1961) Constitution of resorcinol blue [German]. *Angewandte Chemie* 73:434–435.

Nachlas MM, Goldstein TP, Rosenblatt D, Kirsch M, Seligman AM (1959) Influence of chemical structure on the rate of azo coupling and its significance in histochemical methodology. *Journal of Histochemistry and Cytochemistry* 7:50–65.

Naff MB, Naff AS (1963) Thin layer chromatography on microscope slides. *Journal of Chemical Education* 40:534–535.

Naftalin RJ, Pedley KC (1999) Regional crypt function in rat large intestine in relation to fluid absorption and growth of the pericryptal sheath. *Journal of Physiology* 514:211–227.

Nagai H, Murakami Y, Morita Y, Yokoyama K, Tamiya E (2001) Development of a microchamber array for picoliter PCR. *Analytical Chemistry* 73:1043–1047.

Naganuma T, Takasugi H, Kimura H (1998) Abundance of thraustochytrids in coastal plankton. *Marine Ecology – Progress Series* 162:105–110.

Nagata M, Irvine JR (1997) Differential dispersal patterns of male and female masu salmon fry. *Journal of Fish Biology* 51:601–606.

Nakada S, Yamada M, Ito T, Fujimoto M (1980) Purification of methylthymol blue and methylxylenol blue by high-performance liquid chromatography. *Bulletin of the Chemical Society of Japan* 53:3365–3366.

Nakamura F, Akasaki T, Togo M, and Iwamoto K (1998) Determination of piperine in seasoned pork meat by high performance liquid chromatography (HPLC) [Japanese]. *Kanzei Chuo Bunseki Shoho* 37:1–6.

Nakamura Y, Simpo S, Lee M, Oikawa T, Yoshii T, Noda K, Kuwahara Y, Kawasaki K (2000) Histology and tetracycline labeling of a single section of alveolar bone of first molars in the rat. *Biotechnic and Histochemistry* 75:1–6.

Nakane H, Miller FJ, Faraci FM, Toyoda K, Heistad DD (2000) Gene transfer of endothelial nitric oxide synthase reduces angiotensin II-induced endothelial dysfunction. *Hypertension* 35:595–601.

Nakao M, Suita S, Taguchi T, Hirose R, Shima Y (2001) Fourteen-year experience of acetylcholinesterase staining for rectal mucosal biopsy in neonatal Hirschsprung's disease. *Journal of Pediatric Surgery* 36:1357–1363.

Nakos G, Gossrau R (1993) Light microscopic visualization of monoamine oxidase using a cerium method. *Acta Histochemica* 95:203–219.

Narisawa K, Kageyama K, Hashiba T (1996) Efficient root infection with single resting spores of *Plasmodiophora brassicae*. *Mycological Research* 100:855–858.

Nash N, Allan P, Bevenue A, Beckman H (1963) A technique for the recovery of compounds from thin-layer chromatograph strips for infrared analysis. *Journal of Chromatography* 12:421–423.

Nathanson SD, Avery M, Anaya P, Sarantou T, Hetzel FW (1997) Lymphatic diameters and radionuclide clearance in a murine melanoma model. *Archives of Surgery* 132:311–315.

Nauman RV, West PW, Tron F, Gaeke GC (1960) A spectrophotometric study of the Schiff reaction as applied to the quantitative determiniation of sulfur dioxide. *Analytical Chemistry* 32:1307–1311.

NCCLS (1975) Approved Standard: ASM-1. Standard test for labeling efficiency of fluorescein isothiocyanate (FITC). NCCLS (National Committee for Clinical Laboratory Standards), Villanova, PA.

NCCLS (1997) Quality assurance for immunocytochemistry. NCCLS (National Committee for Clinical Laboratory Standards) **17**, 1–45. Wayne, PA.

Nebe-von Caron G, Stephens PJ, Hewitt CJ, Powell JR, Badley RA (2000) Analysis of bacterial function by multi-colour fluorescence flow cytometry and single cell sorting. *Journal of Microbiological Methods* 42:97–114.

Nebel BR (1931) Lacmoid-martius yellow for staining pollen tubes in the style. *Stain Technology* 6:27–29.

Negishi K, Salas R, Laufer M (1997) Origins of horizontal cell spectral responses in the retina of marine teleosts (*Centropomus* and *Mugil* sp.). *Journal of Neuroscience Research* 47:68–76.

Nemcova I, Metal B, Podlaha J (1986) Dissociation constants of arsenazo III. *Talanta (Oxford)* 33:841–842.

Nettleton GS, Carpenter AM (1977) Studies of the mechanism of the periodic acid-Schiff histochemical reaction for glycogen using infrared spectroscopy and model chemical compounds. *Stain Technology* 52:63–77.

Nettleton GS, Johnson LR, Sehlinger TE (1986) Thin layer chromatography of commercial samples of amido black 10B. *Stain Technology* 61:329–336.

Newcomer EHA (1938) A procedure for growing, staining, and making permanent slides of pollen tubes. *Stain Technology* 13:89–91.

Newman GR, Jasani B (1998) Silver development in microscopy and bioanalysis: a new versatile formulation for modern needs. *Histochemical Journal* 30:635–645.

Newton WCF (1925) Chromosome studies in *Tulipa* and some related genera. *Journal of the Linnaean Society, Botany* 47:339–354.

Ni YJ, Iwatani Y, Morimoto K, Kawai A (1996) Studies on unusual cytoplasmic structures which contain rabies virus envelope proteins. *Journal of General Virology* 77:2137–2147.

Nichols DH, Lovas S, Adrian TE, Miller CA, Murphy RF (1998) Peptides bind to eosinophils in the rat stomach. *Anatomical Record* 250:172–181.

Niehaus GD, Mehendale SR (1998) Quantifying rat pulmonary intravascular mono-nuclear phagocytes. *Anatomical Record* 252:626–636.

Nielson AJ, Griffith WP (1979) Tissue fixation by osmium tetroxide. A possible role for proteins. *Journal of Histochemistry and Cytochemistry* 27:997–999.

Niemann CU, Henthorn TK, Krejcie TC, Shanks CA, Enders-Klein C, Avram MJ (2000) Indocyanine green kinetics characterize blood volume and flow distribution and their alteration by propranolol. *Clinical Pharmacology and Therapeutics* 67:342–350.

Nishikawa S, Sasaki F (1996) Internalization of styryl dye FM1–43 in the hair cells of lateral line organs in *Xenopus* larvae. *Journal of Histochemistry and Cytochemistry* 44:733–741.

Nishimura A, Tsuhako M (2000) Single strand conformation polymorphism analysis of Ras oncogene by capillary electrophoresis with laser-induced fluorescence detector. *Chemical and Pharmacological Bulletin* 48:774–778.

Nishimura H, Nelson GH, Rosenblum WI (1989) Chicago sky blue and a helium-neon laser abolish endothelium dependent relaxation *in vivo* in the microcirculation. *Microcirculation, Endothelium and Lymphatics* 5:435–440.

Nishimura S, Ishiko O, Honda K, Hyun Y, Ogita S (2000) New cervicoscopy with tolu-idine blue staining for superficial cervical invasion by endometrial carcinoma. *Gynecologic and Obstetric Investigation* 50:133–135.

Niu WD, Eto JN, Kimura Y, Takeda FH, Matsumoto K (1998) A study on microleakage after resin filling of class V cavities prepared by Er:YAG laser. *Journal of Clinical Laser Medicine and Surgery* 16:227–231.

Niv H, Gutman O, Henis YI, Kloog Y (1999) Membrane interactions of a constitutively active GFP-Ki-ras 4B and their role in signalling. *Journal of Biological Chemistry* 274:1606–1613.

Nocht B (1898) Zur Farbung der Malariaparasiten. *Zentralblatt fur Bakteriologie* 24:839–843.

Nohammer G (1983) A modification of the amidoblack-TCA-staining for quantitative microspectrophotometrical determination of proteins in tissue sections. *Histochemistry* 80:395–400.

Nohammer G (1990) Quantification of the histochemical staining for carbonyls and DNA using 3-hydroxy-2-naphthoic acid hydrazide and fast blue B. *Histochemistry* 94:485–488.

Nohammer G, Desoye G (1997) Mercurochrom can be used for the histochemical demonstration and microphotometric quantitation of both protein thiols and protein (mixed) disulfides. *Histochemistry and Cell Biology* 107:383–390.

Nordstrom T, Knekt M, Nordstrom E, Lindqvist C (1999) A microplate-based fluorometric assay for monitoring human cancer cell attachment to cortical bone. *Analytical Biochemistry* 267:37–45.

Novello AF, Brauer MM (1986) Mercurochrome: a test for chromatin organization at the electron microscope level. *Cellular and Molecular Biology* 32:493–497.

Noyan S, Kahveci Z, Cavusoglu I, Minbay FZ, Sunay FB, Sirmali SA (2000) Effects of microwave irradiation and chemical fixation on the localization of perisinusoidal cells in the liver by gold impregnation. *Journal of Microscopy* 197:101–106.

Nuessly GS, Nagata RT, Skiles ES, Christenson JR, Elliott C (1995) Techniques for differentially staining *Liriomyza trifolii* (Diptera: Agromyzidae) eggs and stipples within Cos lettuce leaves. *Florida Entomologist* 78:258–264.

Nursten HE, Williams KE (1972) The ready desulphonation of two simple azo dyes acids. *Journal of the Society of Dyers and Colourists* 88:148–150.

O'Brien J, Wilson I, Orton T, Pognan F (2000) Investigation of the alamar blue (resazurin) fluorescent dye for the assessment of mammalian cell cytotoxicity. *European Journal of Biochemistry* 267:5421–5426.

O'Brien TP, McCulley ME (1981) *The Study of Plant Structure: Principles and selected methods.* Termarcarphi, Melbourne.

O'Connor JE, Kimler BF, Morgan MC, Tempas KJ (1988) A flow cytometric assay for intracelllular nonprotein thiols using mercury orange. *Cytometry* 9:529–532.

O'Malley JP, Waran MT, Balice-Gordon RJ (1999) *In vivo* observations of terminal Schwann cells at normal, denervated, and reinnervated mouse neuromuscular junctions. *Journal of Neurobiology* 38:270–286.

O'Sullivan NA, Fallon R, Carroll C, Smith T, Maher M (2000) Detection and differentiation of *Campylobacter jejuni* and *Campylobacter coli* in broiler chicken samples using a PCR/DNA probe membrane based colorimetric detection assay. *Molecular and Cellular Probes* 14:7–16.

Oana H, Ueda M, Washizu M (1999) Visualization of a specific sequence on a single large DNA molecule using fluorescence microscopy based on a new DNA-stretching method. *Biochemical and Biophysical Research Communications* 265:140–143.

Ogawa H, Baba Y, Oka K (1999) Dendritic Ca^{2+} transient increase evoked by wind stimulus in the cricket giant interneuron. *Neuroscience Letters* 275:61–64.

Ogawa J, Tsurumi T, Yamada S, Koide S, Shohtsu A (1994) Blood vessel invasion and expression of sialyl Lewis and proliferating cell nuclear antigen in stage 1 non-small cell lung cancer. *Cancer* 73:1177–1183.

Ogawa K, Barka T (1992) *Electron Microscopic Cytochemistry and Immunocytochemistry in Biomedicine.* CRC Press, Boca Raton, Florida.

Ohm TG, Jung E, Schnecko A (1992) A subpopulation of hippocampal glial cells specific for the zinc containing mossy fiber zone in man. *Neuroscience Letters* 145:181–184.

Ohma N, Takagi Y, Takano Y (2000) Distribution of non-collagenous dentin matrix proteins and proteoglycans, and their relation to calcium accumulation in bisphosphonate-affected rat incisors. *European Journal of Oral Science* 108:222–232.

Ohmura N, Lackie SJ, Saiki H (2001) An immunoassay for small analytes with theoretical detection limits. *Analytical Chemistry* 73:3392–3399.

Ohno I, Lea RG, Flanders KC, Clark DA, Banwatt D, Dolovich J, Denburg J, Harley CB, Gauldie J, Jordana M (1992) Eosinophils in chronically inflamed human upper airway tissues express transforming growth factor beta-1 gene (TGF Beta 1). *Journal of Clinical Investigation* 89:1662–1668.

Ohta Y, Iwasaki Y, Abe H, Kato M (1991) Activation of spinal neurons by afferent fibers in the ventral roots of rats. *Neuroscience Letters* 130:137–139.

Okamura N, Tajima Y, Onoe S, Sugita Y (1991) Purification of bicarbonate-sensitive sperm adenylcyclase by 4-acetamido-4'-isothiocyanostilbene-2,2'-disulfonic acid-affinity chromatography. *Journal of Biological Chemistry* 266:17754–17759.

Okpokwasili GC, Odokuma LO (1996) Tolerance of *Nitrobacter* to toxicity of hydrocarbon fuels. *Journal of Petroleum Science and Engineering* 16:89–93.

Oktay K, Nugent D, Newton H, Salha O, Chatterjee P, Gosden RG (1997) Isolation and characterization of primordial follicles from fresh and cryopreserved human ovarian tissue. *Fertility and Sterility* 67:481–486.

Oldmixon EH (1988) Mallory's phloxine B-methylene blue-azure II stain emphasizes elastin and collagen bundles in epoxy embedded lung. *Stain Technology* 63:165–170.

Olerud S, Lorenzi GL (1970) Triple fluorochrome labeling in bone formation and bone resorption. *Journal of Bone Joint Surgery* 52A:274–278.

Olivarius FD, Hansen AR, Karlsmark T, Wulf FC (1996) Water protective effect of barrier creams and moisturizing creams: a new *in vivo* test method. *Contact Dermatitis* 35:219–225.

Oliveira CC, Sartini RP, Reis BF, Zagatto EAG (1996) Multicommutation in flow analysis. 4. Computer-assisted splitting for spectrophotometric determination of copper and zinc in plants. *Analytica Chimica Acta* 332:173–178.

Ollett WS (1951) Further observations on the Gram-Twort stain. *Journal of Pathology and Bacteriology* 63:166.

Olsson Y, Arvidson B, Hartman M, Pettersson A, Tengvar C (1983) Horseradish peroxidase histochemistry. A comparison between various methods used for identifying neurons labeled by retrograde axonal transport. *Journal of Neuroscience Methods* 7:49–59.

Omar AI, Senatorov VV, Hu B (2000) Ethidium bromide staining reveals rapid cell dispersion in the rat dentate gyrus following ouabain-induced injury. *Neuroscience* 95:73–80.

Omar MS, Raoof AMS (1996) *Onchocerca fasciata*: histochemical demonstration of succinate and NADH dehydrogenase. *Journal of Helminthology* 70:47–51.

Opremcak EM, Bruce RA (1999) Surgical decompression of branch retinal vein occlusion via arteriovenous crossing sheathotomy – A prospective review of 15 cases. *Retina* 19:1–5.

Ormerod M (1998) Monitoring electroporation by flow cytometry. In: JE Celis (ed.) *Cell Biology: A Laboratory Handbook*. Academic Press, San Diego. pp. 88–91.

Ormerod MG (1994) *Flow Cytometry. A Practical Approach*. 2nd Edn. IRL Press, Oxford.

Ornstein L, Mautner W, Davis BJ, Tamura R (1957) New horizons in fluorescence microscopy. *Journal of the Mount Sinai Hospital* 24:1066–1078.

Ortiz PP, Sarrat R, Daret D, Whyte J, Torres A, Lamaziere JM (2000) Elastin variations implicating in vascular smooth muscle cells phenotype in human tortuous arteries. *Histology and Histopathology* 15:95–100.

Osaka K, Tyurina YY, Dubey RK, Tyurin VA, Ritov VB, Quinn PJ, Branch RA, Kagan VE (1997) Amphotericin B as an intracellular antioxidant. Protection against 2,2'-azobis(2,4-dimethylvaleronitrile)induced peroxidation of membrane phospholipids in rat aortic smooth muscle cells. *Biochemical Pharmacology* 54:937–945.

Ota D, Marchesan M, Ferrero EA (1996) Sperm release behaviour and fertilization in the grass goby. *Journal of Fish Biology* 49:246–256.

Otto C, Bauer K (1996) Dipeptide uptake: a novel marker for testicular and ovarian macrophages. *Anatomical Record* 245:662–667.

Oud PS, Henderik JBJ, Huysmans ACLM, Pahlplatz MMM, Hermkens H, Tas J, James J, Vooijs GP (1984) The use of light green and orange II as quantitative protein stains, and their combination with the Feulgen method for the simultaneous determination of protein and DNA. *Histochemistry* 80:49–57.

Paddock SW (1982) Incident light microscopy of normal and transformed cultured fibroblasts stained with coomassie blue R250. *Journal of Microscopy* 128:203–205.

Pagano RE, Martin OC (1994) Use of fluorescent analogs of ceramide to study the Golgi apparatus of animal cells. In: JE Celis (ed.) *Cell Biology: A Laboratory Handbook.* Academic Press, New York. pp. 387–393.

Paim APS, Reis BF (2000) An automatic spectrophotometric titration procedure for ascorbic acid determination in fruit juices and soft drinks based on volumetric fraction variation. *Analytical Sciences* 16:487–491.

Pak CC, Puri A, Blumenthal R (1997) Conformational changes and fusion activity of vesicular stomatitis virus glycoprotein:[^{125}I]iodonaphthyl azide photolabeling studies in biological membranes. *Biochemistry* 36:8890–8896.

Palomo MJ, Izquierdo D, Mogas T, Paramio MT (1999) Effects of semen preparation on IVF of prepubertal goat oocytes. *Theriogenology* 51:927–940.

Papanicolaou GN (1941) Some improved methods for staining vaginal smears. *Journal of Laboratory and Clinical Medicine* **26:**1200–1205.

Papkovsky DB (1991) Luminescent porphyrin probes. *Applied Fluorescence Technology* 3:16–24.

Pappenheim A (1912) Zur Blutzellfarbung im klinischen Bluttrockenpraparat und zur histologischen Schnittpraparatfarbung der haematopoetischen Gewebe nach meinen Methoden. *Folia Haematologica* 13:339–344.

Parrington LJ, Sharpe AN (1998) Factors affecting aerobic colony counts for bottled water. *Food Microbiology* 15:79–90.

Parry G, Blenis J, Hawkes SP (1982) Detection of transformed cells using a fluorescent probe: the molecular basis for the differential reaction of fluorescamine with normal and transformed cells. *Cytometry* 3:97–103.

Partearroyo MA, Cabezon E, Nieva J-L, Alonso A, Goni FM (1994) Real-time measurements of chemically induced membrane fusion in cell monolayers, using a resonance energy transfer method. *Biochemica et Biophysica Acta* 1189:175–180.

Patel S, Farrell J, Blades KJ, Grierson DJ (1998) The value of a phenol red impregnated thread for differentiating between the aqueous and non-aqueous deficient dry eye. *Ophthalmic and Physiological Optics* 18:471–476.

Paterakis GS (1996) Reticulocyte counting in thalassaemia. *Clinical and Laboratory Haematology* 18:17–28.

Patro IK, Patro N, Singhal P, Bhardwaj SK (1996) Simultaneous demonstration of microscopic anatomy of brainstem, Nissl substance, lipofucsin and cytological changes using a combination method. *Acta Histochemica et Cytochemica* 29:237–241.

Pattanavibool R, Klimaszewska K, Von Aderkas P (1998) Interspecies protoplast fusion in *Larix*: comparision of electric and chemical methods. *In Vitro Cellular and Developmental Biology – Plant* 34:212–217.

Paul AL, Ferl RL (1993) Osmium tetroxide footprinting of a scaffold attachment region in the maize Adh1 promoter. *Plant Molecular Biology* 22:1145–1151.

Paul RN (1980) The use of thionin and acridine orange in staining semithin sections of plant material embedded in epoxy resin. *Stain Technology* 55:195–196.

Payan C, Coyyin J, Lemarie C, Ramont C (2001) Inactivation of hepatitis B virus in plasma by hospital in-use chemical disinfectants assessed by a modified HepG2 cell culture. *Journal of Hospital Infection* 47:282–287.

Payne JN, Lawes INC, Proctor GB, Horobin RW (1983) Variation between different samples of SITS with respect to axonal transport and toxicity. *Neuroscience Letters* 42:229–234.

Payton BN (1970) Histological staining properties of procion yellow M-4RS. *Journal of Cell Biology* 45:659–662.

Pearce RB (1984) Staining fungal hyphae in wood. *Transactions of the British Mycological Society* 82:564–567.

Pearce RKB, Owen A, Daniel S, Jenner P, Marsden CD (1997) Alternations in the distribution of glutathione in the substantia nigra in Parkinson's disease. *Journal of Neural Transmission* 104:661–677.

Pearse AGE (1957) Solochrome dyes in histochemistry with particular reference to nuclear staining. *Acta Histochemica* 4:95–101.

Pearse AGE (1972) *Histochemistry: Theoretical and Applied.* 3rd Edn. v. 2. Churchill Livingstone, Edinburgh.

Pearse AGE (1980) *Histochemistry: Theoretical and Applied.* 4th Edn. v. 1. Churchill-Livingstone, Edinburgh.

Pearse AGE (1985) *Histochemistry: Theoretical and Applied.* 4th Edn. v. 2. Churchill-Livingstone, Edinburgh.

Pearse AGE, Stoward PJ (1980,1985,1991) *Histochemistry, Theoretical and Applied,* 4th Edn. Vol. 1. Preparative and Optical Technology. Vol. 2. Analytical Technique. Vol. 3. Enzyme Histochemistry. Churchill-Livingstone, Edinburgh.

Pecora JD, Brugnera-Junior A, Cussioli AL, Zanin F, Silva R (2000) Evaluation of dentin root canal permeability after instrumentation and Er:YAG laser application. *Lasers in Surgery and Medicine* 26:277–281.

Pecorella I, McCartney ACE, Lucas S, Michaels S, Ciardi A, DiTondo U, Garner A (1995) Histological study of oxalosis in the eye and adnexa of AIDS patients. *Histopathology* 27:431–438.

Peeters F, Wuest A, Piepke G, Imboden GM (1996) Horizontal mixing in lakes. *Journal of Geophysical Research – Oceans* 101:18361–18375.

Pelander A, Ojanpera I, Sivonen K, Himberg K, Waris M, Niinivaara K, Vuori E (1996) Screening for cyanobacterial toxins in bloom and strain samples by thin-layer chromatography. *Water Research* 30:1464–1470.

Peng Q, Farrants GW, Madslien K, Bommer JC, Moan J, Danielsen HE, Nesland JM (1991) Subcellular localisation, redistribution and photobleaching of sulphonated aluminium phthalocyanins in a human melanoma cell line. *International Journal of Cancer* 49:290–295.

Penner MH (1968) Thin-layer chromatography of certified coal tar color additives. *Journal of Pharmaceutical Sciences* 57:2132–2135.

Penney DP, Powers JM (1995) Report from the Biological Stain Commission Laboratory. Light green SF yellowish. *Biotechnic and Histochemistry* 70:217.

Perera PAC (1991) Periodic acid-basic fuchsin technique for the demonstration of carbohydrates and mucosubstances in paraffin sections – an alternative to the PAS technique. *Journal of Histotechnology* 14:155–158.

Perlnemolnar I, Szakacsnepinter M, Kovago A, Petroczy J, Kralovanszky PU, Matyas J (1985) Investigation of the dye-binding stoichiometry of food and feed-stuff proteins based on reactions with orange G, acid orange 12 and amido black 10B [Hungarian]. *Magyar Kemiai Folyoirat* 91:157–166.

Perry SW, Epstein LG, Gelbard HA (1997a) *In situ* trypan blue staining of monolayer cultures for permanent fixation and mounting. *BioTechniques* **22**:1020.

Perry SW, Epstein LG, Gelbard HA (1997b) Simultaneous *in situ* detection of apoptosis and necrosis in monolayer cultures by TUNEL and Trypan Blue staining. *BioTechniques* 22:1102–1106.

Peters AT, Freeman HS (1991) *Color Chemistry: The Design and Synthesis of Organic Dyes and Pigments.* Elsevier, New York.

Petersen H (1941) Die Probleme der Zellenlehre und die ihrer Geschichte. *Anatomischer Anzeiger* 90:1–42.

Peterson AR, Conn HJ, Melin C (1933a) Methods for the standardization of biological stains: Part I. General considerations. *Stain Technology* 8:87–94.

Peterson AR, Conn HJ, Melin C (1933b) Methods for the standardization of biological stains: Part II. The fluorane derivatives. *Stain Technology* 8:95–99.

Peterson AR, Conn HJ, Melin C (1933c) Methods for the standardization of biological stains: Part III. Certain nitro and azo dyes. *Stain Technology* 8:121–130.

Peterson AR, Conn HJ, Melin C (1934a) Methods for the standardization of biological stains: Part IV. The triphenylmethane derivatives. *Stain Technology* 9:41–48.

Peterson AR, Conn HJ, Melin C (1934b) Methods for the standardization of biological stains: Part V. Miscellaneous dyes. *Stain Technology* 9:147–155.

Peterson CA, Emanuel ME, Humphreys GB (1981) Pathway of movement of apoplastic fluorescent dye tracers through the endodermis at the site of secondary root formation in corn (*Zea mays*) and broad bean (*Vicia faba*). *Canadian Journal of Botany* 59:618–625.

Peterson JM, Perdomo JA, Burnis JS (1995) Influence of kernal position, mechanical damage and controlled deterioration on estimates of hybrid maize seed quality. *Seed Science and Technology* 23:647–657.

Petry HM, Bassi CJ (1991) Green-sensitive cone photoreceptors are selectively labelled by procion yellow dye in goldfish retina. *Visual Neuroscience* 6:15–18.

Pfuller U, Franz H, Preiss A (1977) Sudan black B: chemical structure and histochemistry of the blue main components. *Histochemistry* 54:237–250.

Philip KA, Dascombe MJ, Fraser PA, Pentreath VW (1994) Blood–brain barrier damage in experimental African Trypanosomiasis. *Annals of Tropical Medicine and Parasitology* 88:607–616.

Phimphivong S, Saavedra SS (1998) Terbium chelate membrane label for time-resolved, total internal reflection fluorescence microscopy of substrate-adherent cells. *Bioconjugate Chemistry* 9:350–357.

Pianese G (1896) Beitrag zur Histologie und Aetiologie des Carcinoms. *Beitrage zur pathologischen Anatomie und zur allgemeinen Pathologie* (Suppl. 1). Cited in Lillie (1977).

Picker O, Wietasch G, Scheeren TWL, Arndt JO (2001) Determination of total blood volume by indicator dilution: a comparison of mean transit time and mass conservation principle. *Intensive Care Medicine* 27:767–774.

Piekut DT, Casey SM (1983) Penetration of immunoreagents in vibratome-sectioned brain: a light and electron microscope study. *Journal of Histochemistry and Cytochemistry* 31:669–674.

Pierard GE (1992) Microscopic evaluation of the dansyl chloride test. *Dermatology* 185:37–40.

Pierard-Franchimont C, Arrese JE, Durupt G, Ries G, Cauwenbergh G, Pierard GE (1998) Correlation between *Malassezia* spp load and dandruff severity. *Journal de Mycologie Medicale* 8:83–86.

Pihakaski-Maunsbach K, Walles B (1990) Seasonal variation of storage substances in stem-cells of *Diapensia lapponica* – a light microscopic study. *Nordic Journal of Botany* 10:493–500.

Pilloni L, Lecca S, Van Eyken P, Flore C, Demelia L, Pilleri G, Nurchi AM, Farci AMG, Ambu R, Callea F, Faa G (1998) Value of histochemical stains for copper in the diagnosis of Wilson's disease. *Histopathology* 33:28–33.

Pinna-Senn E, Lisanti JA, Ortiz MI, Dalmasso G, Bella JL, Gosalvez J, Stockert JC (2000) Specific heterochromatic banding of metaphase chromosomes using nuclear yellow. *Biotechnic and Histochemistry* 75:132–140.

Pischinger A (1926) Die Lage des isoelektrischen Punktes histologischer Elemente als Ursache ihrer verschiedenen Farbbarkeit. *Zeitschrift fur Zellforschung* 3:169–197.

Pitman KT, Johnson JT, Edington H, Barnes L, Day R, Wagner RL, Myers EN (1998) Lymphatic mapping with isosulfan blue dye in squamous cell carcinoma of the head and neck. *Archives of Otolaryngology – Head and Neck Surgery* 124:790–793.

Plasek J, Sigler K (1996) Slow fluorescent indicators of membrane potential: a survey of different approaches to probe response analysis. *Journal of Photochemistry and Photobiology B – Biology* 33:101–124.

Platt B, Fiddler G, Riedel G, Henderson Z (2001) Aluminium toxicity in the rat brain: histochemical and immunocytochemical evidence. *Brain Research Bulletin* 55:257–267.

Plieth C, Sattelmacher B, Hansen UP (1997) Cytoplasmic Ca^{2+}-H^+-exchange buffers in green algae. *Protoplasma* 198:107–124.

Pohl R, Kramer PA, Thrall RS (1998) Confocal laser scanning fluorescence microscopy of intact unfixed rat lungs. *International Journal of Pharmaceutics* 168:69–77.

Polak JM, Van Noorden S (1997) *Introduction to Immunocytochemistry* (Royal Microscopical Society Microscopy Handbooks, 37). 2nd Edn. BIOS Scientific Publications, Oxford.

Pond AL, Coyne CP, Chambers HW, Chambers JE (1996) Identification and isolation of two rat serum proteins with A esterase activity toward paraoxon and chlorpyrifos-oxon. *Biochemical Pharmacology* 52:363–369.

Poole WH (1875) A double staining with haematoxylin and aniline. *Quarterly Journal of Microscopical Science* 23:375–377.

Poot M (1998) Staining of mitochondria. In: JE Celis (ed.) *Cell Biology: A Laboratory Handbook.* Academic Press, San Diego. pp. 513–517.

Popescu A, Doyle RJ (1996) The Gram stain after more than a century. *Biotechnic and Histochemistry* 71:145–151.

Popper H (1940) Histological demonstration of vitamin A in the human liver by means of fluorescence microscopy. *Proceedings of the Society for Experimental Biology and Medicine* 43:234–236.

Porro TJ, Morse HT (1965) Fluorescence and absorption spectra of biological dyes (II). *Stain Technology* 40:173–176.

Porro TJ, Dadik SP, Green M, Morse HT (1963) Fluorescence and absorption spectra of biological dyes. *Stain Technology* 38:37–48.

Postlethwait EM, Joad JP, Hyde DM, Schelegle ES, Bric JM, Weir AJ, Putney LF, Wong VJ, Velsor LW, Plopper CG (2000) Three-dimensional mapping of ozone-induced acute cytotoxicity in tracheobronchial airways of isolated perfused rat lung. *American Journal of Respiratory, Cell and Molecular Biology* 22:191–199.

Potvin C (1979) A simple, modified methyl green-pyronin Y stain for DNA and RNA in formalin-fixed tissues. *Laboratory Medicine* 10:772–774.

Powers MM, Clark G, Darrow M, Emmel VM (1960) Darrow red, a new basic dye. *Stain Technology* 35:19–21.

Prat O, Lopez E, Mathis G (1991) Europium(III) cryptate: a fluorescent label for the detection of DNA hybrids on solid support. *Analytical Biochemistry* 195:283–289.

Prenna G, Piva N, Zanotti L (1962) Reazioni di Feulgen fluorescenti e loro possibilita citofluorometriche quantitative. 1. Studio isochimico di alcuni reagenti tipo Schiff fluorescenti nella reazione di Feulgen. *Rivista di Istochimica Normale e Pathologica* 8:427–446.

Prento P (1993) Van Gieson's picrofuchsin. The staining mechanisms for collagen and cytoplasm and an examination of the dye diffusion rate model of differential staining. *Histochemistry* 99:163–174.

Prento P (2001) A contribution to the theory of biological staining based on the principles for structural organization of biological macromolecules. *Biotechnic and Histochemistry* 76:137–161.

Presnell JK, Schreibman MP (1997) *Humason's Animal Tissue Techniques.* 5th Edn. Johns Hopkins University Press, Baltimore.

Price EAC, Hutchings MJ, Marshall C (1996) Causes and consequences of sectoriality in the clonal herb *Glechoma hederacea. Vegetatio* 127:41–54.

Prieto M, Chauvet N, Alonso G (2000) Tanycytes transplanted into the adult rat spinal cord support the regeneration of lesioned axons. *Experimental Neurology* 161:27–37.

Pringle JR, Preston RA, Adams AEM, Stearns T, Drubin DG, Haarer BK, Jones EW (1989) Fluorescent microscopy methods for yeast. *Methods in Cell Biology* 32:357–435.

Proctor GB, Horobin RW (1983) The aging of Gomori's aldehyde-fuchsin: the nature of the chemical changes and the chemical structures of the coloured components. *Histochemistry* 77:255–267.

Proctor GB, Horobin RW (1985a) A widely applicable analytical system for biological stains: reverse-phase thin layer chromatography. *Stain Technology* 60:1–6.

Proctor GB, Horobin RW (1985b) Purification of Oil red O using preparative paper chromatography. *Stain Technology* 60:247–248.

Proctor GB, Horobin RW (1987) Basic fuchsin homologues. Are pure dyes really necessary? *Medical Laboratory Sciences* 44:398–400.

Proctor GB, Horobin RW (1988) Chemical structures and staining mechanisms of Weigert's resorcin-fuchsin and related elastic fiber stains. *Stain Technology* 63:101–111.

Proescher F (1933) Pinacyanol as a histological stain. *Proceedings of the Society for Experimental Biology and Medicine* 31:79–81.

Prudencio C, Sansonetty F, Sousa MJ, Corte-Real M, Leao C (2000) Rapid detection of efflux pumps and their relation with drug resistance in yeast cells. *Cytometry* 39:26–35.

Puchtler H, Sweat F (1962) Amido black as a stain for hemoglobin. *Archives of Pathology* 73:245–249.

Puchtler H, Sweat F (1964) Histochemical specificity of staining methods for connective tissue fibers. Resorcin-fuchsin and van Gieson's picro-fuchsin. *Histochemie* 4:24–34.

Puchtler H, Sweat F, Levine M (1962) On the binding of Congo red by amyloid. *Journal of Histochemistry and Cytochemistry* 10:355–364.

Puchtler H, Rosenthal SI, Sweat F (1964) Revision of the amidoblack stain for hemoglobin. *Archives of Pathology* 78:76–78.

Puchtler H, Waldrop FS, Valentine LS (1973) Polarization microscopic studies of connective tissue stained with picro-Sirius Red F3BA. *Beitrag fur Pathologie* 150:174–187.

Puchtler H, Waldrop FS, Meloan SN (1980) On the mechanism of Mallory's phosphotungstic acid-hematoxylin stain. *Journal of Microscopy* 119:383–390.

Pump B, Hirnle P (1996) Preoperative lymph node staining with liposomes containing patent blue violet. A clinical case report. *Journal of Pharmacy and Pharmacology* 48:699–701.

Purushothaman V, Premkumar B, Venkatesan RA (1998) Resistogram typing of *Salmonella* serotypes of avian origin. *Indian Journal of Animal Sciences* 68:1119–1120.

Putt M (1991) Development and evaluation of tracer particles for use in microzooplankton herbivory studies. *Marine Ecology – Progress Series* 77:27–37.

Pyapali GK, Turner DA, Madison RD (1992) Anatomical and physiological localization of prelabeled grafts in rat hippocampus. *Experimental Neurology* 116:133–144.

Pyle JL, Kavalali ET, Piedras-Renteria ES, Tsien RW (2000) Rapid reuse of readily releasable pool vesicles at hippocampal synapses. *Neuron* 28:221–231.

Quan S, Yamano S, Nakagawa K, Irahara M, Kamada M, Aono T (2001) Penetrating capacity of human spermatozoa cool preserved in electrolyte-free solution. *Journal of Reproductive Medicine* 46:957–961.

Quekett J (1848) *A Practical Treatise on the Use of the Microscope.* Baillière, London.

Quintero-Hunter I, Grier H, Muscato M (1991) Enhancement of histological detail using metanil yellow as counterstain in periodic acid Schiff hematoxylin staining of glycol methacrylate tissue sections. *Biotechnic and Histochemistry* 66:169–172.

Raban P (1963) Adsorption chromatography of direct azo dyes. *Nature* 199:596–597.

Racoosin EL, Swanson JA (1994) Labeling of endocytic vesicles using fluorescent probes for fluid-phase endocytosis. In: JE Celis (ed.) *Cell Biology: A Laboratory Handbook.* Academic Press, New York. pp. 375–380.

Radosevic K, Garritsen HSP, VanGraft M, DeGrooth BG, Greve J (1990) A simple and sensitive flow cytometric assay for the determination of the cytotoxic activity of human natural killer cells. *Journal of Immunological Methods* 135:81–89.

Raffaelli A, Pucci S, Desideri I, Bellina CR, Bianchi R, Salvadori P (1999) Investigation on the iodination reaction of methylene blue by liquid chromatography-mass spectrometry with ionspray ionisation. *Journal of Chromatography* A 854:57–67.

Ragsdale GK, Phelps J, Luby-Phelps K (1997) Viscoelastic response of fibroblasts to tension transmitted through adherens junctions. *Biophysical Journal* 73:2798–2808.

Rahn BA, Perren SM (1970) Calcein blue as a fluorescent label in bone. *Experientia* 26:519–520.

Rahn BA, Perren SM (1971) Xylenol orange, a fluorochrome useful in polychrome sequential labelling of calcifying tissues. *Stain Technology* 46:125–129.

Ramirez-Zacarias JL, Castro-Munozledo F, Kuri-Harcuch W (1992) Quantitation of adipose conversion and triglycerides by staining intracytoplasmic lipids with oil red O. *Histochemistry* 97:493–497.

Ramjiawan B, Maiti P, Aftanas A, Kaplan H, Fast D, Mantsch HH, Jackson M (2000) Noninvasive localization of tumors by immunofluorescence imaging using a single chain Fv fragment of a human monoclonal antibody with broad cancer specificity. *Cancer* 89:1134–1144.

Ramon y Cajal S (1891) Sur la structure de l'ecorce cerebrale de quelques mammiferes. *Cellule* 7:125–176.

Ramsan M, Montresor A, Foum A, Ameri H, DiMatteo L, Albonico M, Savioli L (1999) Independent evaluation of the nigrosin-eosin modification of the Kato-Katz technique. *Tropical Medicine and International Health* 4:46–49.

Ranvier L (1875) *Traite technique d'histologie.* Savy, Paris.

Ranvier LA (1874) Des applications de la purpurine a l'histologie. *Archives de Physiologie* 1:761–773.

Rao NV, Narayama KL, Jairaj MA, Chowdry TV (1985) Thin layer chromatography of azine and oxazine dyes. *Journal of the Indian Chemical Society* 62:173–174.

Rapi S, Ermini A, Bartolini L, Caldini A, Del Genovese A, Miele AR, Buggiani A, Fanelli A (1998) Reticulocytes and reticulated platelets: simultaneous measurement in whole blood by flow cytometry. *Clinical Chemistry and Laboratory Medicine* 36:211–224.

Raser KJ, Posner A, Wang KKW (1995) Casein zymography – a method to study mu-calpain, M-calpain and their inhibitory agents. *Archives of Biochemistry and Biophysics* 319:211–216.

Rashid F, Horobin RW (1990) Interaction of molecular probes with living cells and tissues. 2. A structure-activity analysis of mitochondrial staining by cationic probes, and a discussion of the synergistic nature of image-based and biochemical approaches. *Histochemistry* 94:303–308.

Rasmussen ES (1999) Use of fluorescent redox indicators to evaluate cell proliferation and viability. *In Vitro and Molecular Toxicology* 12:47–58.

Raspail FV (1825) Développement de la fecule dans les organes de la fructification des céréales, et analyse microscopique de la fecule, suivie d'éxperiences propres a en éxpliquer la conversion en gomme. *Annales des Sciences Naturelles* 6:224–239 and 384–427.

Raspail FV (1830) *Essai de chimie microscopique appliquée à la physiologie, ou l'art de transporter le laboratoire sur le porte-objet, dans l'étude des corps organises*. Privately published by the author, Paris.

Rattanachaiyanont M, Weerachatyanukul W, Leveille MC, Taylor T, D'Amours D, Rivers D, Leader A, Tanphaichitr N (2001) Anti-SLIP1-reactive proteins exist on human spermatozoa and are involved in zona pellucida binding. *Molecular Human Reproduction* 7:633–640.

Rawlings PK, Schneider FW (1970) Models for competitive cooperative linear adsorption. The amylose-iodine-iodide complex. *Journal of Chemical Physics* 52:946–952.

Rawlins TE, ed. (1933) *Phytopathological and Botanical Research Methods*. Wiley, New York.

Ray SK, Schaecher KE, Shields DC, Hogan EL, Banik NL (2000) Combined TUNEL and double immunofluorescent labeling for detection of apoptotic mononuclear phagocytes in autoimmune demyelinating disease. *Brain Research Protocols* 5:305–311.

Raybould HE, Gschossman JM, Ennes H, Lembo T, Mayer EA (1999) Involvement of stretch-sensitive calcium flux in mechanical transduction in visceral afferents. *Journal of the Autonomic Nervous System* 75:1–6.

Raza M, Pal S, Rafiq A, DeLorenzo RJ (2001) Long-term alteration of calcium homeostatic mechanisms in the pilocarpine model of temporal lobe epilepsy. *Brain Research* 903:1–12.

Reaven E, Leers-Sucheta S, Nomoto A, Azhar S (2001) Expression of scavenger receptor class B type 1 (SR-BI) promotes microvillar channel formation and selective cholesteryl ester transport in a heterologous reconstituted system. *Proceedings of the National Academy of Sciences of the United States of America* 98:1613–1618.

Reavent E, Tsai L, Azhar S (1996) Intracellular events in the 'selective' transport of lipoprotein-derived cholesteryl esters. *Journal of Biological Chemistry* 271:16208–16217.

Reempts JV, Borgers M (1975) A simple polychrome stain for conventionally fixed epon-embedded tissues. *Stain Technology* 50:19–23.

Rees DD, Rogers RA, Cooley J, Mandle RJ, Kenney DM, Remoldo O'Donnell E (1999) Recombinant human monocyte/neutrophil elastase inhibitor protects rat lungs against injury from cystic fibrosis airway secretions. *American Journal of Respiratory, Cell and Molecular Biology* 20:69–78.

Reichel CG (1758) *De vasis plantarum spiralibus*. Breitkopf, Lipsiae.

Reichenbach A, Grimm D, Mozhaiskaja W, Distler C (1995) Visualisation of Muller (retinal glial) cells by bulk filling with procion yellow. *Journal of Brain Research* 36:305–311.

Reid S, Cross R, Snow EC (1996) Combined Hoechst 33342 and merocyanine 540 staining to examine murine B cell cycle stage, viability and apoptosis. *Journal of Immunological Methods* 192:43–54.

Render J (1997) Cell fate maps in the *Ilyanassa obsoleta* embryo beyond the third division. *Developmental Biology* 189:301–310.

Renier M, Tamanini A, Nicolis E, Rolfini R, Imler JL, Pavirani A, Cabrini G (1995) Use of a membrane potential-sensitive probe to assess biological expression of the cystic fibrosis transmembrane conductance regulator. *Human Gene Therapy* 6:1275–1283.

Rentsch G, Wittekind D (1967) The labelling of proteins with fluorescent fractions of fluorescein isothiocyanate. *Acta Histochemica Suppl.* 7:191–198.

Restaino L, Frampton EW, Irbe RM, Schabert G, Spitz H (1999) Isolation and detection of *Listeria monocytogenes* using fluorogenic and chromogenic substrates for phosphatidylinositol-specific phospholipase C. *Journal of Food Protection* 62:244–251.

Rettie GH, Haynes CG (1964) Thin-layer chromatography and its application to dyes. *Journal of the Society of Dyers and Colourists* 80:629–640.

Reuter K (1901) Uber den farbenden Bestandtell der Romanowsky-Nochtschen Malaria Plasmodienfarbung seine Reindarstellung und praktische Verwendung. *Zentralblatt fur Bakteriologie* 30:248–256.

Reynolds JEF, ed. (1993) *Martindale, The Extra Pharmacopoeia*. 30th Edn. The Pharmaceutical Press, London.

Rhys P, Zollinger H (1972) *Fundamentals of the Chemistry and Application of Dyes*. University of Michigan Press, Ann Arbor.

Riccio ML, Rossolini GM, Lombardi G, Chiesurin A, Satta G (1997) Expression cloning of different bacterial phosphatase-encoding genes by histochemical screening of genomic libraries onto an indicator medium containing phenolphthalein diphosphate and methyl green. *Journal of Applied Microbiology* 82:177–185.

Ricken S, Leipziger J, Greger R, Nitschke R (1998) Simultaneous measurements of cytosolic and mitochrondrial Ca^{2+} transients in HT29 cells. *Journal of Biological Chemistry* 273:34961–34969.

Rinaldi T, Ricci C, Porro D, Bolotin-Fukuhara M, Frontali L (1998) A mutation in a novel yeast proteasomal gene, RPN11/MPR1, produces a cell cycle arrest, overreplication of nuclear and mitochondrial DNA, and an altered mitochondrial morphology. *Molecular Biology of the Cell* 9:2917–2931.

Ringertz NR (1968) Cytochemical demonstration of basic proteins by dansyl staining. *Journal of Histochemistry and Cytochemistry* 16:440–441.

Rinkevich B (2000) Steps towards the evaluation of coral reef restoration by using small branch fragments. *Marine Biology* 136:807–812.

Robbins MH, Drago RS (1997) Activation of hydrogen peroxide for oxidation by copper(II) complexes. *Journal of Catalysis* 170:295–303.

Robert R, Nail S, Marot-Leblond A, Cottin J, Miegeville M, Quenouillere S, Mahaza C, Senet JM (2000) Adherence of platelets to *Candida* species *in vivo*. *Infection and Immunity* 68:570–576.

Roberts W (1863) On peculiar appearances exhibited by blood-corpuscles under the influence of solutions of magenta and tannin. *Proceedings of the Royal Society* 12:481–491.

Robins JH, Abrams GD, Pincock JA (1980) The structure of Schiff reagent aldehyde adducts and the mechanism of the Schiff reaction as determined by nuclear magnetic resonance spectroscopy. *Canadian Journal of Chemistry* 58:339–347.

Roca J, Martinez E, Vazquez JM, Lucas X (1998) Selection of immature pig oocytes for homologous *in vitro* penetration assays with the brilliant cresyl blue test. *Reproduction Fertility and Development* 10:479–485.

Rodgers W, Glaser M (1993) Distribution of proteins and lipids in the erythrocyte membrane. *Biochemistry* 32:12591–12598.

Roding J, Naujok A, Zimmermann HW (1986) Effects of ethidium bromide, tetramethylethidium bromide and betaine B on the ultrastructure of HeLa cell mitochondria *in situ*. *Histochemistry* 85:215–222.

Rodman JS, Lipman R, Brown A, Bronson RT, Dice JF (1998) Rate of accumulation of luxol fast blue staining material and mitochondrial ATP synthase subunit 9 in motor neuron degeneration mice. *Neurochemical Research* 23:1291–1296.

Rodrigues CO, Scott DA, Docampo R (1999) Presence of a vacuolar H+-pyrophosphatase in promastigotes of *Leishmania donovani* and its localisation to a different compartment from the vacuolar H+-ATPase. *Biochemical Journal* 340:759–766.

Rodriguez GG, Phipps D, Ishiguro K, Ridgway HF (1992) Use of a fluorescent redox probe for direct visualization of actively respiring bacteria. *Applied and Environmental Microbiology* 58:1801–1808.

Rodriguez LA (1955) Experiments on the histologic locus of the hematoencephalic barrier. *Journal of Comparative Neurology* 102:27–45.

Rodriguez-Pena JM, Cid VJ, Arroyo J, Nombela C (2000) A novel family of cell wall-related proteins regulated differently during the yeast life cycle. *Molecular and Cellular Biology* 20:3245–3255.

Rodriguez-Peralta LA (1966) Hematic and fluid barriers in the optic nerve. *Journal of Comparative Neurology* 126:109–122.

Roe JN, Szoka FC, Verkman AS (1990) Fiber optic sensor for the detection of potassium using fluorescence energy transfer. *Analyst* 115:353–358.

Roe MA, Lillie RD, Wilcox H (1940) American azures in the preparation of satisfactory Giemsa stains. *Public Health Reports* 55:1272–1278.

Rogers RA, Oldmixon EH, Brain JD (1992) Enhanced contrast within plastic embedded tissue by Lucifer yellow CH – An ideal stain for laser scanning confocal microscopy. *Molecular Biology of the Cell* 3:A185.

Rojanasakul Y, Wang LY, Bhat M, Glover DD, Malanga CJ, Ma JKH (1992) The transport barrier of epithelia – a comparative study of membrane-permeability and charge selectivity in the rabbit. *Pharmaceutical Research* 9:1029–1034.

Romanowsky DL (1891) Zur Frage der Parasitologie und Therapie der Malaria. *St Petersburg Medizinische Wochenschrift* 16:297–315.

Romeis B (1968) *Mikroskopische Technik*. 16th Edn. Oldenbourg Verlag, Munich.

Rong N, Ausman LM, Nicolosi RJ (1997) Oryzanol decreases cholesterol absorption and aortic fatty streaks in hamsters. *Lipids* 32:303–309.

Roque AL, Jafarey NA, Coulter P (1965) A stain for the histochemical demonstration of nucleic acids. *Experimental and Molecular Pathology* 4:266–274.

Ros Barcelo A, Munoz R, Sabater F (1989) Activated charcoal as an adsorbent of oxidized 3,3'-diaminobenzidine in peroxidase histochemistry. *Stain Technology* 64:97–98.

Rosa CG (1953) Preparation and use of aldehyde fuchsin stain in the dry form. *Stain Technology* 28:299–302.

Rosen DL (1999) Bacterial endospore detection using photoluminescence from terbium dipicolinate. *Reviews in Analytical Chemistry* 18:1–21.

Rosen H, Michel BR, Chait A (1991) Phagocytosis of opsonized oil droplets by neutrophils – adaptation to a microtiter plate format. *Journal of Immunological Methods* 144:117–125.

Rosenquist T (1981) Improved visualization of oxytalan fibres with resorcin fuchsin. *Basic and Applied Histochemistry* 25:129–132.

Rosenthal SI, Puchtler H, Sweat F (1965) Paper chromatography of dyes. *Archives of Pathology* 80:190–196.

Ross DW, Bishop C, Henderson A, Kaplow L (1990) Whole blood staining in suspension for nonspecific esterase and alkaline phosphatase analyzed with a Technicon H-1. *Cytometry* 11:552–555.

Rosselet A, Ruch F (1968) Cytofluorometric determination of lysine with dansyl chloride. *Journal of Histochemistry and Cytochemistry* 16:459–466.

Rost FWD (1995) *Fluorescence Microscopy*. v. 2. Cambridge University Press, Cambridge.

Rostoker G, Delchier JC, Chaumette MT (2001) Increased intestinal intra-epithelial T lymphocytes in primary glomerulonephritis – A role of oral tolerance breakdown in the pathophysiology of human primary glomerulonephritides? *Nephrology Dialysis Transplantation* 16:513–517.

Roth J (1982) Applications of immunocolloids in light microscopy. Preparation of protein A-silver and protein A-gold complexes and their application for localization of single and multiple antigens in paraffin sections. *Journal of Histochemistry and Cytochemistry* 30:691–696.

Roth J, Bendayan M, Orci L (1978) Ultrastructural localization of intracellular antigens by the use of protein A–gold complex. *Journal of Histochemistry and Cytochemistry* 26:1074–1081.

Rothbarth PH, Hendriks-Sturkenboom I, Ploem JS (1976) Identification of monocytes in suspensions of mononuclear cells. *Blood* 48:139–149.

Rothe F, Canzler U, Wolf G (1998) Subcellular localization of the neuronal isoform of nitric oxide synthase in the rat brain: a critical evaluation. *Neuroscience* 83:259–269.

Roybal JE, Pfenning AP, Turnipseed SB, Hurlbut JA, Long AR (1996) Dye residues in foods of animal origin. *ACS Symposium Series* 636:169–183.

Ruchel R, Schaffrinski M (1999) Versatile fluorescent staining of fungi in clinical specimens by using the optical brightener Blankophor. *Journal of Clinical Microbiology* 37:2694–2696.

Rugsaseel S, Kirimura K, Usami S (1993) Selection of mutants of *Aspergillus niger* showing enhanced productivity of citric acid from starch in shaking culture. *Journal of Fermentation and Bioengineering* 75:226–228.

Ruiz-Arguello MB, Goni FM, Alonso A (1998) Vesicle membrane fusion induced by the concerted activities of sphingomyelinase and phospholipase C. *Journal of Biological Chemistry* 273:22977–22982.

Rumbaut RE, Harris NR, Sial AJ, Huxley VH, Granger DN (1999) Leakage responses to L-NAME differ with the fluorescent dye used to label albumin. *American Journal of Physiology – Heart and Circulatory Physiology* 276:H333–H339.

Rungby J, Kassem M, Eriksen EF, Danscher G (1993) The von Kossa reaction for calcium deposits: silver lactate staining increases sensitivity and reduces background. *Histochemical Journal* 25:446–451.

Rushing LG, Thompson HC (1997) Simultaneous determination of malachite green, gentian violet and their leuco metabolites in catfish or trout tissue by high performance liquid chromatography with visible detection. *Journal of Chromatography* B **688:**325–330.

Rusmah M, Rahim ZHA (1992) Diffusion of buffered glutaraldehyde and formocresol from pulpotomized primary teeth. *Journal of Dentistry for Children* 59:108–110.

Rutenberg AM, Gofstein R, Seligman AM (1950) Preparation of a new tetrazolium salt which yields a blue pigment on reduction, and its use in the demonstration of enzymes in normal and neoplastic tissues. *Cancer Research* 10:113–121.

Ruth EB (1946) Demonstration of the ground substance of cartilage, bone and teeth. *Stain Technology* 21:27–30.

Ruzin SE (1999) *Plant Microtechnique and Microscopy*. Oxford University Press, New York.

Sabin FR (1923) Studies on living human blood cells. *Bulletin of the Johns Hopkins Hospital* 34:277–288.

Sabin FR (1929) Chemical agents: supravital stains. In: CE McClung (ed.) *Handbook of Microscopical Technique for Workers in both Animal and Plant Tissues.* Hoeber, New York. pp. 81–87.

Sabnis RW, Deligeorgiev TG, Jachak MN, Dalvi TS (1997) DiOC(6)(3): a useful dye for staining the endoplasmic reticulum. *Biotechnic and Histochemistry* 72:253–258.

Sadana US, Claassen N (1996) A simple method to study the oxidizing power of rice roots under submerged soil conditions. *Zeitschrift fur Pflanzenernahrung und Bodenkunde* 159:634–646.

Sadoni N, Sullivan KF, Weinzierl P, Stelzer EH, Zink D (2001) Large-scale chromatin fibers of living cells display a discontinuous functional organization. *Chromosoma* 110:39–51.

Sadurski R, Tsukada H, Ying XY, Bhattacharya S, Bhattacharya J (1994) Diameters of justacapillary venules determined by oil-drop method in rat lung. *Journal of Applied Physiology* 77:718–725.

Saffiotti U, Daniel LN, Mao Y, Shi XG, Williams AO, Kaighn ME (1994) Mechanisms of carcinogenesis by crystalline silica in relation to oxygen radicals. *Environmental Health Perspectives* 102:159–163.

Sagi SR, Rao KA, Rao MSP (1992) A titrimetric method for the determination of oxalic and malonic acids in a mixture. *Journal of the Indian Chemical Society* 69:671–673.

Saha DR, Niyogi SK, Nair GB, Manna B, Bhattacharya SK (2000) Detection of faecal leucocytes and erythrocytes from stools of cholera patients suggesting an evidence of an inflammatory response in cholera. *Indian Journal of Medical Research* 112:5–8.

Sai Y, Kajita M, Tamai I, Wakama J, Wakamiya T, Tsuji A (1998) Adsorptive-mediated endocytosis of a basic peptide in enterocyte-like Caco-2 cells. *American Journal of Physiology* 275:G514–G520.

Sajap AS (1999) Detection of foraging activity of *Coptotermes curvignathus* (Isoptera: Rhinotermitidae) in an *Hevea brasiliensis* plantation in Malaysia. *Sociobiology* 33:137–143.

Sakamoto K, Muratani M, Ogawa T, Nagamachi Y (1993) Evaluation of a new test for colorectal neoplasms – A prospective study of asymptomatic population. *Cancer Biotherapy* 8:49–55.

Sakamoto T, Sun J, Barnes PH, Chung KF (1994) Effect of a bradykinin receptor antagonist, Hoe-140, against bradykinin-induced and vagal stimulation-induced airway responses in the guinea pig. *European Journal of Pharmacology* 251:137–142.

Sakuraba H, Takamatsu Y, Satomura T, Kawakami R, Ohshima T (2001) Purification, characterization, and application of a novel dye-linked L-proline dehydrogenase from a hyperthermophilic archaean, *Thermococcus profundus*. *Applied and Environmental Microbiology* 67:1470–1475.

Salthouse TN, Macdonald BG, Agnello D (1971) Vinyl sulfone dyes: possible applications in histology. *Stain Technology* 46:245–248.

Salthouse TN (1962) Luxol fast blue ARN: A new solvent azo dye with improved staining qualities for myelin and phospholipids. *Stain Technology* 37:313–316.

Salthouse TN (1965) Selective staining of collagen and elastin by luxol fast blue G in methanol: A histochemical study. *Journal of Histochemistry and Cytochemistry* 13:133–140.

Sanchez JC, Wirth P, Jaccoud S, Appel RD, Sarto C, Wilkins MR, Hochstrasser DF (1997) Simultaneous analysis of cyclin and oncogene expression using multiple monoclonal antibody immunoblots. *Electrophoresis* 18:638–641.

Sanchez R, Risopatron J, Sepulveda G, Pena P, Miska W (1995) Evaluation of the acrosomal membrane in bovine spermatozoa – effects of proteinase inhibitors. *Theriogenology* 43:761–768.

Sandell JH, Masland RH (1988) Photoconversion of some fluorescent markers to a diaminobenzidine product. *Journal of Histochemistry and Cytochemistry* 36:555–559.

Sanderson JB (1994) *Biological Microtechnique* (Microscopy Handbooks, 28). BIOS Scientific Publications and Royal Microscopical Society, Oxford.

Sandoval R, Leiser J, Molitoris BA (1998) Aminoglycoside antibiotics traffic to the Golgi complex in LLC-PK1 cells. *Journal of the American Society of Nephrology* 9:167–174.

Sandritter W (ed.) (1964) *100 Years of Histochemistry in Germany*. English edition, edited by F. H. Kasten. Schattauer-Verlag, Stuttgart.

Sandritter W, Kiefer G, Rick W (1966) Gallocyanin chrome alum. In: GL Weid (ed.) *Introduction to Quantitative Cytochemistry*. Academic Press, New York.

Sano A, Kurita N, Coelho R, Takeo K, Nishimura K, Miyaji M (1993) A comparative study of four different staining methods for estimation of live yeast from cells of *Paracocci dioides* Brasiliensis. *Mycopathologia* 124:157–161.

Saoji AM, Jad CY, Kelkar SS (1983) Remazol Billiant Blue as a pre-stain for the immediate visualization of human-serum proteins on polyacrylamide-gel disk-electrophoresis. *Clinical Chemistry* 29:42–44.

Sarasquete C, Munoz-Cueto JA, de Canales MLG, Garcia-Garcia A, Rodriguez-Gomez FJ, Pinuela C, Rendon C, Rodriguez RB (1997) Histochemical and immunocytochemical study of gonadotropic pituitary cells of the killifish, *Fundulus heteroclitus*, during annual reproductive cycle. *Scientia Marina* 61:439–449.

Saris N-EL, Seppala AJ (1969) Binding of bromthymol blue by mitochondrial structural protein. *European Journal of Biochemistry* 7:267–272.

Sasaki CY, Passaniti A (1998) Identification of anti-invasive but noncytotoxic chemotherapeutic agents using the tetrazolium dye MTT to quantitate viable cells in Matrigel. *BioTechniques* 24:1038–1043.

Savarese JJ, Erickson H, Scully SP (1996) Articular chondrocyte tenascin-C production and assembly into de novo extracellular matrix. *Journal of Orthopaedic Research* 14:273–281.

Scala C, Preda P, Cenacchi G, Martinelli GN, Manara GC, Pasquinelli G (1993) A new polychrome stain and simultaneous methods of histological, histochemical and immunohistochemical stainings performed on semithin sections of Bioacryl-embedded human tissues. *Histochemical Journal* 25:670–677.

Schachar RA (1996) Histology of the ciliary muscle-zonular connections. *Annals of Ophthalmology – Glaucoma* 28:70–79.

Schaeffer AB, Fulton M (1933) A simplified method of staining endospores. *Science* 77:194.

Schafer KH, Hansgen A, Mestres P (1999) Morphological changes of the myenteric plexus during early postnatal development of the rat. *Anatomical Record* 256:20–28.

Schafer M, Roffael E (1996) Effect of storage of residues from saw mill processing of pine and spruce and its suitability as a raw-material for particleboards and MDF. 2. Analytical and microscopical investigations of the change of extractive contents of saw mill residues of pine and spruce. [German]. *Holz als Roh- und Werkstoff* 54:157–162.

Schaffer (1888) Die Farberei zum Studium der Knochenentwicklung. *Zeitschrift fur wissenschaftliche Mikroskopie* 5:1–19. Cited in Lillie (1977).

Schaffer GF, Peterson RL (1993) Modifications to clearing methods used in combination with vital staining of roots colonized with vesicular-arbuscular mycorrhizal fungi. *Mycorrhiza* 4:29–35.

Schapiro FB, Grinstein S (2000) Determinants of the pH of the Golgi complex. *Journal of Biological Chemistry* 275:21025–21032.

Schemitsch EH, Turchin DC, Kowalski MJ, Swiontkowski MF (1998) Quantitative assessment of bone injury and repair after reamed and unreamed locked intramedullary nailing. *Journal of Trauma – Injury, Infection and Critical Care* 45:250–255.

Schenk E (1981) A newly certified dye – alcian blue 8GX. *Stain Technology* 56:129–131.

Schep LJ, Tucker IG, Young G, Ledger R, Butt AG (1998) Permeability of the salmon (*Oncorhynchus tshawytscha*) posterior intestine *in vivo* to two hydrophilic markers. *Journal of Comparative Physiology B – Biochemical and Environmental Physiology* 168:562–568.

Schichnes D, Nemson J, Sohlberg L, Ruzin SE (1998) Microwave protocols for paraffin microtechnique and *in situ* localization in plants. *Microscopy and Analysis* 4:491–496.

Schiek RC (1982) Inorganic pigments. In: M Grayson (ed.) *Kirk-Othmer Encyclopedia of Chemical Technology*. Wiley, New York. pp. 788–838.

Schilling WP, Sinkins WG, Estacion M (1999) Maitotoxin activates a nonselective cation channel and a P2Z/P2X(7)-like cytolytic pore in human skin fibroblasts. *American Journal of Physiology, Cell Physiology* 277:C755–C765.

Schipper NW, Baak JPA, Smeulders AWM (1991) Automated selection of the most epithelium-rich areas in gynecologic tumor sections. *Analytical and Quantitative Cytology and Histology* 13:395–402.

Schmid I, Ferbas J, Uittenbogaart CH, Giorgi JV (1999) Flow cytometric analysis of live cell proliferation and phenotype in populations with low viability. *Cytometry* 35:64–74.

Schmitt O, Eggers R (1999) Flat-bed scanning as a tool for quantitative neuroimaging. *Journal of Microscopy* 196:337–346.

Schmorl G (1934) *Die pathologisch-histologischen Untersuchungsmethoden.* Vogel, Berlin.

Schmued LC, Fallon JH (1985) Selective neuronal uptake of 4-acetamido-4′-isothio-cyanatostilbene-2,2′-disulfonic acid (SITS) and other related stilbenes *in vivo*: a fluorescent whole cell staining technique. *Brain Research* 346:124–129.

Schmued LC, Snavely LF (1993) Photoconversion and electron microscopic localization of fluorescent axon tracer fluoro-ruby (rhodamine-dextran-amine). *Journal of Histochemistry and Cytochemistry* 41:777–782.

Schnedl W, Wachtler F, Abraham R, Dann O (1982) Fluorescent staining of *Trypanosoma rhodesiense* by trypanocides and related drugs. *Mikroskopie (Wien)* 39:139–142.

Schnell SA, Wessendorf MW (1995) Bisbenzimide: a fluorescent counterstain for tissue autoradiography. *Histochemistry and Cell Biology* 103:111–114.

Schoenwaelder MEA, Clayton MN (1999) The presence of phenolic compounds in isolated cell walls of brown algae. *Phycologia* 38:161–166.

Schott I, Hartmann D, Gieselmann V, Lullmann-Rauch R (2001) Sulfatide storage in visceral organs of arylsulfatase A-deficient mice. *Virchows Archiv fur pathologische Anatomie und Physiologie und fur klinische Medizin* 439:90–96.

Schottelius J, Kuhn EM, Enriquez R (2000) *Microsporidia* and *Candida* spores: their discrimination by calcofluor, trichrome-blue and methylene-blue combination staining. T*ropical Medicine and International Health* 5:453–458.

Schrader S (1994) Influence of earth worms on the pH conditions of their environment by cutaneous mucus secretion. *Zoologischer Anzeiger* 233:211–219.

Schrader S, Joschko M (1991) A method for studying the morphology of earthworm burrows and their function in respect to water-movement. *Pedobiologica* 35:185–190.

Schraermeyer U, Polyanovsky A, Pivovarova N, Zierold K, Stieve H, Gribakin F (1999) Extracellular compartments of the blowfly eye: ionic content and topology. *Visual Neuroscience* 16:461–474.

Schulte E, Wittekind D (1987) Microwave-stimulated staining of histological material with the standard azure B-eosin Y stain. *Acta Anatomica* 130:83.

Schulte E, Wittekind D (1989) Standardization of the Feulgen-Schiff technique. Staining characteristics of pure fuchsin dyes; a cytophotometric investigation. *Histochemistry* 91:321–331.

Schulte E, Wittekind DH (1989) A quick and standardized Giemsa stain for wet-fixed cytological material. *Stain Technology* 64:253–254.

Schulte E, Wittekind D, Kretschmer V (1988) Victoria blue B – a nuclear stain for cytology. A cytophotometric study. *Histochemistry* 88:427–433.

Schulte EKW (1994) Improving biological dyes and stains: Quality testing versus standardization. *Biotechnic and Histochemistry* 69:7–17.

Schulte EKW, Lyon H, Prento P (1991) Gallocyanin chromalum as a nuclear stain in cytology. I. A cytophotometric comparison of the Husain-Watts gallocyanin chromalum staining protocol with the Feulgen procedure. *Histochemical Journal* 23:241–245.

Schulte EKW, Lyon HO, Hoyer PE (1992) Simultaneous quantification of DNA and RNA in tissue sections. A comparative analysis of the methyl green-pyronin technique with the gallocyanin chromealum and Feulgen procedures using image cytometry. *Histochemical Journal* 24:305–310.

Schultz G (1931–1939) *Farbstofftabellen*. 7th Edn. v. 1, 2 and Suppl. Akademische Verlagsgesellschaft, Leipzig.

Schumacher TE, Smucker AJM, Eshel A, Curry RB (1983) Measurement of short term root-growth by prestaining with neutral red. *Crop Science* 23:1212–1214.

Schumacher U, Adam E (1994) Standardization of staining in glycosaminoglycan histo-chemistry: alcian blue, its analogues, and diamine methods. *Biotechnic and Histochemistry* 69:18–24.

Schurr U, Schuberth B, Aloni R, Pradel KS, Schmundt D, Jahne B, Ullrich CI (1996) Structural and functional evidence for xylem-mediated water transport and high transpiration in *Agrobacterium tumefaciens*-induced tumors of *Ricinus communis*. Botanica Acta **109**:405–411.

Schwartz E (1867) Ueber eine methode doppelter Farbung mikroskopischer Objecte und ihre Anwendung zur Untersuchung der Muskulatur des Milz, Lymphdrusen und anderer Organe. *Akademische Sitzungberichte (Wien)* 55:671–691.

Schwarzacher T, Heslop-Harrison P (2000) *Practical* in situ *Hydridization*. Bios, Oxford.

Schweizer D, Ambros PF (1994) Chromosome banding. Stain combinations for specific regions. *Methods in Molecular Biology* 29:97–112.

Schwenke R, Geyer G (1966) Zur Anwendung von Reaktivfarbstoffen. *Acta Histochemica* 24:366–375.

Scopsi L, Larsson L-I (1986) Bodian's silver impregnation of endocrine cells. A tentative explanation to the staining mechanism. *Histochemistry* 86:59–62.

Scorilas A, Bjartell A, Lilja H, Moller C, Diamandis EP (2000) Streptavidin-polyvinyl-amine conjugates labeled with a europium chelate: Applications in immunoassay, immunohistochemistry, and microarrays. *Clinical Chemistry* 46:1450–1455.

Scott DA, Docampo R, Benchimol M (1998) Analysis of the uptake of the fluorescent marker 2',7'-bis-(2-carboxyethyl)-5(and 6)-carboxyfluorescein (BCECF) by hydrogeno-somes in *Trichomonas vaginalis*. *European Journal of Cell Biology* 76:139–145.

Scott JE (1967) On the mechanism of the methyl green-pyronin stain for nucleic acids. *Histochemie* 9:30–47.

Scott JE (1973) Affinity, competition and specific interactions in the biochemistry and histochemistry of polyelectrolytes. *Biochemical Society Transactions* 1:787–806.

Scott JE (1996) Alcian blue – now you see it, now you don't. *European Journal of Oral Sciences* 104:2–9.

Scott JE, Dorling J (1965) Differential staining of acid glycosaminoglycans (mucopolysac-charides) by alcian blue in salt solutions. *Histochemistry* 5:221–233.

Scott JE, Thomlinson AM (1998) The structure of interfibrillar proteoglycan bridges (shape modules) in extracellular matrix of fibrous connective tissues and their stability in various chemical environments. *Journal of Anatomy* 192:391–405.

Scott JE, Willet IH (1966) Binding of cationic dyes to nucleic acids and other biological polyanions. *Nature* 209:985–986.

Scott RE (1924a) Standardization of biological stains. *Military Surgeon* 55:229–243.

Scott RE (1924b) Standardization of biological stains. II. Methylene blue. *Military Surgeon* 55:337–352.

Seeley R, Vandemark PJ, Lee JJ (1991) *Microbes in Action. A Laboratory Manual of Microbiology*. 4th Edn. Freeman, New York.

Seidler E (1980) New nitro-monotetrazolium salts and their use in histochemistry. *Histochemical Journal* 12:619–630.

Seidler E (1991) The tetrazolium-formazan system – design and histochemistry. *Progress in Histochemistry and Cytochemistry* 24:1–86.

Seki M (1932) Substantive (direkte) Farbung der histologisch fixierten Preparate. *Folia Anatomica Japonica* 10:635–654.

Sekiguchi T, Nakamura H, Banscho Y (1969) Organic pigments of the phthalocyanine series. XIX. Synthesis and separations of the 4-sulfonated copper phthalocyanines. [Japanese]. *Kogyo Kagaku Zasshi* 72:115–160.

Selbo PK, Sandvig K, Kirveliene V, Berg K (2000) Release of gelonin from endosomes and lysosomes to cytosol by photochemical internalization. *Biochimica et Biophysica Acta* 1475:307–313.

Sells MA, Boyd JT, Chernoff J (1999) p21-Activated kinase 1 (pak1) regulates cell motility in mammalian fibroblasts. *Journal of Cell Biology* 145:837–849.

Senseman DM (1996) High-speed optical imaging of afferent flow through rat olfactory bulb slices: voltage-sensitive dye signal reveal periglomerular cell activity. *Journal of Neuroscience* 16:313–324.

Serino A, Kan K, Graves K, Kule C, Anthony A (2000) Age, strain, and semi-chronic hydergine treatment effects on motor activity and neuronal nucleic acid-protein metabolism in male mice. *Life Sciences* 67:1489–1505.

Serpe M, Joshi A, Kosman DJ (1999) Structure-function analysis of the protein-binding domains of Mac1p, a copper-dependent transcriptional activator of copper uptake in *Saccharomyces cerervisiae. Journal of Biological Chemistry* 274:29211–29219.

Serrato-Valenti G, Cornara L, Modenesi P, Piana M, Mariotti MG (2000) Structure and histochemistry of embryo envelope tissues in the mature dry seed and early germination of *Phacelia tanacetifolia. Annals of Botany* 85:625–634.

Servant G, Weiner OD, Neptune ER, Sedat JW, Bourne HR (1999) Dynamics of chemo-attractant receptor in living neutrophils during chemotaxis. *Molecular Biology of the Cell* 10:1163–1178.

Shah IA, Miller I (1991) Simplified luxol fast blue stain for selective visualization and quantification of gastric parietal cells. *Journal of Histotechnology* 14:247–249.

Shan WS, Tanaka H, Phillips GR, Arndt K, Yoshida M, Colman DR, Shapiro L (2000) Functional cis-heterodimers of N- and R-cadherins. *Journal of Cell Biology* 148:579–590.

Shapeiro XX (1961) Freezing out, a safe technique for concentration of dilute solutions. *Science* 133:2063–2064.

Shapiro H (2001) Multiparameter flow cytometry of bacteria: implications for diagnostics and therapeutics. *Cytometry* 43:223–226.

Shapiro HM (1988) *Practical Flow Cytometry.* 2nd Edn. Liss, New York.

Shapiro HM (1994) Cell membrane potential analysis. *Methods in Cell Biology* 41:121–133.

Shapiro HM (2000) Membrane potential estimation by flow cytometry. *Methods* 21:271–279.

Shapiro SH (1991) Keratin as a factor in histopathologic diagnosis. *Journal of Histotechnology* 28:51–55.

Sharp RR, Cunningham AB, Komlos J, Billmayer J (1999) Observation of thick biofilm accumulation and structure in porous media and corresponding hydrodynamic and mass transfer effects. *Water Science and Technology* 39:195–201.

Shaw SR, Varney LP (1999) Primitive, crustacean-like state of blood–brain barrier in the eye of the apterygote insect *Petrobius* (Archaeognatha) determined from uptake of fluorescent tracers. *Journal of Neurobiology* 41:452–470.

Sheehan DC, Hrapchak BB (1987) *Theory and Practice of Histotechnology.* 2nd Edn. Battelle, Columbus, Ohio.

Sheen B (1971) China clay – a sorbant for thin-layer chromatography. *Journal of Chromatography* 60:363–370.

Shelef LA, Firstenberg-Eden R (1997) Novel selective and non-selective optical detection of microorganisms. *Letters in Applied Microbiology* 25:202–206.

Shelly WB (1969) Fluorescent staining of elastic tissue with rhodamine B and related xanthene dyes. *Histochemie* 20:244–249.

Shen YL, Gao JB, Xu KW, Xue LG, Zhang YM, Shi BK, Li DC, Wei XB, Higuchi S (1997) Babesiasis in Nanjing area, China. *Tropical Animal Health and Production* 29:S19–S22.

Shepherd VA, Goodwin PB (1992) Seasonal patterns of cell-to-cell communication in Characorallina Klein ex Willd. 2. Cell-to-cell communication during the development of antheridia. *Plant Cell and Environment* 15:151–162.

Shibusawa Y, Chiba T, Matsumoto U, Ito Y (1995) Countercurrent chromatographic isolation of high density lipoprotein fractions from human serum. *American Chemical Society Symposium Series* 593:119–128.

Shikata T, Uzawa T, Yoshiwara N, Akatsuka T, Yamazaki S (1974) Staining methods of Australia antigen in paraffin sections: detections of cytoplasmic inclusion bodies. *Japanese Journal of Experimental Medicine* 44:25–36.

Shilina NM, Koterov AN, Kon II (1997) The effect of gamma-irradiation on the level of transferrin in the plasma of mice and the degree of glycosylation [Russian]. *Biull. Eksp. Biol. Med.* 123:46–50.

Shimoni M, Reuveni R, Bar Zur A (1996) Relation between peroxidase, beta-1,3-glucanase, the se gene and partial resistance of maize to *Exserohilum turcicum*. *Canadian Journal of Plant Pathology* 18:403–408.

Shirai K, Matsuoka M (1996) Structure and properties of hematein derivatives. *Dyes and Pigments* 32:159–169.

Shoemaker K, Rubin J, Zumbro GI, Tackett R (1996) Evans blue and gentian violet – Alternatives to methylene blue as a surgical marker dye. *Journal of Thoracic and Cardiovascular Surgery* 112:542–544.

Shorr E (1941) A new technic for staining vaginal smears. III. A single differential stain. *Science* 94:545.

Siboni G, Rothmann C, Ehrenberg B, Malik Z (2001) Spectral imaging of MC540 during murine and human colon carcinoma cell differentiation. *Journal of Histochemistry and Cytochemistry* 49:147–154.

Sick TJ, Perez-Pinzon MA (1999) Optical methods for probing mitochondrial function in brain slices. *Methods* 18:104–108.

Siegel I (1967) Toluidine blue 0 and naphthol yellow S; a highly polychromatic general stain. *Stain Technology* 42:29–30.

Sievers E, Santer R, Oldigs HD, Haase S, Schleyerbach U, Schaub J (1995) Gastrointestinal passage time in preterm infants. *Monatsschrift Kinderheilkunde* 143:S76–S80.

Silvander M, Johnsson M, Edwards K (1998) Effects of PEG-lipids on permeability of phosphatidylcholine/cholesterol liposomes in buffer and in human serum. *Chemistry and Physics of Lipids* 97:15–26.

Silver RB (1998) Ratio imaging: practical considerations for measuring intracellular calcium and pH in living tissue. *Methods in Cell Biology* 56:237–251.

Simmons A (1997) *Hematology. A Combined Theoretical and Technical Approach.* Butterworth-Heinemann, Boston.

Simpson ME (1921) Vital staining of human blood with special reference to the separation of the monocytes. *University of California Publications in Anatomy* 1:1–9.

Simpson WJ, Fernandez JL, Hammond JRM (1992) Differentiation of brewery yeasts using a disk-diffusion test. *Journal of the Institute of Brewing* 98:33–36.

Sims PJ, Waggoner AS, Wang CH, Hoffman JF (1974) Studies on the mechanism by which cyanine dyes measure membrane potential in red blood cells and phosphatidyl choline vesicles. *Biochemistry* 13:3315–3330.

Simson JAV (1977) The influence of fixation on the carbohydrate cytochemistry of rat salivary gland secretory granules. *Histochemical Journal* 9:645–657.

Singer JJ (1976) Organic pigments. In: M Grayson (ed.) *Kirk-Othmer Encyclopedia of Chemical Technology.* Wiley, New York. pp. 838–870.

Singer M (1952) Staining of tissue sections with acid and basic dyes. *International Review of Cytology* 1:211–255.

Sipka S, Antal-Szalmas P, Szollosi I, Csipo I, Lakos G, Szegedi G (2000) Ceramide stimulates the uptake of neutral red in human neutrophils, monocytes and lymphocytes. *Annals of Hematology* 79:83–85.

Sire MF, Vernier JM (1980) Lipid staining on semithin sections with Sudan black B or Nile blue sulphate. Application to intestinal fat absorption. *Acta Histochemica et Cytochemica* 13:193–201.

Sirivaidyapong S, Cheng FP, Marks A, Voorhout WF, Bevers MM, Colenbrander B (2000) Effect of sperm diluents on the acrosome reaction in canine sperm. *Theriogenology* 53:789–802.

Sivashanmugam P, Rajalakshmi M (1997) Sperm maturation in rhesus monkey. Changes in ultrastructure, chromatin condensation, and organisation of lipid bilayer. *Anatomical Record* 247:25–32.

Sivitz M, Kallen RG, Latties AM (1973) Procion yellow and catecholamine derivatives. Chemical relationships. *Journal of Histochemistry and Cytochemistry* 21:87–92.

Sjolund KF, Von Heijne M, Hao JX, Xu XJ, Sollevi A, Wiesenfeld-Hallin Z (1998) Intrathecal administration of the adenosine A1 receptor agonist R-phenylisopropyl adenosine reduces presumed pain behaviour in a rat model of central pain. *Neuroscience Letters* 243:89–92.

Skopp G, Potsch L, Eser HP, Moller MR (1996) Preliminary practical findings on drug monitoring by a transcutaneous collection device. *Journal of Forensic Sciences* 41:933–937.

Skripchenko A, Robinette D, Wagner SJ (1997) Comparison of methylene blue and methylene violet for photoinactivation of intracellular and extracellular virus in red cell suspensions. *Photochemistry and Photobiology* 65:451–455.

Skubatz H, Kunkel DD, Meeuse BJD (1993) Ultrastructural changes in the appendix of the sauromatum-guttatum inflorescence during anthesis. *Sexual Plant Reproduction* 6:153–170.

Slaninova I, Sestak S, Svoboda A, Farkas V (2000) Cell wall and cytoskeleton reorganization as the response to hyperosmotic shock in *Saccharomyces cerevisiae. Archives of Microbiology* 173:245–252.

Slavik J (1994) *Fluorescent Probes in Cellular and Molecular Biology.* CRC Press, Boca Raton, FL.

Sloper JC (1954) Histochemical observations on the neurohypophysis in dog and cat, with reference to the relationship between neurosecretory material and posterior lobe hormone. *Journal of Anatomy* 88:576–577.

Sloper JC (1955) Hypothalamic neurosecretion in the dog and cat, with particular reference to the identification of neurosecretory material with posterior lobe hormone. *Journal of Anatomy* 89:301–316.

Smid JR, Monsour PA, Rousseau EM, Young WG (1992) Cytochemical localization of dipeptidyl peptidase II activity in rat incisor tooth ameloblasts. *Anatomical Record* 233:493–503.

Smiley JF, Goldman-Racic PS (1993) Silver-enhanced diaminobenzidine-sulfide (SEDS): a technique for high resolution immunoelectron microscopy demonstrated with monoamine immunoreactivity in monkey cerebral cortex and caudate. *Journal of Histochemistry and Cytochemistry* 41:1393–1404.

Smital T, Kurelec B (1998) The activity of multixenobiotic resistance mechanism determined by rhodamine B-efflux method as a biomarker of exposure. *Marine Environmental Research* 46:443–447.

Smith CF, Townsend DE (1999) A new medium for determining the total plate count in food. *Journal of Food Protection* 62:1404–1410.

Smith JC, Russ P, Cooperman BS (1976) Synthesis, structure determination, spectral properties, and energy-linked spectral responses of the extrinsic probe oxonol V in membranes. *Biochemistry* 15:5094–5105.

Smith RE, Reynolds CJ, Elder EA (1992) The evolution of proteinase substrates with special reference to dipeptidylpeptidase IV. *Histochemical Journal* 24:637–647.

Society of Dyers and Colourists. *Colour Index International*. 4th revision of 3rd Edn. Vols 1–9 (1971–1992); CD-ROM Version 2 with additions to 1996. Bradford, UK.

Sone Y, Nicolaysen A, Staub NC (1997) Effect of particles on sheep lung hemodynamics parallels depletion and recovery of intravascular macrophages. *Journal of Applied Physiology* 83:1499–1507.

Song W, Warren A, Hill BF (1998) Description of a new freshwater ciliate, *Euplotes shanghaiensis* nov. spec. from China (Ciliophora, Euplotidae). *European Journal of Protistology* 34:104–110.

Song WB, Petz W, Warren A (2001) Morphology and morphogenesis of the poorly-known marine urostylid ciliate, *Metaurostylopsis marina* (Kahl, 1932) nov. gen., nov. comb. (Protozoa, Ciliophora, Hypotrichida). *European Journal of Protistology* 37:63–76.

Song YC, Liu LH, Ding Y, Tian XB, Yao Q, Meng L, He CR, Xu MS (1994) Comparisons of G-banding patterns in six species of the Poaceae. *Hereditas* 121:31–38.

Sorisky A, Pardasani D, Gagnon A, Smith TJ (1996) Evidence of adipocyte differentiation in human orbital fibroblasts in primary culture. *Journal of Clinical Endocrinology and Metabolism* 81:3428–3431.

Soros CL, Dengler NG (1996) Leaf morphogenesis and growth in *Cyperus eragrostis* (Cyperaceae). *Canadian Journal of Botany* 74:1753–1765.

Southwell BR, Furness JB (2001) Immunohistochemical demonstration of the NK(1) tachykinin receptor on muscle and epithelia in guinea pig intestine. *Gastroenterology* 120:1140–1151.

Sparks DL, Lue LF, Martin TA, Rogers J (2000) Neural tract tracing using Di-I: a review and a new method to make fast Di-I faster in human brain. *Journal of Neuroscience Methods* 103:3–10.

Speel EJM, Ramaekers FCS, Hopman AHN (1997) Sensitive multicolor fluorescence *in situ* hybridization using catalyzed reporter deposition (CARD) amplification. *Journal of Histochemistry and Cytochemistry* 45:1439–1446.

Speel EJM, Hopman AHN, Komminoth P (1999) Amplification methods to increase the sensitivity of *in situ* hybridization: Play CARD(S). *Journal of Histochemistry and Cytochemistry* 47:281–288.

Spencer CI, Berlin JR (1995) A method for recording intracellular $[Ca^{2+}]$ transients in cardiac myocytes uisng calcium green-2. *Pflugers Archiv – European Journal of Physiology* 430:579–583.

Spicer SS, Lillie RD (1961) Histochemical identification of basic proteins with biebrich scarlet at alkaline pH. *Stain Technology* 36:365–370.

Spiekermann P, Rehm BHA, Kalscheuer R, Baumeister D, Steinbuchel A (1999) A sensitive, viable-colony staining method using Nile red for direct screening of bacteria that accumulate polyhydroxyalkanoic acids and other lipid storage compounds. *Archives of Microbiology* 171:73–80.

Sporri S, Bell B, Dreher E, Schneider H, Matamedi M (2000) Tubal sterilization by means of endoluminal coagulation: an *in vivo* study in rabbits. *Contraception* 62:141–147.

Srividya K, Balasubramanian N (1996) Indirect spectrophotometric determination of ascorbic acid in pharmaceutical samples and fruit juices. *Analyst* 121:1653–1655.

Staehle HJ, Spiess V, Heinecke A, Muller HP (1995) Effects of root-canal filling materials containing calcium hydroxide on the alkalinity of root dentin. *Endodontics and Dental Traumatology* 11:163–168.

Stalinski J (1994) Digestion, defecation and food passage rate in the insectivorous bat *Myotis myotis*. *Acta Theriologica* 39:1–11.

Stamp GW, Wright NA (1990) *Advanced Histopathology*. Springer-Verlag, New York.

Stamps WT, Linit MJ (1995) A rapid and simple method for staining lipid in fixed nematodes. *Journal of Nematology* 27:244–247.

Stankiewicz M, Shaw RJ, Jonas WE, Cabaj W, Grimmett DJ, Douch PGC (1994) A technique for the isolation and purification of viable mucosal mast cells/globule leukocytes from the small intestine of parasitised sheep. *International Journal for Parasitology* 24:307–309.

Stapelfeldt H, Jun H, Skibsted LH (1993) Fluorescence properties of carminic acid in relation to aggregation, complex-formation and oxygen activation in aqueous food models. *Food Chemistry* 48:1–11.

Stark MJ, Smalley KM, Rowe EC (1969) Methylene blue staining of axons in the ventral nerve cord of insects. *Stain Technology* 44:97–102.

Stefanini M, De Martino C, Zamboni L (1967) Fixation of ejaculated spermatozoa for electron microscopy. *Nature* 216:173–174.

Stein GM, Edlund U, Pfuller U, Bussing A, Schietzel M (1999) Influence of polysaccharides from *Viscum album* L – on human lymphocytes, monocytes and granulocytes *in vitro*. *Anticancer Research* 19:3907–3914.

Stellmach J (1984) Fluorescent redox dyes. 1. Production of fluorescent formazan by unstimulated and phorbol ester stimulated or digitonin-stimulated Ehrlich ascites tumor cells. *Histochemistry* 80:137–143.

Stellmach J, Severin E (1987) A fluorescent redox dye – Influence of several substrates and electron carriers on the tetrazolium salt formazan reaction of Ehrlich ascites tumor cells. *Histochemical Journal* 19:21–26.

Steward WW (1978) Functional connections between cells as revealed by dye-coupling with a highly fluorescent naphthalimide tracer. *Cell* 14:741–759.

Steward WW (1981) Lucifer dyes. Highly fluorescent dyes for biological tracing. *Nature* 292:17–21.

Stewart AK, Boyd CAR, Vaughan Jones RD (1999) A novel role for carbonic anhydrase: cytoplasmic pH gradient dissipation in mouse small intestine enterocytes. *Journal of Physiology* 516:209–217.

Stiller D, Katenkamp D, Thoss K (1972) Staining mechanism of thioflavine T with special reference to the localization of amyloid. [German]. *Acta Histochemica* 42:234–245.

Stockert JC (1977) Osmium tetroxide/p-phenylenediamine staining of nucleoli and Balbiani rings in *Chironomus* salivary glands. *Histochemistry* 53:43–56.

Stockert JC, Trigoso CI (1993) Fluorescence of eosinophil leukocyte granules induced by the fluorogenic reagent 2-methoxy-2,4-diphenyl-3 (2H)-furanone. *Blood Cells* 19:423–430.

Stockert JC, Trigoso CI (1994) Selective fluorescence reaction of indigocarmine stained eosinophil leukocyte granules induced by alkaline reduction of the bound dye to its leuco derivative. *Acta Histochemica* 96:8–14.

Stockert JC, Del Castillo P, Armas-Portela R (1989) Alcian yellow as a fluorescent dye. *Acta Histochemica* (Jena) 85:59–64.

Stockert JL (1975) Uranyl-EDTA-hematoxylin: a new selective staining technique for nucleolar material. *Histochemistry* 43:313–322.

Stockinger L (1964) Vitalfarbung und Vitalfluorochromierung Tierischer Zellen. In: L Heilbrunn (ed.) *Protoplasmatologia. Handbuch der Protoplasmaforschung.* Springer-Verlag, Wien. p. 1096.

Stohr M, Vogt-Schaden M, Knobloch M, Vogel R, Futterman G (1978) Evaluation of eight fluorochrome combinations for simultaneous DNA-protein flow analysis. *Stain Technology* 53:205–215.

Stoughton RH (1930) Thionin and orange G for the differential staining of bacteria and fungi in plant tissue. *Annals of Applied Biology* 17:162–164.

Stoward PJ (1967) Studies in fluorescence histochemistry. 1. The demonstration of sulphomucin. *Journal of the Royal Microscopical Society* 87:215–235.

Stoward PJ, Pearse AGE (1991) *Histochemistry, Theoretical and Applied*, 4th Edn. Vol. 3. Enzyme Histochemistry. Churchill-Livingstone, Edinburgh.

Straub SG, Kornreich B, Oswald RE, Nemeth EF, Sharp GWG (2000) The calcimimetic R-467 potentiates insulin secretion in pancreatic beta cells by activation of a nonspecific cation channel. *Journal of Biological Chemistry* 275:18777–18784.

Stretton AOW, Kravitz EA (1968) Neuronal geometry: determination with a technique of intracellular dye injection. *Science* 162:132–134.

Stricker SA, Silva R, Smythe T (1998) Calcium and endoplasmic reticulum dynamics during oocyte maturation and fertilization in the marine worm *Cerebratulus lacteus. Developmental Biology* 203:305–322.

Stromberg A, Karlsson A, Ryttsen F, Davidson M, Chiu DT, Orwar O (2001) Microfluidic device for combinatorial fusion of liposomes and cells. *Analytical Chemistry* 73:126–130.

Sturmer D (1977) Syntheses and properties of cyanine and related dyes. In: A Weissberger, EC Taylor (eds) *Chemistry of Heterocyclic Compounds.* Wiley, New York. pp. 441–587.

Stutz MH, Ludemann WH, Saas S (1968) Improved method for preparative chromatography. *Analytical Chemistry* 40:258–259.

Sugihara T, Sawada S, Hakura A, Hori Y, Uchida K, Sagami F (2000) A staining procedure for micronucleus test using new methylene blue and acridine orange: specimens that are supravitally stained with possible long-term storage. *Mutation Research – Genetic Toxicology and Environmental Mutagenesis* 470:103–108.

Suller MTE, Lloyd D (1999) Fluorescence monitoring of antibiotic-induced bacterial damage using flow cytometry. *Cytometry* 35:235–241.

Sumi Y, Inoue T, Muraki T, Suzuki T (1983a) A highly sensitive chelator for metal staining, bromopyridylazo-diethylaminophenol. *Stain Technology* **58**:325–328.

Sumi Y, Inoue T, Muraki T, Suzuki T (1983b) The staining properties of pyridylazophenol analogs in histochemical staining of a metal. *Histochemistry* 77:1–7.

Sumi Y, Itoh MT, Yoshida M, Akama Y (1999) Highly sensitive chelating agents for histochemical staining of rare earth metals. *Histochemistry and Cell Biology* 112:179–182.

Sumner AT (1980) Dye binding mechanisms in G-banding of chromosomes. *Journal of Microscopy* 119:397–406.

Sumner AT (1990) *Chromosome Banding.* Unwin Hyman, London.

Sumner AT (1994) Chromosome banding and identification. *Fluorescence. Methods in Molecular Biology* 29:83–96.

Sumner BEH (1965) Experiments to determine the composition of aldehyde-fuchsin solutions. *Journal of the Royal Microscopical Society* 84:181–187.

Sun JY, Chen XG, Hu ZD (1997) Spectrophotometric flow injection determination of trace iodide in table salt and laver through the reaction of iodate with 3,5-bromo-2-PADAP and thiocyanate. *Fresenius' Journal of Analytical Chemistry* 357:1002–1005.

Sun YX, Ye BX, Wang Y, Tang XR, Zhou XY (1998) Study on the determination of neuro-transmitters using poly(neutral red) coated carbon fiber microelectrodes. *Microchemical Journal* 58:182–191.

Suveges I, Modis L (1970) A new fluorescence microscopic technic for the investigation of the cornea. *Acta Histochemica* 35:85–89.

Suyama K, Yamamoto H, Naganawa T, Iwata T, Komada H (1993) A plate-count method for aerobic cellulose decomposers in soil by Congo red staining. *Soil Science and Plant Nutrition* 39:361–365.

Suzuki T, Sumi Y, Miyazaki K, Muraki T, Nokubi K, Kimura M, Kata M (1978) Histochemical staining of chromium by chrome azurol S. *Acta Histochemica et Cytochemica* 11:46–51.

Swain D, De DN (1990) Differential staining of the cell cycle of plant cells using safranin and indigo-picrocarmine. *Stain Technology* 65:197–204.

Swank L, Davenport HA (1935) Chlorate-osmic-formalin method for degenerating myelin. *Stain Technology* 10:87–90.

Sweat F, Puchtler H, Rosenthal SI (1964) Sirius red F3BA as stain for connective tissue. *Archives of Pathology* 78:69–72.

Swei A, Lacy F, DeLano FA, Schmid Schonbein GW (1997) Oxidative stress in the Dahl hypertensive rat. *Hypertension* 30:1628–1633.

Sykes G (1958) *Disinfection and Sterilization*. Spon, London.

Szeimies RM, Lorenzen T, Karrer S, Abels C, Plettenberg A (2001) Photochemotherapy of cutaneous AIDS-related Kaposi sarcoma with indocyanine green and laser light [German]. *Hautarzt* 52:322–326.

Szentkuti L (1997) Light microscopical observations on luminally administered dyes, dextrans, nanospheres and microspheres in the pre-epithelial mucus gel layer of the rat distal colon. *Journal of Controlled Release* 46:233–242.

Szichy L, Mityko J, Gajdacs (1974) Application of thin-layer chromatography to mill-scale analysis of dyeing techniques. [Hungarian]. *Kolorisztikai Ertesito* 16:31–39.

Szucs S, Vamosi G, Poka R, Sarvary A, Bardos H, Balazs M (1998) Single-cell meas-urement of superoxide anion and hydrogen peroxide production by human neutrophils with digital imaging fluorescence microscopy. *Cytometry* 33:19–31.

Tachibana H, Kobayashi S, Kaneda Y, Takeuchi T, Fujiwara T (1997) Preparation of a monoclonal antibody specific for *Entamoeba dispar* and its ability to distinguish *E. dispar* from *E. histolytica*. *Clinical and Diagnostic Laboratory Immunology* 4:409–414.

Tago H, Kimura H, Maeda T (1986) Visualization of detailed acetylcholinesterase fiber and neuron staining in rat brain by a sensitive histochemical procedure. *Journal of Histochemistry and Cytochemistry* 34:1431–1438.

Takahashi A, Camacho P, Lechleiter JD, Herman B (1999) Measurement of intracellular calcium. *Physiological Reviews* 79:1089–1125.

Takahashi A, Kono K, Amemiya H, Iizuka H, Fujii H, Matsumoto Y (2001) Elevated caspase-3 activity in peripheral blood T cells coexists with increased degree of T-cell apoptosis and down-regulation of TCR zeta molecules in patients with gastric cancer. *Clinical Cancer Research* 7:74–80.

Takahashi Y, Nilsson S, Berggren B (1995) Aeroallergen immunoblotting with human IgE antibody. *Grana* 34:357–360.

Takalo H, Mukkala VM, Mikola H, Liitti P, Hemmila I (1994) Synthesis of europium(III) chelates suitable for labeling of bioactive molecules. *Bioconjugate Chemistry* 5:278–282.

Takamatsu T, Nakanishi K, Fukuda M, Fujita S (1981) Cytofluorometry on cells isolated from paraffin sections after blocking of the background fluorescence by azocarmine G. *Histochemistry* 71:161–170.

Takaya K (1967) Luxol fast blue MBS and phloxine; a stain for mitochondria. *Stain Technology* 42:207–211.

Takeuchi S, Maeda A (1979) Fluorescein mercuric acetate as a probe of the dynamic structure of double-helical DNA. *Biochimica et Biophysica Acta* 563:365–374.

Tamaoki J, Chiyotani A, Takemura H, Konno K, Matsumoto T, Ashida Y (1997) Stimulation of airway mucociliary transport and epithelial ciliary motility by the triazolopyridazin derivative TAK-225. *Journal of Pharmacology and Experimental Therapeutics* 281:1186–1190.

Tamse A, Katz A, Kablan F (1998) Comparison of apical leakage shown by four different dyes with two evaluating methods. *International Endodontic Journal* 31:333–337.

Tamura Z, Abe S, Ito K, Maedo M (1996) Spectrophotometric analysis of the relationship between dissociation and coloration, and of the structural formulas of phenolphthalein in aqueous solution. *Analytical Sciences* 12:927–930.

Tanabe K, Kawasaki T, Maeda M, Tsuji A, Yabuuchi M (1987) Chemiluminescence assay for dihydronicotinamide-adenine dinucleotide and its application to determination of biological substances using 1-methoxyphenazine methosulfate and isoluminol [Japanese]. *Bunseki Kagaku* 36:82–87.

Tanaka KR, the Subcommittee on Cellular Enzymology. (1984) *National Committee for Clinical Laboratory Standards* 4:346–354.

Tandon P, Ishikawa M, Komamine A, Fukuda H (1999) Incorporation of fluorescein-conjugated anti-mouse immunoglobulin G into permeabilized *Nicotiana tabacum* BY-2 cells. *Plant Science* 140:63–69.

Tang DG, Tokumoto YM, Raff MC (2000) Long-term culture of purified postnatal oligo-dendrocyte precursor cells. Evidence for an intrinsic maturation program that plays out over months. *Journal of Cell Biology* 148:971–984.

Tanke HJ, De Haas RR, Sagner G, Ganser M, van Gijlswijk RPM (1998) Use of platinum coproporphyrin and delayed luminescence imaging to extend the number of targets FISH karyotyping. *Cytometry* 33:453–459.

Tantular IS, Iwai K, Lin K, Basuki S, Horie T, Htay HH, Matsuoka H, Marwoto H, Wongsrichanalai C, Dachlan YP, Kojima S, Ishii A, Kawamoto F (1999) Field trials of a rapid test for G6PD deficiency in combination with a rapid diagnosis of malaria. *Tropical Medicine* 4:245–250.

Tas J, van der Ploeg M, Mitchell JP, Cohn NS (1980) Protein staining methods in quanti-tative cytochemistry. *Journal of Microscopy* 119:295–311.

Tas J, Mendelson D, Noorden CJF (1983) Cuprolinic blue: a specific dye for single-stranded RNA in the presence of magnesium chloride. I. Fundamental aspects. *Histochemical Journal* 15:801–804.

Tato A, Ferrer JM, Quintana E, Romero JB, Del Castillo P, Stockert JC (1990) Observations on the contrasting reaction of some electron dense stains applied on epoxy-embedded tissue sections. *Zeitschrift fur mikroskopische-anatomische Forschung* 104:337–348.

Taylor CR (1992a) Quality assurance and standardization in immunohistochemistry. A proposal for the annual meeting of the Biological Stain Commission, June, 1991. *Biotechnic and Histochemistry* 67:110–117.

Taylor CR (1992b) Report of the Immunohistochemistry Steering Committee of the Biological Stain Commission; 'Proposed format: Package insert for immunohisto-chemical products'. *Biotechnic and Histochemistry* 67:323–338.

Taylor CR (1994) *Immunomicroscopy: A Diagnostic Tool for the Surgical Pathologist.* 2nd Edn. Saunders, Philadelphia.

Taylor DL, Wang YL, eds (1989) *Fluorescence Microscopy of Living Cells in Culture. Part B. Quantitative Fluorescence Microscopy – Imaging and Spectroscopy.* Academic Press, San Diego.

Teichman RJ, Fujimoto M, Yanagimachi R (1972) A previously unrecognised material in mammalian spermatozoa as revealed by malachite green and pyronine. *Biology of Reproduction* 7:73–81.

Tepperman BL, Kiernan JA, Soper BD (1989) The effect of sialoadenectomy on gastric mucosal integrity in the rat: roles of epidermal growth factor and prostaglandin E2. *Canadian Journal of Physiology and Pharmacology* 67:1512–1519.

Terasaki M (1993) Probes for the endoplasmic reticulum. In: WT Mason, WT Real (eds) *Fluorescent and Luminescent Probes for Biological Activity: A Practical Guide to Technology for Quantitative Analysis (Biological Techniques)*. Academic Press, London. pp. 120–123.

Terasaki M (1994) Redistribution of cytoplasmic components during germinal vesicle breakdown in starfish oocytes. *Journal of Cell Science* 107:1797–1805.

Terasaki M (1998) Labeling of the endoplasmic reticulum with DiOC6(3). In: JE Celis (ed.) *Cell Biology:A Laboratory Handbook*. Academic Press, London. pp. 501–506.

Terasaki M (2000) Dynamics of the endoplasmic reticulum and golgi apparatus during early sea urchin development. *Molecular Biology of the Cell* 11:897–914.

Terner JY, Gurland J, Gaer F (1964) Phosphotungstic acid-hematoxylin; spectrophotometry of the lake in solution and in stained tissue. *Stain Technology* 39:141–154.

Terra WR, Regel R (1995) pH buffering in *Musca domestica* midguts. *Comparative Biochemistry and Physiology A – Physiology* 112:559–564.

Tettey JNA, Skellern GG, Midgley JM, Grant MH, Wilkinson R, Pitt AR (1999) Intracellular localization and metabolism of the phenanthridium trypanocide, ethidium bromide, by isolated rat hepatocytes. *Xenobiotica* 29:349–360.

Thanawongnuwech R, Brown GB, Halbur PG, Roth JA, Royer RL, Thacker BJ (2000) Pathogenesis of porcine reproductive and respiratory syndrome virus-induced increase in susceptibility to *Streptococcus suis* infection. *Veterinary Pathology* 37:143–152.

Thanos S, Kacza J, Seeger J, Mey J (1994) Old dyes for new scopes – The phagocytosis-dependent long-term fluorescence labeling of microglial cells *in vivo*. *Trends in Neurosciences* 17:177–182.

Thanos S, Fischer D, Pavlidis M, Heiduschka P, Bodeutsch N (2000) Glioanatomy assessed by cell–cell interactions and phagocytic labelling. *Journal of Neuroscience Methods* 103:39–50.

Thatte U, Bagadey S, Dahanukar S (2000) Modulation of programmed cell death by medicinal plants. *Cellular and Molecular Biology* 46:199–214.

Thom SM, Horobin RW, Seidler E, Barer MR (1993) Factors affecting the selection and use of tetrazolium salts as cytochemical indicators of microbial viability and activity. *Journal of Applied Bacteriology* 74:433–443.

Thomas M, Nicklee T, Hedley DW (1995) Differential effects of depleting agents on cytoplasmic and nuclear nonprotein sulfhydryls – a fluorescence image cytometry study. *British Journal of Cancer* 72:45–50.

Thomas PGA, Ball BA (1996) Cytofluorescent assay to quantify adhesion of equine spermatozoa to oviduct epithelial cells *in vitro*. *Molecular Reproduction and Development* 43:55–61.

Thomas RL, Matsko CM, Lotze MT, Amoscato AA (1999) Mass spectrometric identification of increased C16 ceramide levels during apoptosis. *Journal of Biological Chemistry* 274:30580–30588.

Thompson SW (1966) *Selected Histochemical and Histopathological Methods*. Thomas, Springfield, IL.

Thomsen JS, Ebbesen EN, Mosekilde L (1998) Relationships between static histomorphometry and bone strength measurements in human iliac crest bone biopsies. *Bone* 22:153–163.

Thrane C, Olsson S, Nielsen TH, Sorenson J (1999) Vital fluorescent stains for detection of stress in *Pythium ultimum* and *Rhizoctonia solani* challenged with viscosinamide from *Pseudomonas fluorescens* DR54. *FEMS Microbiology and Ecology* 30:11–23.

Tian H, Huhmer AFR, Landers JP (2000) Evaluation of silica resins for direct and efficient extraction of DNA from complex biological matrices in a miniaturized format. *Analytical Biochemistry* 283:175–191.

Tian WH, Festoff BW, Blot S, Diaz J, Hantai D (1995) Synaptic transmission blockade increases plasminogen-activator activity in mouse skeletal-muscle poisoned with botulinum toxin type-A. *Synapse* 20:24–32.

Tiano L, Ballarini P, Santoni G, Wozniak M, Falcioni G (2000) Morphological and functional changes of mitochondria from density separated trout erythrocytes. *Biochemica et Biophysica Acta* 1457:118–128.

Tilak BD (1971) Naphthoquinonoid dyes and pigments. Ch. 1 In: K Venkataraman (ed.) *The Chemistry of Synthetic Dyes*. Academic Press, New York. pp. 1–55.

Timmers ACJ, Reiss HD, Bohsung J, Traxel K, Schel JHN (1996) Localization of calcium during somatic embryogenesis of carrot (*Daucus carota* L). *Protoplasma* 190:107–118.

Tisserant B, Gianinazzi-Pearson V, Gianinazzi S, Gollotte A (1993) Implanta histochemical staining of fungal alkaline phosphatase activity for analysis of efficient arbuscular mycorrhizal infections. *Mycological Research* 97:245–250.

Tojyo Y, Tanimura A, Nezu A, Matsumoto Y (1998) Activation of beta-adrenoceptors does not cause any change in cytosolic Ca^{2+} distribution in rat parotid acinar cells. *European Journal of Pharmacology* 360:73–79.

Tolivia J, Navarro A, Tolivia D (1994) Polychromatic staining of epoxy semithin sections. A new and simple method. *Histochemistry* 101:51–55.

Tolstoouhov AV (1928) The effect of preliminary treatment (fixing fluid) on staining properties of the tissues. *Stain Technology* 3:49–56.

Tominaga T, Tominaga Y, Yamada H, Matsumoto G, Ichikawa M (2000) Quantification of optical signals with electrophysiological signals in neural activities of Di-4-ANEPPS stained rat hippocampal slices. *Journal of Neuroscience Methods* 102:11–23.

Tormey WP, O'Brien PA (1993) Clinical associations of an increased transthyretin band in routine serum and urine protein electrophoresis. *Annals of Clinical Biochemistry* 30:550–554.

Torres JJ, Soler A, Saez J, Ortuno JF (1997) Determination of low concentration of sulphorhodamine B in wastewaters and stabilization ponds. *Water Research* 31:3183–3186.

Tran D, Golick M, Rabinovitz H, Rivlin D, Elgart G, Nordlow B (2000) Hematoxylin and safranin O staining of frozen sections. *Dermatologic Surgery* 26:197–199.

Tran NN, Leroy P, Bellucci L, Robert A, Nicolas A, Atkinson J, Capdeville-Atkinson C (1995) Intracellular concentrations of fura-2 and fura-2/am in vascular smooth muscle cells following perfusion loading of fura-2/am in arterial segments. *Cell Calcium* 18:420–428.

Trawoger R, Kolobow T, Cereda M, Giacomini M, Usuki J, Horiba K, Ferrans VJ (1997) Clearance of mucus from endotracheal tubes during intratracheal pulmonary ventilation. *Anesthesiology* 86:1367–1374.

Trevan DJ, Sharrock A (1951) A methyl green-pyronin-orange G stain for formalin-fixed tissues. *Journal of Pathology and Bacteriology* 63:326–329.

Trevor AJ, Katzung BG, Masters SB (2002) K*atzung and Trevor's Pharmacology: Examination and Board Review*. 6th Edn. Lange Medical/McGraw-Hill, New York.

Trigoso CI, Stockert JC (1995) Fluorescence of the natural dye saffron – Selective reaction with eosinophil leukocyte granules. *Histochemistry and Cell Biology* 104:75–77.

Trigoso CI, Espada J, Stockert JC (1995) Fluorescence of eosinophil leucocyte granules induced by 1-hydroxy-3,6,8-pyrenetrisulfonate. Visualization of differences in protein isoelectric points. *Histochemistry and Cell Biology* 104:69–73.

Trindade AV, Brandao APS, de Sousa AF, Farias CF, de Brito-Gitirana L (1998) Enhancement of the paraldehyde-fuchsin staining. *Journal of Histotechnology* 21:147–150.

Trojanowski JQ (1983) Native and derivatized lectins for *in vivo* studies of neuronal connectivity and neuronal cell biology. *Journal of Neuroscience Methods* 9:185–204.

Trosko JE, Chang CC, Wilson MR, Upham B, Hayashi T, Wade M (2000) Gap junctions and the regulation of cellular functions of stem cells during development and differentiation. *Methods* 20:245–264.

Trotter PJ (2000) A novel pathway for transport and metabolism of a fluorescent phosphatidic acid analog in yeast. *Traffic* 1:425–434.

Tsai MY, Morfini G, Szebenyi G, Brady ST (2000) Release of kinesin from vesicles by hsc70 and regulation of fast axonal transport. *Molecular Biology of the Cell* 11:2161–2173.

Tsakalidou E, Manolopoulou E, Tsilibari V, Georgalaki M, Kalant-Zopoulos G (1993) Esterolytic activities of *Enterococcus durans* and *Enterococcus faecium* strains isolated from Greek cheese. *Netherlands Milk and Dairy Journal* 47:145–150.

Tsau Y, Wenner P, O'Donovan MJ, Cohen LB, Loew LM, Wuskell JP (1996) Dye screening and signal-to-noise ratio for retrogradely transported voltage-sensitive dyes. *Journal of Neuroscience Methods* 70:121–129.

Tseng WC, Haselton FR, Giorgio TD (1999) Mitosis enhances transgene expression of plasmid delivered by cationic liposomes. *Biochemica et Biophysica Acta – Gene Structure and Expression* 1445:53–64.

Tsubota K, Kaido M, Yagi Y, Fujihara T, Shimmura S (1999) Diseases associated with ocular surface abnormalities: the importance of reflex tearing. *British Journal of Ophthalmology* 83:89–91.

Tsuji A, Koshimoto H, Sato Y, Hirano M, Sei-Iida Y, Kondo S, Ishibashi K (2000) Direct observation of specific messenger RNA in a single living cell under a fluorescence microscope. *Biophysical Journal* 78:3260–3274.

Tsuji A, Sato Y, Hirano M, Suga T, Koshimoto H, Taguchi T, Ohsuka S (2001) Development of a time-resolved fluorometric method for observing hybridization in living cells using fluorescence resonance energy transfer. *Biophysical Journal* 81:501–515.

Tsuji H, Kohli Y, Fukumitsu S, Morita K, Kaneko H, Ohkawara T, Minami M, Ueda K, Sawa Y, Matsuzaki H, Morinaga O, Ohkawara Y (1999) *Helicobacter pylori*-negative gastric and duodenal ulcers. *Journal of Gastroenterology* 34:455–460.

Tsukise A, Fujimori O, Yamada K (1990) An efficient histochemical method for deoxyribonucleic acids using a silver enhancement procedure. *Histochemical Journal* 22:409–415.

Tsutsumi Y, Onada N, Osamura RY (1990) Victoria blue-hematoxylin and eosin staining – a useful routine stain for demonstration of invasion by cancer cells. *Journal of Histotechnology* 13:271–274.

Tu SI, Patterson D, Shen SY, Brauer D, Hsu AF (1996) NADH-linked electron-transfer induces Cd^{2+} movement in corn root plasma membrane vesicles. *Plant and Cell Physiology* 37:141–146.

Tucker JD, Breneman JW, Lee DA, Ramsey MJ, Swiger RR (1994) Fluorescence *in situ* hybridization of human and mouse DNA probes to determine the chromosomal contents of cell lines and tumors. In: JE Celis (ed.) *Cell Biology: A Laboratory Handbook.* Academic Press, New York. pp. 450–458.

Turner HA (1955) Theories of dyeing. 1. Dyeing processes at completion. *Journal of the Society of Dyers and Colourists* 71:29–46.

Tusl T (1968) The separation of alizarin complexan from impurities by paper chromatography. *Journal of Chromatography* 37:546–548.

Tuvia S, Garver TD, Bennett V (1997) The phosphorylation state of the FIGQY tyrosine of neurofascin determines ankyrin-binding activity and patterns of cell segregation. *Proceedings of the National Academy of Sciences of the United States of America* 94:12957–11962.

Uga S, Wan Q, Hatono N, Ishikawa S (1994) Response of mouse lens to central needle injury. *Ophthalmic Research* 26:181–188.

Ukhanov K, Ukhanova M, Taylor CW, Payne R (1998) Putative inositol 1,4,5-triphosphate receptor localized to endoplasmic reticulum in *Limulus* photoreceptors. *Neuroscience* 86:23–28.

Ullrich S, Karrasch B, Hoppe HG, Jeskulke K, Mehrens M (1996) Toxic effects on bacterial metabolism of the redox dye 5-cyano-2,3-ditolyl tetrazolium chloride. *Applied and Environmental Microbiology* 62:4587–4593.

Uncini A, Di Muzio A, Di Guglielmo G, De Angelis MV, De Luca G, Lugaresi A, Gambi D (1999) Effect of rhTNF-alpha injection into rat sciatic nerve. *Journal of Neuroimmunology* 94:88–94.

Unkenholz EG, Brown ML, Pope KL (1997) Oxytetracycline marking efficacy for yellow perch fingerlings. *Progressive Fish-Culturist* 59:280–284.

Unna PC (1891) Uber die Reifung unsrer Farbstoffe. *Zeitschrift fur wissenschaftliche Mikroskopie* 8:475–487.

Utakoji JR, Masukuma S (1974) Fluorescent staining of L cell centromeres and chromocenters with 1-dimethylaminonaphthalene-5-sulfonyl chloride and G-bandings. *Experimental Cell Research* 87:111–119.

Vacca LL (1985) *Laboratory Manual of Histochemistry.* Raven Press, New York.

Vachier I, LeDoucen C, Loubatiere J, Damon M, Terouanne B, Nicolas JC, Chanez P, Godard P (1994) Imaging reactive oxygen species in asthma. *Journal of Bioluminescence and Chemiluminescence* 9:171–175.

van Amstel TNM, Kengen HMP (1996) Callose deposition in the primary wall of suspension cells and regenerating protoplasts, and its relationship to patterned cellulose synthesis. *Canadian Journal of Botany* 74:1040–1049.

van Beneden E, Julin C (1884) La spermatogenese chez l'Ascaride megalocephale. *Memoires de l'Academie Royale des Sciences, des Lettres et de Beaux-Arts de Belgique* 7:312–342.

Van den Munckhof RJM (1996) *In situ* heterogeneity of peroxisomal oxidase activities: an update. *Histochemical Journal* 28:401–429.

van der Loos CM (1999) *Immunoenzyme Multiple Staining Methods.* BIOS Scientific Publishers, Oxford.

Van der Reijden ED, Monchamp JD, Lewis SM (1997) The formation, transfer, and fate of spermatophores in *Photinus* fireflies (Coleoptera: Lampyridae). *Canadian Journal of Zoology* 75:1202–1207.

Van der Veen H, Hoekstra OS, Paul MA, Cuesta MA, Meijer S (1994) Gamma probe-guided sentinel node biopsy to select patients with melanoma for lymphadenectomy. *British Journal of Surgery* 81:1769–1770.

Van Driel BEM, De Goeij AFPM, Song JY, De Bruine AP, Van Noorden CJF (1997) Development of oxygen insensitivity of the quantitative histochemical assay of G6PDH activity during colorectal carcinogenesis. *Journal of Pathology* 182:398–403.

Van Duijn P (1956) A histochemical specific thionine-SO_2 reagent and its use in a bi-color method for deoxyribonucleic acid and periodic acid Schiff positive substances. *Journal of Histochemistry and Cytochemistry* 4:55–63.

van Gieson J (1889) Laboratory notes of technical methods for the nervous system. *New York Medical Journal* 50:57–60.

Van Ginneken CJ, De Smet MJ, Van Meir FJ, Weyns AA (1998) Microwave staining of intestinal whole-mount preparations with cuprolinic blue. *Histochemical Journal* 30:703–709.

Van Ginneken CJ, De Smet MJ, Van Meir FJ, Weyns AA (1999) Microwave staining of enteric neurons using cuprolinic blue (quinolinic phthalocyanin) combined with enzyme histochemistry and peroxidase immunohistochemistry. *Journal of Histochemistry and Cytochemistry* 47:13–21.

Van Holde KE (1985) *Physical Biochemistry.* 2nd Edn. Prentice-Hall, Englewood Cliffs, NJ.

van Landeghem GF, DHaese PC, Lamberts LV, Djukanovic L, Pejanovic S, Goodman WG, DeBroe ME (1998) Low serum aluminum values in dialysis patients with increased bone aluminum levels. *Clinical Nephrology* 50:69–76.

van Leeuwenhoek A. (1719) *Epistolae physiologicae super compluribus naturae arcanis* (In volume III of *Opera omnia.* Leiden; Langerak. (Republished in 1722)).

Van Noorden CJF (1983) Sensitivity to light of solutions of phenazine methosulfate. *Histochemical Journal* 15:275–276.

Van Noorden CJF, Frederiks WM (1992) *Enzyme Histochemistry: A Laboratory Manual of Current Methods* (Royal Microscopical Society Handbooks, 26). Oxford University Press, Oxford.

Van Noorden CJF, Gossrau R (1991) Quantitative histochemical and cytochemical assays. In: PB Bach, JR Baker (eds) *Histochemical and Immunohistochemical Techniques: Applications to Pharmacology and Toxicology.* Chapman and Hall, London. pp. 114–145.

Vandaele DJ, Perlman AL, Cassell MD (1995) Intrinsic fiber architecture and attachments of the human epiglottis their contributions to the mechanism of deglutition. *Journal of Anatomy* 186:1–15.

Vandelft JL, Ossterhuis JA, Barthen ER, Eijkelenboom AM (1983) Janus green vital staining of the cornea. *Documenta Ophthalmologica* 55:47–50.

Vaney DI (1992) Photochromic intensification of diaminobenzidine reaction-product in the presence of tetrazolium salts – applications for intracellular labelling and immuno-histochemistry. *Journal of Neuroscience Methods* 44:217–223.

Vargic T, Mrsa V (1994) Detection of exo-beta-1,3-glucanase activity in polyacrylamide gels after electrophoresis under denaturing or nondenaturing conditions. *Electrophoresis* 15:902–906.

Vaughan REA (1914) A method for the differential staining of fungus and host cells. *Annals of the Missouri Botanical Garden* 1:241–242.

Vazquezduhalt R, Semple KM, Westlake DWS, Fedorak PM (1993) Effect of water-miscible organic solvents on the catalytic activity of cytochrome C. *Enzyme and Microbial Technology* 15:936–943.

Vega SY, Rust MK (2001) Developing marking techniques to study movement and foraging of Argentine ants (Hymenoptera: Formicidae). *Sociobiology* 37:27–39.

Vejar L, LeCerf P (1997) Sputum lipid laden macrophages in the diagnosis of pulmonary aspiration in children [Spanish]. *Revista Medica de Chile* 125:191–194.

Velazquez JLP, Frantseva MV, Carlen PL (1997) *In vitro* ischemia promotes glutamate-mediated free radical generation and intracellular calcium accumulation in hippocampal pyramidal neurons. *Journal of Neuroscience* 17:9085–9094.

Vellar ID (2001) Preliminary study of the anatomy of the venous drainage of the intra-hepatic and extrahepatic bile ducts and its relevance to the practice of hepatobiliary surgery. *ANZ Journal of Surgery* 71:418–422.

Venkataraman K, ed. (1952–1978) *The Chemistry of Synthetic Dyes.* v. 1–8. Academic Press, New York.

Venkataraman K (1977) *The Analytical Chemistry of Synthetic Dyes.* Wiley, Chichester, UK.

Verbelen JP, Stickens D (1995) *In vivo* determination of fibril orientation in plant cell walls with polarization CSLM. *Journal of Microscopy* 177:1–6.

Vercelli A, Innocenti GM (1993) Morphology of visual callosal neurons with different locations, contralateral targets or patterns of development. *Experimental Brain Research* 94:393–404.

Vercelli A, Repici M, Garbossa D, Grimaldi A (2000) Recent techniques for tracing pathways in the central nervous system of developing and adult mammals. *Brain Research Bulletin* 51:11–28.

Verheyen G, Joris H, Crits K, Nagy Z, Tournaye H, Van Steirteghem A (1997) Comparison of different hypo-osmotic swelling solutions to select viable immotile spermatozoa for potential use in intracytoplasmic sperm injection. *Human Reproduction Update* 3:195–203.

Verity PG, Beatty TM, Williams SC (1996) Visualization and quantification of plankton and detritus using digital confocal microscopy. *Aquatic Microbial Ecology* 10:55–67.

Verna C, Zaffe D, Siciliani G (1999) Histomorphometric study of bone reactions during orthodontic tooth movement in rats. *Bone* 24:371–379.

Vertutdoi A, Ohnishi SI, Bolard J (1994) The endocytic process in CHO cells, a toxic pathway of the polyene antibiotic Amphotericin B. *Antimicrobial Agents and Chemotherapy* 38:2373–2379.

Vickerstaff T (1954) *The Physical Chemistry of Dyeing*. 2nd Edn. Oliver and Boyd, London.

Vico P, Coessens B, Heymans O, Vandeweyer E (1994) New mixture for simultaneous anatomical and radiological cadaver studies. *European Journal of Plastic Surgery* 17:17–19.

Vidal BC (1980) Aorta elasticae and tendon collagen reactivity to 8-anilino-1-naphthalene sulphate (ANS) and dansylchloride. *Cellular and Molecular Biology* 26:583–588.

Vidal M, Mangeat P, Hoekstra D (1997) Aggregation reroutes molecules from a recycling to a vesicle-mediated secretion pathway during reticulocyte maturation. *Journal of Cell Science* 110:1867–1877.

Villaneuva AR, Longo JA, Weiner G (1994) Staining and histomorphometry of micro-cracks in the human femoral head. *Biotechnic and Histochemistry* 69:81–88.

Villatte F, Bachman TT, Hussein AS, Schmid RD (2001) Acetylcholinesterase assay for rapid expression screening in liquid and solid media. *BioTechniques* 30:81–84,86.

Villaverde S, Fernandez MT (1997) Non-toluene-associated respiration in a *Pseudomonas putida* 54G biofilm grown on toluene in a flat-plate vapor-phase bio-reactor. *Applied Microbiology and Biotechnology* 48:357–362.

Visser R (1967) Continuous preparative thin-layer chromatography. *Analytica Chimica Acta* 38:157–162.

Vit JP, Moustacchi E, Rosselli F (2000) ATM protein is required for radiation-induced apoptosis and acts before mitochrondrial collapse. *International Journal of Radiation Biology* 76:841–851.

Vogt W (1925) Gestaltungsanalyse am Amphibienkeim mit ortlicher Vitalfarbung. 1. Methodik. *Roux Archiv fur Entwicklungs-mechanik* 106:542–610.

Vogtle F, Knops P (1991) Dyes for visual distinction between enantiomers – crown ethers as optical sensors for chiral compounds. *Angewandte Chemie* (International Edition in English) 30:958–960.

Volk B, Popper H (1944) Histological demonstration of fat in urine and stool by means of fluorescence microscopy. *American Journal of Clinical Pathology* 14:234–238.

Volkmann R, Strauss F (1934) Ein Ersatz fur die Hornowskysche Kombination zur Darstellung von Elastin, Muskulatur und Kollagen auf der Basis der Azanmethode. Zeitschrift fur wissenschaftliche Mikroskopie 51:244–249 (Abstract in *Stain Technology* 10:70).

von Bohlen und Halbach O, Kiernan JA (1999) Diaminobenzidine induces fluorescence in nervous tissue and provides intrinsic counterstaining of sections prepared for peroxidase histochemistry. *Biotechnic and Histochemistry* 74:236–243.

von Gerlach J (1858) *Mikroskopische Studien aus dem Gebiete der menschlichen Morphologie.* Enke, Erlangen.

von Recklinghausen F (1860) Eine Methode mikroskopische hohle und solide Gebilde von einander zu unterschneiden. *Virchows Archiv fur pathologische Anatomie und Physiologie und fur klinische Medizin* 19:451–452.

Vongsavan N, Matthews B (1991) The permeability of cat dentin *in vivo* and *in vitro*. *Archives of Oral Biology* 36:641–646.

Vroom JM, DeGrauw KJ, Gerritsen HC, Bradshaw DJ, Marsh PD, Watson GK, Birmingham JJ, Allison C (1999) Depth penetration and detection of pH gradients in biofilms by two-photon excitation microscopy. *Applied and Environmental Microbiology* 65:3502–3511.

Wagner L, Base W, Wiesholzer M, Sexl V, Bognar H, Worman CP (1991) Detection of BLT substrate specific proteases in individual human peripheral blood leukocytes and bone marrow cells – Application of the method to the classification of leukemia. *Journal of Immunological Methods* 142:147–155.

Wainwright M (1996) Non-porphyrin photosensitizers in biomedicine. *Chemical Society Reviews* 25:351–359.

Wainwright M, Phoenix DA, Marland J, Wareing DRA, Bolton FJ (1997) *In vitro* photobactericidal activity of aminoacridines. *Journal of Antimicrobial Chemotherapy* 40:587–589.

Wainwright M, Phoenix DA, Laycock SL, Wareing DRA, Wright PA (1998) Photobactericidal activity of phenothiazinium dyes against methicillin-resistant strains of *Staphylococcus aureus*. *FEMS Microbiology Letters* 160:177–181.

Walczysko P, Wagner E, Albrechtova JTP (2000) Use of co-loaded fluo-3 and fura red fluorescent indicators for studying the cytosolic Ca^{2+} concentrations distribution in living plant tissue. *Cell Calcium* 28:23–32.

Waldeyer W (1863) Untersuchungen uber den Ursprung und den Verlauf des Axencylinders bei Wirbelllosen und Wirbelthieren, sowie uber dessen Endverhalten in den quergestreiften Muskelfaser. *Henle und Pfeufers Zeitschrift fur rationale Medizin* 20:193–256.

Waldrop FS, Puchtler H (1975) Luxol fast blue MBSN-Levafix red violet E-2BL, a combined stain for myelin sheaths and glia fibers. *Archives of Pathology* 99:529–532.

Walker JM (1994) The dansyl-Edman method for peptide sequencing. *Methods in Molecular Biology* 32:329–334.

Wall J, Solomon A (1999) Flow cytometric characterization of amyloid fibrils. *Methods in Enzymology* 309:460–466.

Wallace DJ (1989) The use of quinacrine (Atabrine) in rheumatic diseases – a reexamination. *Seminars in Arthritis and Rheumatism* 18:282–296.

Wang BH, Lu ZX, Polya GM (1997) Inhibition of eukaryote protein kinases by isoquinoline and oxazine alkaloids. *Planta Medica* 63:494–498.

Wang H, Zhu XW, Wang YZ (1999) Fluorescence measurement of proton pumping of tonoplast H+-APTase [Chinese]. *Progress in Biochemistry and Biophysics* 26:184–187.

Wang Q, Ko KS, Kapus A, McCulloch CAG, Ellen RP (2001) A spirochete surface protein uncouples store-operated calcium channels in fibroblasts: a novel cytotoxic mechanism. *Journal of Biological Chemistry* 276:23056–23064.

Wang SL, Li J, Zhu XZ, Zhao ZT, Sun T, Dong H, Zhang YG (1998) Gland atrophy following retrograde injection of methyl violet as a treatment in chronic obstructive parotitis. *Oral Surgery, Oral Medicine, Oral Pathology, Oral Radiology and Endodontics* 85:276–281.

Wang WH, Abeydeera LR, Han YM, Prather RS, Day BN (1999) Morphologic evaluation and actin filament distribution in porcine embyros produced *in vitro* and *in vivo*. *Biology of Reproduction* 60:1020–1028.

Wang XY, Wong WC, Ling EA (1995) Localization of NADPH-diaphorase activity in the submucous plexus of the guinea-pig intestine: light and electron-microscopic studies. *Journal of Neurocytology* 24:271–281.

Wang YL, Taylor DL, eds (1989) *Fluorescence Microscopy of Living Cells in Culture. Part A. Fluorescent Analogs, Labeling Cells, and Basic Microscopy.* Academic Press, San Diego.

Ward RK, Nation PN, Maxwell M, Barker CL, Clothier RH (1997) Evaluation *in vitro* of epidermal cell keratinization. *Toxicology in Vitro* 11:633–636.

Waring DR, Hallas G, eds (1990) *The Chemistry and Application of Dyes.* Plenum Press, New York.

Warnakulasuriya KAAS, Johnson NW (1996) Sensitivity and specificity of OraScan(R) toluidine blue mouthrinse in the detection of oral cancer and precancer. *Journal of Oral Pathology and Medicine* 25:97–103.

Wasiluk KR, Fulcher RG, Jones RJ, Gengenbach BG (1994) Characterization of starch granules in maize using microspectrophotometry. *Starch* 46:369–373.

Watanabe J, Mondo H, Takeda K, Takamori Y, Kanamura S (1999) Amplification of immunostaining intensity by the peroxidase-antiperoxidase, avidin-biotin-peroxidase complex or catalyzed signal amplication method gives erroneous antigen content in sections. *Acta Histochemica et Cytochemica* 32:407–413.

Watanabe N, Tsukada N, Smith CR, Phillips MJ (1991) Motility of bile canaliculi in the living animal: implications for bile flow. *Journal of Cell Biology* 113:1069–1080.

Watanabe Y, Ito T, Harada T, Kobayashi S, Ozaki T, Nimura Y (1995) Spatial-distribution and pattern of extrinsic nerve strands in the aganglionic segment of congenital aganglionosis – stereoscopic analysis in spotting lethal rats. *Journal of Pediatric Surgery* 30:1471–1476.

Watanabe Y, Konishi M, Shimada M, Ohara H, Iwamoto S (1998) Estimation of age from the femur of Japanese cadavers. *Forensic Science International* 98:55–65.

Webster J, Stone BA (1994) Isolation, histochemistry monosaccharide composition of the walls of roots hairs from *Heterozostera tasmanica* (Martens-ex-Aschers) Denhartog. *Aquatic Botany* 47:29–37.

Weenk GH, Vandenbrink JA, Struijk CB, Mossel DAA (1995) Modified methods for the enumeration of spores of mesophilic *Clostridium* species in dried foods. *International Journal of Food Microbiology* 27:185–200.

Wehby RG, Frank ME (1999) NOS- and non-NOS NADPH diaphorases in the insular cortex of the Syrian golden hamster. *Journal of Histochemistry and Cytochemistry* 47:197–207.

Weidner WJ, Sillman AJ (1997) Low levels of cadmium chloride damage the corneal endothelium. *Archives of Toxicology* 71:455–460.

Weigele M, DeBernardo SL, Tengi JP, Leimgruber W (1972) A novel reagent for the fluoro-metric assay of primary amines. *Journal of the American Chemical Society* 94:5927–5928.

Weigert C (1878) Bismarckbraun als Farbemittel. *Archiv fur mikroskopische Anatomie* 15:258–260.

Weinberg PD, Winlove CP, Parker KH (1994) Measurement of absolute tracer concentra-tions in tissue sections using digital imaging fluorescence microscopy. Application to the study of plasma protein uptake by the arterial wall. *Journal of Microscopy* 173:127–141.

Weinhaus AJ, Bhagroo NV, Brelje TC, Sorenson RL (2000) Dexamethasone counteracts the effect of prolactin on islet function: implications for islet regulation in late pregnancy. *Endocrinology* 141:1384–1393.

Weisbart RH, Baldwin R, Huh B, Zack DJ, Nishimura R (2000) Novel protein transfection of primary rat cortical neurons using an antibody that penetrates living cells. *Journal of Immunology* 164:6020–6026.

Weise DA (1980) A histochemical study of protein-bound SH and SS in chromosomes of hyacinth and fava bean root tips. *Histochemistry* 70:29–32.

Weiss M, Pick U (1991) Uptake of the fluorescent indicator atebrin into acidic vaculoles in the halotolerant alga *Dunaliella salina*. *Planta* 185:494–501.

Weissenberg R (1937) Bismark brown as a vital dye for localised staining on the egg of lamphrey and the opportunity of preserving it for paraffin sections. *Collecting Net* 12:131.

Welcher FJ (1949) *Organic Analytical Reagents*. Van Nostrand, New York.

Wendland B, McCaffery JM, Xiao Q, Emr SD (1996) A novel fluorescence-activated cell sorter-based screen for yeast endocytosis mutants identifies a yeast homologue of mammalian eps15. *Journal of Cell Biology* 135:1485–1500.

Wenyou N, Junzhu H, Mau ZX (1991) A new staining method for constriction marks in skin. *Forensic Science International* 50:147–152.

Werner T, Klimant I, Huber C, Krause C, Wolfbeis OS (1999) Fiber optic ion-microsensors based on luminescence lifetime. *Mikrochimica Acta* 131:25–28.

Wessendorf MW (1991) Fluorogold – composition, and mechanism of uptake. *Brain Research* 553:135–148.

Westermark GT, Johnson KH, Westermark P (1999) Staining methods for identification of amyloid in tissue. *Methods in Enzymology* 309:3–25.

Whitaker JE, Haugland RP, Prendergast FG (1991) Spectral and photophysical studies of benzo[c]xanthene dyes: dual emission pH sensors. *Analytical Biochemistry* 194:330–344.

White WL, Harbuck SC, Noble LW (1994) Demonstration of nucleolar organizer region associated protein – utility of the Bielschowsky stain and counterstains. *Journal of Histotechnology* 17:99–103.

Whitman GF, Boldt JP, Martinez JE, Pantazis CG (1991) Flow cytometric analysis of induced human Graafian follicles. 1. Demonstration and sorting of two luteinized cell-populations. *Fertility and Sterility* 56:259–264.

Wiatr SM (1976) Acridine orange fluorescence in plastic-embedded root tips of *Vicia faba*. *Mikroskopie* 32:313–317.

Wiederkehr A, Meier KD, Riezman H (2001) Identification and characterization of *Saccharomyces cerevisiae* mutants defective in fluid-phase endocytosis. *Yeast* 18:759–773.

Wiedorn KH, Kuhl H, Galle J, Caselitz J, Vollmer E (1999) Comparison of in-situ hybridization, direct and indirect in-situ PCR as well as tyramide signal amplification for the detection of HPV. *Histochemistry and Cell Biology* 111:89–95.

Wilhelm D, Mansmann U, Neudeck H, Matejevic D, Vetter K, Graf R (1999) Increase of segments of elastic-type blood vessel walls in fetal placental stem villi during pre-eclampsia at term. *Anatomy and Embryology* 200:597–605.

Willeford KO, Parker TA, Diehl SV (1998) Efficacy of phloxine B as a bacteriocidal agent in plants. *Journal of Agricultural and Food Chemistry* 46:1637–1641.

Williams BL, Wilson K (1976) Chromatographic techniques. In: BL Williams, K Wilson (eds) *A Biologist's Guide to Principles and Techniques of Practical Biochemistry*. Edward Arnold, London. pp. 52–98.

Williams JM, Uebelhart D, Thonar EJMA, Kocsis K, Modis L (1996) Alteration and recovery of the spatial orientation of the collagen network of articular cartilage in adolescent rabbits following intraarticular chymopapain. *Connective Tissue Research* 34:105–117.

Williams JT, Southerland SS, Souza J, Calcutt AJ, Cartledge RG (1999) Cells isolated from adult skeletal muscle capable of differentiating into multiple mesodermal phenotypes. *American Surgeon* 65:22–26.

Williams TG, Colman B (1994) Rapid separation of carbonic anhydrase isozymes using cellulose acetate membrane electrophoresis. *Journal of Experimental Botany* 45:153–158.

Williamson B, Duncan GH, Harrison JG, Harding LA, Elad Y, Zimand G (1995) Effect of humidity on infection of rose petals by dry-inoculated conidia of *Botrytis cinerea*. *Mycological Research* 99:1303–1310.

Willis P, Caudle AB, Fayrerhosken RA (1994) Fine-structure of equine oocytes matured in-vitro for 15 hours. *Molecular Reproduction and Development* 37:87–92.

Wilman DEV, Connors TA (1983) Molecular structure and anti-tumour activity of alkylating agents. Ch. 8 In: S Neidle, M Waring (eds) *Molecular Aspects of Anti-cancer Drug Action*. Macmillan, London. pp. 233–282.

Wilson CM (1983) Staining of proteins on gels: comparisons of dyes and procedures. *Methods in Enzymology* 91:236–247.

Wilson CM (1992) An update on protein stains: amido black, coomassie blue G, and coomassie blue R. *Biotechnic and Histochemistry* 67:224–234.

Wilson JG (1910) Intra vitam staining with methylene blue. *Anatomical Record* 4:267–277.

Wilson JM, Laurent P, Tufts BL, Benos DJ, Donowitz M, Vogl AW, Randall DJ (2000) NaCl uptake by the branchial epithelium in freshwater teleost fish: An immunological approach to ion-transport protein localization. *Journal of Experimental Biology* 203:2279–2296.

Wilson WD, Ezrin C (1954) Three types of chromophil cells of the adenohypophysis demonstrated by a modification of the periodic acid-Schiff technique. *American Journal of Pathology* 30:891–899.

Windham GL, Williams WP (1994) Reproduction of *Meloidogyne incognita* and *M. arenaria* on tropical corn hydrids. *Journal of Nematology* 26:753–755.

Wirth SJ (1992) Water-soluble, dye-labeled fatty-acid derivatives for preliminary detection of lipolytic microorganisms in agar media. *FEMS Microbiology Letters* 95:77–80.

Wismar BL (1966) Quad-type stain for the simultaneous demonstration of intracellular and extracellular tissue components. *Stain Technology* 41:309–313.

Wissowzky A (1876) Ueber das Eosin als reagenz auf Hamoglobin und die Bildung von Blutgefassen und Blutkorperchen bei Saugetier und Huhnerembryonen. *Archiv fur mikroskopische Anatomie* 13:479–496.

Wittekind D (1985) Standardization of dyes and stains for automated cell pattern recognition. *Analytical and Quantitative Cytology and Histology* 7:6–30.

Wittekind DH (1983) On the nature of Romanowsky-Giemsa staining and its significance for cytochemistry and histochemistry: an overall review. *Histochemical Journal* 15:1029–1047.

Wittekind DH (1991) Die Romanowsky-Giemsa (RG)-Farbung (1891–1991). Ihr Wesen, Moglichkeiten ihrer Anwendung, ihre Bedeutung. *Verhandlungen Deutsche Gesselschaft fur Zytologie* 17:66–75.

Wittekind D, Kretschmer V (1972) Lichtmikroskopische Untersuchungen zur Strukturerhaltung farbstoffinduzierter autophagischer Vakuolen durch verschiedene Fixativen. *Zeitschrift fur Zellforschung* 126:518–535.

Wittekind D, Schulte E (1987a) Die Bedeutung der Standardisierung der Zell- und Gewebspraparaten fur bildanalytische Operationen. In: S Eins (ed.) *Kvantitative und strukturelle Bildanalyse in der Medizin*. GIT-Verlag, Darmstadt. pp. 5–12.

Wittekind D, Schulte E (1987b) Standardized azure-B as a reticulocyte stain. *Clinical and Laboratory Haematology* **9**:395–398.

Wittekind D, Schulte E, Schmidt G, Frank G (1991) The standard Romanowsky-Giemsa stain in histology. *Biotechnic and Histochemistry* 66:282–295.

Wittekind DH, Gehring T (1985) On the nature of Romanowsky-Giemsa staining and the Romanowsky-Giemsa effect. 1. Model experiments on the specificity of azure B-eosin Y stain as compared with other thiazine dye-eosin Y combinations. *Histochemical Journal* 17:263–296.

Wittekind DH, Kretschmer V, Lohr W (1976) Kann Azur B-Eosin die May-Grunwald-Giemsa Farbung ersetzen? *Blut* 32:71–78.

Wittekind DH, Kretschmer V (1987) On the nature of Romanowsky-Giemsa staining and the Romanowsky-Giemsa effect. II. A revised Romanowsky-Giemsa staining procedure. *Histochemical Journal* 19:399–401.

Wohlrab F, Schwarz J (1988) Spectrophotometric investigations on the binding of Victoria blue 4R to oxidized insulin. *Acta Histochemica* 84:187–194.

Wojcik C, Schroeter D, Wilk S, Lamprecht J, Paweletz N (1996) Ubiquitin mediated proteolysis centers in Hela cells – indication from studies of an inhibitor of the chymotrypsin-like activity of the proteasome. *European Journal of Cell Biology* 71:311–318.

Wolff RA, Chien GL, van Winkle DM (2000) Propidium iodide compares favourably with histology and triphenyl tetrazolium chloride in the assessment of experimentally-induced infarct size. *Journal of Experimental and Cellular Cardiology* 32:225–232.

Wolman M, Bubis JJ (1965) The cause of the green polarization colour of amyloid stained with Congo red. *Histochemie* 4:351–356.

Wolosker H, de Souza DO, de Meis L (1996) Regulation of glutamate transport into synaptic vesicles by chloride and proton gradient. *Journal of Biological Chemistry* 271:11726–11731.

Wolters GHJ, Pasma A, Konijnendijk W, Bouman PR (1979) Evaluation of the glyoxal-bis-(2-hydroxyanil)-method for staining of calcium in model gelatin films and pancreatic islets. *Histochemistry* 62:137–151.

Wood ML, Green AJ (1958) Studies on textile dyes for biological staining. I. Pontacyl blue black SX, pontacyl violet 6R and luxol fast yellow TN. *Stain Technology* 33:279–281.

Wrenn BA, Venosa AD (1996) Selective enumeration of aromatic and aliphatic hydrocarbon degrading bacteria by a most-probable-number procedure. *Canadian Journal of Microbiology* 42:252–258.

Wright JH (1902) A rapid method for the differential staining of blood films and malaria parasites. *Journal of Medical Research* 7:138–144.

Wright KM, Horobin RW, Oparka K (1996) Phloem mobility of fluorescent xenobiotics in *Arabidopsis* in relation to their physicochemical properties. *Journal of Experimental Botany* 47:1779–1788.

Wright PR, Richfield Fratz N, Rasooly A, Weisz A (1997) Quantitative analysis of components of the color additives DandC red Nos 27 and 28 (Phloxine B) by thin-layer chromatography and video densitometry. *Journal of Planar Chromatography* 10:157–162.

Wu FJ, Friend JR, Remmel RP, Cerra FB, Hu WS (1999) Enhanced cytochrome P450IA1 activity of self-assembled rat hepatocyte spheroids. *Cell Transplantation* 8:233–246.

Wu KC, Kim RJ, Bluemke DA, Rochitte CE, Zerhouni EA, Becker LCJA (1998) Quantification and time course of microvascular obstruction by contrast-enhanced echocardiography and magnetic resonance imaging following acute myocardial infarction and reperfusion. *Journal of the American College of Cardiology* 32:1756–1764.

Wu M, Kiernan JA (2001) A new method for surface staining large slices of fixed brain using a copper phthalocyanine dye. *Biotechnic and Histochemistry* 76:253–255.

Wu R, Hoshino T, Nagura M (1999) Endocochlear potential in focal lesions of the guinea pig cochlea. *Hearing Research* 128:102–111.

Wyatt JH (1972) The demonstration of berylium oxide in paraffin sections with chrom-oxane stains. *Stain Technology* 47:33–36.

Xia P, Bungay PM, Gibson CC, Kovbasnjuk ON, Spring R (1998) Diffusion coefficients in the lateral intercellular spaces of Madin-Darby canine kidney cell epithelium determined with caged compounds. *Biophysical Journal* 74:3302–3312.

Yack JE (1993) Janus green B as a rapid, vital stain for peripheral nerves in chordotonal organs in insects. *Journal of Neuroscience Methods* 49:17–22.

Yamada K (1969) Combined histochemical staining of 1,2-glycol and sulfate groupings of mucopolysaccharides in paraffin sections. *Histochemie* 20:271–276.

Yamada K (1970) Dual staining of some sulfated mucopolysaccharides with alcian blue (pH 1.0) and ruthenium red (pH 2.5). *Histochemie* 23:13–20.

Yamada K (1978) Concanavalin A-peroxidase-diaminobenzidine-periodic acid-m-aminophenol-fast black K: a method for the dual staining of neutral complex carbohydrates. *Histochemical Journal* 10:573–584.

Yamada M, Kawai M, Mochizuki H, Hata Y, Mashima Y (1998) Fluorophotometric measurement of the buffering action of human tears *in vivo*. *Current Eye Research* 17:1005–1009.

Yamada S, Itano HA (1966) Phenanthrenequinone as an analytical reagent for arginine and other monosubstituted guanidines. *Biochimica et Biophysica Acta* 130:538–540.

Yamaguchi K, Tamura Z, Maeda M (1997) Molecular structure of bromophenol blue having a gamma-sultone ring. *Analytical Sciences* 13:1057–1058.

Yamaguchi K, Chijiiwa K, Shimizu S, Yokohata K, Tanaka M (1998) Litmus paper helps detect potential pancreatoenterostomy leakage. *American Journal of Surgery* 175:227–228.

Yamamoto M, Hibi H, Miyake K (1995) Effects of alpha-blocker on daily testicular sperm production and sperm concentration, motility, intraluminal pressure and fluid movement in the rat epididymis. *Tohoku Journal of Experimental Medicine* 177:25–37.

Yamazaki Y, Furukawa F, Nishikawa A, Takahashi M, Oka S (1998) Histochemical determination of stereoselectivity of esterases in normal pancreas and pancreatic tubular adenocarcinoma of hamsters. *Biotechnic and Histochemistry* 73:23–31.

Yan SM, Finato N, Artico D, Di Loreto C, Cataldi P, Bussani R, Silvestri F, Beltrami CA (1998) DNA content, apoptosis and mitosis in transplanted human hearts. *Advances in Clinical Pathology* 2:205–219.

Yang DH, Tsuyama S, Ohmori J, Murata F (1998) Sulphated glycosaminoglycans in guinea pig basophils studied by means of cationic colloidal gold. *Histochemistry and Cell Biology* 109:189–194.

Yang J, Tezel G, Patil RV, Wax MB (2000) Flow cytometry for quantification of retrogradely labeled retinal ganglion cells by fluoro-gold. *Current Eye Research* 21:981–985.

Yao DL, Komoly S, Zhang QL, Webster HD (1994) Myelinated axons demonstrated in the CNS and PNS by antineurofilament immunoreactivity and luxol fast blue counterstaining. *Brain Pathology* 4:97–100.

Yates R, Moran J, Addy M, Mullan PJ, Wade WG, Newcombe R (1997) The comparative effect of acidified sodium chlorite and chlorhexidine mouthrinses on plaque regrowth and salivary bacterial counts. *Journal of Clinical Periodontology* 24:603–609.

Yatome C, Ogawa T (1995) Reduction of methylene blue by *Bacillus* sp. *Journal of Environmental Science and Health Part A – Environmental Science and Engineering and Toxic and Hazardous Substance Control* 30:31–39.

Ye B, Gitler C, Gressel J (1997) A high sensitivity, single-gel, polyacrylamide gel electrophoresis method for the quantitative determination of glutathione reductases. *Analytical Biochemistry* 246:159–165.

Yim KO, Bradford KJ (1998) Callose deposition is responsible for apoplastic semipermeability of the endosperm envelope of muskmelon seeds. *Plant Physiology* 118:83–90.

Yokoyama H, Kim JH, Sato J, Sano M, Hirano K (1996) Fluorochrome Uvitex 2B stain for detection of the Microsporidian causing beko disease of yellowtail and goldstriped amberjack juveniles. *Fish Pathology* 31:99–104.

Yoo JJ, Lee I, Atala A (1998) Cartilage rods as a potential material for penile reconstruction. *Journal of Urology* 160:1164–1168.

Yoon SH, Lee CH, Kim DY, Kim KW, Park KH (1994) Time–temperature indicator using phospholipid–phospholipase system and application to storage of frozen pork. *Journal of Food Science* 59:490–493.

Yoshida M, Nagatsuka Y, Muramatsu S, Niijima K (1991) Differential roles of the caudate nucleus and putamen in motor behavior of the cat as investigated by local injection of GABA antagonists. *Neuroscience Research* 10:34–51.

Yoshika S, Umeda T, Kurahashi Y (1972) An effective reactivation of alkaline phosphatase in hard tissue completely decalcified for light and electron microscopy. *Histochemie* 29:296–306.

Yu W, Dodds WK, Banks MK, Skalsky J, Strauss EA (1995) Optimal staining and sample storage for direct microscopic enumeration of total and active bacteria in soil with two fluorescent dyes. *Applied and Environmental Microbiology* 61:3367–3372.

Zacharias E (1881) Uber die chemische Beschaffenheit des Zellkerns. *Botanische Zeitung* 39:169–176.

Zagon IS, Vavra J, Steele I (1970) Microprobe analysis of protargol stain deposition in two protozoa. *Journal of Histochemistry and Cytochemistry* 18:559–564.

Zajic G, Forge A, Schacht J (1993) Membrane stains as an objective means to distinguish isolated inner and outer hair-cells. *Hearing Research* 66:53–57.

Zanker V (1973) Methode und Ergebnisse der Farbstoffreinigung. 6. Gang der Reinigung eines Farbstoffes, demonstriert am Beispiel des Acridinorange. *Acta Histochemica Suppl.* XIII-B:281–290.

Zanker V (1981) Grundlagen der Farbstoff-Substrat-Bezeihungen in der Histochemie. *Acta Histochemica Suppl.* 24:151–168.

Zarfati D, Harris A, Garzozi HJ, Zacish M, Kagemann L, Jonescu-Cuypers CP, Martin B (2000) A review of ocular blood flow measurement techniques. *Neuroophthalmology* 24:401–409.

Zbaeren J, Zbaeren D, Geiser T, Haeberli A (1998) A new method for GMA sections – Immunofluorescence with the ELF-precipitate combined with a classical hematoxylin/eosin-phloxine stain. *Journal of Histotechnology* 21:25–27.

Zeindl-Eberhart E, Jungblut PR, Rabes HM (1997) A new method to assign immunodetected spots in the complex two-dimensional electrophoresis pattern. *Electrophoresis* 18:799–801.

Zeng S, Yi FX, Guo ZG (1999) Regulatory role of protein tyrosine phosphorylation in platelet activating factor-induced signal transduction in platelets. *Acta Pharmacologica Sinica* 20:157–161.

Zhang J, Davidson RM, Wei MD, Loew LM (1998) Membrane electric properties by combined patch clamp and fluorescence ratio imaging in single neurons. *Biophysical Journal* 74:48–53.

Zhang YH, Lu J, Elmquist JK, Saper CB (2000) Lipopolysaccharide activates specific populations of hypothalamic and brainstem neurons that project to the spinal cord. *Journal of Neuroscience* 20:6578–6586.

Zhao QY, Zhou RZ, Temsamani J, Zhang ZW, Roskey A, Agrawal S (1998) Cellular distribution of phosphorothioate oligonucleotide following intravenous administration in mice. *Antisense Nucleic Acid Drug Development* 8:451–458.

Zhou HY, Salih E, Glimcher MJ (1998) Isolation of a novel bone glycosylated phosphoprotein with disulphide cross-links to osteonectin. *Biochemical Journal* 330:1423–1431.

Zhou JG, Meno JR, Hsu SSF, Winn HR (1994) Effects of theophylline and cyclohexyladenosine on brain injury following normoglycemic and hyperglycemic ischemia – a histopathologic study in the rat. *Journal of Cerebral Blood Flow and Metabolism* 14:166–173.

Zhou T, Paulitz TC (1993) *In vitro* and *in vivo* effects of *Pseudomonas* spp. on *Pythium aphanidermatum*: zoospore behavior in exudates and on the rhizoplane of bacteria-treated cucumber roots. *Phytopathology* 83:872–876.

Zhu HP, Clark SM, Benson SC, Rye HS, Glazer AN, Mathies RA (1994) High-sensitivity capillary electrophoresis of double-stranded DNA fragments using monomeric and dimeric fluorescent intercalating dyes. *Analytical Chemistry* 66:1941–1948.

Zieglansberger W, Reiter C (1974) Interneuronal movement of procion yellow in cat spinal neurones. *Experimental Brain Research* 20:527–530.

Zimmermann HW (1983) Physikalisch-chemische Grundlagen der Farbung fur manuelle und apparative Zytodiagnostik. *Microscopica Acta Suppl.* 6:45–58.

Zipfel E, Grzes JR, Naujok A, Seiffert W, Wittekind DH, Zimmermann HW (1984) Uber Romanowsky Farbstoffe und den RG-effekt. **3**. Mitteilung: Mikrospektralphotometrische Untersuchung der RG-Farbung. Spectralphotometrischer Nachweis eines DNA-Azur B-Eosin Y Komplexes, der den RG-Effekt verursacht. *Histochemistry* 84:337–351.

Ziv NE, Smith SJ (1996) Evidence for a role of dendrite filopodia in synaptogenesis and spine formation. *Neuron* 17:91–102.

Zochowski M, Wachowiak M, Falk CX, Cohen LB, Lam YW, Antic S, Zecevic D (2000) Imaging membrane potential with voltage-sensitive dyes. *Biological Bulletin* 198:1–21.

Zollinger H (1961) *Azo and Diazo Chemistry. Aliphatic and Aromatic Compounds* (Transl. H. E. Nursten). Interscience Publishers, London.

Zollinger H (1991) *Color Chemistry: Synthesis, Properties and Applications of Organic Dyes and Pigments.* 2nd Edn. VCH Verlag, Weinheim and New York.

Zollinger H (1999) *Color. A Multidisciplinary Approach.* Wiley-VCH, Weinheim.

Zozulya V, Blagoi Y, Lober G, Voloshin I, Winter S, Makitruk V, Shalamay A (1997) Fluorescence and binding properties of pheanzine derivatives in complexes with polynucleotides of various base compositions and secondary structures. *Biophysical Chemistry* 65:55–63.

Zucker-Franklin D (1968) Electron microscopic studies of human granulocytes: structural variations related to function. *Seminars in Hematology* 5:109–133.

Zuppinger H (1874) Eine methode, Axencylinderfortsatze der Ganglienzellen des Ruckenmarkes zu demonstrieren. *Archiv fur mikroskopische Anatomie* 10:255–256.

Index

A **bold** page number indicates the beginning of the principal entry for that compound.

539